EFFECTS OF ACID RAIN ON FOREST PROCESSES

WILEY SERIES IN

ECOLOGICAL AND APPLIED MICROBIOLOGY

RECENT TITLES

MICROBIAL LECTINS AND AGGLUTININS:
Properties and Biological Activity
David Mirelman, Editor, 1986

THERMOPHILES: General,
Molecular, and Applied Microbiology
Thomas D. Brock, Editor, 1986

INNOVATIVE APPROACHES
TO PLANT DISEASE CONTROL
Ilan Chet, Editor, 1987

PHAGE ECOLOGY
Sagar M. Goyal, Charles P. Gerba, and
Gabriel Bitton, Editors, 1987

BIOLOGY OF ANAEROBIC
MICROORGANISMS
Alexander J.B. Zehnder, Editor, 1988

THE RHIZOSPHERE
J.M. Lynch, Editor, 1990

BIOFILMS
William G. Characklis and
Kevin C. Marshall, Editors, 1990

ENVIRONMENTAL MICROBIOLOGY
Ralph Mitchell, Editor, 1992

BIOTECHNOLOGY IN PLANT
DISEASE CONTROL
Ilan Chet, Editor, 1993

ANTARCTIC MICROBIOLOGY
E. Imre Friedmann, Editor, 1993

WASTEWATER MICROBIOLOGY
Gabriel Bitton, 1994

EFFECTS OF ACID RAIN ON
FOREST PROCESSES
Douglas L. Godbold and Aloys Hüttermann,
Editors, 1994

EFFECTS OF ACID RAIN ON FOREST PROCESSES

Edited by

DOUGLAS L. GODBOLD
ALOYS HÜTTERMANN

Forstbotanisches Institut
Universität Göttingen
Göttingen, Germany

A JOHN WILEY & SONS, INC., PUBLICATION
New York • Chichester • Brisbane • Toronto • Singapore

Address All Inquiries to the Publisher
Wiley-Liss, Inc., 605 Third Avenue, New York, NY 10158-0012

Printed in the United States of America.

Library of Congress Cataloging-in-Publication Data

Effects of acid rain on forest processes / editors, Douglas Godbold, Aloys Hüttermann.
p. cm. – (Wiley series in ecological and applied microbiology)
Includes index.
ISBN 0-471-51768-2
1. Forest ecology. 2. Forest plants—Effect of acid deposition on. 3. Acid deposition—Environmental aspects. 4. Microbial ecology. I. Godbold, Douglas. II. Hüttermann, Aloys. III. Series.
QH541.5.F6E44 1994
574.5′2642—dc20 93-41974

The text of this book is printed on acid-free paper.

CONTENTS

CONTRIBUTORS

Christopher S. Cronan, Department of Botany and Plant Pathology, University of Maine, Orono, ME 04469

Johannes Eichorn, Abt. Waldschäden, Hessische Forstliche Versuchsanstalt, D-34346 Hann. Münden, Germany

Douglas L. Godbold, Forstbotanisches Institut, Universität Göttingen, Büsgenweg 2, D-37077 Göttingen, Germany

Franz Gruber, Forstbotanisches Institut, Universität Göttingen, Büsgenweg 2, D-37077 Göttingen, Germany

Aloys Hüttermann, Forstbotanisches Institut, Universität Göttingen, Büsgenweg 2, D-37077 Göttingen, Germany

Georg Jentschke, Forstbotanisches Institut, Universität Göttingen, Büsgenweg 2, D-37077 Göttingen, Germany

Christine Rapp, Institut für Waldbau, Universität Göttingen, Büsgenweg 1, D-37077 Göttingen, Germany

Andreas Roloff, Institut für Forstbotanik, Technische Universität Dresden, Pienerstr. 7, D-01737 Tharandt, Germany

Matthias Schaefer, Zoologisches Institut, Universität Göttingen, Berlinerstr. 28, D-37073 Göttingen, Germany

Bernhard Ulrich, Institut für Bodenkunde und Waldernährung, Universität Göttingen, Büsgenweg 2, D-37077 Göttingen, Germany

Volkmar Wolters, Institut für Zoologie, Universität Mainz, Saarstr. 21, D-55099 Mainz, Germany

James J. Worrall, College of Environmental Science and Forestry, State University of New York, Syracuse, NY 13210

PREFACE

Throughout the history of science many discoveries have been made by chance, by workers intending to investigate something other than the discovery they eventually made. The history of investigations of forest ecosystems is no different.

The Solling project, which ran from 1966 to 1973 as a German contribution to the International Biological Programme, was a pioneering project on forest ecosystems. The Solling project ran parallel to investigations of forest ecosystems in Belgium, Denmark, England, Hungary, Poland, and what was then Czechoslovakia. The project involved scientists from many disciplines, and addressed factors influencing primary and secondary production, biodiversity, and biogeochemical cycles. Even after the official end of the Solling project, ecosystem research was continued on the site. This allowed long-term monitoring of changes in flora and element budgets that later became critical factors in the prediction and investigation of forest decline in the area.

The original objective of the Solling project was an investigation of the functioning of a forest ecosystem, whereby the only human effects were those of past and present forestry practices. However, it soon became apparent that even this isolated forest area was being influenced by pollutants that had been transported over long distances. The concept of a clean air zone in German forests was slowly being eroded. The alarm was sounded originally by Ulrich in publications in 1979. It had become obvious to Ulrich, through element budget studies, that the input of sulfur and acid components was affecting severely the nutrient budgets in the ecosystem, and that these would have far-reaching consequences for tree health and ecosystem structure.

Ulrich had the foresight and conviction to publish his forebodings long before the symptoms of forest decline became apparent in Germany. As is often the case, his ideas were at times met with derision. But the appearance of symptoms of forest decline in Europe and North America also produced a great deal of interest in the media, and as a consequence, a push in forest ecosystem research. The number of forest sites investigated in Germany was expanded greatly, and research of forest ecosystems was put on a firmer basis by the founding of the Forest Ecosystem Research Centre in Göttingen, with which a number of authors in this book are affiliated.

Much of the work carried out in Europe and North America was concerned with finding the causes of forest decline, and a number of theories were developed to explain this phenomenon. These theories included soil acidification, the direct effects of gaseous pollutants, and attack by pathogens. Many of these theories were pursued on a basis of mutual exclusion and the desire to be proven right often obscured the basic quest for well-founded scientific knowledge. Today we now know that forest ecosystems are inordinately complex and may be affected by a large number of factors. One factor that stands out is acid deposition, and it is this subject that this book addresses.

The beginning of much of the data in support of the major role of soil acidification in forest decline was collected in the Solling project. This book reflects the more recent developments in investigations of the effects of acid deposition. It has become increasingly clear that the major influence of acid deposition is indirect, i.e., the effects are mediated by change in soil chemistry that influences all other processes. Thus this book begins where the original evidence for changes in soil chemistry started—in investigations of element budgets. Element budgets and changes in soil chemistry are discussed for both European and North American forest ecosystems. As well as providing the rooting area for forest trees, forest soils support numerous life forms that in turn directly or indirectly influence the vitality of the trees. The book addresses the effects of acid deposition not only on root and canopy growth of forest trees, but also on the micro- and macroscopic flora and fauna. Similarly, pathogenic organisms are also influenced by the chemical and physical properties of the soil, and are addressed in one chapter of the book. The final chapter tries to simplify the complex interaction in the forest using an integrative theory of forest ecosystems.

Due to the dominance of authors from the Göttingen Forest Ecosystem Research Centre, both the structure of the book (from the soil to the canopy) and the approach taken in the investigations tend to be bottom-up. The book undoubtedly reflects a Göttingen point of view, for which we offer no apologies. The chapters have been written by experts in their field, and the text represents a balance of their research and review of literature. The authors have presented their opinions; as has been seen so often in ecosystem research, only time will tell if they are the right ones.

Douglas L. Godbold
Aloys Hüttermann

NUTRIENT AND ACID–BASE BUDGET OF CENTRAL EUROPEAN FOREST ECOSYSTEMS

BERNHARD ULRICH

Institut für Bodenkunde und Waldernährung, Universität Göttingen, Büsgenweg 2, D-37077 Göttingen, Germany

1. INTRODUCTION

Anthropogenic emissions include acid precursors (SO_2 and NO_x) as well as nutrients (especially nitrogen, N). By reactions with dust particles (e.g., soil dust) the acids formed from the acid precursors can release nutrients such as Ca, Mg, and K from the solid phases into solution. The reaction can take place in air or with deposited particles at plant surfaces. This implies an increase of the input of dissolved nutrient cations into ecosystems due to the emission of acid precursors. The effects of the input of acidity and nutrients on the input and output fluxes in the forest canopy, in the ecosystem as a whole, and in the soil, give insights into which processes in the ecosystem are affected. Based on our existing knowledge of the processes, we can draw conclusions on how the processes are affected. The instruments are proton and acid–base budgets (Ulrich, 1978–1980, 1991a). The macronutrients are an indispensible part of these budgets, since they can exist as bases (K^+, Mg^{2+}, and Ca^{2+}) or as acids (NO_3^-, SO_4^{2-}, and PO_4^{3-}). The anions can react as bases, some cations (NH_4^+, Al^{3+}, Fe and Mn ions, and trace metals) can react as acids (cf. Ulrich, 1991a; van Breemen, 1991; Bruggenwert et al., 1991). In this chapter, only major ions that contribute significantly to the cation–anion budget are taken into account.

Effects of Acid Rain on Forest Processes, pages 1–50

Being part of the International Biological Program (IBP), ecosystem research in the Solling, a mountain area in West Germany, has been ongoing since 1966. Early measurements of the element (ion) cycling revealed high rates of deposition of various air pollutants, especially sulfuric acid (H_2SO_4), in this remote area (Mayer and Ulrich, 1974). Based on available data on ecosystem budgets for the period 1969–1975, the detrimental effects of acid deposition on soils and forest ecosystems have been pointed out (Ulrich et al., 1979). The major conclusions drawn at that time were the evident acidification of soils, increasing risk of root damage due to Al toxicity, and the prognosis of forest decline caused by acid deposition.

In the 1980s, the fluxes in forest ecosystems were assessed for a number of ecosystems in central Europe (FBW, 1989; Ulrich, 1989, 1991a) and North America (Johnson and Lindberg, 1992). In most cases the measurements cover only a time period of 2–3 years. How well short-term budgets reflect the long-term behavior of the ecosystem is still an unanswered question. Therefore, the long-term data set available from the Solling is analyzed in this chapter with respect to temporal trends and their causes.

2. ASSESSMENT OF ION FLUXES AND OF ACID–BASE BUDGETS

2.1. Ion Fluxes

The widely used term "base cations" is not used in this section, since it is not applicable to the ions Na, K, Mg, and Ca in soil solutions. In soil solutions of pH < 8 these ions do not exist as bases, that is, they cannot react as proton acceptors. A base can only be the combination of these ions with anions of weak acids like H_2CO_3 or, in the solid soil phase, substances with cation exchange properties. In acid deposition and soil solution these ions exist as neutral salts with the anions of strong acids (SO_4, NO_3, and Cl). The term "base cations" is incorrect and is misleading.

The assessment of ion fluxes is based on data of bulk deposition (BD_X), of stand precipitation [CP_X, represents the sum of throughfall (TF) and stemflow], and of outflow with water (SO_X), where X represents the following ions:

M_b cations: Na^+, K^+, Mg^{2+}, Ca^{2+} (either neutral salts or weak bases).
M_a cations: H^+, Mn^{2+}, Al^{3+}, Fe^{2+}, NH_4^+ (potential proton donors).
Anions: SO_4^{2-} and NO_3^- (potential proton acceptors), Cl^- (neutral or acid salts).

Organic bound nitrogen is estimated as the difference between Kjeldahl $-$N and NH_4^+, but is not used to calculate the charge balance. All data are expressed in kilomole ion equivalents (IE) per hectare per year (kmol$_c$ ha^{-1} yr^{-1}).

In solutions of pH < 5.0, the acid neutralization capacity (ANC, alkalinity) is negative. Such solutions possess a base neutralization capacity (BNC, acidity), which is defined as the sum of M_a cations. As a relative measure of acidity, the acidity degree (AD) is used.

$$AD\ (\text{in }\%) = [M_a/(M_b + M_a)] \times 100 \tag{1.1}$$

Total deposition (TD_X) is composed of bulk deposition (BD_X) and interception deposition (ID_X).

$$TD_X = BD_X + ID_X \tag{1.2a}$$

The interception deposition ID_X is composed of the impacted particles (particle interception ID_{part}) and the adsorption of gases (gas interception ID_{gas}).

$$ID_X = (ID_{part} + ID_{gas})_X \tag{1.3}$$

$$TD_X = BD_X + ID_{part,X} + ID_{gas,X} \tag{1.2b}$$

The canopy and the stem can act as sinks or sources of ions with respect to the precipitation passing through the canopy (TF) and along the stem (stemflow). The canopy precipitation CP is thus composed of

$$CP_X = BD_X + ID_X + Q_X \tag{1.4}$$

where Q represents the source and sink function, respectively, of the canopy and of the stem.

For the assessment of the interception deposition the assumption is made that Q_X is negligible for Na, SO_4, and Cl. This assumption was also found valid for North American forests (Johnson et al., 1993). Thus it follows:

$$ID_X = CP_X - BD_X \tag{1.5}$$

for X = Na, SO_4, and Cl.

In remote forest areas of central Europe, ID_{part} is mainly due to the interception of cloud water (occult deposition). For this case we assume that the relationships between the ions in cloud water are similar to those in bulk deposition. Generally, this assumption seems valid for the northwestern lowland and the German uplands. It does not apply to coastal forests (incomplete equilibration of sea water particles with air pollutants), to forests in the vicinity of communities and arable lands (ID_{part} due to dry deposition of dust particles), and to some regions in southern Germany (occurrence of NaCl free cloud water?). For the cases where the assumption mentioned above is justified, ID_{part} (occult deposition) can be estimated from the following relation (Ulrich, 1983).

$$(ID/ND)_{Na} = (ID_{part}/ND)_X \tag{1.6a}$$

for X = H, Na, K, Mg, Ca, Mn, Al, Fe, NH_4, SO_4, Cl, and NO_3. The ratio $(ID/ND)_{Na}$ may be called the sodium factor f_{Na}.

$$ID_{part,X} = f_{Na} \times ND_X \quad (\text{X as in Eq. 1.6a}) \tag{1.6b}$$

The parameter ID_{gas} is necessarily zero for Na, K, Mg, Ca, Mn, Al, and Fe (no gas phase).

Applying Eqs. 1.5 and 1.3, the interception deposition of SO_2, HCl, NH_4, and NO_3 can be estimated.

$$ID_{gas,X} = CP_X - BD_x - ID_{part,X} \quad (X = SO_4, Cl, NH_4, \text{ and } NO_3) \quad (1.7)$$

For NH_3 and HNO_3 a gaseous deposition and an uptake into the leaf is possible (canopy as a sink, Q negative). If in the case of NH_4 or NO_3, the ID_{part} exceeds the difference CP − BD, then ID_{gas} cannot be estimated.

The gaseous deposition of SO_2, HCl, and HNO_3 is accompanied by an equivalent deposition of H^+ and the deposition of NH_3 by an equivalent consumption of H^+. Based on this, the gaseous deposition budget of protons is calculated as follows:

$$ID_{gas,H} = ID_{gas,SO_4} + ID_{gas,Cl} + ID_{gas,NO_3} - ID_{gas,NH_4} \quad (1.8)$$

A proton buffering by NH_3 is thus considered as a NH_4 input. The input of acidity (acid load by deposition, AL_{dep}) is given by $TD_{Ma\ cations}$.

For all fluxes the cation–anion budget theoretically has to be balanced (sum of cations minus sum of anions equals zero). Deviations from zero include analytical errors as well as the contribution of ions that were not analyzed. Usually, bicarbonate (HCO_3^-) and organic anions are not determined. In most cases the ion budgets of CP, BD, TD, and SO are positive (anion deficiency) or balanced (see the data sets presented in FBW, 1989). In the calculation of the acid–base budgets the following assumptions are made: anion deficits at $pH < 5.0$ are caused by organic anions and at $pH > 5.0$ by HCO_3^-. As the cation–anion budget of a flux also includes all of its errors, the interpretation of the difference is flawed with uncertainties. The data available for fluxes of N_{org} and of C_{org} (DOC) can substantiate the presumption of organic anions.

The calculation algorithms result in quantitative hypotheses concerning the deposition of the ions specified. The model assumes occult (cloud water) deposition as the main pathway of particle deposition. The hypotheses on the deposition flux rates have to be tested. Tests can be experiments to exclude other pathways (Ulrich et al., 1973; Fassbender, 1977; Lindberg, 1989) and direct assessment of the fluxes by micrometeorological methods. From 1989–1990 micrometeorological measurements were performed concerning the exchange between atmosphere and canopy of the spruce stand F1. The results on particle deposition agree with the ID_{part} calculated according to Section 2. These results showed that the efficiency of particle deposition increases during fog events by an order of magnitude. According to these measurements, 50% of particle deposition in 1990 was due to fog deposition (Gravenhorst, 1992). Deposition of cloud water can also occur during rain events, such events can hardly be separated from wet deposition.

A further qualitative test is the plausibility of the data (see Section 3).

2.2. Acid–Base Budgets

For any compartment of the ecosystem, acid–base budgets can be assessed, provided that the input and output of cations and anions is known. Three compartments will be considered: the forest canopy, the ecosystem as a whole, and the soil.

For any ion, the budget B is defined as the difference between input I and output O.

$$B_X = I_X - O_X \tag{1.9}$$

Input I and output O of ions occur usually with a transport medium. The transport medium may be water (see Section 2.1) and organic matter. The compartment considered is treated as a blackbox. Positive budgets show an element increase in the compartment due to input (the compartment acts as a sink), negative budgets show element losses from the compartment (the compartment acts as a source).

From the negative or positive signs of the compartment budget the triggering processes can often be concluded on the basis of existing knowledge, the budget reflects the net rate of the process.

Only for Cl may we preclude that changes in the storage are due to the accumulation or loss of neutral salts, respectively:

Accumulation of neutral salts: Positive values of $(I - O)$ for Cl
Loss of neutral salts: Negative values of $(I - O)$ for Cl

as far as equivalent changes in the storage of M_b cations can be related to changes in the storage of Cl.

The *acid load from the ion input–output budget* (AL_{I-O}) of the compartment is the sum of

1. The sink function of the compartment for acid input. This corresponds to the sum of positive budgets $[(I–O) > 0]$ of H, Mn, Al, Fe, and NH_4.
2. The source function of the compartment for anions. This corresponds to the sum of negative budgets $[(I–O) < 0]$ of SO_4, NO_3, and HCO_3 or C_{org}, respectively. For the estimation of HCO_3 and C_{org} see above. Chloride is only taken into account if its budget is not balanced by M_b cations. Processes are the dissolution of aluminum sulfates, nitrification (formation of HNO_3), dissociation of dissolved organic acids, or H_2CO_3, respectively. The dissolution of aluminum sulfates represents the mobilization of an input of H_2SO_4 that was stored in the solid soil phase previously (Prenzel, 1983; cf. 4, 5).
3. The sink function of the compartment for M_b cations, corrected for neutral salt accumulation (cf. above). This value corresponds to the sum of positive budgets $[(I–O) > 0]$ of Na, K, Mg, and Ca. It is presumed that the M_b cations are bound to acidic groups releasing M_a cations (e.g., protons). In an ecosystem this may be due to a net accumulation of phytomass.

The following *acid–base reactions* balance the total acid load in the compartment:

4. The consumption of protons through the dissolution of the oxides of Mn, Al and Fe, and of silicates, results in a source function of the compartment for these ions. This result corresponds to the sum of the negative budgets [(I–O) < 0] of Mn, Al, and Fe. A negative budget of Al can also be caused by the dissolution of aluminum sulfates. In this case the release of Al represents the consumption of protons when aluminum sulfate has been formed (cf. 2). For H and NH_4, negative budgets have not yet been observed.
5. The consumption of protons connected with the formation of uncharged compounds from anions. Processes are the formation of aluminum sulfates (reaction of H_2SO_4 with AlOOH), the biogenic transformation of HNO_3 into N_{org} or uncharged nitrogen species (denitrification), the formation of H_2S, the accumulation of organic acids. The result of these processes is a sink function of the compartment for these anions. This result corresponds to the sum of positive budgets [(I–O) > 0] of SO_4, NO_3, and C_{org}. Chloride is only taken into account if its budget is not balanced by M_b cations. The HCO_3^- can be transformed to H_2CO_3 if the pH changes from greater than 5.0 to less than 5.0.
6. The consumption of protons connected with the release of M_b cations. The processes can be cation exchange by M_a cations, silicate weathering, and mineralization. These processes result in a source function of the compartment for M_b cations, corrected for the leaching of neutral salts (cf. above). This result corresponds to negative budgets [(I–O) < 0] of Na, K, Mg, and Ca.

According to the ecosystem definition given elsewhere (Ulrich, Chapter 10), the silicate minerals belong to the environment of the ecosystem. If silicate weathering can be assessed separately, the M_b cations released can be taken into account as an input and the protons consumed as an output.

The acid load from the ion input–output budget (AL_{I-O}) and the sum of the acid–base reactions balance each other.

In the *soil budgets* (surface humus plus mineral soil) additional ecosystem internal fluxes and changes in the storage have to be included. In the following equation, only aerated soils are considered. Proton production due to oxidation reactions [e.g., oxidation of sulfides or of Fe(II)], and proton consumption due to reduction processes [e.g., reduction of Fe(III)] are not taken into account.

To calculate the soil input I_{soil}, the litter fall (LF) must be known.

$$\text{Soil input} \qquad I_{soil,X} = CP_X + LF_X \qquad (1.10)$$

To calculate the plant uptake from the soil into the shoot, U, the forest (wood) increment W is needed.

$$\text{Plant uptake from soil} \qquad U_{soil,X} = LF_X + W_X + CP_X - TD_X \quad (1.11)$$

The cation excess in the forest increment (cation excess equals surplus of nutrient cations above mobile inorganic anions in the organic matter) is equivalent to the acid load of the soil due to biomass increment (AL_W).

The soil budget (SB) is defined as:

$$SB_X = I_{soil,X} - O_{soil,X} = (TD - SO)_X - W_x \tag{1.12}$$

The acid load of the soil (AL_{soil}) is equal to:

$$AL_{soil} = AL_{I-O} + AL_W \tag{1.13}$$

3. TEMPORAL TRENDS IN ION FLUXES AND BUDGETS IN A NORWAY SPRUCE ECOSYSTEM IN THE SOLLING

3.1. Site and Methods

The ecosystem research of the Solling project focused on two adjacent forest ecosystems represented by a 140-year-old stand of European beech (*Fagus sylvatica*) (B1) and by a 105-year-old stand of Norway spruce (*Picea abies* Karst.) (F1). Both stands are located on the plateau of the Solling mountains in northwest Germany (9°30′ east, 51°40′ north) at about 500-m elevation. The beech stand developed from natural regeneration, the spruce stand was planted. Mean annual temperature is 6.4°C, average annual precipitation is 1088 mm. The soil is classified as podzolic brown earth (typical dystrochrept), it was developed from weathered triassic sandstone (Buntsandstein) covered by loess sediments, partly intermixed by solifluction. The soils are strongly and deeply acidified, pH($CaCl_2$) reaching from 2.9 to 4.2, and the base saturation of the effective cation exchange capacity (CEC) is less than 5% in the whole soil profile. Detailed chemical characteristics including changes in the chemical soil state are given by Ulrich et al. (1979) and Matzner (1988, 1989). Detailed data on the forest stands as well as other results from the Solling project were given in Ellenberg et al. (1986).

The field and laboratory methods used for assessing the element fluxes are described by Meiwes et al. (1984a, b) and König et al. (1989). Precipitation was collected with bulk samplers. A comparison between wet only samplers collected on a daily basis, and bulk samplers collected every week, was made at the spruce site F1 (Fig. 1.1, measuring period June 1991–March 1992, Ibrom, 1993). The bulk samplers collected generally higher amounts of ions, the deviation is larger for bulk precipitation than for throughfall (exceptions: trace elements). For the main ions, the fluxes found with wet only samplers in throughfall can be expected to amount to 90–100% of those found with bulk samplers. The difference is larger for Na, K, NH_4, Cl, and NO_3 in precipitation collected on a tower above the forest. This may indicate that the increased flux in bulk collectors is mainly due to wet particles (fog and cloud water). The distinction between wet and occult deposition seems to be difficult.

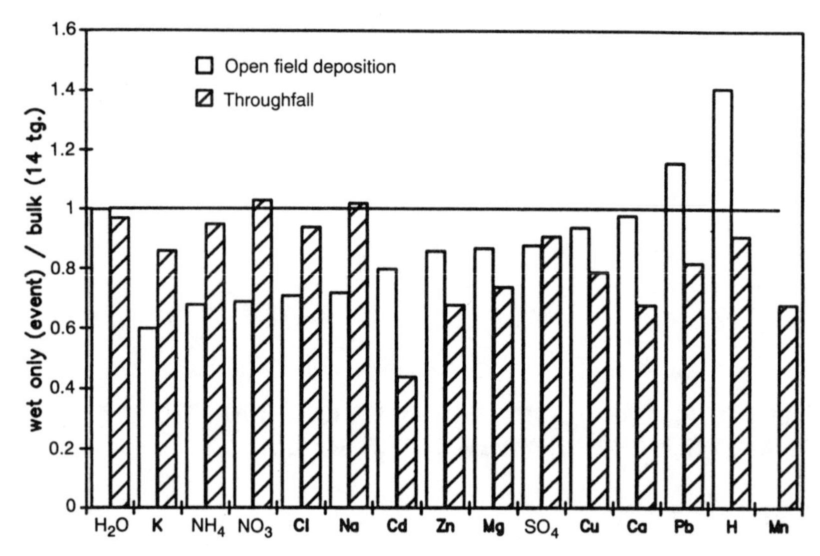

Fig. 1.1. Solling F1: Throughfall in 1990. Comparison of bulk and wet only samplers. Wet only samplers on a daily basis, bulk samplers on a fortnight basis. (From Ibrom, 1993.)

The spruce stand has a high roughness (z_0 = 1.5–3 m, turbulent diffusion coefficient K = 5–10 m^2/s with a maximum during midday, atmospheric resistance R_a = 2–5 s/m^{-1}, Gravenhorst, 1992).

The water outflow is calculated by simulation models that were further developed with time (van der Ploeg and Benecke, 1974; Hauhs, 1986; Manderscheid, 1991).

The primary data (flux rates of bulk deposition, TF, and outflow) are taken from the following sources:

Water fluxes: Benecke, 1984 (up to 1975); Matzner, 1988 (1982–1985); Wenzel, 1989 (1986–1987); Manderscheid (personal communication); Manderscheid et al., 1993 (1988–1991).

Ion fluxes: Matzner et al., 1982 (up to 1979); Matzner and Ulrich, 1983 (1980–1981); Matzner, 1988 (up to 1985); Wenzel, 1989 (1986–1987); Meiwes (personal communication, 1988–1991).

The lysimeter plates for the collection of percolating soil water at a 90-cm depth were renewed in December 1976.

3.2. Mean Values and Temporal Trends

In Table 1.1, the arithmetic mean values of the ion fluxes in the spruce ecosystem together with the standard deviation (SD) between years are compiled. Bulk deposition and TF fluxes cover the time period 1969–1991, with the exception of NH_4 and NO_3, which were analyzed since 1971. The outflow fluxes cover the period 1973–

TABLE 1.1. **Solling F1: Mean Values and Standard Deviation (SD) of the Ion Fluxes**

	Na	K	Mg	Ca	H	Mn	Al	NH_4	SO_4	Cl	NO_3
						(kmol$_c$ ha^{-1} yr^{-1})					
BD[a] 1969/1971–1991	0.36	0.09	0.13	0.42	0.73	0.01	0.10	0.84	1.31	0.45	0.61
SD	0.09	0.02	0.05	0.20	0.28	0.01	0.06	0.15	0.29	0.11	0.11
r	ns	ns	−**	−***	−*	ns	−*	ns	−***	ns	ns
TF[b] 1969/1971–1991	0.78	0.71	0.38	1.48	2.82	0.19	0.29	1.23	4.83	1.09	1.16
SD	0.18	0.12	0.07	0.30	1.06	0.05	0.11	0.32	1.11	0.21	0.23
r	ns	ns	ns	−***	−*	ns	ns	+***	−***	ns	+*
ID^{c}_{part} 1969/1971–1991	0.43	0.11	0.16	0.51	0.91	0.02	0.13	1.01	1.64	0.55	0.75
SD	0.13	0.04	0.09	0.24	0.46	0.01	0.09	0.34	0.71	0.19	0.26
r	ns	ns	−*	−***	−*	ns	−*	ns	−*	ns	ns
ID^{d}_{gas} 1969/1971–1991	—	—	—	—	1.98	—	—	0.05	1.90	0.10	0.03
SD					0.65			0.25	0.64	0.10	0.07
r					ns			ns	ns	+**	—
TD[e] 1969/1971–1991	0.78	0.19	0.30	0.93	3.63	0.03	0.24	1.90	4.83	1.09	1.39
SD	0.18	0.05	0.14	0.41	1.03	0.02	0.14	0.34	1.11	0.21	0.28
r	ns	ns	−*	−***	−*	ns	−*	ns	−***	ns	ns
TD–TF[f] 1969/1971–1991	0	−0.52	−0.09	−0.55	0.81	−0.16	−0.05	0.68	0	0.02	0.23
SD		0.12	0.11	0.29	0.57	0.04	0.09	0.34		0.04	0.21
r		ns	−*	ns	ns	ns	ns	−**		—	ns

(continued)

TABLE 1.1. Solling F1: Mean Values and Standard Deviation (SD) of the Ion Fluxes *(Continued)*

	Na	K	Mg	Ca	H	Mn	Al	NH_4	SO_4	Cl	NO_3
SO[g] 1973–1991	0.85	0.09	0.46	0.51	0.36	0.32	6.26	0.02	6.07	1.18	1.08
SD	0.30	0.04	0.19	0.28	0.17	0.15	3.21	0.02	2.93	0.41	0.74
r	ns	ns	ns	ns	ns	ns	ns	ns	ns	ns	ns
TD–SO[h] 1973–1991	−0.03	0.10	−0.19	0.25	3.25	−0.28	−6.03	1.93	−1.29	−0.06	0.34
SD	0.36	0.04	0.20	0.39	1.02	0.15	3.22	0.31	3.03	0.46	0.71
r	ns	ns	ns	ns	−**	ns	ns	ns	−*	ns	ns

	Na	K	Mg	Ca	H	Mn	Al	N_t	SO_4	$H_2PO_4^{-1}$
LF[i]	0.02	0.13	0.06	0.54	–	0.16	0.13	3.31	0.37	0.10
W[j]	0.01	0.17	0.10	0.44		0.13	0.01	0.87	0.07	0.04
TD–SO–W[k]	−0.04	−0.07	−0.29	−0.19		−0.41	−6.04	1.53	−1.36	−0.03

[a] Bulk deposition (BD).
[b] Throughfall (TF).
[c] Deposition of particles = dry + occult deposition (ID_{part}).
[d] Gaseous deposition (SO_2, HNO_3, HCl, and NH_3) (ID_{gas}).
[e] Total deposition (TD).
[f] Canopy budget (sink/source function of the canopy) (TD–TF).
[g] Outflow of water and of ions in the water (SO).
[h] Ecosystem budget (sink/source function of the ecosystem) (TD–SO).
[i] Litter fall (LF).
[j] Actual biomass (Wood) increment (W).
[k] Soil budget (sink/source function of the soil) (= I_{soil}–SO–U) (TD–SO–W).
Confidence level that *r* is different from zero: *, 5%; **, 1%; ***, 0.1%.

1991. As an indication of changes in the fluxes during the measuring period, the correlation coefficient r is given with the year as the independent variable. The direction of change (− tendency for decrease, + tendency for increase during the measuring period) is indicated together with the significance of r (confidence level that r is different from zero at 5%: *, 1%: **, 0.1%: ***). Table 1.2 lists the mean values of the water fluxes, of pH (weighted mean), of total N (N_t), of ion sums, and of the cation–anion budget. In BD, TF, and in the SO, total N corresponds to Kjeldahl N + NO_3. The difference between Kjeldahl N and NH_4 represents analytical errors in organic bound N (N_{org}). The parameter N_t in total deposition (TD) represents the sum of total deposition of NH_4 and NO_3 and of N_{org} in bulk deposition.

In *bulk deposition* (BD), Mn and Al have low values. Together these elements contribute 6.5% to the acidity (sum of M_a cations). This contribution should not be neglected, the deposition of Mn and Al should therefore be included in ion budgets. For all other ions the variation between years lies between 11 and 38%, with maximum values for Ca and Mg. Precipitation is quite high (1042 mm), its coefficient of variation is 19% (cf. Fig. 1.8). The cation–anion budget indicates a small cation excess (C − A), which is of the same order of magnitude as N_{org}. The standard deviation indicates that the variation is larger than for N_{org}, thus in single years errors can play a significant role. Both H^+ and NH_4 are equally significant for the acid load. During the measuring period, there is a tendency for a decrease in bulk deposition of SO_4, Ca, Mg, H and Al, and for an increase of pH. Wesselink et al. (1993) report significant decreasing time trends for the concentrations in bulk precipitation for H, SO_4 (***), Mg, Al, Cl (**), and Ca (*) between 1976 and 1991.

A comparison of the TF data with BD reveals that the pH in *throughfall* is much lower, this indicates the interception deposition of protons (H^+, strong acidity) by the canopy. On the other hand the acidity degree is lower (57%), indicating that buffering reactions during the canopy passage take place. The increase in flux rate for NH_4 and NO_3 is lower than for Na and Cl. This result indicates that the canopy may act as a sink for deposited nitrogen. A comparable increase in flux rate is shown for Ca, H, and SO_4. The highest increases are reached by K and Mn, indicating a source function of the canopy due to the leaching of these ions from needles. During the measuring period, a decreasing tendency for the fluxes with TF exists for SO_4, Ca, H, and N_{org}, whereas the fluxes of NH_4 and NO_3, as well as pH, tend to increase. Wesselink et al. (1993) report significant decreasing time trends for the concentrations in TF for H, SO_4 (***), Al (**), and Ca (*) between 1976 and 1991.

The calculated *particle deposition* (ID_{part}) is in the mean slightly higher than bulk deposition (factor 1.25, SD 0.45). This factor varies annually between 0.44 and 2.18, indicating a great annual variability of the conditions for particle (cloud water) deposition. The annual variability lies between 33 and 42%, it is higher only for Mg. Particle deposition shows a decreasing tendency especially in the case of Ca, but also for H, SO_4, Al, and Mg.

The calculated *deposition of gases* (ID_{gas}) is significant only for SO_2. The calculated deposition of HCl contributes only 1.3% to the acid load by deposition (total deposition of M_a cations; cf. Table 1.2), but it shows a tendency to increase,

TABLE 1.2. Solling F1: Mean Values and Standard Deviation (SD) of Water Fluxes, pH, and Ion Budgets

	H_2O (mm)	pH	N_{org}	N_t	M_b[a]	M_a[b] (kmol$_c$ ha^{-1} yr^{-1})	Cation Sum[c]	Anion Sum[d]	C–A[f]	Acidity degree[e] (%)
BD[g] 1969/1971–1991	1041	4.20	0.30	1.74	0.99	1.68	2.66	2.34	0.32	63
SD	209	0.26	0.13	0.28	0.24	0.40	0.53	0.37	0.32	6
r	ns	+**	ns	ns	−**	ns	−**	−**	ns	ns
TF[h] 1969/1971–1991	766	3.46	0.63	2.92	3.34	4.50	7.84	7.01	0.83	57
SD	172	0.21	0.33	0.58	0.51	1.10	1.41	1.30	0.69	5
r	ns	+***	−*	+*	−*	ns	−*	−*	ns	ns
TD[i] 1971–1991	1041	–	–	3.59	2.19	5.75	7.93	7.24	0.70	72
SD				0.55	0.53	1.30	1.56	1.36	0.70	5
r				ns	−***	ns	−**	−*	ns	ns
TD–TF[j] 1969/1971–1991	275	–	–	0.58	−1.15	1.25	0.10	0.23	−0.13	–
SD	74			0.49	0.45	0.64	0.59	0.21	0.55	
r	ns			ns	ns	ns	ns	ns	ns	

SO[k] 1973–1991	438	4.10	0.16	1.26	1.91	6.96	8.87	8.22	0.65	78
SD	166	0.15	0.06	0.79	0.71	3.46	4.10	3.71	0.63	6
r	ns	+*	ns	ns	ns	ns	ns	ns	ns	+***
TD–SO[l] 1973–1991	597	–	–	2.40	0.13	−1.13	−1.00	−1.01	0.01	–
SD	99			0.81	0.87	3.41	4.19	3.73	0.89	
r	ns			ns	ns	−*	−*	−*	ns	

[a] M_b = Na, K, Mg, and Ca.
[b] M_a = H, Mn, Al, and NH_4.
[c] Cation sum = $M_a + M_b$.
[d] Anion sum = SO_4, Cl, and NO_3.
[e] Acidity degree in % = (M_a/cation sum) × 100.
[f] Cation sum minus anion sum = C–A.
[g] Bulk deposition = BD.
[h] Throughfall = TF.
[i] Total deposition = TD.
[j] Canopy budget (sink/source function of the canopy) = TD–TF.
[k] Outflow of water and of ions in the water = SO.
[l] Ecosystem budget (sink/source function of the ecosystem) = TD–SO.
Confidence level that *r* is different from zero: *, 5%; **, 1%, ***, 0.1%.

however. Deposition of HNO_3 seems to be negligible. This number may be an erroneous result of the calculation, since Lindberg et al. (1990) found a high removal rate for HNO_3 in the adjacent beech stand. The HNO_3 deposition seems therefore to be underestimated. The high deposition of SO_2 can be explained by relative high SO_2 concentrations (e.g., in 1984: annual mean 28 μg of SO_2 m^{-3}, now it is ~19) and a long wetness duration of the canopy. Since the surface uptake of SO_2 is followed by dissolution and oxidation to H_2SO_4, it is connected with an equivalent production of protons. Sulfur dioxide deposition is in the mean responsible for 34% of the acid load by deposition.

The annual variability of *total deposition* (TD) is for most ions small (17–25%). The higher variability of Ca, Mg, Al, and Mn may be due to acid–base reactions of deposited protons with dry or occult deposited particles (which probably originate primarily from soil dust) at the needle surface, and to needle leaching. The cation–anion budget (C–A) reveals an anion deficit that accounts for 9% of the total deposition, compared to 12% of bulk deposition. The acidity degree of total deposition (72%) is higher than that of bulk deposition, indicating that interception deposition contributes more acidity than alkalinity.

The *canopy budget* (TD–TF) is zero by definition for Na. It is also zero for SO_4^{2-} and only slightly different from zero with high variation for Cl, Al, and the cation–anion budget (C–A). A source function of the canopy is calculated for K, Ca, Mn [coefficients of variation (CV) 23 − 53%], and Mg (CV = 120%). This finding can be due to the dissolution of dry deposited particles, and of leaching from the needles (especially K and Mn). In 1969, 1971, and 1972 the canopy acted as sink for Mg, but since then it has acted as a source. The stand shows periodic yellowing as a sign of latent Mg deficiency. The reason for the high variability of Mg is discussed in Section 3.3. A sink function of the canopy is calculated for protons, NH_4 (with decreasing tendency), and NO_3. This reflects an acid buffering capacity of the canopy and the uptake of nitrogen from deposition. The calculated nitrogen uptake corresponds to 22% of total N uptake in the above-ground part of the stand and to 25% of the totally deposited N. The uptake of NH_4 and NO_3 from deposition has been experimentally demonstrated for beech (Brumme et al., 1992) and spruce (Eilers et al., 1992), both tree species take up NH_4 preferentially. Johnson et al. (1993) report that on the average, in North American case studies with considerably lower N deposition (<1.3 kmol ha^{-1} $year^{-1}$), 40% of the deposited inorganic N was retained by the canopy.

According to the acid–base budget (Table 1.3), the acid load of the canopy results from the sink function of H^+ and NH_4. The acid buffering is due mainly to the leaching of M_b cations and of Mn, the assimilation of nitrate plays a minor role. Twenty eight percent of the acidity deposited is buffered during the canopy passage.

The *outflow* (SO, in 80–100-cm soil depth) of the transport medium water is relatively high (438 mm). A variation coefficient of 38% indicates substantial outflow rates in each year (cf. Fig. 1.8). This finding may not be the case in ecosystems with lower precipitation where the variation coefficients can be much higher. The variation coefficients for the output fluxes of most ions are at 35–55%. Exceptions are NH_4, NO_3, and the anion deficit (C–A, 69–100%). In case of NH_4, the output

TABLE 1.3. Solling F1: Canopy Acid–Base Budget

	$kmol_c$ ha^{-1} yr^{-1}	% of $AL_{I\text{–}O}$
Acid Load from Input–Output Budget ($AL_{I\text{–}O}$) (see p. 5)		
1. Retention of acid input (H^+: 50%, NH_4: 42% of $AL_{I\text{–}O}$)	1.49	92
2. Release of anions (C_{org}: 8%)	0.13	8
3. Retention of deposited M_b cations	0	
Sum = $AL_{I\text{–}O}$	1.62	100
Acid–Base Reactions (see p. 6)		
4. Release of M_a cations (Mn: 10%, Al: 3%)	0.21	13
5. Retention of deposited anions (NO_3: 14%, Cl: 1%)	0.25	15
6. Release of M_b cations (K: 32%, Mg: 6%, Ca: 34%)	1.16	72

flux is always very low. The cation excess (anion deficit) amounts to 7% of the cation sum; it is much higher than the flux of N_{org}. Dominating ions in the outflow are Al and sulfate (70 and 68% of cation sum, respectively). The concentration of dissolved organic carbon (DOC) is low (~5 mg Cl^{-1}), Al is almost totally inorganic bound (98%) (Dietze and Ulrich, 1992), a considerable fraction exists as $AlSO_4^+$. The acidity degree is higher than in total deposition (TD) and is increasing, this indicates the dissolution of acidity (aluminum sulfates) from soil. The pH lies with 4.1 in the Al buffer range, it shows a weak tendency to increase. Wesselink et al. (1993) report significant increasing time trends for the concentrations in outflow water for H, Mg (***), Na, Al, SO_4 (**), and Ca (*) between 1973 and 1978, as well as decreasing time trends between 1978 and 1991 for Ca, Mg, H, SO_4 (***), Na, and Al (**). This finding indicates that the variation of the outflow fluxes includes opposite time trends.

The *ecosystem budget* (TD–SO) is slightly negative for Na and Cl. The large annual variability of the budget, due to the variability in precipitation and outflow, make it very uncertain to use Cl or Na output of a single year or the mean of a few years to calculate the outflow rate of water. Even after a period of 19 years of measurement of the outflow of Na and Cl, the mean budget deviates from zero due to variations in input and output. Since the Na and Cl budget are similar, there is no indication for the release of Na by weathering. The sulfate budget is negative showing a high variability. The budgets of Mn and Al are negative (leaching). Aluminum shows a similar standard deviation as sulfate. Both H^+ and NH_4 are strongly retained (to 90 and 99%, respectively), while NO_3 is slightly retained (to 24%). Of the M_b cations, the budget of Mg is negative, while K and Ca are positive.

The acid–base budget given in Table 1.4 is calculated as a mean of the different years and not from the mean fluxes compiled in Table 1.1. Therefore, it shows higher values for such ions that have positive and negative budgets in different years, such events balance each other in calculating mean fluxes. The acid–base budget shows a mean annual acid load of 8.04 $kmol_c$ ha^{-1}, 64% is due to the sink function of the ecosystem for acid input (H^+ and NH_4), 31% is due to the source function for anions (mainly SO_4), and 5% is due to the sink function for M_b cations (Ca and K). In the acid–base reactions, the source function for M_a cations plays the dominant role (78% of acid buffering). This finding is characteristic for ecosystems with acid subsoils (Al buffer range) and indicates that the acute and historical input of strong acidity is transformed into weaker acidity and transferred with the seepage water to the seepage conductor. The acute strong acidity corresponds to the deposition of H^+ and NH_4, the historical strong acidity corresponds to the source function of SO_4. Nineteen percent of acid buffering occurs from the retention of sulfate (as aluminum sulfate) and the transformation of nitrate to uncharged compounds, 3% by the net leaching of M_b cations, mainly Mg.

The *soil budget* (TD–SO–W) is given in Table 1.1. It is negative for the nutrients K, Mg, Ca, for the cation acids Mn, Al, and for SO_4, and it is positive for N. The acid load of the soil due to forest increment amounts to 0.72 $kmol_c$ $ha^{-1}year^{-1}$. Thus the total acid load of the soil is 8.76 $kmol_c$. The forest increment contributes

TABLE 1.4. Solling F1: Ecosystem Acid–Base Budget

	$kmol_c$ ha^{-1} yr^{-1}	% of AL_{I-O}
Acid Load from Input–Output Budget (AL_{I-O})		
1. Retention of acid input (H^+: 40%, NH_4: 24% of AL_{I-O})	5.18	64
2. Release of anions (SO_4: 25%, NO_3 2%, C–A 4%)	2.47	31
3. Retention of deposited M_b cations (Ca: 4%, K: 1%)	0.39	5
Sum = AL_{I-O}	8.04	100
Acid–Base Reactions		
4. Release of M_a cations (Mn: 3%, Al: 75%)	6.31	78
5. Retention of deposited anions (SO_4: 9%, NO_3: 6%, C–A: 4%)	1.52	19
6. Release of M_b cations (Mg: 2%, Ca: 1%)	0.22	3

9%. In other spruce, pine, and beech ecosystems of northwest Germany, the contribution of forest increment to the total acid load of the soil was found to be less than 30% (Bredemeier, 1987). In the spruce ecosystem in the Solling there is also an accumulation of an organic top layer from decomposition residues of fine root litter (Ulrich, 1989). Nitrogen is totally accumulated in the organic top layer. The budget of the mineral soil is even more negative for the M_b cations since they are also accumulated in the organic top layer. A separation is very difficult, however. If we take into consideration the storage changes of exchangeable cations in soil, Matzner (1988, 1989) calculated a release by silicate weathering to a soil depth of 1 m of 0.08 and 0.36 $kmol_c\ ha^{-1}\ year^{-1}$ for K and Mg, respectively. The release of K reflects the weathering of illites to vermiculites (illites are the dominating clay minerals). The release of Mg reflects the destruction of 2:1 layer silicates in the most acid top soil (Al–Fe buffer range) (Flehmig et al., 1990). There is no indication of the release of Ca by silicate weathering.

3.3. Temporal Patterns and Their Causes

For many fluxes, the linear regression coefficients given in Tables 1.1 and 1.2 indicate a temporal trend. In this section, the temporal patterns are discussed on the basis of the annual fluxes.

In Figure 1.2 the deposition of S is shown. Sulfur input has had maximum values from the mid 1970s to the early 1980s. In this period the deposition of SO_2 as a gas was highest. The lowest values were reached after 1989. This development can be

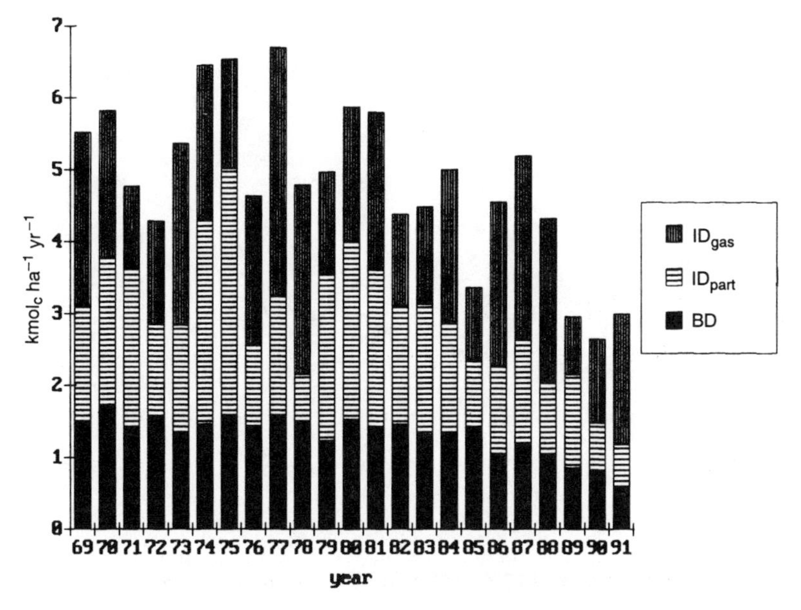

Fig. 1.2. Solling F1: Bulk deposition (BD), particle deposition (ID_{part}), and gaseous deposition (ID_{gas}) of S in $kmol_c\ ha^{-1}\ yr^{-1}$.

compared with the development of the emission of SO_2 as compiled in Table 1.5. The comparison shows that S deposition in the spruce forest in Solling is considerably higher than the emission density in West Germany, but the time trends are similar. The high deposition of S in the Solling can be traced back to the high emission rates in East Germany. The sharp falling-off of S deposition since 1989, which shows up in bulk deposition, as well as in particle (cloud water) and SO_2 deposition, is due to emission control measures in West Germany. But it is also due to the decline of industrial activity in East Germany after the reunification in 1990.

This change in the emission scenario is reflected in the temporal development of weighted mean annual pH values in bulk deposition, throughfall, and even in the outflow (Fig. 1.3). The pH in bulk deposition increased since 1986 and had for the first time reached a value above 5 (5.02) in 1991. At the same time the pH in throughfall also increases but remains below pH 4. This finding shows that there is still interception deposition of strong acidity due to dry deposition of SO_2. The pH in outflow passed through a minimum between 1975 and 1981, since then there is no marked tendency for increase.

In Figure 1.4 total N deposition and the fluxes of NH_4 + NO_3, as well as that of NH_4 alone in throughfall, are shown together with the 3 years sliding mean. The variability in the deposition of total N is high (cf. Table 1.2), but there is no time trend recognizable (Manderscheid et al., 1993). Between the periods of 1971 and 1975 and 1986 and 1990, Matzner and Meiwes (1993) calculated a percent increase in bulk deposition of 11 and 24% for NH_4 and NO_3, and in throughfall of 40 and 30%. According to the emission data (Table 1.5), an increase in deposition of NO_3 since 1975 is not to be expected. The emission of NH_3 from agriculture (feedlots) is

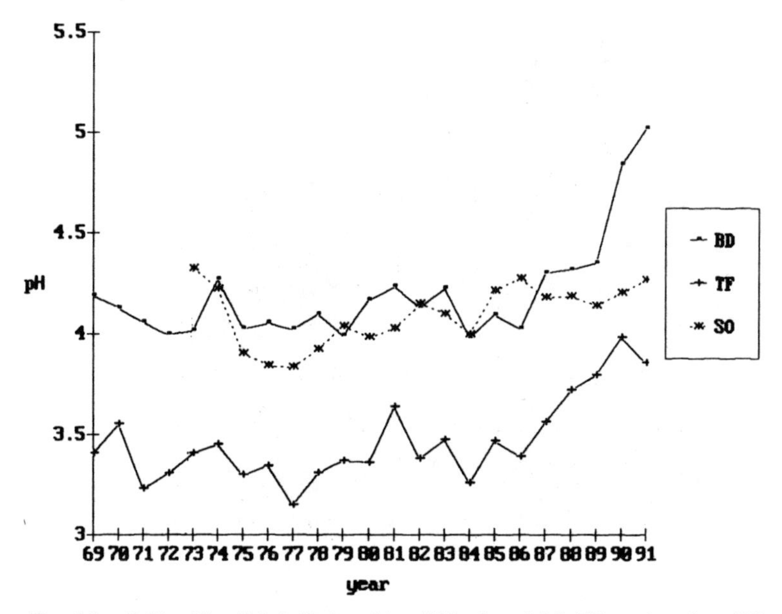

Fig. 1.3. Solling F1: pH in bulk deposition (BD), throughfall (TF), and outflow (SO).

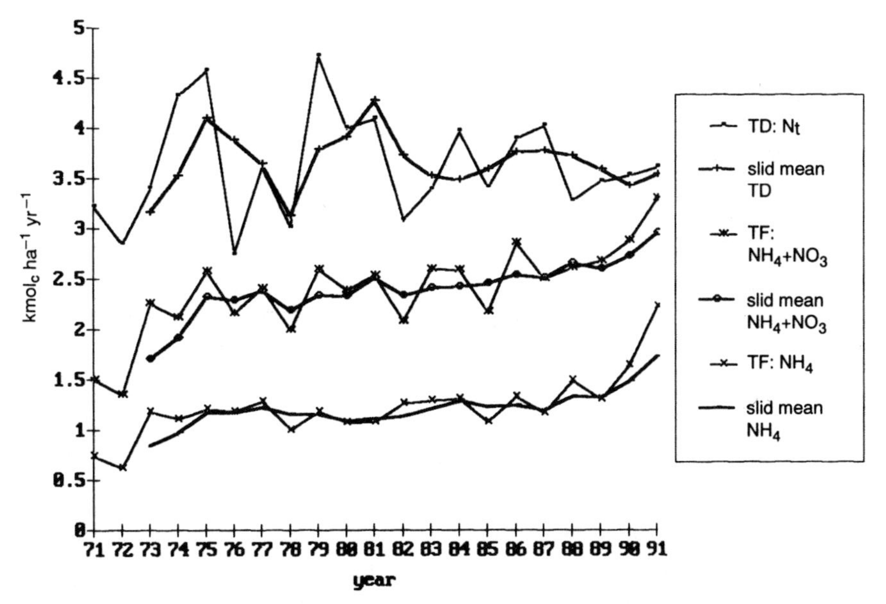

Fig. 1.4. Solling F1: Total deposition (TD) of total nitrogen (N_t), as well as fluxes of NH_4 and NH_4 + NO_3 in throughfall in kmol$_c$ ha^{-1} yr^{-1}.

TABLE 1.5. Emission of SO_2 and NO_x in West, East, and Total Germany

		1966	1970	1975	1980	1985	1990
SO_2	West	4.3	4.7	4.2	4.0	3.0	1.2
	East	?	?	11.9	12.5	15.6	13.8
	Total	—	—	6.9	6.6	6.8	5.0
NO_x	West	1.7	2.1	2.2	2.6	2.6	2.3
	East	?	?	1.2	1.2	1.3	1.3
	Total	—	—	1.9	2.2	2.2	2.0
Acidity	West	6.0	6.8	6.4	6.6	5.6	3.5
	East	—	—	13.1	13.7	16.9	15.1
	Total	—	—	8.8	8.7	9.0	7.0
NH_3	West	?	1.2	1.3	1.3	1.4	1.3
	East	?	1.2	1.4	1.4	1.5	1.1
	Total	—	1.2	1.3	1.4	1.4	1.3
NO_x + NH_3	West	—	3.3	3.5	3.9	4.0	3.6
	East	—	—	2.6	2.6	2.8	2.6
	Total	—	—	3.2	3.5	3.6	3.3

[a]Data expressed as emission density in kmol$_c$ ha^{-1} yr^{-1} of the area of West, East, and total Germany, respectively.
Source: BMU, 1992; Umweltbundesamt, 1993.

not well documentated, but may be at a high level since the mid 1970s. The increase of NH_4 and NO_3 fluxes in throughfall thus do not reflect increases in deposition, but decreases in canopy uptake. This result is confirmed by the data on canopy uptake presented in Fig. 1.5. There was no nitrate uptake by the canopy from 1976–1978. This finding correlates with a nitrification pulse in the warm–dry years 1975–1976 and high nitrate outputs in 1977–1978. Thus, canopy uptake of NO_3 and NH_4 decreased during this period when the NO_3 concentration in soil solution was high. The decreasing trend of N uptake by the canopy from the early 1980s onward can be interpreted as approaching N saturation of the stand. In 1991, the calculation yielded no N uptake by the canopy. An increase in nitrate deposition during the coming years is to be expected in contrast to the prediction for West Germany (Table 1.5) as a consequence of increasing NO_x emissions in East Germany resulting from an increase in car traffic.

Since SO_2 and NO_x are acid precursors, their emission results in the formation of an equivalent amount of acidity (cf. Tables 1.1 and 1.2). In Figure 1.6 the total deposition of acidity is compiled. The deposition of acidity in the spruce forest in the Solling is of comparable magnitude to the emission density in West Germany. Up to the mid 1980s the variation in the emission of acid precursors in West Germany was low, whereas the variation in deposition of acidity was high (see Fig. 1.6). This variation is not coupled to precipitation, bulk deposition shows only minor variation. It is mainly due to changes in either particle or gaseous deposition and does probably reflect changes in the climatic conditions for cloud water impaction and canopy wetness (dry deposition of SO_2). As for S, 1989–1991 are the 3 years with the lowest values. The decrease in these years is evident for protons in bulk, particulate, and gaseous deposition. The first year where the calculation yields a gaseous input of NH_3 is 1991. This could be a consequence of the pH of bulk

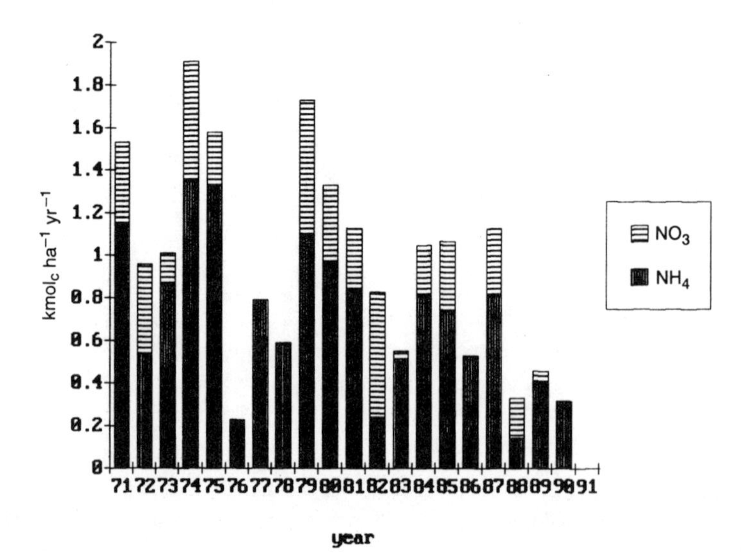

Fig. 1.5. Solling F1: Canopy budget (TD–TF) of NH_4 and NO_3 in kmol$_c$ ha^{-1} yr^{-1}.

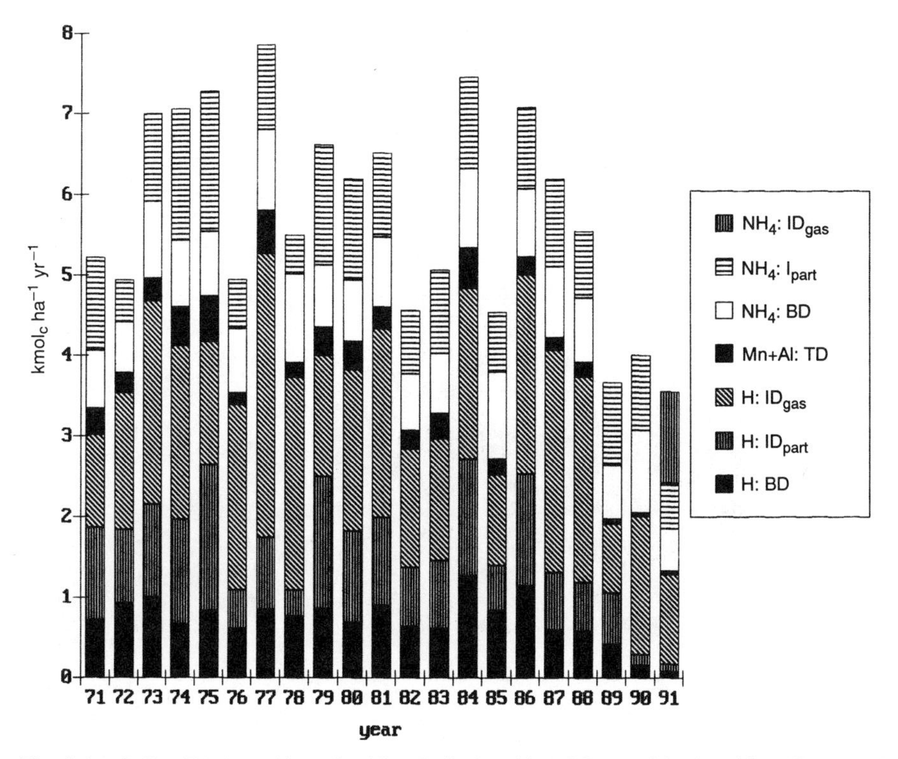

Fig. 1.6. Solling F1: Deposition of acidity: Bulk deposition (BD), particle deposition (ID_{part}), and gaseous deposition (ID_{gas}) of protons (H) and ammonium (NH_4), as well as total deposition (TD) of manganese and aluminum (Mn + Al) in kmol$_c$ ha^{-1} yr^{-1}.

deposition (1991: 5.02), which indicates that rain water is no longer a strong sink for NH_3. Thus, NH_3 from feedlots may be distributed over large regions and captured by wet canopy surfaces where SO_2 is dissolving and creating low pH. Measurements of NH_3 concentration in air showed a concentration increase in the canopy (Gravenhorst, 1992), this could indicate that from dry canopy surfaces NH_3 is emitted. Further measurements are necessary to quantify the processes and fluxes of NH_3 under the changed air quality conditions.

The decrease in the pH of bulk deposition may be the reason for a strong decrease in the deposition of dissolved Al since 1989 (Fig. 1.7). From this development we conclude that the deposition of Al in earlier years was a consequence of reactions between strong acids and Al compounds (primarily soil particles) in the canopy.

To a large extent, precipitation determines the temporal pattern of fluxes in the ecosystem, especially of the outflow. The temporal pattern of precipitation (BD) and outflow (SO) is represented in Fig. 1.8. The outflow of most ions is correlated with the outflow of the transport medium water. For Na, K, Mg, Al, SO_4, Cl, and NO_3 the correlation coefficients r lie between 0.79*** and 0.89. The correlation is weaker for H and Mn (r = 0.72*** and 0.77, respectively), and is not significant for Ca and the cation–anion budget C–A. This finding is reflected in the variability

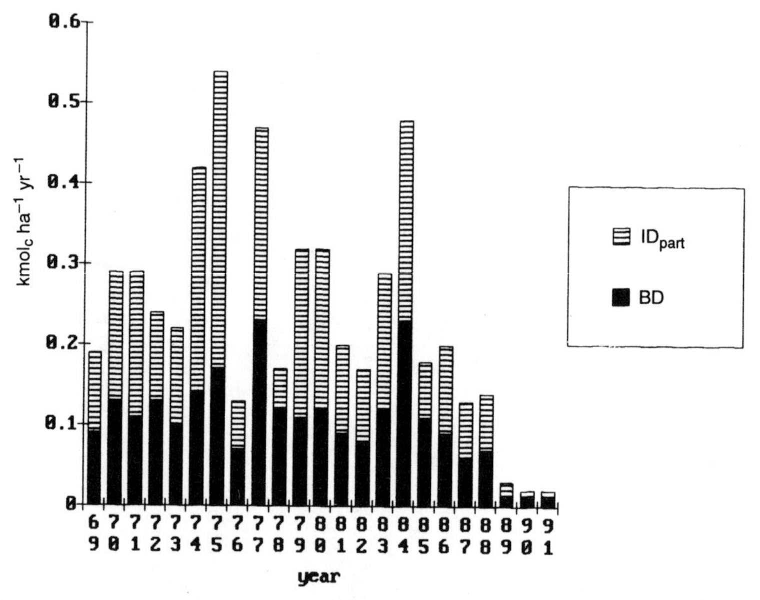

Fig. 1.7. Solling F1: Bulk deposition (BD) and particle deposition (ID_{part}) of Al in kmol$_c$ ha^{-1} yr^{-1}.

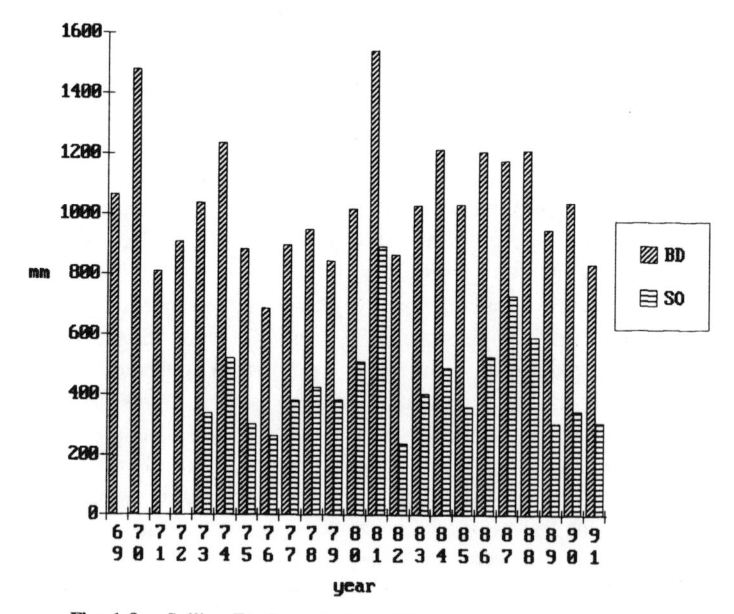

Fig. 1.8. Solling F1: Precipitation (BD) and outflow (SO) of water.

of the weighted mean annual concentrations: for K, Cl, Mg, and Na the CV lies between 20 and 25%, for SO_4, Mn, H, and Al between 27 and 38%, for NO_3 and Ca it is 40 and 48%, respectively. Significant tendencies for a decrease in the concentrations of the outflow exist for K (***) as well as for H, Mn, and Al (*). For ecological parameters that are dependent on climatic and biological variability, coefficients of variation below 25% are quite low.

The close correlation with water outflow results in large differences of the annual ecosystem budget for Cl and Na (Fig. 1.9). Generally, both fluxes are correlated, but the budget of both ions can also be strongly discoupled on an annual basis. Although the coefficient of variation of chloride concentration in water outflow is low (22%), the maximal deviations from the annual mean (0.275 mmol L^{-1}) are substantial (min 0.17, max 0.40). This variation in Cl concentration is too high in order to use annual chloride budgets for assessing annual water outflows. The Cl budget can be positive in years with high outflow (e.g., 1988), and negative in years with low outflow (e.g., 1980 and 1991). It is remarkable that there may be periods of 5 years where the Cl and Na budget deviate in the same direction.

The nitrate outflow is well correlated with water outflow (r = 0.85; cf. Fig. 1.10), but the nitrate concentration in outflow water shows a higher variability (CV = 40%). Nitrate is the only ion where the concentration is correlated with water outflow: there is a tendency for the nitrate concentration to increase with an increasing outflow rate (r = 0.53*). The highest nitrate concentration was reached in 1987, the year with the second highest outflow rate. There is also a tendency that the nitrate output in years with high outflow rates is increasing with time (cf. Fig. 1.10). According to this temporal pattern, N saturation of the ecosystem may result in

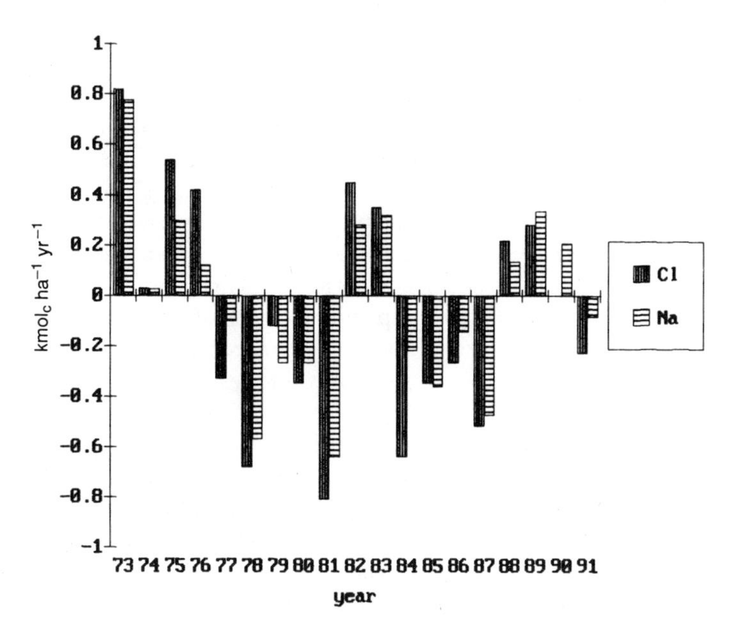

Fig. 1.9. Solling F1: Ecosystem budget (TD–SO) of chloride (Cl) and sodium (Na) in $kmol_c\ ha^{-1}\ yr^{-1}$.

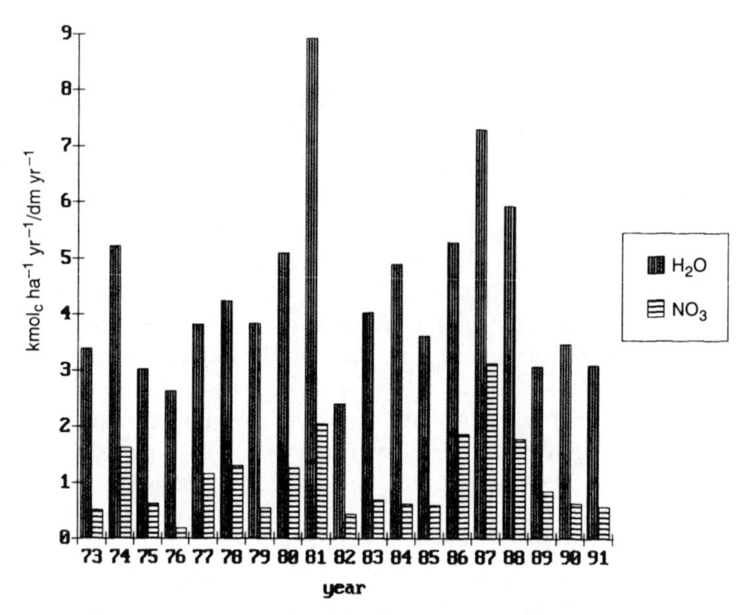

Fig. 1.10. Solling F1: Outflow of water (H_2O, in dm yr^{-1}) and of nitrate (NO_3, in kmol$_c$ ha^{-1} yr^{-1}).

increasing leaching rates of nitrate especially in wet years with high outflow rates. In 1987, nitrate leaching amounted to 78% of the total N deposition.

Sulfate and aluminum are the dominant ions in the outflow (68 and 71% of the cation sum, respectively). Their close correlation with water outflow is depicted in Figure 1.11. Sulfate outflow strongly follows the outflow of water. The fluxes of SO_4 and Al are strongly coupled ($r = 0.96$). These relationships indicate that the outflow regime has been dominated by acid–base reactions of protons that originate from the input of H^+ and NH_4, and by the transport of sulfate through the soil column. The protons react mainly in the top soil and release Al from clay minerals that are destructed. In the strongly acidified top mineral soil (Fe–Al or Fe buffer range according to Ulrich, 1981) aluminum chlorites, which have been considered as the final stage of clay mineral development in acid soils, are no longer stable. The aluminum hydroxy layers are dissolved, resulting in freely expanding smectites (Frank and Gebhardt, 1991). Further steps in the degradation of clay minerals are the loss of K from vermiculites and of Mg from expanded smectites (Flehmig et al., 1990). The final reaction products are amorphous silicates that are rich in Si and that cover the surfaces of minerals and aggregates (Veerhoff and Brümmer, 1991). The Al concentrations of the soil solution in the upper horizons indicate undersaturation with respect to gibbsite, alunite, and jurbanite (Matzner and Prenzel, 1992). The release of Al under these conditions seems to be restricted by kinetic constraints. These constraints can be due to diffusion (from pore walls into aggregates and back) and to the dissolution rate of minerals. The statement of Binkley (1992) that the export of acidity in the form of Al results in the reduction of the net H^+ load to the ecosystem is not correct: the release of Al is the result of an acid–base reaction

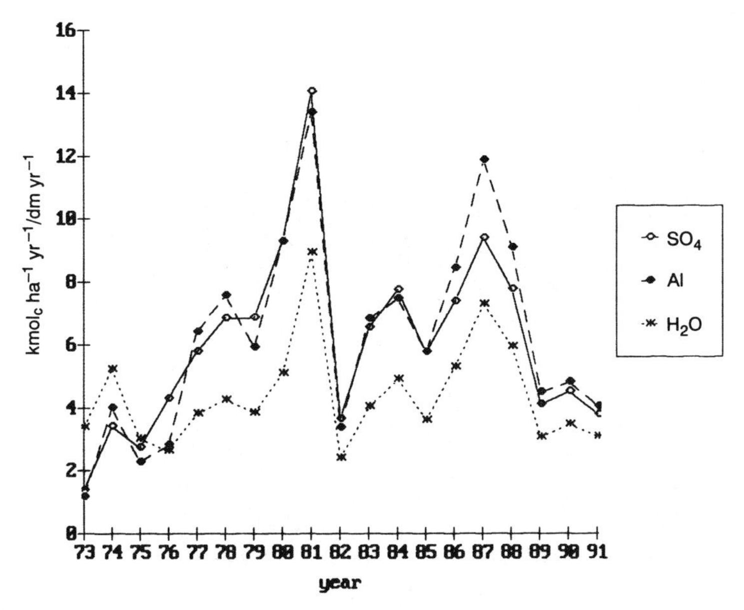

Fig. 1.11. Solling F1: Outflow of sulfate (SO_4) and aluminum (Al) in kmol$_c$ ha^{-1} yr^{-1} and outflow of water (H_2O, in dm yr^{-1}).

between protons and silicate (clay) minerals. The ecosystem and soil budget clearly show that the protons are retained in the soil and that they ultimately destabilize the clay minerals and lead to an irreversible reduction of cation exchange capacity.

In Figure 1.12 the ecosystem budget (TD–SO) of sulfur is shown, together with the water outflow. Between 1973 and 1977 sulfate was retained in the soil. Since then the sulfate budget was negative with the exception of 1982, the year with the lowest water outflow. High negative S budgets correlate with high water outflows. In Figure 1.13 the mean annual concentrations of sulfate, aluminum, and protons in the outflow water are represented (for ease of comparison, the proton concentrations were multiplied by 10). These data show a distinct time trend. Proton concentration increased from 1973 to 1977, corresponding to a pH change from 4.33 to 3.83 (weighted annual mean), since then the time trend mirrors a decrease with fluctuations (increase of pH, Fig. 1.3). The fluctuations (pH decrease) follow warm–dry years (1975–1976 and 1982–1983) and reflect nitrification pulses (acidification pushes) in the topsoil. The increase in proton concentration at the beginning of the measuring period was followed with a time delay of 1–2 years by a strong increase in SO_4 and Al concentrations. This time delay may mainly reflect the effect of soil chemical heterogeneity: The reactions at the surfaces of the soil pores where the acid water is percolating (e.g., dissolution of aluminum sulfates) are diffusion limited. The concentrations of Al and SO_4 are more or less equivalent, a difference in concentrations occurs in years with higher nitrate output (1986–1988) and reflects Al leaching as nitrate. The main process responsible for the negative S budget can be traced back to the dissolution of an aluminum sulfate with the mole ratio

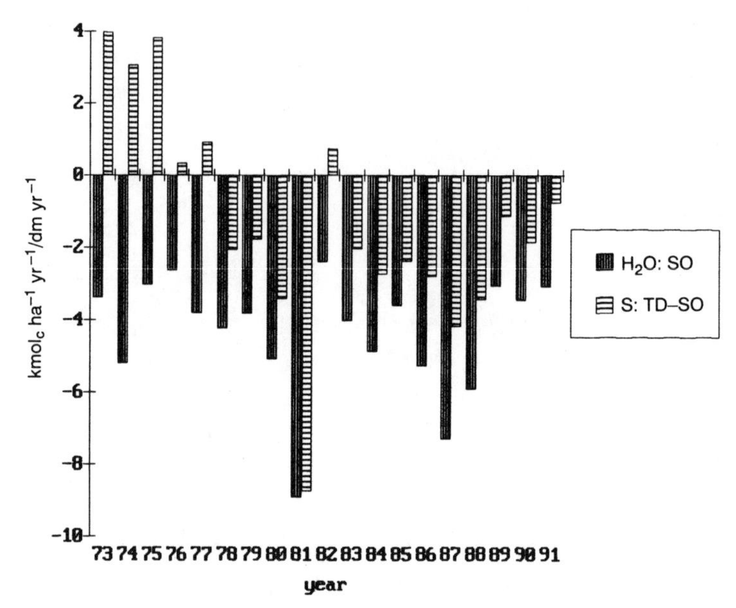

Fig. 1.12. Solling F1: Ecosystem budget (TD–SO) of sulfur [(I − O) SO_4] in kmol$_c$ ha^{-1} yr^{-1} and outflow of water (in dm yr^{-1}).

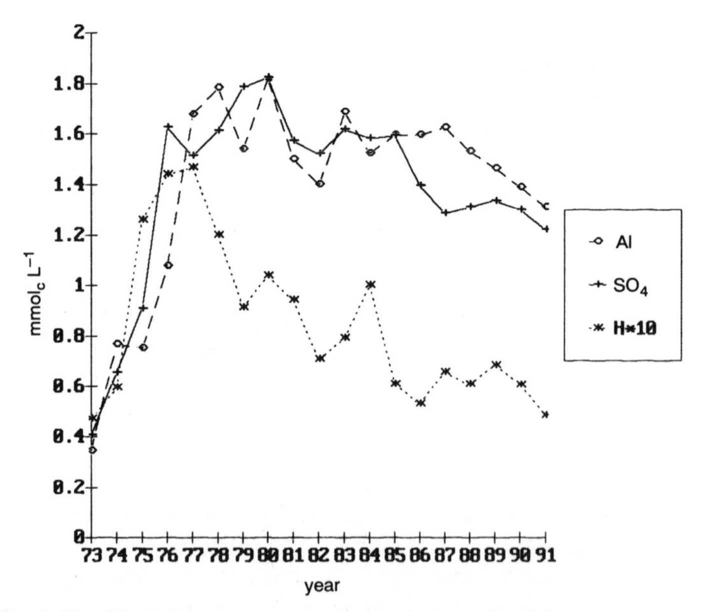

Fig. 1.13. Solling F1: Weighted mean annual concentrations of aluminum (Al), sulfate (SO_4), and protons (H) in the outflow water in mmol$_c$ L^{-1} (for ease of comparison the proton concentration has been multiplied by 10).

$Al:SO_4 = 1$. This result is in agreement with the findings of Prenzel (1983) and Förster (1986). Matzner and Prenzel (1992) found a relative constant ion product $pAl + pOH + pSO_4 = 17.2$ (SD = 0.2). The solubility of such aluminum sulfates is pH dependent. The pH decrease is not correlated with the deposition of acidity in these years. It has to be traced back to ecosystem internal processes (acidification pulses) that can no longer be buffered due to the decrease in base saturation caused by long-term acid deposition.

The output flux of Ca is only weakly correlated with water outflow. The Ca concentration shows a relative high variation. Figure 1.14 shows the time trend in input (TD) and output (SO) for Ca. The input shows maximum values between 1970 and 1972 and an almost steady decrease since 1982 (Manderscheid et al., 1993), which is also reflected in throughfall (Matzner and Meiwes, 1993). Since the early 1980s the decrease in Ca deposition is related to the decrease in SO_2 emission and in deposition of acidity. It seems to reflect a decreasing dissolution of Ca bearing particles in the air. The output was low from 1973 to 1977, during the phase of increasing proton, sulfate, and aluminum concentrations (cf. Figure 1.13). It remained at a high level till 1984 with the exception of 1982, the year with the lowest water outflow. The weighted mean annual Ca concentration in the outflow water (Fig. 1.15) varied greatly till 1983 with maximum values from 1978 to 1980. Since then there is a decreasing tendency. This time pattern reflects a change in input only since 1983, and is not correlated with the fluctuation of water outflow. Until the early 1980s the fluctuations with high Ca concentrations seem to reflect breakdowns in the retention ability of the ecosystem for deposited Ca. This fluctuation is also

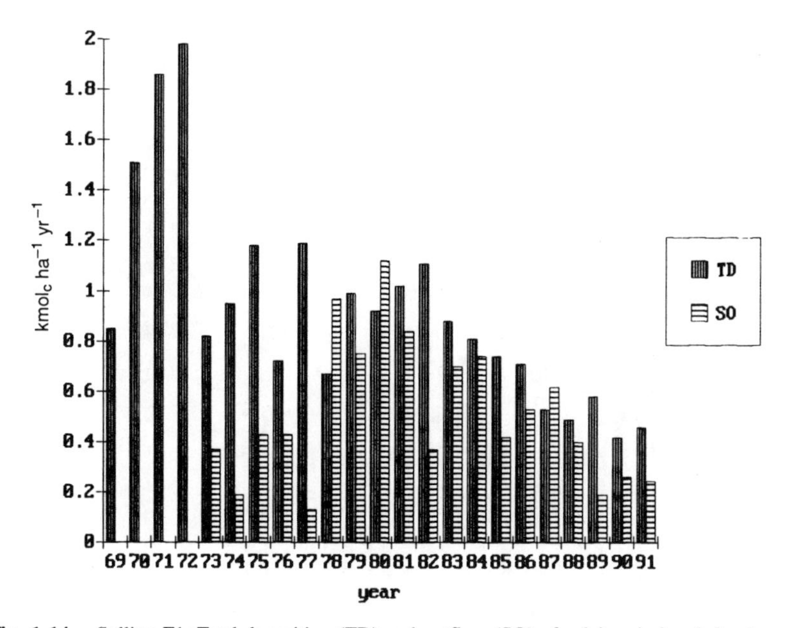

Fig. 1.14. Solling F1: Total deposition (TD) and outflow (SO) of calcium in $kmol_c\ ha^{-1}\ yr^{-1}$.

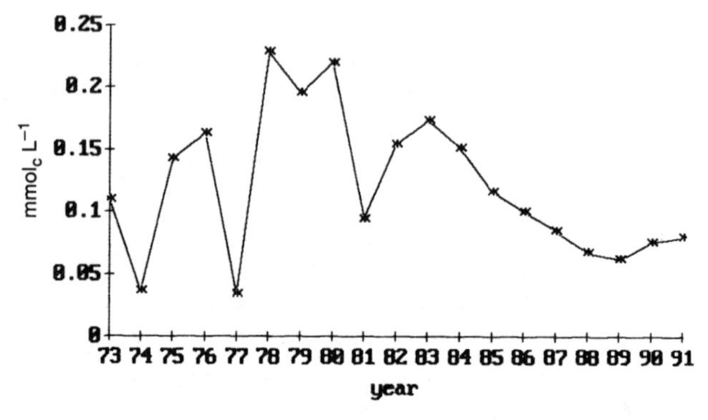

Fig. 1.15. Solling F1: Weighted mean annual Ca concentration in outflow water in mmol$_c$ L^{-1}.

reflected in a decrease of the exchangeable Ca storage in the soil (Matzner, 1988). The development of the outflow since the early 1980s indicates that changes in the input of dissolved Ca rapidly show up in the outflow indicating that the storage and retrieval function for Ca in the soil is negligible: The soil has lost a property that is important for ecosystem stability.

Magnesium has a high variability in particle deposition (cf. Table 1.1, CV = 56%) and in the canopy budget (TD–TF), where negative values (uptake by canopy) were calculated for the years 1969, 1971, and 1972. Using the ratio of Na/Mg in sea water, the contribution of sea spray to Mg deposition can be calculated. In Figure 1.16, Mg deposition is compiled separately for the sea born and land born fraction

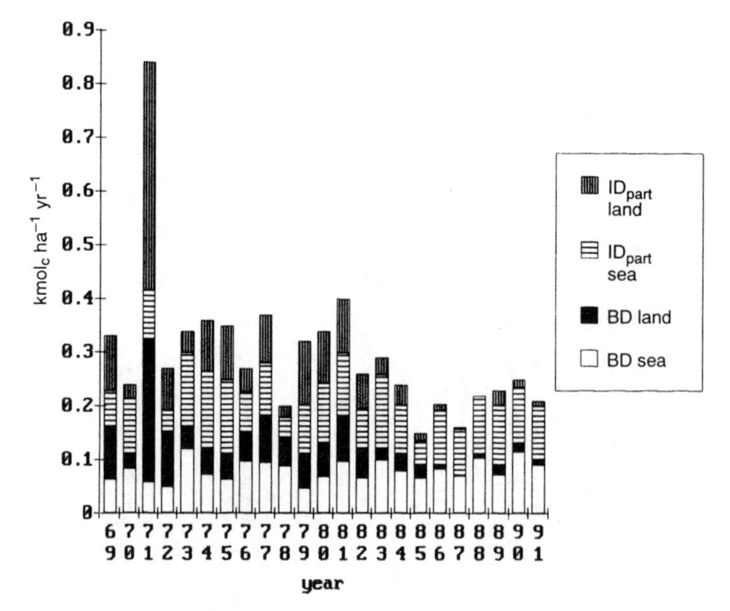

Fig. 1.16. Solling F1: Sea born (sea) and land born fractions (land) of bulk deposition (BD) and particle deposition (ID_{part}) of Mg in kmol$_c$ ha^{-1} yr^{-1}.

in bulk and particle deposition. The years with negative canopy budgets show the largest land born fractions of deposition. These data may reflect a local source of Mg bearing dry particles at the beginning of the measuring period that were collected in the samplers of bulk deposition. This finding may also explain the high Ca input from 1970 to 1972 (cf. Fig. 1.14). Under this condition the calculation algorithm may produce erraneous results (overestimation of Mg deposition in 1969, 1971, and 1972). The decreasing tendency in deposition of Mg (see Fig. 1.16) is also significant for the period from 1973 to 1991. It begins as for Ca in 1982 and may have the same reasons. Since only the land born fraction of Mg deposition is subjected to changes, this decrease is less pronounced than for Ca.

The ratios of Ca and Mg to Al concentrations have proven to be useful stress indicators. In Figure 1.17 the weighted mean annual molar Ca/Al and Mg/Al ratios are shown. Their decrease with time is significant ($r = 0.60$ and 0.82, respectively). The fluctuations at the beginning of the measuring period are much larger if a higher time resolution is considered (data presented by Matzner, 1988, 1989). This fluctuations can be explained by soil chemical heterogeneity in the soil depth where the lysimeters were placed: Slowly percolating water was still influenced by acid–base reactions that took place in the interior of the aggregates, whereas rapidly percolating water mirrors the chemical state of the pore walls. The fluctuations of high frequency and amplitude of the Ca(Mg)/Al ratio (see Matzner, 1988, 1989) represent the exhaustion of the buffer capacity of the soil and the break of the process hierarchy (cf. Ulrich, Chapter 10): the chemical state of the soil solution is no longer determined by the activities of mineralizers and the ion uptake of fine roots, but determines (limits) the activities of mineralizers and fine roots. For Norway spruce the Ca/Al ratios above 1 indicate negligible Al stress, whereas at ratios below 0.2–0.1 a fine root system can no longer be maintained or established (see

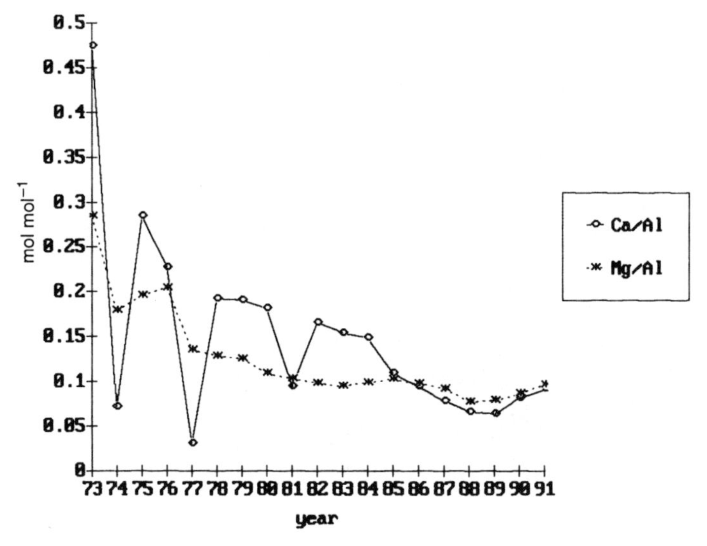

Fig. 1.17. Solling F1: Temporal trend of the molar Ca/Al and Mg/Al ratios in water outflow.

review in Ulrich, 1989). At Mg/Al ratios below 0.2–0.3, Mg uptake is strongly reduced, resulting in decreasing Mg contents of the needles and the development of Mg deficiency. The data show that the conditions for fine root development in the subsoil were drastically impaired during the measuring period. This development started much earlier, the fluctuations of the outflow concentrations represent its final phase. It passed over in a phase of steady decrease during the 1980s. This increasing trend since 1989 may reflect the effect of decreasing deposition of acidity and sulfate: The release of Al in the soil and the outflow of sulfates are decreasing. However, since the capacity of the soil to buffer acidity by cation exchange is exhausted, the recovery of the Ca(Mg)/Al ratios will be very limited.

In Figure 1.18 the acid load from the ion input–output budget of the ecosystem is compiled. Due to ecosystem internal reactions on acid deposition, the acid load can be more than twice as large as the deposition of acidity. The figure shows the large variations (by a factor of almost 4) in acid load as a consequence of internal ecosystem dynamic processes. The largest effect has the release of sulfate that reflects a pH decrease due to unbuffered nitrification–acidification pulses. Three periods can be distinguished:

1973–1977: The acid load is due to deposition of protons and NH_4 with a small contribution due to retention of M_b cations.

1978–1985: The acid load due to deposition of protons and NH_4 remains at the previous level, it is superimposed by the leaching of sulfate, which fluctuates according to water outflow. The contribution of M_b cations becomes significant only in years with low water outflow.

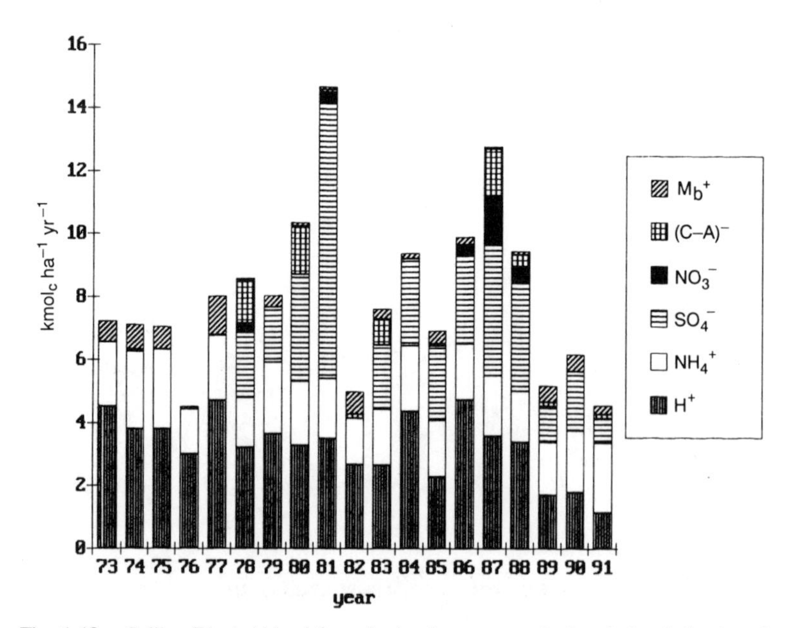

Fig. 1.18. Solling F1: Acid load from the ion input–output budget in kmol$_c$ ha^{-1} yr^{-1}.

1986–to present: In years with high water outflow the leaching of nitrate becomes significant.

1989–1991: Together with a decrease in the acid load by protons and sulfate from acid deposition, the leaching of sulfate decreases. This finding indicates that the ecosystem responds immediately on changes in acid and sulfate input.

The acid–base reactions that balance the acid load are compiled in Figure 1.19. Similar periods as above can be distinguished:

1973–1975: The buffering of the acid load occurs mainly by sulfate retention. The acid input is mainly accumulated in the solid soil phase as aluminum sulfate and in the organic matter as organic bound nitrogen (retention of nitrate). A variable part of the acidity deposited is transferred through the unsaturated zone as dissolved Al and, to a much lesser degree, Mn.

1976–1977: The buffering by sulfate retention decreases, the role of Al leaching increases.

1978–1988: The buffering of deposited acidity is mainly due to Al leaching. The contribution of nitrate retention varies. The leaching of M_b cations in some years becomes a significant contribution.

1989–1991: The acid buffering is almost exclusively due to Al leaching and nitrate retention.

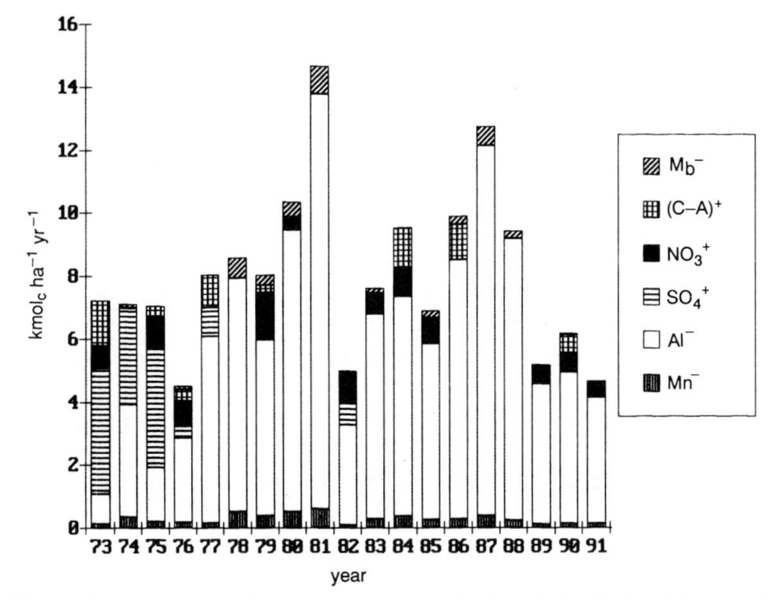

Fig. 1.19. Solling F1: Acid–base reactions balancing the acid load in kmol$_c$ ha^{-1} yr^{-1}.

A comparison of Figures 1.18 and 1.19 shows that the cation–anion budget (C–A) contributes in some years to acid load and in other years to acid buffering. This finding is interpreted as leaching or retention of organic anions in the ecosystem. The outflow data on organic bound N and C do not back this interpretation. One would assume that the leaching of organic anions from soil increases with an increasing water outflow rate, but no correlation was found for C–A. It is highly questionable whether the cation–anion budget represents a real effect in acid load or acid buffering in this ecosystem, it seems to reflect errors.

3.4. Driving Forces for the Fluctuation in the Ion Fluxes

Temporal variation can be caused by climatic variability, by fluctuation of ecosystem internal processes, and by changes in the chemical climate due to changes in emission of air pollutants.

The annual variability of the climate concerns precipitation, transpiration, water outflow, cloud water impaction, and canopy wetness (dry deposition of SO_2), that is, outputs and inputs. The climate induces short-term (seasonal and annual) fluctuations in the fluxes (high frequency) of great amplitude. The general pattern of correlation of the output of anions with the water outflow, or in other words of relatively low variation of concentrations in the outflow, creates a considerable annual variability in the output fluxes, depending on precipitation.

Ecosystem internal processes are a further cause of variation. The variations are generally due to a decoupling of the ion cycle in the ecosystem (see Ulrich, Chapter 10). Such decouplings begin at a small spatial scale with high frequency and amplitude of the fluctuations. The more they extend to a large spatial scale, the more frequency and amplitude decrease. In a special case, a decoupling could be due to forest increment as well as to different rates of ion uptake and mineralization:

- The forest represents a timber tree in the phase of steady forest increment, the growth conditions are maintained by thinning. The fluctuations caused by forest growth are therefore small and not detectable.
- In warm–dry years or after rewetting of a dry soil the mineralization rate can greatly exceed the rate of ion uptake. If the N mineralized is nitrified, this corresponds to a nitrate pulse that indicates an acidification push due to the formation of HNO_3. Such events are bound to the presence of easily decomposable organic matter, that is, to the rhizosphere and surface horizons with high biological activity. These events can show up in the leaching of nitrate (cf. Fig. 1.10), if the nitrate is not taken up by organisms or denitrified during its passage through the soil column. The last two processes correspond to a proton consumption at the locations where they take place.

The third cause of variation are changes in emissions. Under the chemical conditions of the soil since 1976, deposited sulfate is not retained in the soil. It thus determines the output of M_a cations (Al). Changes in S deposition thus immediately affect the output. This result applies also to Ca deposition. In the strongly acidified subsoil, exchangeable Ca is bound on relatively strong acidic groups. Under the condition of constant acid input, a decrease in Ca input results therefore in a decrease in exchangeable Ca due to changes in the ion ratios in soil solution. This finding is reflected in the Ca concentrations (Fig. 1.15) and the Ca/Al ratio (Fig. 1.17) in the outflow water. Until 1977 Ca concentrations could be quite low despite high inputs due to the retention capacity of the root system. The Ca concentration was found to be decreasing since 1982. The exchangeable Ca storage in 0–50-cm soil depth decreased from 13 to 5 $\text{kmol}_c\ \text{ha}^{-1}$ from 1969 to 1983 (Matzner, 1988). The slight increase in Ca concentration in 1990–1991 may be due to low water outflow. The fluctuations of these processes cause fluctuations in the fluxes that overlap each other.

On the basis of soil solution data (equilibrium soil solution according to Ulrich, 1966), an acidification push has been identified in this ecosystem in 1969 (Ulrich, 1980). Such events take place in the macropores where the roots are located. These events are primarily buffered by diffusion of the acidity into the fine pores where cation exchange and aluminum sulfate precipitation can occur (Förster, 1987). For this reason, equilibrium soil solution data does not represent the real acidification of the soil solution in the rhizosphere. The depth gradient and structure of the lignified root system and of the fine roots indicate that during the decades before 1970 the fine root system was shifted towards the surface. The fine root distribution in the 1980s is represented in Figure 1.20 (from Murach and Ulrich, 1988). The main fine root mass is now located in the surface humus layer and the uppermost 20 cm of the mineral soil where Al is mainly organic bound. Decomposition and soil animal activity is almost completely restricted to the surface humus layer (Schauermann, 1986). This development represents the withdrawal of life activities from the mineral soil. The last step in this development seems to be connected with the acidification push in 1969, representing a large scale fluctuation of several years duration

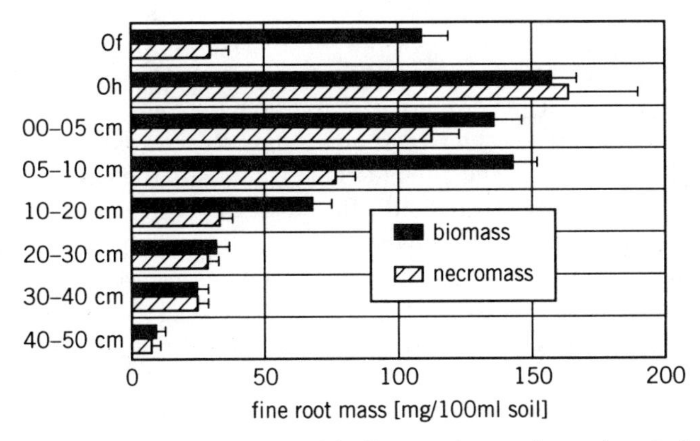

Fig. 1.20. Solling F1: Vertical distribution of the fine roots (mean values and standard deviation of 12 samplings between 1981 and 1985).

and low amplitude. The temporal trend of the fluxes between 1973 and 1977 represents the final phase of this development. These fluctuations are superimposed on the short-term fluctuations due to the variation of precipitation and water outflow. The decrease in Ca retention (see above) may be due to the withdrawal of the fine root system from the subsoil. Together with the life processes of ion uptake and mineralization, the effects of decouplings are now restricted to the topsoil. Nitrification pulses with the consequence of acidification of the soil solution (acidification pushes) were also observed in 10-cm soil depth in the years in 1982 and 1983 (Matzner, 1989). The potential for ecosystem internal processes in the subsoil is now exhausted. The time trend of ecosystem internal acid load (Fig. 1.18) and of acid–base reactions (Fig. 1.19) therefore reflects the tendency of the ecosystem to approach a flux equilibrium (attractor, see Ulrich, Chapter 10) between the input and output of acidity, sulfate, nitrogen, and nutrient cations.

The dependency of forest growth from deposition is evident in respect to N. From 1973 to 1991, 970 kg of N ha^{-1} were deposited, 290 kg were leached. During this time span 230 kg of N were sequestered in the forest increment (annually 12 kg). According to measurements by Brumme and Beese (1992) up to 3 kg of N ha^{-1} can be lost annually as N_2O by denitrification, which amounts to 50 kg of N. The losses as N_2 (denitrogen) are still unknown. Up to 400 kg of N ha^{-1} were sequestered in the soil during this time period (2 kmol of N annually) according to this budget. This finding is confirmed by soil inventories that showed a marked increase in the amount of surface humus (humus form: moder) and in the N storage in this humus (Matzner, 1988). From these data we must conclude that the growth of this forest depends on the N input. The internal N cycle is decoupled. The course of this decoupling is the accumulation of decomposition products of fine root litter in the Oh horizon of the moder. The decomposition products of the fine root litter are stabilized by complexation with Al and Fe (Ulrich, 1989), which was accumulated in the apoplast during the lifetime of the fine roots. Needle litter decomposition is much less affected. The consequences of this effect of acid deposition on the N cycle and N uptake are compensated or even overcompensated by the N input. Without N deposition the forest stand would show retarded growth. Now the stand depends on anthropogenic N emissions for further growth. The present deposition rate is too high, however. It bears the risks that the nitrate load of ground water, as well as the emission of the greenhouse gas N_2O from the forest, will increase. The emission of NO_x and NH_3 must therefore be strongly reduced. For West Germany a short-term reduction of 55% and a long-term reduction of 80% has been requested on the basis of critical loads (Ulrich, 1991b). Since the beginning of industrialization this forest ecosystem has become totally dependent on anthropogenic emissions. This finding represents a very unstable situation for the whole ecosystem, not only for tree growth. The organic top layer, which increases, and where the nitrogen is accumulated, is also very unstable and can be subjected to decomposition and forest fire. There is no guarantee that considerable fractions of the N accumulated in the forest floor will not suddenly appear again in the atmosphere as NH_3 and NO_x or in the ground water (as nitrate).

4. SITE DEPENDING VARIATION OF ION FLUXES AND BUDGETS EXEMPLIFIED WITH CASE STUDIES

4.1. Site Description

The relationship existing between inputs and outputs (outflows) of ions can be illustrated in case studies of ion budgets that were compiled over several years. In the following presentation, the case studies are arranged

1. According to the emission scenario: For the case studies 1–23 the data were collected in the phase of high emission of SO_2 till 1988, for the case studies 24 and 25 only in the phase of declining emission of SO_2 from 1988 to 1991.
2. According to the chemical soil state (buffer range according to Ulrich, 1981).
3. According to a north–south gradient.

Information and data on the case studies 1–9 were published by Ulrich (1989, 1991a), on the case studies 1–22 by FBW (1989). For this reason, only a short description of the case studies will be given. Only those case studies are included, where the water outflow was assessed by a mathematical model.

1 Göttinger Wald: Calcicolous beech wood, age 110 years.
Soil: Rendzina-terra fusca (rendoll) developed on limestone.
Buffer range: Topsoil: Cation exchange, subsoil: Carbonate.
Period: 1981–1987 (Meiwes and Beese, 1988; Sah, 1990).

2 Harste (near Göttingen): Mesotrophic beech wood, age 95 years.
Soil: Loess over triassic limestone, Parabrown earth (alfisol).
Buffer range: Cation exchange.
Period: 1980–1985 (Cassens-Sasse, 1987; Bredemeier, 1987).

3 Lüneburger Heide: Oak wood, age 103 years.
Soil: Podzol-brown earth developed on sandy glacial deposits.
Buffer range: Topsoil: Al, subsoil: Cation exchange/Al.
Period: 1980–1985 (Bredemeier, 1987).

14 Wülfersreuth (Fichtelgebirge): (Norway) spruce forest, age 30 years.
Soil: Podzol-brown earth developed on phyllite.
Buffer range: Topsoil:Fe/Al, subsoil: Cation exchange/Al.
Period: 1984–1986 (Hantschel, 1987; Schulze et al., 1989).

For all further case studies, the buffer range of the soil is the same: Topsoil: Fe/Al, subsoil: Al. According to a first evaluation of a forest soil inventory in Germany (Ulrich and Puhe, 1993), two thirds of the forest soils have a base saturation of less than 15% down to 60-cm soil depth, which corresponds to the Al buffer range. Only in 12% of the forest soils does the base saturation exceed 60%. With respect to the chemical soil state, case study 1 represents a small percentage of forests in Germany and the case studies 2, 3, and 14 are around 30%.

6 Solling B1: Acidophilous beech wood, age 135 years.
Soil: Podzolic brown earth from loess over triassic sandstone (typic dystrochrept).
Period: 1969–1985 (Bredemeier, 1987; Matzner, 1988; Sah, 1990).

5 Lüneburger Heide: Scots pine forest, age 98 years.
Soil: Podzol brown earth from sandy glacial deposits.
Period: 1980–1985 (Bredemeier, 1987).

11 Wingst distr. 28: Spruce forest, age 95 years.
Soil: Podzol (spodosol) from sandy glacial deposits.
Period: Deposition 1983–1990, outflow 1988–1989 (Büttner et al., 1986; Büttner, 1992; Meyer, 1992).

8 Grünenplan distr. 79 (Hils): Spruce forest 107 years.
Soil: Podzolic brown earth from cretacous sandstone.
Period: 1984–1986 (Wiedey and Raben, 1989).

10 Grünenplan distr. 71 (Hils): Spruce forest, 69 years.
Soil: Podzolic brown earth from cretaceous Hils sandstone.
Period: 1984–1987 (Wiedey and Raben, 1989).

12 Lange Bramke (Harz): Spruce forest, age 39 years.
Soil: Podzolic brown earth from devonic Kahleberg sandstone.
Period: 1981–1988 (Hauhs, 1985, 1989).

7 Solling F1: Spruce forest, 100 years (see Section 3).
Soil: Podzolized brown earth from loess over triassic sandstone (typic dystrochrept).
Period: 1969–1991 (Bredemeier, 1987; Matzner, 1988, 1989).

4 Spanbeck (near Göttingen): Spruce forest, 85 years.
Soil: Podzolic brown earth from loess over triassic sandstone.
Period: 1982–1985 (Cassens-Sasse, 1987; Bredemeier, 1987).

23 Elberndorf (Sauerland): Spruce forest, age 65 years.
Soil: Podzolic brown earth from devonic clay schist.
Period: 1985–1988 (Manderscheid, 1992).

15 Oberwarmensteinach (Fichtelgebirge): Spruce forest, age 30 years.
Soil: Podzol (spodosol) from phyllite.
Period: 1984–1986 (Hantschel, 1987; Schulze et al., 1989).

22 Bavarian forest: Spruce forest, age 100 years.
Soil: Podzol from granite.
Period: 1988 (Kreutzer and Heil, 1989).

24 Merzalben (Pfälzer Wald): Mixed oak–beech forest, age of oak 170 years; age of beech 85 years.
Soil: Podzolic brown earth on triassic sandstone.
Period: 1988–1991 (Manderscheid, 1992; Block, 1993).

25 Idar-Oberstein (Hunsrück): Spruce forest, age 114 years.
Soil: Podzolic brown earth from loess over devonic clay schist.
Period: 1988–1991 (Manderscheid 1992; Block, 1993).

A comparison between the actual storages of exchangeable Ca + Mg (M_b) in 0–50 cm of soil depth and the cumulative acid deposition since the beginning of industrialization (see Ulrich, 1989) is given in Figure 1.21 for case studies 1–23. The storage of exchangeable Ca + Mg in soil decreases strongly from the carbonate buffer range (case study 1: 1100 $kmol_c$ ha^{-1}) and the cation exchange buffer range (10–1000 $kmol_c$ ha^{-1}, depending on clay content, humus content, stone content, and base saturation; 200 $kmol_c$ in case study 2) to the Al buffer range (<15 $kmol_c$ ha^{-1}). The calculated cumulative acid deposition amounts to 60–260 $kmol_c$ ha^{-1}. These figures are a low estimate. In the case study Solling F1, the acid deposition monitored from 1969 to 1991 amounts to 132 $kmol_c$ ha^{-1}. It was shown that in forests of low altitudes like in southern Lower Saxony, considerable losses of exchangeable Ca, Mg, and K took place between 1954 and 1986 (Ulrich et al., 1989). The comparison given in Figure 1.21 demonstrates that during the last decades in many forests the chemical soil state was changed drastically by acid deposition (loss of nutrient cations and decrease in base saturation).

4.2. Ion Fluxes

In Figure 1.22 the input of acidity is compiled. Acid deposition is relatively low for the deciduous forests 1–3, for the healthy spruce stand in the transition to Al buffer range (14), for the pine forest (5), and for the phase of reduced SO_2 emission (24 and 25). Solling F1 (7) has the highest input, but there are several other spruce

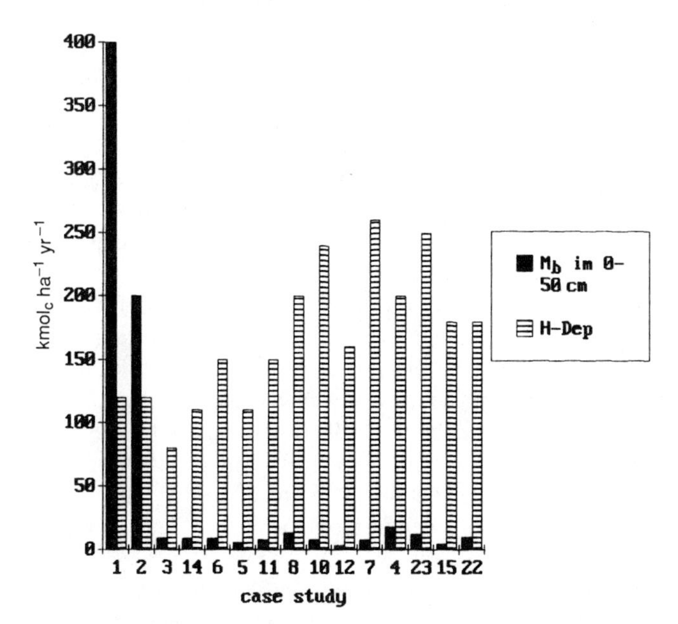

Fig. 1.21. Case studies: Comparison between cumulative acid deposition since start of industrialization (H-Dep) and the storages of exchangeable Ca + Mg in the soil depth 0–50 cm (M_b) in $kmol_c$ ha^{-1} yr^{-1} (M_b storage in case study 1 is 1100 (keq).

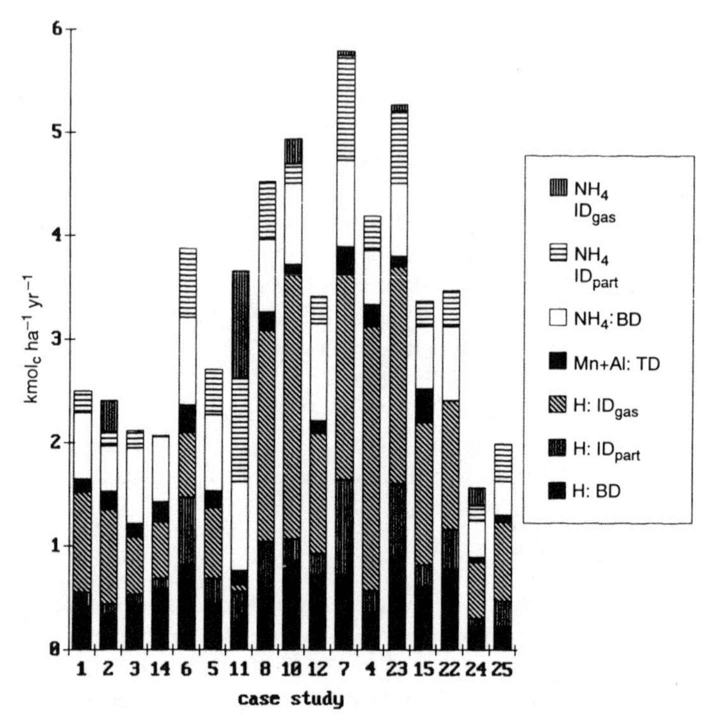

Fig. 1.22. Case studies: Deposition of acidity in kmol$_c$ ha^{-1} yr^{-1}.

forests in the northern parts of the central European mountain ranges that are comparable (8, 10, and 23). The contribution of ammonium to acid deposition is relatively large in all case studies. A special case is study 11, which lies in a region of intensive animal breeding and high NH_3 emission. This forest is subjected to high NH_3 deposition, which reacts in the canopy with SO_2 (cf. Fig. 1.23). From the data on anion deposition compiled in Fig. 1.23 it can be concluded that acid deposition is mostly due to SO_2, since sulfate is the dominating anion.

In Table 1.6 the mean value of the canopy budget (TD–TF) is given for the case studies, together with the temporal mean of case study 7 (Solling F1) for the period 1973–1991. No canopy uptake of NH_4 or NO_3 is found in 7 and 10 of the case studies, respectively. This explains the low mean value and the high variability. For the other ions the mean values between the time series and the site series are quite similar. The coefficients of variation show that there are site specific effects on the canopy budget, but they are rather small compared to the temporal variation, with the exception of Mn. The Mn leaching from the canopy depends on the availability of Mn in the soil.

The deposition of nutrients plays an important role for forest growth and ecosystem stability for forest ecosystems subjected to acid deposition. The fertilizing effect of deposition can be assessed from deposition of nutrients (TD) and the ecosystem budget (TD–SO). Both data sets are given in Figures 1.24–1.27 for K,

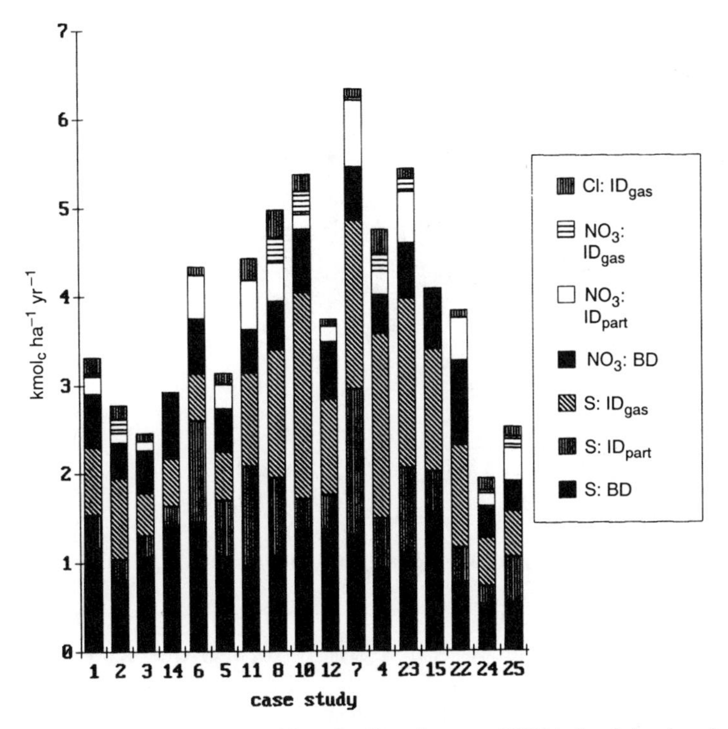

Fig. 1.23. Case studies: Deposition of sulfur, nitrate, and HCl in kmol$_c$ ha^{-1} yr^{-1}.

Mg, Ca, and N (sum of NH_4 and NO_3), respectively. Positive values of the ecosystem budget indicate a storage increase in the ecosystem. Nutrient uptake by the vegetation is the main process responsible for this increase in storage. The sequestration of the nutrients may occur in forest growth and in increasing soil organic matter storage. The importance of this process for forest growth can be evaluated by comparison with nutrient uptake data of the tree stand. For case study 7, which can be considered as representative, such data are given in Table 1.1. Negative values

TABLE 1.6. Case Studies: Mean Values of the Canopy Budget TD–TF

	K	Mg	Ca	H	Mn	NH_4	NO_3
			(kmol$_c$ ha^{-1} yr^{-1})				
No. 7 (F1) (1973–1991)	−0.52	−0.11	−0.63	0.86	−0.16	0.66	0.21
SD	0.12	0.05	0.18	0.55	0.04	0.38	0.21
CV(%)	23	42	29	64	26	58	101
No. 1–25	−0.42	−0.11	−0.48	0.87	−0.09	0.24	0.08
SD	0.15	0.05	0.20	0.52	0.06	0.25	0.10
CV(%)	36	47	41	59	65	104	136

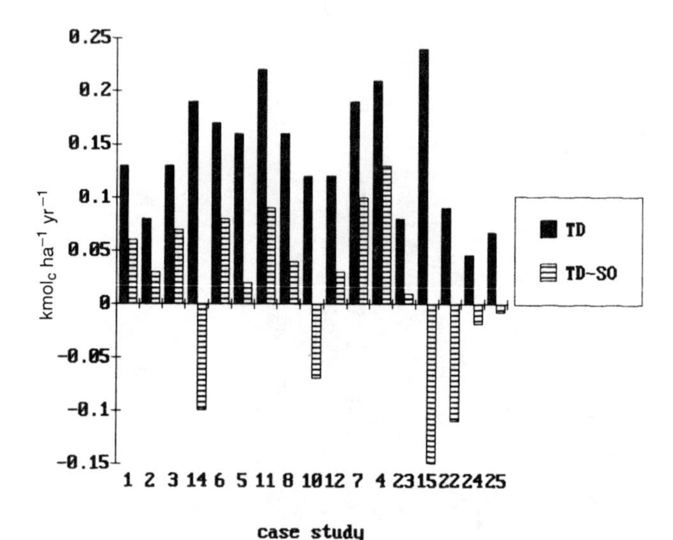

Fig. 1.24. Case studies: Total deposition (TD) and ecosystem budget (TD–SO) of K in kmol$_c$ ha^{-1} yr^{-1}.

indicate that the ecosystem cannot make use of the nutrients deposited, and that additional nutrients are lost.

Potassium deposition (Fig. 1.24) is in most of the case studies below the need for forest growth. The budget is negative for one third of the case studies. This result shows that K leaching due to acid deposition can decrease soil storages and result in K deficiency. In many soils of Central Europe, which developed from glacial and periglacial sediments, the clay minerals are derived from mica and are rich

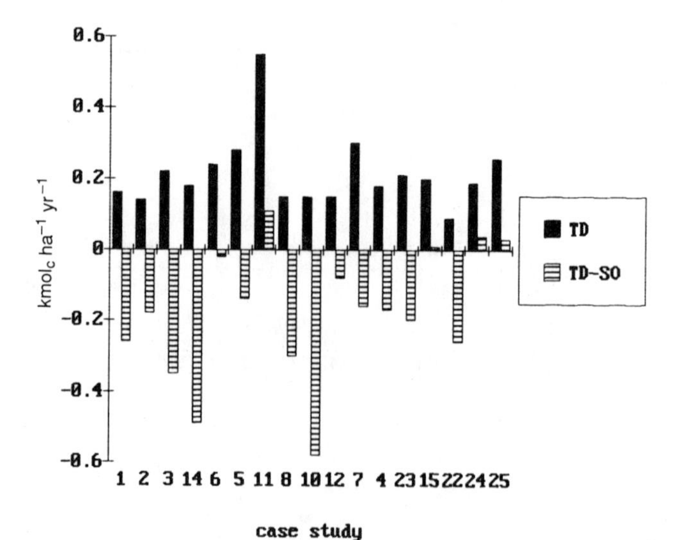

Fig. 1.25. Case studies: Total deposition (TD) and ecosystem budget (TD–SO) of Mg in kmol$_c$ ha^{-1} yr^{-1}.

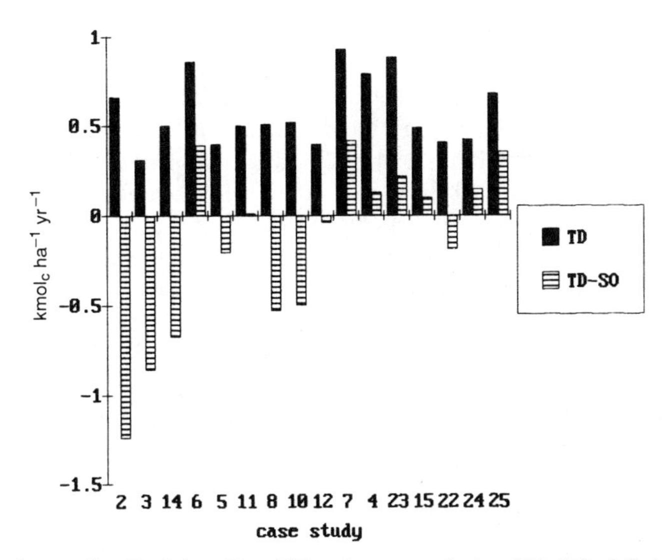

Fig. 1.26. Case studies: Total deposition (TD) and ecosystem budget (TD–SO) of Ca in kmol$_c$ ha^{-1} yr^{-1}.

in K, thus K release by weathering is of significance and can balance K losses. Even under these conditions K deficiency can develop as a consequence of acid deposition by the impoverishment of pore walls and aggregate surfaces in soil (Hildebrand, 1990).

Magnesium deposition (Fig. 1.25) covers or exceeds in all case studies the need for forest growth. In most case studies the ecosystem budget is negative, however. In ecosystems where the soil is in the Al buffer range, the source of the Mg leached

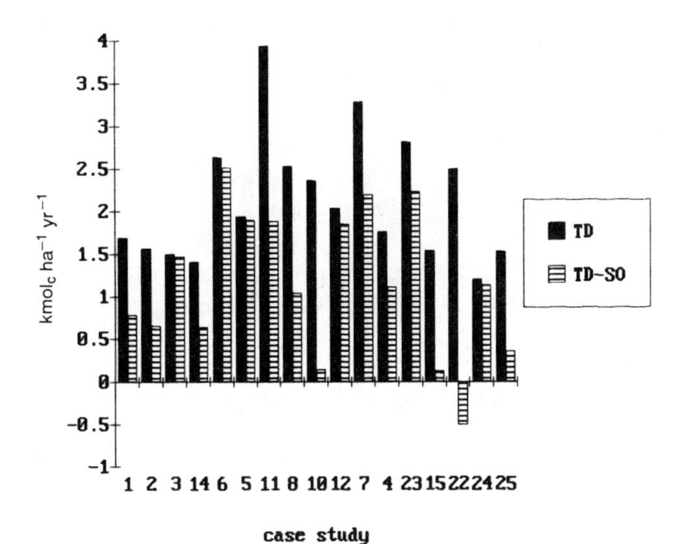

Fig. 1.27. Case studies: Total deposition (TD) and ecosystem budget (TD–SO) of N (NH_4 + NO_3) in kmol$_c$ ha^{-1} yr^{-1}.

may not be the exchangeable storage, but the weathering of clay minerals as discussed in Section 3.3. In the most acid top soils, this process goes on for decades. Magnesium nutrition was impaired by the shift of the Mg/Al ratio in the rooted soil to values below 0.3–0.2 (cf. Fig. 1.17). Magnesium deficiency is a significant phenomenon in the actual forest decline in central Europe (FBW, 1989; Schulze et al., 1989; Blanck et al., 1988; Stock, 1988), which developed during the last decades (review by Ulrich, 1989). As the trees adjust to the stronger acidification of the subsoil by shifting their fine root system to the top soil, they can take up deposited Mg in the humus rich top soil where Al is organically complexed. Magnesium deficiency is therefore most severe in aggrading tree stands that started on soils where the base saturation in the subsoil was high enough to develop a deep reaching root system. If due to acid deposition and cation uptake in the aggrading biomass the exchangeable Mg pool became exhausted, the molar Mg/Al ratio in soil solution falls below the critical value (0.3–0.2) and Mg uptake ceases. If the trees can shift the fine root system to the top soil rapidly enough, they may get enough Mg from deposition to maintain vitality and growth. This example shows how complex the interaction between deposition, soil change, and tree adjustment can be.

Calcium deposition (Fig. 1.26) is of the same order of magnitude as the need for forest growth. Leaching losses play a significant role, however, and the sequestration of deposited Ca is limited. The tree organ where Ca deficiency shows up is the fine root, not the leaf. Its uptake and binding in the fine root apoplast depends on the Ca/Al ratio. The data available on the elemental composition of fine roots indicate that Ca deficiency and Al surplus is a common feature in acid soils (Al buffer range) and determines the fine root development (review in Ulrich, 1989). This feature indicates that a break in the process hierarchy took place, and that the ecosystem is developing towards a new state (cf. Ulrich, Chapter 10).

Only in the case of N (Fig. 1.27) does the ecosystem budget of most case studies considerably exceed the need for forest growth. According to N fertilization experiments in forests in the 1950s, this should result in increased forest growth. Many investigations found that forest growth increased during the last one-to-two decades (Becker et al., 1990; Kenk et al., 1991). There are case studies, however, where despite high C/N ratios in the soil organic matter and retardation of litter decomposition, deposited N is almost not retained (10, 15, and 25). The negative budget of case study 22 represents only 1 year and may not characterize the long-term behavior.

In the case studies where the measurements are continued (1, 3, 5, 6, 11, and 12), similar temporal trends to those discussed in Sections 2 and 3 were found (Meiwes et al., 1993).

The ecosystem budgets (TD–SO) depend greatly on the chemical soil state. This allows the distinction of different ion budget typs.

4.3. Ion Budget Types

Ecosystems with calcareous subsoils: In Table 1.7 the ecosystem acid–base budget is given for case study 1, a beech forest on calcareous soil. This budget is totally

TABLE 1.7. Göttinger Wald: European Beech Forest on Soil Derived from Limestone: Ecosystem Acid–Base Budget[a]

	$kmol_c$ ha^{-1} yr^{-1}	% of AL_{I-O}
Acid Load from Input–Output Budget (AL_{I-O})		
1. Retention of acid input (H+: 9.8%, NH_4: 5.3%, MN + Al: 0.7% of AL_{I-O})	2.46	16
2. Release of anions [C–A (HCO_3)]	13.0	84
3. Retention of deposited M_b cations (K)	0.06	0.4
Sum = AL_{I-O}	15.52	100
Acid–Base Reactions		
4. Release of M_a cations	0	0
5. Retention of deposited anions	0	0
6. Release of M_b cations (Mg: 1.7%, Na: 2.3%, Ca: 96%)	15.52	100

[a]Mean values 1981–1987.

different from that given in Table 1.4 for a spruce forest on acid soil. The acid load is mainly due to the dissociation of ecosystem internally generated H_2CO_3, and the only acid–base reaction is the release of M_b cations, which is almost exclusively Ca. The Ca originates from the dissolution of $CaCO_3$. This reaction implies that the chemical soil state, (e.g., base saturation) is not changed. If the fine earth of the topsoil is free of carbonate, however, acid deposition results in a decrease of base saturation as demonstrated for case study 1 (Meiwes and Beese, 1988). The effect of acid deposition becomes particularly evident if the chemical heterogeneity on a microscale (surface and internal part of soil aggregates) is considered (Hantschel and Pfirrmann, 1990). A typical feature of these ecosystems, which are characterized by rapid litter decomposition, is the fact that nitrate is not retained. This finding is a consequence of a high turnover by mineralization, which provides a sufficient supply of nitrogen.

Ecosystems with subsoils in the cation exchange buffer range: In Fig. 1.28 the acid load from the input–output budget is compiled for the other case studies. Case study 2 represents a beech ecosystem where the rooted soil is still in the cation exchange buffer range. The acid load is due to H and NH_4 retention from deposition with a small contribution by nitrate leaching. The main acid–base reaction (see Fig. 1.29) is the release of M_b cations (Ca and Mg), where the retention of sulfate contributes significantly. The M_b cations originate from the exchangeable pool, thus base saturation is decreasing. This actual process of soil acidification shows up in a great spatial heterogeneity: On a small spatial scale base saturation varies between 80 and 15% (Cassens-Sasse, 1987). It is also reflected in the soil solution, where in addition to the spatial variation high temporal variation also occurs: For example,

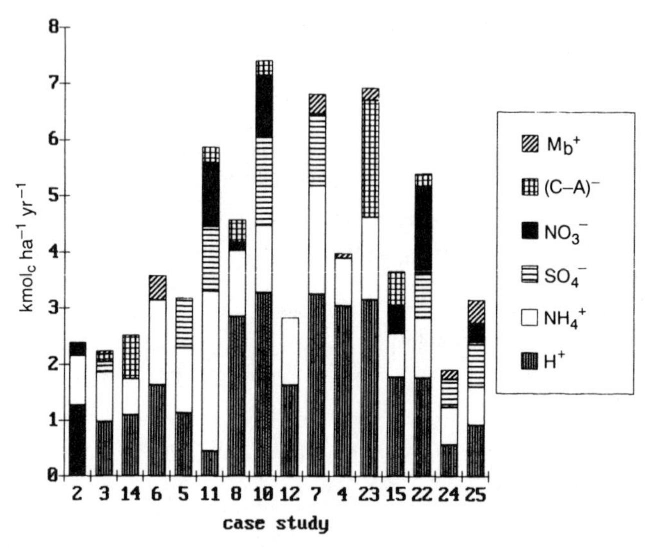

Fig. 1.28. Case studies: Acid load from the ion input–output budget in kmol$_c$ ha^{-1} yr^{-1}.

fluctuations of pH between 5 and 3.8, of the Al concentration between 0.1 and 0.8 mmol L^{-1}. The acidification affects the entire rooted soil. In case study 2, there is only a weakly pronounced gradient of the fine root mass up to a 50-cm depth. The fraction of dead roots ranging between 40 and 50% indicates stress conditions. The fine root dynamics is more pronounced in the stronger acidified horizons; there is a correlation in the seasonal development between sharp reductions of the fine root biomass and those of pH values (Cassens-Sasse, 1987).

In the case studies 3 and 14, the base saturation in the subsoil is already low, but

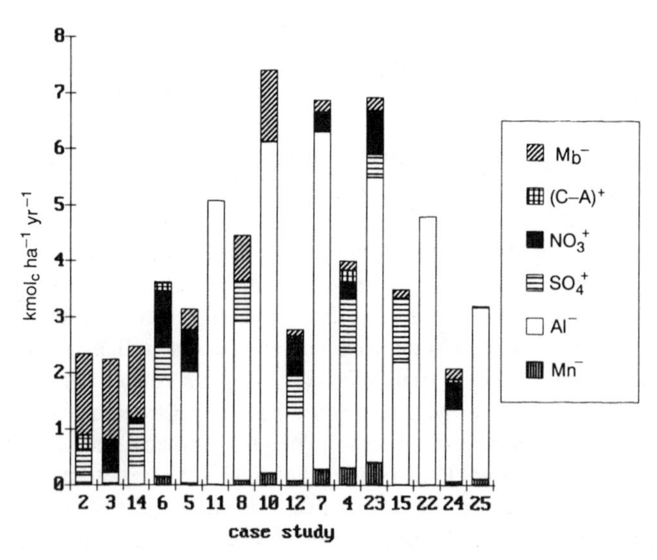

Fig. 1.29. Case studies: Acid–base reactions balancing the acid load in kmol$_c$ ha^{-1} yr^{-1}.

in the depth where the lysimeters are installed ($\sim$ 1 m) there is still some exchangeable Ca left to buffer the Al in the seepage water. Some Al is leached, however, indicating the final phase of the cation exchange buffer system.

Ecosystems with subsoils in the aluminum buffer range: In the case studies with subsoils in the Al buffer range (Figs. 1.28 and 1.29 from No. 6 to the right) leaching of sulfate and nitrate can contribute significantly to the acid load. The variation is quite large, however, and may not only be site specific. The net leaching of sulfate and nitrate can reflect events due to ecosystem internal processes and may not characterize the site specific long-term behavior of the ecosystem. This conclusion must be drawn due to the relative short measuring periods. In any case, these events also show that ecosystem internal processes can increase the acid load considerably. The buffering of the acid load occurs predominantly by the leaching of Al. The contribution of leaching of M_b cations is missing or small. This points to the fact that the exchangeable M_b cations left in the Al buffer range are bound on relatively strong acidic groups and do not act as bases at the pH predominating (3.8–4.2). This finding is an important aspect if the remaining buffer capacity of the soil is calculated from soil chemical data. The exchangeable M_b cations in the Al buffer range seem to be in a flux equilibrium with the deposition. A decrease in deposition of these ions should therefore result in a decrease of exchangeable storage. Deposition decreases also if the tree stand becomes open or if the trees are cut. This decrease could impair the development of regeneration on these strongly acidified soils.

A further trend in the soils of the Al buffer range that are under the influence of acid deposition is the extension of the Fe–Al buffer range in the upper soil layer (A horizon) into greater depth. The Fe–Al buffer range is characterized by pH values below 3.8, and the infiltration of humic acids from the humus cover with the seepage water into the top mineral soil.

In Figure 1.30 a generalization of the depth gradient of the soil acidification including the surface humus is presented. Leaves and litter have a high base saturation as indicated by the fraction of M_b cations. The high base saturation of the leaves already decreases significantly in the L and Of horizon.

Starting from the Oh horizon the base saturation is lower than indicated, because a considerable amount of the cations are bound to strong acidic groups and represent neutral salts. The Ah_e horizon is in the Fe–Al buffer range. Its thickness increases with continuous acid deposition. In the Oh and Ah_e horizon the fine roots, depending on the pH and the Ca and Mg concentration of the soil solution, were found to be suffering proton stress. Further below, the soil is in the Al buffer range. This range may extend beyond the rooting area into the seepage water aquifer. Here, high Al concentrations may still occur until, in the event of acidification pushes, aluminum hydroxysulfates are completely depleted.

In the case of vertical transport of seepage water, further below the zone of the Al buffer range (Fig. 1.30) a zone of variable thickness follows, which is in the cation exchange buffer range (Ulrich and Malessa, 1989). With respect to the pH values (4.0–4.4), this zone does not differ from the Al buffer range. The low pH values indicate steep gradients in the chemical potentials at pore walls and aggregate surfaces. In this zone the Al^{3+} ions transported with the seepage water through the

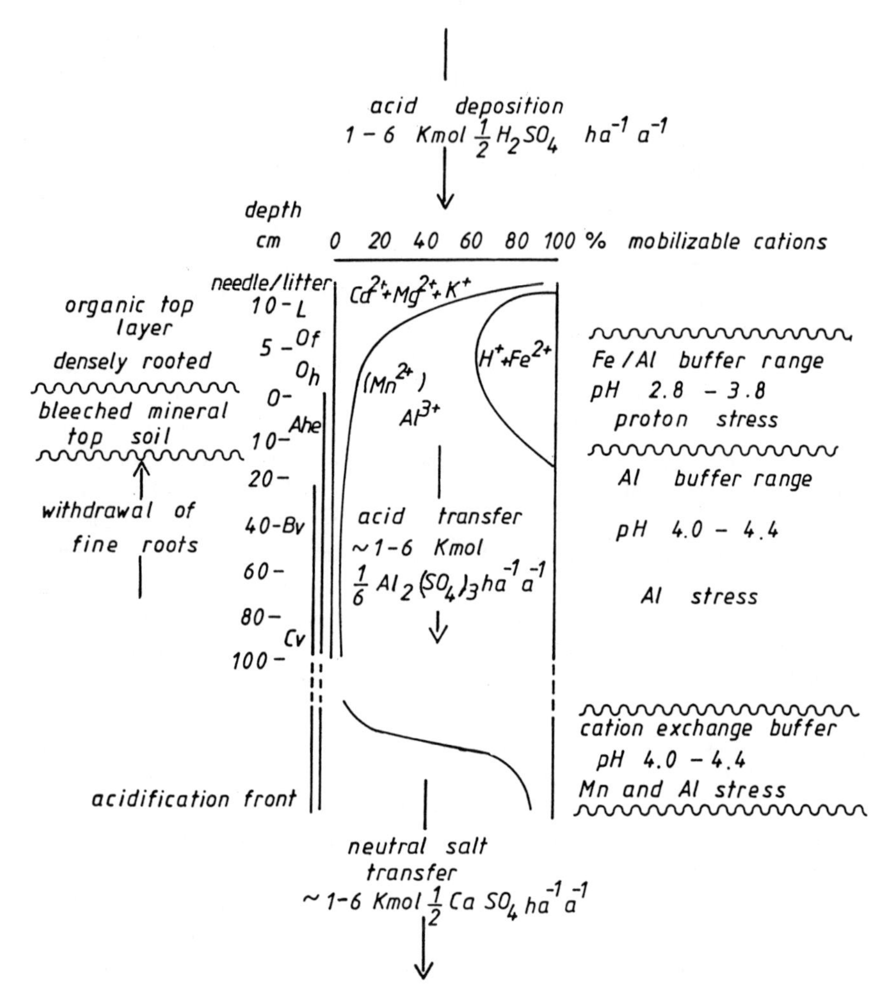

Fig. 1.30. Depth gradient of soil acidity under the influence of acid deposition.

zone of Al buffer range are exchanged against Ca, Mg, and K. These cations are then transported with the seepage water into greater depths.

The acidification front spearheads this zone into deeper reaches. The speed of this movement is influenced by the acid output from the ecosystem and by the storage of exchangeable bases in the area of the acidification front.

REFERENCES

Becker M, Bräker O-U, Kenk G, Schneider O, Schweingruber FH (1990): Kronenzustand und Wachstum von Waldbäumen im Dreiländereck Deutschlanf-Frankreich-Schweiz in den letzten Jahrzehnten. Allg Forstz 45:263–274.

Benecke P (1984): Der Wasserumsatz eines Buchen- und eines Fichtenwaldökosystems im Hochsolling. Schriften Forstl Fak Univ Göttingen 77:158 pp Frankfurt: Sauerländer's Verlag.

Binkley D (1992): H^+ Budgets. In Johnson DW, Lindberg SE (Eds): Atmospheric deposition and forest nutrient cycling. New York, Berlin, Heidelberg: Springer, pp 450–463.

Blanck K, Matzner E, Stock R, Hartmann G (1988): Der Einfluss kleinstandörtlicher bodenchemischer Unterschiede auf die Ausprägung von Vergilbungssymptomen an Fichten im Harz. Forst Holzwir 43:288–292.

Block J (1993): Verteilung und Verlagerung von Radiocäsium in zwei Waldökosystemen in Rheinland-Pfalz insbesondere nach Kalk und Kaliumdüngungen. Ph.D. Thesis Universitat Göttingen, Göttingen, Germany, 287 pp.

Bredemeier M (1987): Stoffbilanzen, interne Protonenproduktion und Gesamtsäurebelastung des Bodens in verschiedenen Waldökosystemen Norddeutschlands. Ber Forschungszantrum Waldökosysteme Universität Göttingen A 33:183 pp.

Bruggenwert MGM, Hiemstra T, Bolt GH (1991): Proton sinks in soil controlling soil acidification. In Ulrich B, Sumner ME (eds) "Soil acidity." Berlin, Heidelberg, New York: Springer, pp 8–27.

Brumme R, Beese F (1992): Effects of liming and nitrogen fertilization on emissions of CO_2 and N_2O from a temperate forest. J Geophys Res 97:12851–12858.

Brumme R, Leimcke U, Matzner E (1992): Interception and uptake of NH_4 and NO_3 from wet deposition by aboveground parts of young beech (Fagus silvatica L.) trees. Plant Soil 142:273–279.

BMU (Bundesumweltministerium) (1992): 5. Immissionschutzbericht, Bonn.

Büttner G, Lamersdorf N, Schultz R, Ulrich B (1986): Deposition und Verteilung chemischer elemente in küstennahen Waldstandorten. Ber Forschungszentrum Waldökosysteme Universität Göttingen B 1: 172 pp.

Büttner G (1992): Stoffeinträge und ihre Auswirkungen in Fichtenökosystemen im nordwestdeutschen Küstenraum. Ber Forschungszentrum Waldökosysteme Universität Göttingen A 84:191 pp.

Cassens-Sasse E (1987): Witterungsbedingte saisonale Versauerungsschübe im Boden zweier Waldökosysteme. Ber Forschungszentrum Waldökosysteme Universität Göttingen A 30:287 pp.

Dietze G, Ulrich B (1992) Aluminum speciation in acid soil water and ground water. In Matthes G, Frimmel F, Hirsch P, Schulz HD, Usdowski H-E (eds): Progress in hydrogeochemistry. Berlin, Heidelberg, New York: Springer, pp 269–281.

Eilers G, Brumme R., Matzner E (1992): Above-ground N-uptake from wet deposition by Norway spruce (Picea abies Karst.). For Ecol Manage 51:239–249.

Ellenberg H, Mayer R, Schauermann J (1986): Ökosystemforschung. Ergebnisse des Sollingprojekts 1966–1986. Stuttgart: Ulmer.

Fassbender HW (1977): Modellversuch mit jungen Fichten zur Erfassung des internen Nährstoffumsatzes. Oecolog Plant 12:263–272.

FBW (Forschungsbeirat Waldschäden/Luftverunreinigungen) (1989): 3. Bericht. Kernforschungszentrum Karlsruhe.

Flehmig W, Fölster H, Tarrah J (1990): Stoffbilanzierung in einer Pseudogley-Parabraunerde aus Löss unter Anwendung der IR-Phasenanalyse. Z Pflanzenernaehr Bodenkd 153:149–155.

Förster R (1986): A multicomponent transport model. Geoderma 38:261–278.

Förster R (1987): Ein Konvektions–Diffusions–Transportmodell mit Multispezies-Kationenaustausch, Ionenkomplexierung und Aluminiumhydroxosulfat zur Simualtion der Sulfatspeicherung in sauren Waldböden. Ber Forschungszantrum Waldökosysteme Universität Göttingen A 28:162 pp.

Frank U, Gebhardt H (1991): Transformation and destruction of clay minerals caused by recent strong acidification. Proceedings of the 7th Euroclay Conference, Vol. 1, Dresden, 369–374.

Gravenhorst G (1992): Stoffaustausch zwischen Atmosphäre und Baumkronen (Forschungsbericht). Ber Forschungszentrum Waldökosysteme Universität Göttingen B 31:1–12.

Hantschel R (1987): Wasser- und elementbilanzen von geschädigten, gedüngten Fichtenökosystemen im

Fichtelgebirge unter Berücksichtigung von physikalischer und chemischer Bodenheterogenität. Bayreuther Bodenkdl Ber 3:219 pp.

Hantschel R, Pfirrmann T (1990): Bodenkundliche Untersuchungen am Forschungsschwerpunkt Wank—Bedeutung für die Waldschäden in den Kalkalpen. Beih Forstwiss Centralbl 40:87–101.

Hauhs M (1985): Wasser- und Stoffhaushalt im Einzugsgebiet der Langen Bramke (Harz). Ber Forschungszentrum Waldökosysteme Universität Göttingen 17:206 pp.

Hauhs M (1986): A model of ion transport through a forested catchment at Lange Bramke, West Germany. Geoderma 38:97–113.

Hildebrand EE (1990): Die Bedeutung der Bodenstruktur für die Waldernährung, dargestellt am Beispiel des Kaliums. Forstwiss Centralbl 109:2–12.

Ibrom A (1993): Die Deposition und die Pflanzenauswaschung (Leaching) von Pflanzennährstoffen in einem Fichtenbestand im Solling. Ph.D. Thesis Universität Göttingen, Göttingen, Germany.

Johnson DW, Lindberg SE (eds) (1992): Atmospheric deposition and forest nutrient cycling. Ecological Studies 91:707 pp.

Johnson DW, Lindberg SE, van Migroet H, Lovett GM, Cole DW, Mitchell MJ, Binkley D (1993): Atmospheric deposition, forest nutrient status, and forest decline: implications of the integrated forest study. In Huettl R, Mueller-Dombois (eds): Forest Decline in the Atlantic and Pacific Region. Berlin, Heidelberg: Springer Verlag, pp. 66–81.

Kenk G, Spiecker H, Diener G (1991): Referenzdaten zum Waldwachstum. Forschungsbericht KfK–PEF 82:1–59. Kernforschungszentrum Karlsruhe.

König N, Loftfield N, Lüter K-L (1989): Atom-Absorptionsspektrographische Bestimmungsmethoden für Haupt-und Spurenelemente in Probelösungen aus Waldökosystem-Untersuchungen. Ber Forschungszentrum Waldökosysteme Universitat Göttingen B 13:84 pp.

Kreutzer K, Heil K (1989): Untersuchungen zum Stoffhaushalt in einem Fichtenbestand der Hochlagen des Bayerischen Waldes. GSF-Bericht 6/89:145-154, Neuherberg.

Lindberg SE (1989): Application of surface analysis methods to studies of atmospheric deposition in forests. In Ulrich B (ed): International Congress on Forest Decline Research. Lectures Vol. 1, pp 269–283. Kernforschungszentrum Karlsruhe.

Lindberg SE, Bredemeier M, Schaefer DA, Qi, L (1990): Atmospheric concentrations and deposition of nitrogen and major ions in conifer forests in the United States and Federal Republic of Germany. Atmos Environ 24A:2207–2220.

Manderscheid B (1991): Die Simulation des Wasserhaushalts als Teil der Ökosystemforschung. Ber Forschungszentrum Waldökosysteme Universität Göttingen B 24:211–242.

Manderscheid B (1992): Modellentwicklung zum Wasser- und Stoffhaushalt am Beispiel von vier Monitoring-Flächen. Ber Forschungszentrum Waldökosysteme Universität Göttingen A 87:233 pp.

Manderscheid B, Matzner E, Meiwes K-J, Xu, Y (1994): Long term development of element budgets in a Norway spruce (Picea abies Karst.) forest of the German Solling area. Water, Air, Soil, Pollut, in press.

Matzner, E (1988): Der Stoffumsatz zweier Waldökosysteme im Solling. Ber Forschungszentrum Waldökosysteme Universität Göttingen A 40:217 pp.

Matzner E (1989): Acidic precipitation: Case study Solling. In Adriano DC, Havas M (eds): Acidic Precipitation Vol. I, New York, Berlin, Heidelberg: Springer, pp 39–83.

Matzner E, Khanna PK, Meiwes K-J, Lindheim M, Prenzel J, Ulrich B (1982): Elementflüsse in Waldökosystemen im Solling—Datendokumentation. Göttinger Bodenkd Ber 71:267 pp.

Matzner E, Meiwes K-J (1994): Long-term development of element fluxes with bulk precipitation and throughfall in two German forests. J Environ Qual, in press.

Matzner E, Prenzel J (1992): Acid deposition in the German Solling area: effects on soil solution chemnistry and Al mobilization. Water, Air, Soil, Pollut 61:221–234.

Matzner E, Ulrich B (1983): Abiotische Folgewirkungen der weiträumigen Ausbreitung von Luftverunreinigungen—Datenband. Bundesministerium des Innern, Luftreinhaltung, Forschungsbericht 104 02 615, 141 pp.

Mayer R, Ulrich B (1974): Conclusions on the filtering action of forests from ecosystems analysis. Ecol Plant 9:157–168.

Meiwes K-J, Beese F (1988): Ergebnisse der Untersuchung des Stoffhaushalts eines Buchenwaldökosystems auf Kalkgestein. Ber Forschungszentrum Waldökosysteme Universität Göttingen B 9:1–141.

Meiwes K-J, König N, Khanna PK, Prenzel J, Ulrich B (1984a): Chemische Untersuchungsverfahren für Mineralboden, Auflagehumus und Wurzeln zur Charakterisierung und Bewertung der Versauerung in Waldböden. Ber Forschungszentrum Waldökosysteme Universität Göttingen 7:1–67.

Meiwes K-J, Hauhs M, Gerke H, Asche N, Matzner E, Lamersdorf N (1984b): Die Erfassung des Stoffkreislaufs in Waldökosystemen—Konzept und Methodik. Ber Forschungszentrum Waldökosysteme Universität Göttingen 7:68–142.

Meiwes K-J, Meesenburg H, Schultz-Sternberg R (1994): Long term development of atmospheric deposition in selected forest ecosystems in northern Germany. Comm Europ Communities, Air Poll Res Report, in press.

Meyer M (1992): Untersuchungen zur Restabilisierung geschädigter Waldökosysteme im norddeutschen Küstenraum (Fallstudie Wingst II). Ber Forschungszentrum Waldökosysteme Universistät Göttingen A 94:306 pp.

Murach D, Ulrich B (1988): Destabilization of forest ecosystems by acid deposition. GeoJournal 17: 253–260.

Prenzel J (1983): A mechanism for storage and retrieval of acid in acid soils. In Ulrich B, Pankrath J (eds): Effects of Accumulation of Air Pollutants in Forest Ecosystems. Dordrecht: Reidel, pp 157–170.

Sah SP (1990): Vergleich des Stoffhaushalts zweier Buchenwaldökosysteme auf Kalkgestein und auf Buntsandstein. Ber Forschungszentrum Waldökosysteme Universität Göttingen A 59:140 pp.

Schauermann J (1986): Verteilung der Bodenfauna. In Ellenberg H, Mayer R, Schauermann J (eds): Ökosystemforschung—Ergebnisse des Sollingprojekts 1966–1986. Stuttgart: Ulmer, pp 208–219.

Schulze ED, Lange OL, Oren R (1989): Forest decline and air pollution. Ecological Sudies 77: 475 pp, Berlin, Heidelberg, New York: Springer.

Stock R (1988): Aspekte der regionalen Verbreitung "neuartiger Waldschäden" an Fichte im Harz. Forst Holz 43:283–286.

Ulrich B (1966): Kationenaustausch-Gleichgewichte in Böden. Z Pflanzenernaehr Düngung Bodenkd 113:141–159.

Ulrich B (1978–1980): Production and consumption of hydrogen ions in the ecosphere. In Hutchinson TC, Havas M (eds): Effects of acid precipitation on terrestrial ecosystems. New York, London: Plenum Press, pp 255–282.

Ulrich B (1980): Die Wälder in Mitteleuropa: Messergebnisse ihrer Umweltbelastung, Theorie ihrer Gefährdung, Prognose ihrer Entwicklung. Allg Forstz 35:1198–1202.

Ulrich B (1981): Ökologische Gruppierung von Böden nach ihrem chemischen Bodenzustand. Z Pflanzenernaehr Bodenkd 144:289–305.

Ulrich B (1983): Interaction of forest canopies with atmospheric constituents: SO_2, alkali and earth alkali cations and chloride. In Ulrich B, Pankrath J (eds): Effects of Accumulation of Air Pollutants in Forest Ecosystems. Dordrecht: Reidel, pp 33–45.

Ulrich, B (1989): Effects of Acidic Precipitation on Forest Ecosystems in Europe. In Adriano DC, Johnson AH (eds): Acidic Precipitation Vol. 2 Biological and Ecological Effects. New York, Berlin, Heidelberg: Springer pp 189–272.

Ulrich B (1991a): An ecosystem approach to soil acidification. In Ulrich B, Sumner ME (eds): Soil acidity. Berlin, Heidelberg, New York: Springer pp 28–79.

Ulrich B (1991b): Introduction to acidic deposition effects: critical deposition rates and emission densities. In Chadwick MJ, Hutton M (eds): Acid deposition in Europe. Stockholm Environment Institute, pp 2–16.

Ulrich B, Malessa V (1989): Tiefengradienten der Bodenversauerung. Z Pflanzenernaehr Bodenkd 152: 81–84.

Ulrich B, Steinhardt U, Müller-Suur A (1973): Untersuchungen über den Bioelementgehalt in der Kronentraufe. Göttinger Bodenkd Ber 29:133–192.

Ulrich B, Mayer R, Khanna PK (1979): Deposition von Luftverunreinigungen und ihre Auswirkungen in Waldökosystemen im Solling. Schriften Forstl Fak Universität Göttingen 58: Frankfurt: Sauerländer Verlag 291 pp.

Ulrich B, Meyer H, Jänich K, Büttner G (1989): Basenverluste in den Böden von Hainsimsen-Buchenwäldern in Südniedersachsen zwischen 1954 und 1986. Forst Holz 44:251–253.

Ulrich B, Puhe J (1994): Auswirkungen der zukünftuigen Klimaveränderung auf mitteleuropäische Waldökosysteme und deren Rückkopplungen auf den Triebhauseffekt. Veröffentlichungen der Enquete-Kommission Schutz der Erdatmosphäre des Deutschen Bundestags, in press.

Umweltbundesamt (1993): Daten zur Umwelt 1991/9F2. E Schmidt Berlin, Verlag 675 pp.

van der Ploeg RR, Benecke P (1974): Simulation of one-dimensional moisture transfer in unsaturated, layered, field soils. In Ulrich B, Mayer R, Heller H (Eds): Data analysis and data synthesis of forest ecosystems. Göttinger Bodenkd Ber 30:150–169.

van Breemen N (1991): Soil acidification and alcalinization. In Ulrich B, Sumner ME (eds): Soil acidity. Berlin, Heidelberg, New York: Springer pp 1–7.

Veerhoff M, Brümmer GW (1991): Mineralogische und chemische Charakterisierung von Abbauprodukten der Silikatverwitterung unter stark sauren Bedingungen. Mitteilungen Dt Bodenkundl Ges 66 (II):1123–1126.

Wenzel B (1989): Kalkungs- und Meliorationsexperimente im Solling: Initialeffekte auf Boden, Sickerwasser und Vegetation. Ber Forschungszentrum Waldökosysteme Universität Göttingen A 51:274 pp.

Wesselink LG, Meiwes K-J, Matzner E, Stein A (1994): Long term changes in water and soil chemistry in a spruce and beech forest, Solling, Germany. Env Sci Technol, submitted.

Wiedey GA, Raben GH (1989): Datendokumentation zur Waldschadensforschung im Hils. Ber Forschungszentrum Waldökosysteme Universität Göttingen B 12:305 pp.

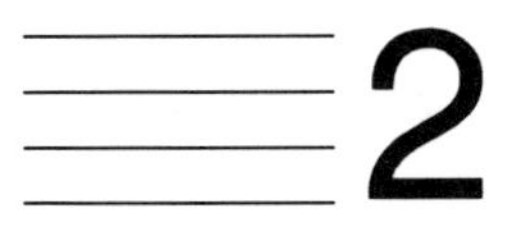

ALUMINUM BIOGEOCHEMISTRY IN THE ALBIOS FOREST ECOSYSTEMS: THE ROLE OF ACIDIC DEPOSITION IN ALUMINUM CYCLING

CHRISTOPHER S. CRONAN

Department of Plant Biology and Pathology, University of Maine, Orono, Maine 04469

1. INTRODUCTION

Aluminum is a nonessential element for living organisms, but is a critical geochemical component of most natural ecosystems, including forested watersheds. Distributed throughout the biosphere, aluminum is more abundant in the Earth's crust than all other elements except oxygen and silicon. Most soils are rich in aluminum oxides, aluminosilicate clays, and unweathered aluminosilicate minerals. These aluminous soil components are an important part of the soil physical matrix and also participate in a number of important geochemical processes, including ion exchange reactions involving major cations and anions, acid–base buffering, phosphorus immobilization, and pedogenetic weathering and translocation of aluminum. These pH-sensitive soil aluminum fractions are also a potential source of colloidal and soluble aluminum capable of interacting directly with living cells through a variety of antagonistic or toxic mechanisms.

It has been hypothesized that one of the major ecological consequences of atmospheric sulfur and nitrogen deposition is an increased mobilization and transport of

Effects of Acid Rain on Forest Processes, pages 51–81

aluminum in forest soils (Johnson, 1979; Ulrich et al., 1980). Because of the potential toxicity of this metal (Fig. 2.1), it was proposed that aqueous aluminum could be a critical biogeochemical link between atmospheric pollution and sensitive biotic communities in terrestrial and aquatic environments exposed to acidic deposition (Cronan and Schofield, 1979). Early studies showed that free ionic aluminum at concentrations of less than 10 μmol L^{-1} in acidified surface waters produced toxic conditions for some aquatic organisms (Baker and Schofield, 1982; Muniz and Leivestad, 1980). Also reported were some instances of increased root dieback and forest decline in Germany that were associated with both high levels of acidic deposition and elevated concentrations of aluminum sulfate and nitrate in soil solutions (Ulrich et al., 1980; Godbold et al., 1988). Based on that evidence, there was a need to define more clearly the nature and extent of the interactions between acidic deposition and aluminum in the environment.

The ALBIOS (Aluminum in the Biosphere) project was designed to examine the general patterns of aluminum (Al) biogeochemistry and Al toxicity in forested ecosystems exposed to acidic deposition. Primary study objectives were (1) to quantify regional patterns of Al mobilization, chemistry, and transport in areas receiving atmospheric deposition of strong acids; (2) to evaluate the potential for Al toxicity in the forests of eastern North America and northern Europe; and (3) to determine the influence of acidic deposition on aluminum cycling and toxicity in forested watershed ecosystems.

The geochemical and plant response components of ALBIOS were organized into complementary field and laboratory studies focused on the following two hypotheses: (1) acidic deposition increases the concentrations of labile Al in soil solutions and surface waters of forested watersheds; and (2) in forest ecosystems, acidic deposition increases bioavailable Al to levels that are toxic to trees and aquatic biota, causing growth reductions, nutritional deficiencies, or mortality. This chapter presents a summary of previous ALBIOS reports by Cronan et al. (1989), Kelly et al. (1990), Raynal et al. (1990), Sucoff et al. (1990), and Cronan and Schofield (1990), along with more recent findings and analysis of the ALBIOS results.

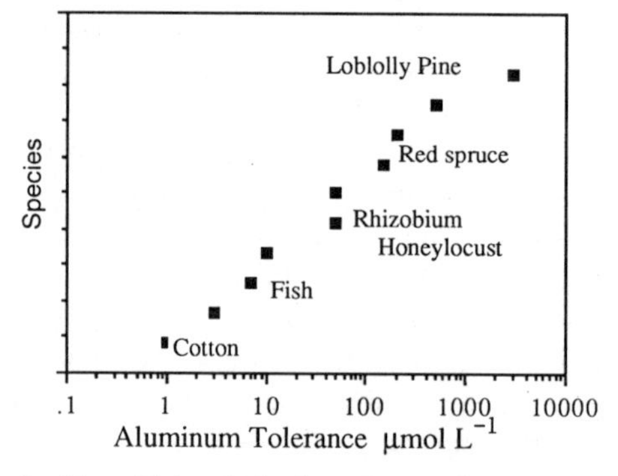

Fig. 2.1. Variations in the Al sensitivity of microbes, plants, and animals. Data from references cited in Cronan et al. (1986, 1989) and Cronan and Schofield (1990).

1.1. Background Literature

Previous studies of Al transport focused primarily on individual watersheds from northern temperate regions characterized by high loading of acidic deposition and elevated concentrations of Al in surface waters. These investigations included work in the Adirondack region of New York (David and Driscoll, 1984; Cronan, 1985; Driscoll et al., 1987), the New England region of the eastern United States (Likens et al., 1977; Cronan, 1980; Johnson et al., 1981; Driscoll et al., 1985), the Canadian Shield region (Campbell et al., 1986; Hendershot et al., 1986; Manley et al., 1987), the Cascade Mountains of Washington (Ugolini et al., 1977; Dahlgren and Ugolini, 1989), southwestern Sweden (Nilsson and Bergkvist, 1983), Denmark (Rasmussen, 1986), Norway (Sullivan et al., 1986), the Netherlands (van Breemen et al., 1982; Van Grinsven et al., 1987), and West Germany (Pillai-Nair, 1978; Ulrich et al., 1980; Hauhs, 1985). Other recent studies also extended to such geographic regions as the Pine Barrens of New Jersey (Budd et al., 1981; Turner et al., 1985), the Shenandoah Mountains of Virginia (Cozzarelli et al., 1987), the Smoky Mountains of Tennessee (Joslin et al., 1987; Johnson et al., 1991), the Sangre de Cristo Mountains of New Mexico (Graustein, 1981), and the Rocky Mountains of Colorado (Litaor, 1987). These biogeochemical investigations contributed to our improved understanding of the processes that regulate the chemistry and transport of aqueous Al.

Previous investigations of Al toxicity to trees included studies of at least 33 species in the following genera: *Abies, Acer, Alnus, Betula, Eleaegnus, Fagus, Fraxinus, Picea, Pinus, Populus, Prunus, Pseudotsuga, Quercus, Thuja, and Tsuga* (e.g., McCormick and Steiner, 1978; Rost-Siebert, 1983; Schier, 1985; Van Praag et al., 1985; Hutchinson et al., 1986; Paganelli et al., 1987; Hecht-Bucholz et al., 1987). These studies indicated that the visible or measurable symptoms of Al toxicity in trees include reduced root elongation and branching, a generalized thickening of the roots, and necrosis of cells near the root and shoot meristems. Other possible effects of Al toxicity and antagonism are summarized in Table 2.1.

Plants are adversely affected by exposure to ionic Al either through antagonistic interference with cation uptake and metabolism or through direct irreversible damage to plant cells from Al interactions with orthophosphate groups of biomolecules (Haug, 1984; Sucoff et al., 1990). These interactions may have the following consequences: (a) Al can bind to nucleotides and nucleic acids, inhibiting cell division; (b) Al can bind to adenosine triphosphate (ATP), adenosine diphosphate (ADP), or membrane-bound ATPases, interfering with energy transfer; (c) Al may interfere with enzyme systems like acid phosphatases; (d) Al may bind to calmodulin, seriously disrupting cellular control and contractile processes; (e) Al can bind to phospholipid groups and alter membrane permeability; and (f) Al can bind to apoplast surface adsorption sites and interfere with plant uptake or selectivity for nutrient ions such as calcium (Ca) or magnesium (Mg). Thus, Al toxicity can potentially affect energy transformations, cell division, membrane transport, nutrient accumulation, and various activities regulated by calmodulin, a multifunctional Ca-dependent regulatory protein.

TABLE 2.1. Observed Symptoms Associated With Aluminum Toxicity or Antagonism[a]

Cellular Responses
Inhibition of cell division and mitotic activity
Decreased water permeability in cell membranes
Shoot Responses
Delay in budbreak and leaf expansion
Decreased stem and leaf production
Decreased shoot biomass
Decreased tissue Mg and Ca concentrations
Increased tissue Al concentration
Altered tissue P concentrations
Root Responses
Increased tissue Al concentration
Reduced plant water uptake
Decreased root biomass
Decreased fine root branching
Decreased terminal root elongation
Decreased tissue Mg and Ca concentrations
Altered tissue P concentrations

[a] From Cronan et al., 1989.

2. STUDY SITES

The ALBIOS investigation involved a comparison of Al biogeochemistry in 10 North American and 4 European watersheds (Table 2.2). These forested study sites were selected to provide contrasts in key environmental parameters, including atmospheric deposition, climate, soil properties, bedrock and surficial geology, and forest cover. Estimated annual sulfur (S) deposition at these watersheds ranges 20-fold from approximately 4 kg of S ha^{-1} at the Experimental Lakes Area, Ontario, to greater than 80 kg of S ha^{-1} at Solling, Germany. For the North American sites, estimated annual S inputs from wet plus dry deposition range from about 4 kg of S ha^{-1} in the upper Midwest, to about 11 kg of S ha^{-1} at Plastic Lake in eastern Ontario, to about 18 kg of S ha^{-1} in the Adirondack Park of New York, to about 20 kg of S ha^{-1} at Camp Branch watershed in central Tennessee (Cronan and Goldstein, 1989).

3. METHODS

At each catchment, four to seven study plots were established along topographic drainage gradients and soils were sampled by horizon for detailed physical–chemical analysis. Samples were refrigerated at field moisture until analysis. Soil chemical parameters were determined by selective extraction using neutral 1 *M* potassium chloride (KCl) for exchangeable acidity, 1 *M* ammonium chloride

TABLE 2.2. The ALBIOS Watersheds

Watershed	Map Coordinates
Southern and Midwestern Region	
Round Lake, Wisconsin	46°30′N,92°0′W
University Forest, Missouri	36°54′N,90°19′W
Oxford Pine, Mississippi	34°23′N,89°27′W
Camp Branch, Tennessee	35°38′N,85°18′W
Coweeta WS 40, North Carolina	35°04′N,83°28′W
Old Rag Mt., Virginia	38°33′N,78°19′W
Northern Region	
Experimental Lakes Area, Ontario	49°42′N,93°44′W
Plastic Lake, Ontario	45°11′N,78°49′W
Huntington Forest, New York	44°01′N,74°16′W
Big Moose, New York	43°50′N,74°52′W
Birkenes, Norway	58°23′N,08°14′E
Lake Gardsjon, Sweden	58°04′N,12°03′E
West German Region	
Lange Bramke, Germany	51°52′N,10°25′E
Solling, Germany	51°40′N,09°30′E

(NH_4Cl) for exchangeable base cations, 0.2 *M* copper acetate at pH 4.8 for adsorbed organic Al, and 0.2 *M* ammonium oxalate at pH 3.0 for amorphous Al. Cation exchange capacity (CEC) at field pH was estimated as the sum of exchangeable base cations (Ca + Mg + K + Na) and acidity (Al + H^+). Base saturation was calculated as the sum of base cations divided by the CEC. Soil mineralogy was determined by X-ray diffraction analysis, while bulk soil chemistry was estimated by X-ray fluorescence. These background results are reported in a separate project summary by Cronan and Goldstein (1989).

At each plot, solutions were sampled along hydrologic flow gradients from rainfall inputs, through soils, and into surface waters. Paired sets of plastic tension and tension-free lysimeters were used to collect soil water at each plot from the O, B2, and BC or C horizons. The sampling surface of the tension lysimeters consisted of an 0.2-μm Nylon-66 filter membrane sandwiched between two disks of 6-mm thick 35-μm porous polyethylene. Most watersheds were sampled on a storm event basis during the spring or fall between 1983 and 1986. The intent of the sampling program was to develop an extensive analysis of spatial variations in aqueous Al. In this respect, the study goals were similar to those of the Eastern Lake Survey and the National Stream Survey conducted by the U.S. Environmental Protection Agency (EPA). Two of the study sites (Camp Branch, TN and Big Moose, NY) were studied on an intensive basis to evaluate temporal and spatial variations in Al chemistry (Cronan et al., 1990).

After collection, field pH and temperature were measured, and water samples were immediately processed to separate Al species into the following operationally defined categories based on charge and lability: charged labile Al (= labile inorganic); uncharged labile Al (= labile organic); and total acid soluble Al. Total labile Al was estimated by a 10-s reaction with 8-hydroxyquinoline at pH 8.3, followed by extraction into MIBK in the field (Driscoll, 1984). The MIBK extracts were analyzed for Al spectrophotometrically with correction for iron (Fe) interference. Charged and uncharged Al fractions were separated by rapid elution through Rexyn 101 cation exchange resin adjusted to the sample pH. Total acid soluble Al was determined in subsamples containing 0.01% *o*-phenanthroline, 1.0% hydroxylamine hydrochloride, and 0.2% hydrochloric acid (HCl).

These separate Al fractions were discriminated, because of their potential biological and geochemical significance. The charged labile inorganic Al was assumed to represent the most biologically active and potentially toxic fraction. The uncharged labile Al was assumed to include aluminum complexed with organic ligands or bound to silica. Finally, the acid soluble pool of Al was taken as an estimate of the total Al in solution and colloidal suspension for geochemical mass balance purposes.

Other major ions in filtered subsamples were analyzed by standard methods of ion chromatography (Cl^-, SO_4^{2-}, NO_3^-), atomic absorption spectroscopy (Ca, Mg, K, and Na), and electrochemical analysis (F^- and lab air-equilibrated pH). Analytical precision and accuracy were evaluated with interlaboratory calibration tests, analysis of standard reference materials, and repeated sample measurements. Further details on the study methodology have been reported elsewhere by Cronan and Goldstein (1989).

Controlled laboratory experiments and field studies were used to examine Al toxicity responses for one indicator species (honeylocust, *Gleditsia triacanthos*) and several major commercial tree species (red spruce, *Picea rubens;* sugar maple, *Acer saccharum;* red oak, *Quercus rubra;* American beech, *Fagus grandifolia;* European beech, *Fagus sylvatica;* and loblolly pine, *Pinus Taeda*). Efforts to include Norway spruce (*Picea abies*) in the studies were unsuccessful, because of problems with pathogens and disease among the seedlings. The plant response investigations included hydroponic studies of Al toxicity thresholds, greenhouse soil culture experiments, and field root ingrowth core experiments (Sucoff et al., 1990; Kelly et al., 1990; Raynal et al., 1990). During the course of the field studies, tree root and foliar tissue samples were also collected at most of the ALBIOS watersheds as a means of examining correlations between soil and tissue Al chemistry (Joslin et al., 1988).

4. RESULTS AND DISCUSSION

4.1. Patterns of Aluminum Geochemistry and Cycling in the Forest Landscape

4.1.1. Comparative Aluminum Chemistry in Soils

Because soils are major sources and sinks for Al, the 14 ALBIOS watersheds were analyzed in detail for soil chemical characteristics that might relate to the patterns of Al cycling observed across the geographic sampling gradient. At a general level, there were important differences between soil horizons and among watersheds in terms of exchangeable cation chemistry and acidity (Fig. 2.2). Surface O horizons

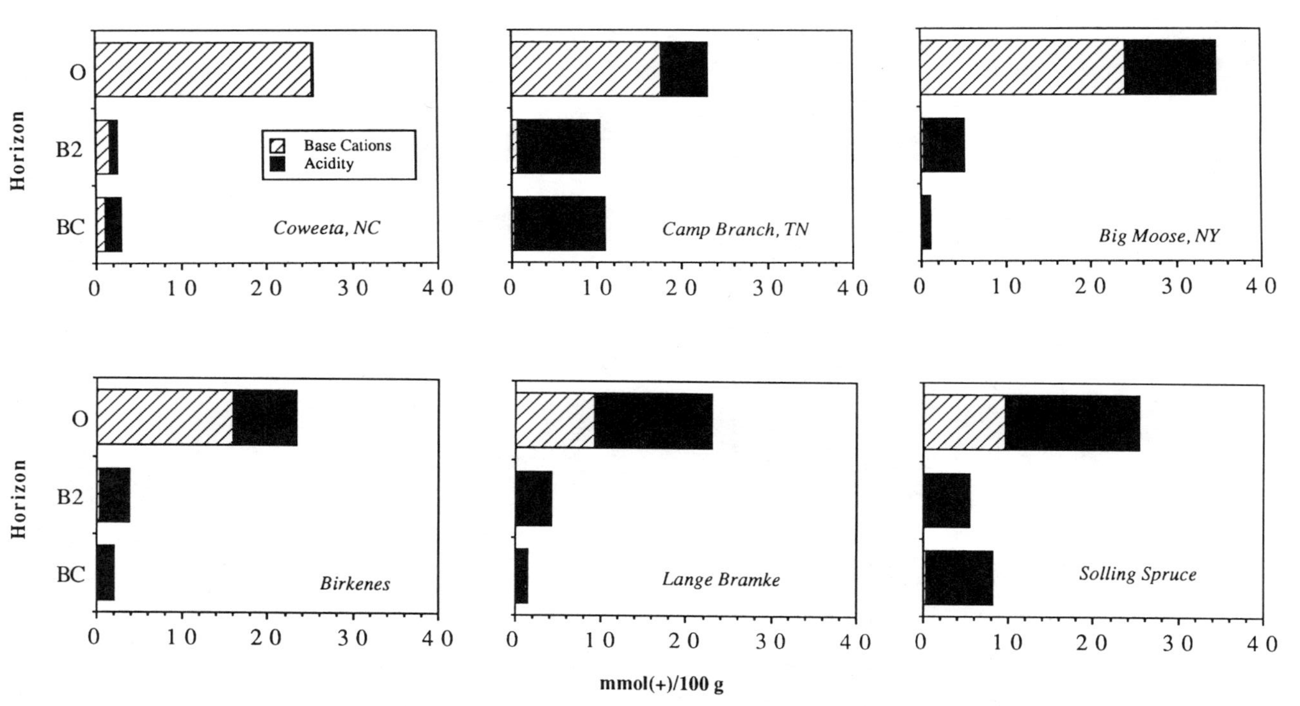

Fig. 2.2. Comparison of soil exchange chemistry for three North American and three northern European ALBIOS forest ecosystems. Base cations refer to the sum of exchangeable Ca + Mg + K + Na, whereas exchangeable acidity includes Al^{3+} + H^{+}.

typically exhibited high exchange capacities and a high percentage of exchangeable base cations (i.e., Ca, Mg, K, and Na). In contrast, mineral B horizons had lower exchange capacities and higher relative amounts of exchangeable acidity. The striking divergence of organic and mineral soil horizons is illustrated in Figure 2.3, where the O horizon samples are shown plotted in a region of lower pH and higher percent base saturation as compared to the B horizon samples from the ALBIOS forest sites.

Selective extractions were used to quantify and to compare several operationally defined, partially overlapping soil Al pools in the ALBIOS study soils. Figure 2.4 presents a comparison of the profile distribution of these soil Al fractions for three contrasting North American and three of the European ALBIOS sites. In virtually every case, the amorphous Al hydroxy pool extracted with acid ammonium oxalate was the largest soil Al fraction. For most soils, this fraction reached maximum concentrations in the illuvial B horizon. One exception to that pattern was found at Coweeta Hydrologic Lab in North Carolina, where downslope erosion enriched the surface soil horizons with high concentrations of this Al fraction. The complexed organic Al pool extracted with copper acetate was the next most abundant soil Al fraction in the northern soils, and reached its highest concentrations in the spodic B horizons of these soils. Finally, the exchangeable Al fraction extracted with neutral KCl was a relatively small proportion of the potentially reactive soil Al reservoir in these soils.

Total amounts of amorphous, organic, and exchangeable Al varied greatly between watersheds, but tended to be concentrated in the B horizons of most forest soils. In contrasting the soil Al pools at Camp Branch, TN versus Big Moose, NY, Cronan et al. (1990) found that the amorphous, organic, and exchangeable pools in the rooting zone of the northern Spodosol were 6×, 3×, and 0.33× as large as the respective pools in the southern Ultisol. Although Al pools in surface O horizons of the ALBIOS soils tended to be relatively small, the amounts of accumulated forest floor Al were sufficiently large that one might wonder about the source of that Al. In an attempt to address that question for the ALBIOS project, Rustad (1988) exam-

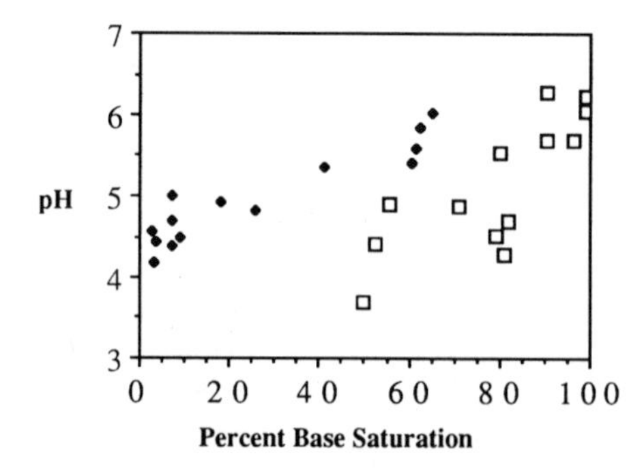

Fig. 2.3. Relationship between soil pH in water and percent base saturation (% BS = sum of base cations/effective cation exchange capacity) for ALBIOS O horizons (open boxes) and B horizons (dark diamonds).

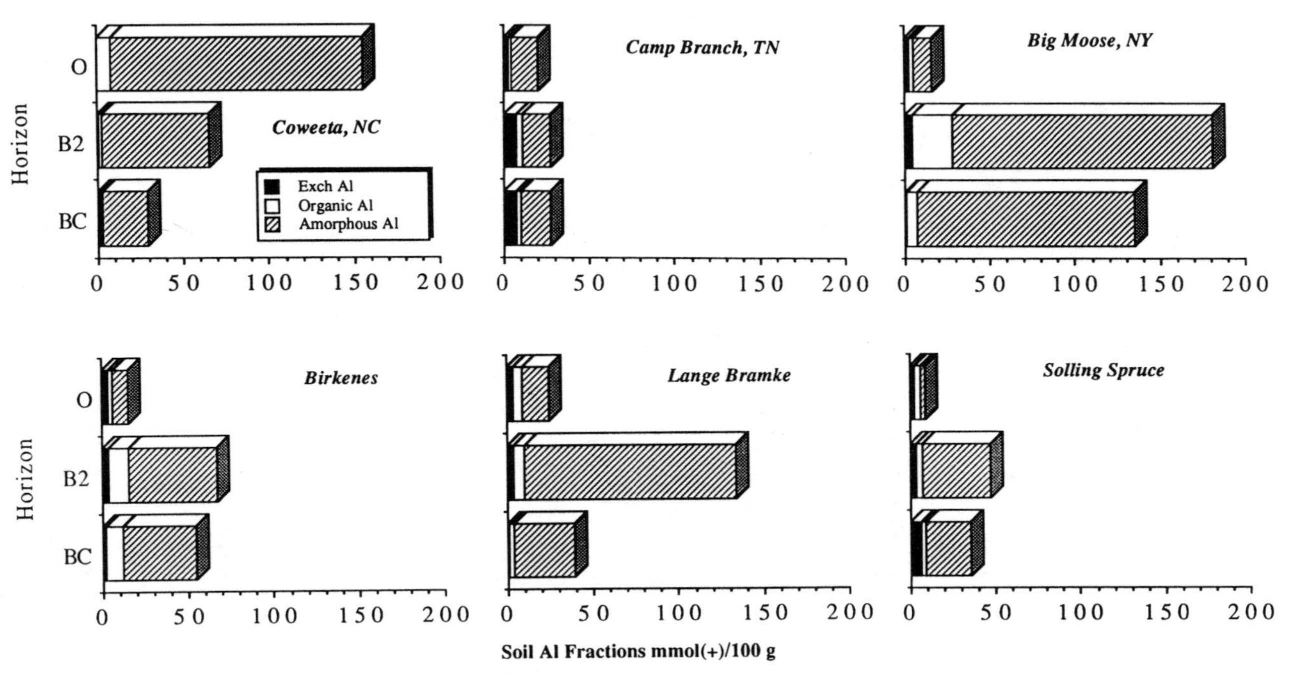

Fig. 2.4. Interregional differences in soil Al fractions for organic and mineral horizons at the ALBIOS study sites. The operationally defined Al pools are partially overlapping, and are depicted using the actual concentration of Al extracted by each salt.

ined Al cycling in the forest floor of a northern spruce-fir forest in Maine, and found that plant cycling of Al was insufficient to account for observed pools and leaching fluxes of Al. It was concluded that much of the Al must have originated through windthrow mixing of mineral soil into the O horizon, followed by in situ weathering.

The Al fractionation data provided further confirmation of the sharp contrasts in chemical properties observed between the surface organic and subsurface mineral soil horizons in the ALBIOS watersheds. As shown on the log–log plot in Figure 2.5, the O horizon cluster was characterized by a low ratio of exchangeable Al/exchangeable Ca and a small concentration of amorphous Al. In comparison, the B2 horizon samples from the ALBIOS sites clustered in a separate region of the graph characterized by a high ratio of exchangeable Al/Ca and high concentrations of amorphous Al.

4.1.2. Aqueous Aluminum Chemistry and Transport

Concentrations of aqueous Al (Fig. 2.6*a*) and strong acid anions (Fig. 2.7) varied considerably in soil drainage and surface waters among the ALBIOS study watersheds (Cronan and Schofield, 1990). Maximum observed concentrations of labile Al in soil waters ranged from 1 to 10 μmol L^{-1} in Ultisols, Alfisols, and some Inceptisols of the southeastern and midwestern United States, to 15 to 70 μmol L^{-1} in northern Spodosols, to 80 to 240 μmol L^{-1} in spruce forest Inceptisols in Germany. In general, the highest concentrations of labile Al and the lowest Ca/Al molar ratios (Fig. 2.6*b*) were found in mineral soil horizons of the northern Spodosols and German Inceptisols. These sensitive soil horizons shared the following characteristics: (a) base saturation was usually less than 10–15% of effective cation exchange capacity at field pH; (b) soil pH in water (1:1) was typically less than 4.9; and (c) soil solution sulfate concentration was generally 80 μmol L^{-1} or greater (Cronan and Schofield, 1990).

For surface organic soil horizons, maximum concentrations of labile Al were

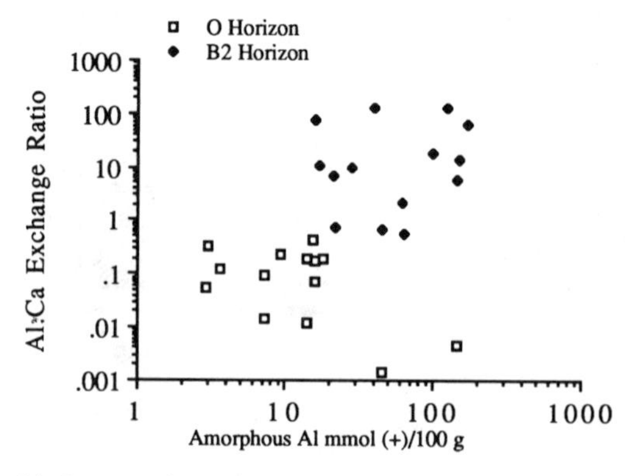

Fig. 2.5. Relationship between the exchangeable Al/Ca ratio and the concentration of extractable amorphous aluminum for O horizon and B2 horizon samples from the ALBIOS study sites.

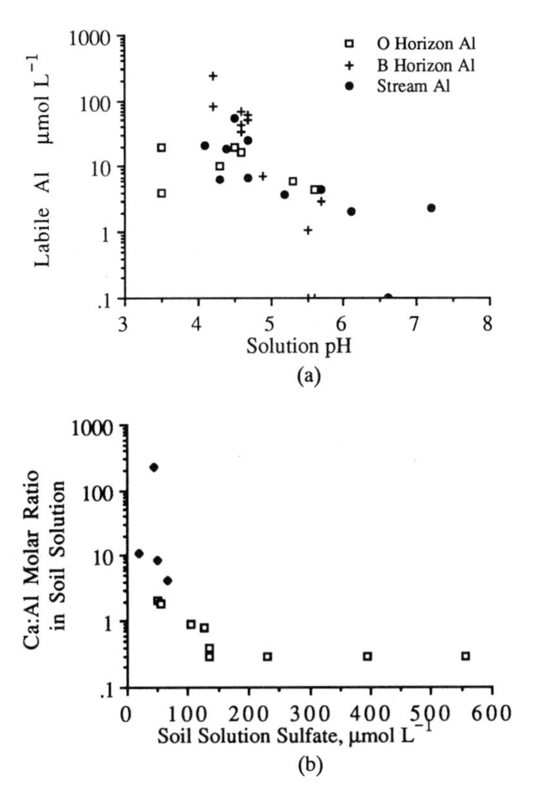

Fig. 2.6. (*a*) Range of variation for peak concentrations of labile aluminum and associated pH in soil solutions and streamwater at the ALBIOS watersheds. (*b*) The Ca/Al molar concentration ratios as a function of soil solution sulfate concentration for B horizons in the ALBIOS North American and European watersheds (Cronan and Goldstein, 1989; Cronan and Schofield, 1990).

typically less than 10–20 µmol L^{-1}, in spite of the low pH and high concentrations of organic ligands characteristic of forest floor leachates. The highest concentrations of labile Al in forest floors occurred where base saturation was less than 20–30% and the bound Al ratio was high (Cronan et al., 1986).

Headwater streams in the study watersheds (Fig. 2.6) contained labile Al concentrations that ranged from less than 1 µmol L^{-1} to peak values of approximately 55 µmol L^{-1} in one German stream (Cronan et al., 1989; Cronan and Schofield, 1990). These differences in stream chemistry were a function of a number of hydrogeochemical factors, including flow paths, soil chemistry of source areas, flow conditions, surficial geology and mineralogy, and mobile anion chemistry. As shown in Figure 2.8, there was a reasonably strong inverse relationship between stream pH and the ratio of strong acid anions (SAA) to Ca. Thus, for those streams with a high ratio of SAA/Ca, much of the cationic charge was contributed by H^+ and Al^{n+} ions.

Some of the general patterns of solution chemistry within individual watersheds

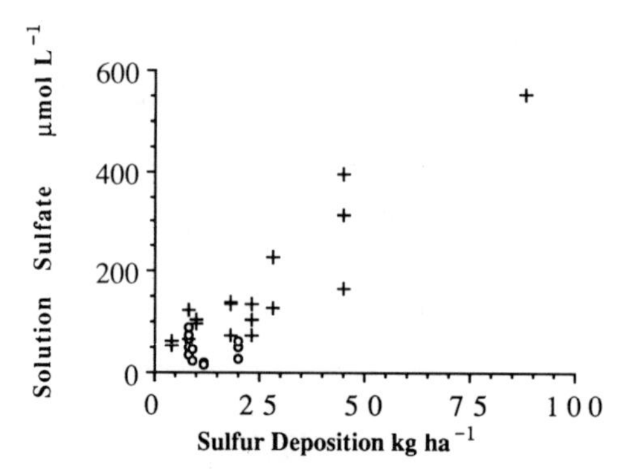

Fig. 2.7. Relationship between estimated total S deposition and soil solution sulfate for the ALBIOS watersheds (Cronan and Schofield, 1990). Open circles indicate southern forest sites where soil sulfate adsorption is an important process limiting soil solution sulfate concentration. Crosses are other ALBIOS sites.

and among the different North American and European study sites are illustrated in Figure 2.9. For most watersheds, sulfate was the dominant anion in soil solutions and streamwater. Exceptions to this were found in a few cases where bicarbonate (e.g., Coweeta, NC) or chloride (e.g., Birkenes, Norway) concentrations were high enough to dominate anion charge equivalents. There were also cases where nitrate concentrations were high enough to provide an important contribution to cationic charge balance (e.g., Big Moose, NY; Lange Bramke, Germany; and Solling, Germany).

Charge contributions from aqueous Al varied widely across the different ALBIOS study sites (Fig. 2.9). For example, ionic Al was virtually absent from soil solutions and streamwater at the southern Coweeta, NC watershed, despite the large pools of amorphous Al hydroxides throughout the soil profile (Fig. 2.4). At the southern Camp Branch, TN watershed, ionic Al concentrations were also low, despite the acidic soil conditions and large concentrations of exchangeable Al. However, large concentrations of free and complexed Al were observed in the mineral soil solutions and headwater streams of several northern and German study

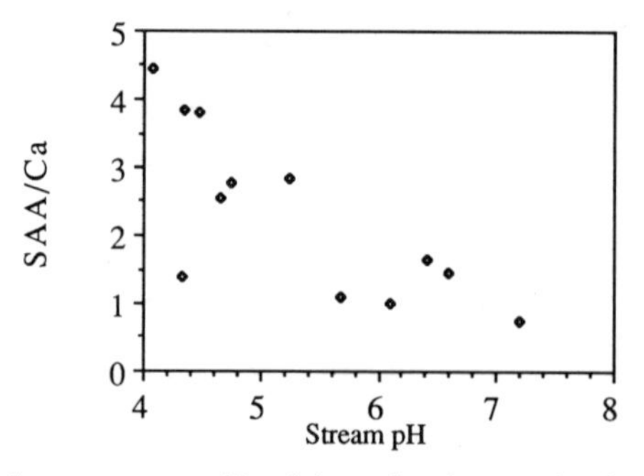

Fig. 2.8. Relationship between stream pH and the molar charge ratio of strong acid anions (SAA = SO_4^{2+} and NO_3^-) to calcium for streams in the ALBIOS watersheds.

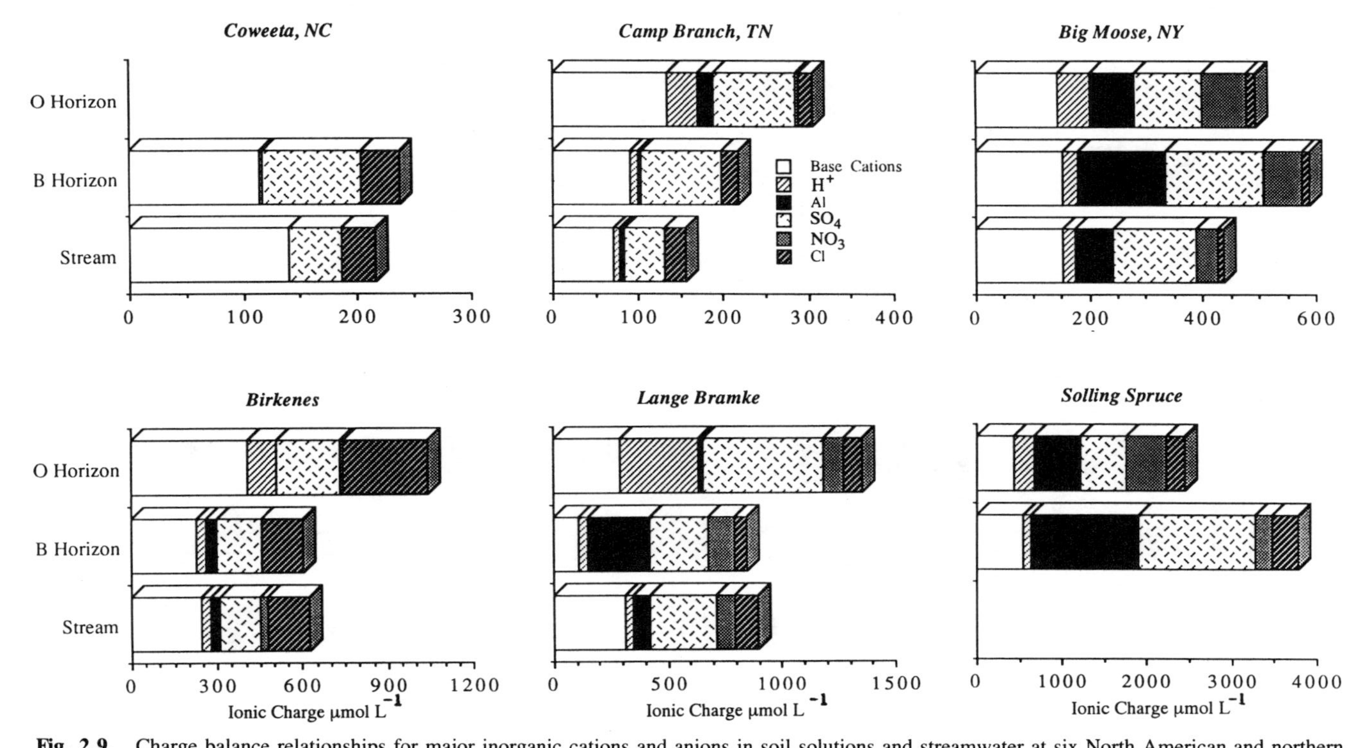

Fig. 2.9. Charge balance relationships for major inorganic cations and anions in soil solutions and streamwater at six North American and northern European ALBIOS watersheds. Data were not available for the Coweeta O horizon and Solling stream. Because of limited collections at the German sites, data from Hauhs (1989) were used in the Lange Bramke B horizon stacked bar, and data from Matzner (1989) and Bredemeier et al. (1990) were used in the Solling graph.

sites. In fact, Al was the major cation contributing to charge balance in the B horizon soil solutions at the Big Moose, Lange Bramke, and Solling spruce ecosystems.

Figure 2.10 presents a closer look at the range of variation observed for soil solutions from three of the mineral B horizons described in Figures 2.2 and 2.4. As indicated in those earlier figures, the Camp Branch, Big Moose, and Solling mineral soils are all very acidic and contain relatively large pools of reactive Al. According to Reuss (1983), one would expect significant Al leaching from soils such as these, where the base saturation is less than 10–15% of effective cation exchange capacity. However, as shown in Figure 2.10, Al concentrations varied by two orders of magnitude between the three forest soil profiles. As discussed by Cronan et al. (1990) and others, these differences can be accounted for in large part by the major differences in mobile strong acid anion flux through the separate soils. This phenomenon will be discussed in more detail elsewhere in this chapter.

Results from the ALBIOS study and other investigations have shown that Al transport through soils involves (a) dynamic concentration changes in aqueous Al associated with movement through different source and sink horizons, and (b) qualitative shifts in the distribution of ionic, complexed, and colloidal Al fractions in soil water during infiltration. As an example, Figure 2.11 illustrates the observed changes in operationally defined aqueous Al fractions for water samples collected along drainage gradients in contrasting southern and northern hardwood watersheds in eastern North America. At the northern drainage basin, Al concentrations were very high throughout the drainage profile (Fig. 2.11). Organic horizon leachates contained up to 30–40 μmol L^{-1} or more of labile uncharged Al, labile charged Al, and acid soluble Al. Much of this Al was presumed to be complexed or bound by soluble or colloidal organic matter (Ugolini et al., 1977; Graustein et al., 1977; Graustein, 1981; David and Driscoll, 1984). As drainage water migrated from the Oa to the Bs horizon, large increases in labile charged Al were evident and some reduction in acid soluble Al occurred (Fig. 2.11). With further transport to the BC horizon and stream, concentrations of labile Al decreased and pH increased.

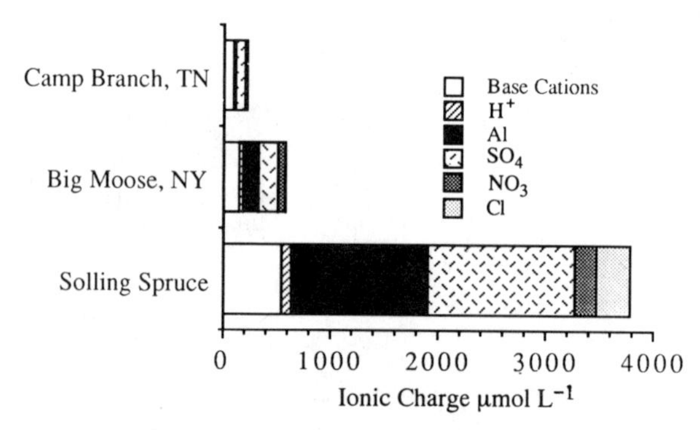

Fig. 2.10. Comparison of soil solution chemistries for acidic B horizons from a Tennessee Ultisol, a New York Spodosol, and a German Inceptisol exposed to respective sulfur deposition inputs of 20, 18, and 88 kg ha^{-1} $year^{-1}$.

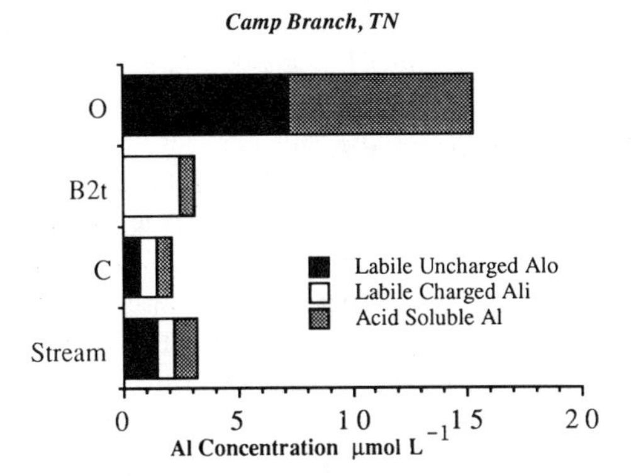

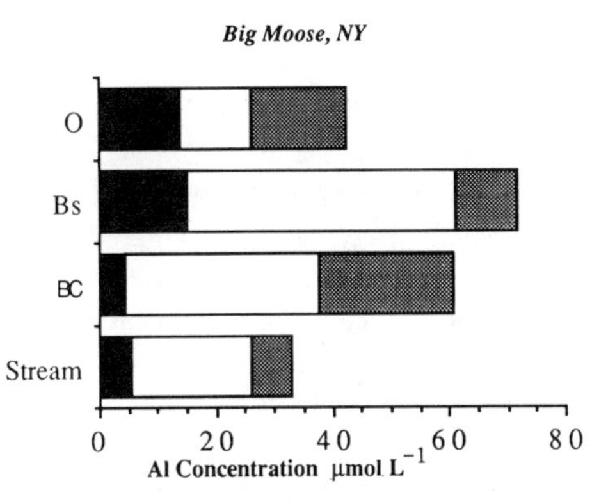

Fig. 2.11. Distribution of operationally defined aqueous Al fractions in drainage waters from a southern watershed at Camp Branch, TN and a northern watershed at Big Moose, NY (Cronan et al., 1990). The Al_o fraction is assumed to represent organic (O) complexes of Al, except in the Camp Branch C horizon, where it may actually be an uncharged Al–silicate complex.

Based upon thermodynamic calculations, the inorganic speciation of Al at the northern site was dominated by aquo aluminum (Al^{3+}) (Cronan et al., 1990). This speciation resulted from the low pH values and high concentrations of labile charged Al, coupled with low concentrations of inorganic complexing ligands. Some complexation by hydroxide (4–11% Al—OH) and fluoride (6–17% Al—F) ligands was evident, while binding by sulfate was low (1–5% $Al—SO_4$). Complexation by hydroxide increased as solution acidity decreased through the soil profile.

In contrast to conditions in the northern watershed, concentrations of labile and acid soluble Al were low throughout the southern ecosystem (Fig. 2.11). Labile charged Al, the most biologically active fraction (Driscoll et al., 1980), ranged from 0 to 2.3 $\mu mol\ L^{-1}$ in soil and stream solutions. Surface soil horizon leachates at the southern watershed clearly exhibited the highest Al concentrations and were dominated by labile uncharged and acid soluble Al. These Al fractions, again, coincided with elevated dissolved organic carbon (DOC) concentrations and were probably associated with the release of organic ligands and humic colloids from the forest floor (Ugolini et al., 1977). With water movement into the B horizon, concentrations of total Al decreased substantially, but concentrations of labile charged Al increased to maximum values (2.3 μmol L^{-1}) in this horizon. Concentrations of Al were very low in C horizon solutions and streamwater at Camp Branch, coincident with higher pH values and low concentrations of SO_4^{2-} and DOC.

At the southern watershed, the calculated speciation of inorganic Al was dominated (>85%) by fluoride complexes (i.e., AlF^{2+} and AlF_2^{+}). Drainage water concentrations of fluoride at the southern watershed were similar to values observed at the northern site; however, these molar concentrations of fluoride generally exceeded concentrations of labile charged Al at the southern watershed. This fact, coupled with the strong affinity of Al for fluoride in acidic solutions (Hem and Roberson, 1967; Plankey et al., 1986), resulted in the predominance of alumino–fluoride species at the southern site.

4.1.3. Relationships Between Acidic Deposition and Aqueous Aluminum

Having established the general range of environmental variation in Al concentrations for the ALBIOS watersheds (Figs.2.6, 2.9, and 2.11), this study next focused on two important questions: (1) can the interregional variations in aqueous Al chemistry be accounted for; and (2) is there a relationship between acidic deposition and the patterns of Al chemistry in soil solutions and streamwater. As predicted by Reuss (1983), our results showed that one of the major controls on intersoil differences in concentrations of labile Al was the stoichiometry of the soil exchange complex. Labile Al was greatest for soils with very low base saturation and a high ratio of exchangeable Al to exchangeable base cations (Fig. 2.12). Conversely, soils with higher base saturation exhibited low-to-nondetectable concentrations of labile

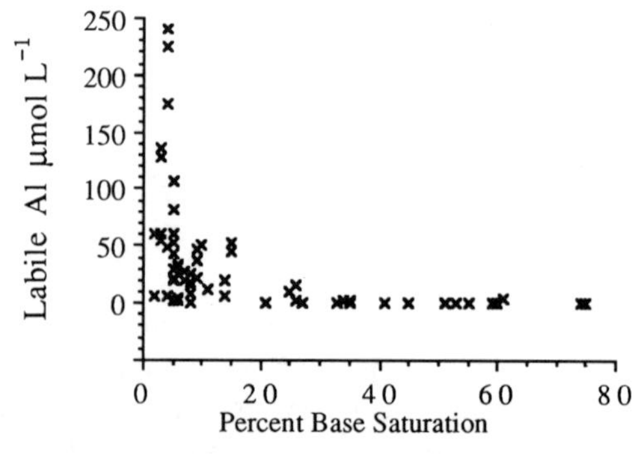

Fig. 2.12. Relationship between soil percent base saturation and labile Al in soil solutions from mineral horizons in the ALBIOS watersheds (Cronan and Schofield, 1990). The *y* axis is raised to provide a clearer view of the points clustered along the $y = 0$ line.

Al. This simple relationship is in agreement with previous thermodynamic predictions, and provides an important tool for stratifying forest soils and watersheds into systems that are more or less sensitive to Al mobilization.

Despite the significance of soil exchange chemistry as a determinant of aqueous Al concentrations, base saturation differences explained only part of the variation in labile Al (Cronan and Schofield, 1990). Although high concentrations of labile Al were only observed in soils with less than 10–15% base saturation, some of the lowest concentrations of labile Al were also found in soils with the same base saturation characteristics. This evidence indicated that additional factors were involved in controlling intersite differences in labile Al concentrations.

Because of our interest in acidic deposition as a source of mobile strong acid anions (Fig. 2.7), we next examined the relationships between labile Al and the sum of strong acid anions in our field samples. Results showed that for those watersheds containing mineral soils with less than 10% base saturation, increased concentrations of SO_4^{2-} plus NO_3^- in soil drainage waters were accompanied by increased concentrations of labile Al (Fig. 2.13). This result is just what Reuss and Johnson (1986) predicted from ion exchange principles for an Al-saturated soil exchange complex exposed to increasing concentrations of acid or salt.

For soils such as those included in Figure 2.13, any addition of mobile anions from acidic deposition, fertilization, nitrification associated with disturbance, or marine aerosol inputs would lead to a general Al leaching response of the type illustrated in Figure 2.13 (Cronan and Schofield, 1990). In the case of this study, it so happens that the major strong acid anion at the ALBIOS watersheds was SO_4^{2-} and the primary source for SO_4^{2-} was acidic deposition. The gradient of increasing strong acid anions indicated in Figure 2.13 was a function of intersite differences in the following ecosystem mass balance relationship:

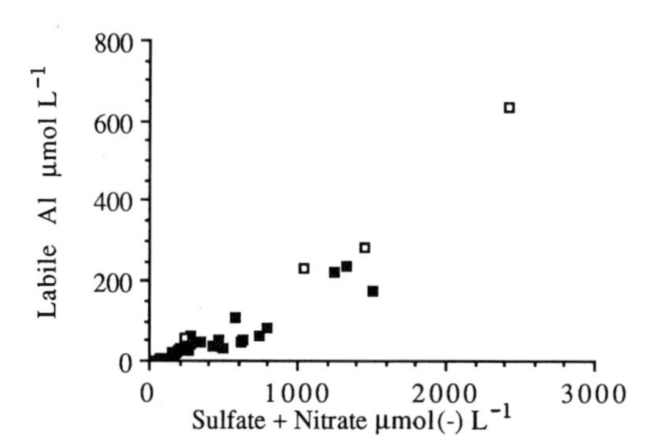

Fig. 2.13. Relationship between strong acid anions and labile Al in soil solutions collected from mineral horizons with a base saturation of 10% or less (Cronan and Schofield, 1990). The points include samples from 7 of the 14 ALBIOS watersheds: Plastic Lake, Big Moose, Camp Branch, Birkenes, Lake Gardsjon, Lange Bramke, and Solling. The open boxes are additional comparative data from an oak woodland in the Netherlands (Van Breemen et al., 1989) and from the southern Appalachian Mountains (Johnson et al., 1991; Joslin and Wolfe, 1992). The filled boxes are the ALBIOS sites.

Strong acid anion leaching = [atmospheric strong acid anion inputs] − [SO_4^{2-} adsorption] − [NO_3^- uptake] + [internal release of strong acid anions]

Thus, Camp Branch, TN, with both high atmospheric inputs (~20 kg of S ha^{-1} $year^{-1}$ and ~7 kg of N $ha^{-1}year^{-1}$) and high ecosystem retention of strong acid anions (~70% sulfate retention and ~85% nitrate retention), plotted at the lower left corner of Figure 2.13. In contrast, Solling, Germany, with high inputs of acidic deposition (~88 kg of S $ha^{-1}year^{-1}$ and ~30 kg of N $ha^{-1}year^{-1}$) and low ecosystem retention of strong acid anions, plotted in the upper right corner of the graph. The open boxes in the graph show additional results from an oak woodland in the Netherlands (van Breemen et al., 1989) and from high elevation spruce stands in the southern Appalachian Mountains (Johnson et al., 1991; Joslin and Wolfe, 1992). It should be noted that some of the scatter in the diagram results from the omission of Cl, which is an important strong acid anion at some of the sites.

Based upon these results, it was concluded that for a specific subset of the ALBIOS study watersheds (i.e., those containing soils with <10–15% base saturation), the prediction stated in the first hypothesis was supported (Cronan and Schofield, 1990). In other words, much of the variation in soil solution Al concentrations could be accounted for by changes in the concentrations of the two major strong acid anions derived from atmospheric deposition: sulfate and nitrate. For systems such as those included in Figure 2.13, one would expect changes in acidic deposition to result in direct changes in the concentrations and transport of strong acid anions and labile Al. The magnitude of this response would be influenced by the amount of net ecosystem retention of sulfate and nitrate by soil adsorption and biological assimilation. Any significant changes in labile Al concentrations would produce corresponding changes in the potential for Al toxicity (Haug, 1984).

Surface waters dominated by hydrologic inputs from the sensitive soil horizons described above would also be expected to exhibit a positive relationship between sulfate plus nitrate and labile Al. As shown in Figure 2.14, some of the ALBIOS streams did, indeed, conform to that general prediction. However, a number of the streams were dominated by groundwater inflows containing neutral salts of sulfate and nitrate; as a consequence, those streams did not fit the pattern observed for most of the more acidic streams.

One of the central questions for the ALBIOS investigation was the following: Is aluminum mobilization in soils and transport to streams a characteristic response observed in all watersheds exposed to acidic deposition? As implied in the preceding paragraph, the answer to this question can be understood in part by considering what happens to strong acids after they are deposited from the atmosphere to the surface of the earth. In general, there are three major fates for inputs of strong acid to a forested watershed: (1) the strong acid may be neutralized by the release of base cations from cation exchange and weathering reactions; (2) the strong acid may be partially neutralized by the release of a mixture of cations, including Al, derived from cation exchange or dissolution reactions; or (3) the strong acid may be neutralized by the alkalinity generation associated with anion adsorption on soil colloids or biological uptake of sulfate and nitrate (Cronan and Schofield, 1990).

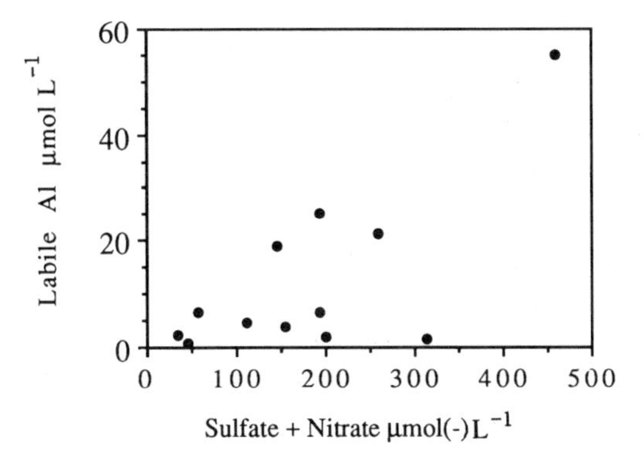

Fig. 2.14. Relationship between strong acid anions and labile Al for streams at the ALBIOS watersheds.

Because of these varied types of responses, there is no simple universal relationship between atmospheric inputs of strong acid and Al transport in the environment.

As an example, we can examine the contrasting Camp Branch, TN and Big Moose, NY watersheds from the ALBIOS study (Fig. 2.15). In an earlier paper, Cronan et al., (1990) reported that the major differences in Al chemistry and transport between the two watersheds were related to different patterns of alkalinity generation and mobile anion transport in the two forest ecosystems. At the northern watershed, atmospheric inputs of acidity were partially neutralized through the release of mixed cations (Ca, Al, Mg, K, and Na) from soils and detritus. Because of the high mobility of sulfate and nitrate in the northern watershed, there was significant transport of Al through the soil profile and into stream water. In the southern watershed, soil sulfate adsorption, biological retention of nitrate, and base cation release were the major sources of acid neutralizing capacity for soil drainage waters and surface waters. Because these processes resulted in both the release of alkalinity and the removal of mobile strong acid anions, concentrations of labile Al remained low throughout most of the drainage profile in the southern watershed.

4.1.4. Pools and Fluxes of Aluminum in Forest Ecosystems

Efforts have only recently been made to quanify the various pools and transfers associated with Al cycling in forest ecosystems. As shown in Table 2.3, previous studies provided diverse pieces of the Al cycle, but there are few if any comprehensive data sets on this topic. In the ALBIOS study, portions of the Al cycle were examined in detail in two contrasting forest ecosystems: the southern Camp Branch watershed in eastern Tennessee, and the northern Big Moose watershed in the Adirondack Park of New York. As indicated in Table 2.3, there were several distinct contrasts between the northern and southern sites. At the southern ecosystem, the estimated pool of KCl exchangeable Al in the upper 50 cm of soil was approximately three times larger than the exchangeable Al pool at the northern ecosystem. In contrast, the pools of cooper (Cu)-extractable "adsorbed organic Al" and ammonium (NH_4)-oxalate extractable "amorphous Al" at the southern site were roughly

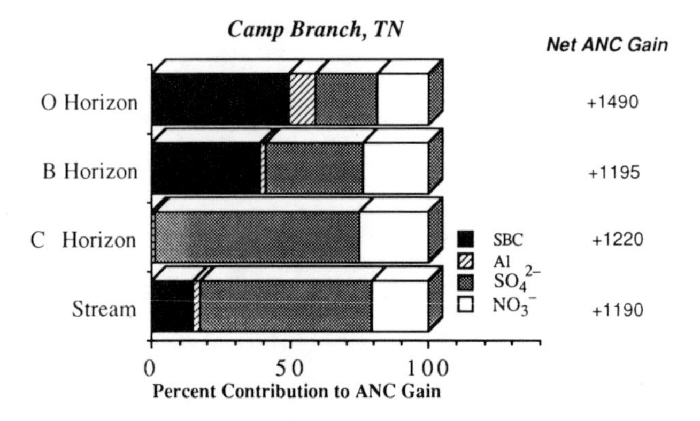

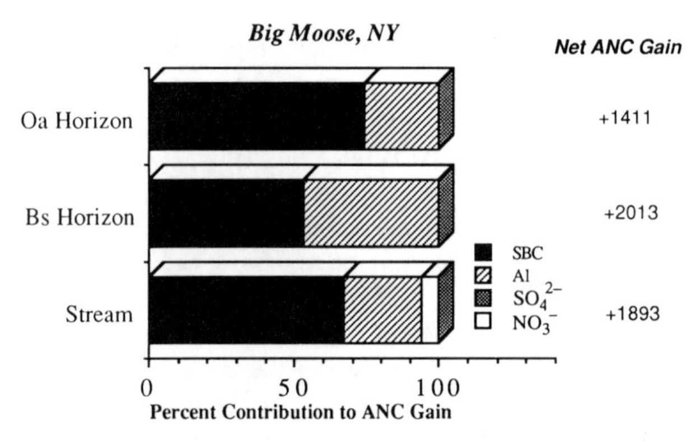

Fig. 2.15. Components of acid neutralizing capacity (ANC) generation at the southern Camp Branch and northern Big Moose watersheds (Cronan et al., 1990). For each ecosystem level, the net ANC gain compared to atmospheric deposition inputs is shown in $mol_c\ ha^{-1}$ at the right of the graph. The bar graph indicates the percentage of that ANC gain that is contributed by base cation release (SBC = sum of base cations), Al release, sulfate adsorption, and nitrate uptake and accumulation.

one third and one sixth the size of the respective pools at the northern drainage basin.

Transfers of Al out of each soil horizon were also distinctly different between watersheds, both in terms of magnitude and major source and sink horizons. At the southern site, the surface horizons were the major source for Al export in the soil profile, while the B horizon appeared to act as the major sink for Al. By comparison, the B horizon was the major source for mobile Al at the northern site, and the major Al sink was either in the C horizon or in the soil–stream interface (Driscoll et al., 1989). Fluxes of labile Al from individual soil horizons were much greater at the northern watershed (where values averaged 17 kg of Al ha^{-1} $year^{-1}$ in B horizons) than at the southern watershed (where maximum exports were <2 kg of Al ha^{-1} $year^{-1}$). By comparison, estimated annual Al fluxes from the lower B

TABLE 2.3. Comparative Aluminum Cycling Data for Different Forest Ecosystems[a]

Component	Site 1	Site 2	Site 3	Site 4	Site 5	Site 6	Site 7
				Al Pools (kg ha^{-1})			
Aboveground	2.6				29.3	8.4	
Root system	14.1				44.5		
Forest floor	838				255	110	370
Mineral soil							
Exchangeable		4,670	1,510			530	
Humic		2,290	6,880				
Amorphous		11,800	72,900				
Total					18,950	270,000	
				Al Transfers (kg ha^{-1} yr^{-1})			
Throughfall	0.3			0.2	0.3	1.4	2.9
Litter fall	0.65	4.7	0.4		1.2	2	
Root turnover					58.6		
O Leaching	2.1	1.8	8.6	26.7		3.1	
B Leaching		0.5	17	22.7		3.6	51.5
C Leaching		0.2	0.5			3.6	
Stream	2.65	0.5	7.3				

[a] Sites 1–7 include Tunk Mt., ME; Camp Branch, TN; Big Moose, NY; Great Smoky Mt., NC; Findley Lake, WA; Pine Barrens, NJ; and Solling Spruce, Germany (Rustad, 1988; Cronan et al., 1990; Johnson et al., 1991; Vogt et al., 1987; Turner et al., 1985).

horizon at the Solling spruce site were about 52 kg ha^{-1} (Table 2.3). Finally, headwater stream exports of labile Al were almost 15 times greater at the northern watershed as compared to the southern ecosystem.

Although it was not possible to develop complete budgets for Al cycling by the plant and microbial communities at the Camp Branch and Big Moose watersheds, two of the important biological transfers—litter fall and decomposition—were examined. Results of litterfall analyses (Rustad and Cronan, 1989) indicated that litterfall cycling of Al was approximately 0.4 kg of Al ha^{-1} $year^{-1}$ at the northern drainage basin, but was as much as 10 times higher at the southern watershed. It is important to note that these litter inputs may not represent an immediate source of Al to the forest floor. Instead, the decomposing litter may act as a sink for Al (Rustad and Cronan, 1988), and the decaying litter may accumulate two to four times the original content of Al. At the southern and northern watersheds, litter decomposition resulted in the immobilization of roughly 8.9 and 1.5 kg of Al ha^{-1}, respectively, during the first 24 months following each input of fresh litter. At the ecosystem level, results from the intensive studies indicated that Al cycling at the southern site was controlled more by biotic transfers (e.g., plant uptake and litterfall), while solution transfers (e.g., soil leaching) dominated Al cycling at the northern site (Cronan et al., 1990).

4.2. Evidence for Aluminum Stress and Toxicity in the Forest Landscape

The plant response components of the ALBIOS investigation were intended to address the following questions: (a) what are the patterns of tree sensitivity to Al; (b) what are the apparent mechanisms and symptoms of Al toxicity in trees; (c) do field conditions exist where Al stress or toxicity are likely to occur; (d) is there a relationship between acidic deposition and Al toxicity to trees; and (e) is Al toxicity an important stress factor in forest decline? These questions were examined with a combination of controlled laboratory and inferential field studies (Kelly et al., 1990; Ryanal et al., 1990; Sucoff et al., 1990).

4.2.1. Patterns of Tree Sensitivity to Aluminum

Tree sensitivity to Al varies as a function of such factors as solution pH, the chemical speciation of Al, calcium concentration in the medium, temperature, overall ionic strength of the medium, the form of inorganic nitrogen in the soil solution, mycorrhizal interactions, soil moisture, plant nutrient status, initial vigor, and the species and genetic stock of the plant (Sucoff et al., 1990). Because of these interacting parameters, the actual threshold concentrations for Al toxicity often vary substantially between different trees and sets of environmental conditions. In fact, Al toxicity thresholds for various tree species have been reported over the concentration range of less than 50 μmol L^{-1} to greater than 3000 μmol L^{-1}.

One of the ALBIOS objectives was to compare tree sensitivities to Al under standard solution and soil culture conditions (Cronan et al., 1989). For each tree species, response curves were developed relating various plant growth and nutrient parameters to changes in soluble Al in the growth medium. Figure 2.16*a* illustrates that in hydroponic experiments with red spruce seedlings, changes in soluble Al were accompanied by decreases in root and total plant biomass (Thornton et al., 1987). Similar kinds of responses were observed in solution and/or soil culture for honeylocust, sugar maple, European beech, and some groups of red oak seedlings (Sucoff et al., 1990; Kelly et al., 1990; Raynal et al., 1990). Most of the test species also exhibited a strong negative relationship between increasing root tissue Al concentration and plant growth (Fig. 2.16*b*). For perspective, Joslin and Wolfe (1988) found that field populations of red and Norway spruce at five of the ALBIOS study sites exhibited fine root concentrations of Al (Table 2.4) that were well within the range associated with growth reductions in the lab seedlings (Fig. 2.16*b*).

Seedlings of the six major tree species and one indicator species (honeylocust) exhibited a range of different thresholds for growth and nutrient responses to Al (Cronan et al., 1989). There were significant reductions in root and/or shoot growth of red spruce, honeylocust, European beech, sugar maple, and some groups of red oak seedlings (Fig. 2.17) at soluble Al concentrations ranging from less than 200 to 800 μmol L^{-1} (Thornton et al., 1986a,b, 1987, 1989; Sucoff et al., 1990; DeWald et al., 1990; Joslin and Wolfe, 1988, 1989). In comparison, American beech, loblolly pine, and some of the cohorts of red oak tolerated up to 3000 μmol L^{-1} of soluble Al (Thornton et al., 1989; Tepper et al., 1989; Raynal et al., 1990). On the

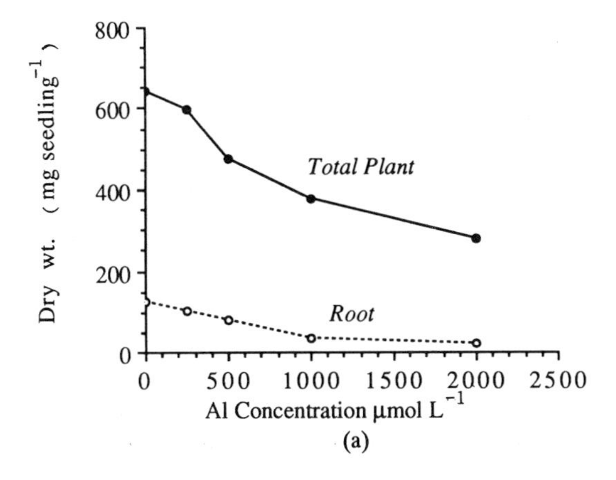

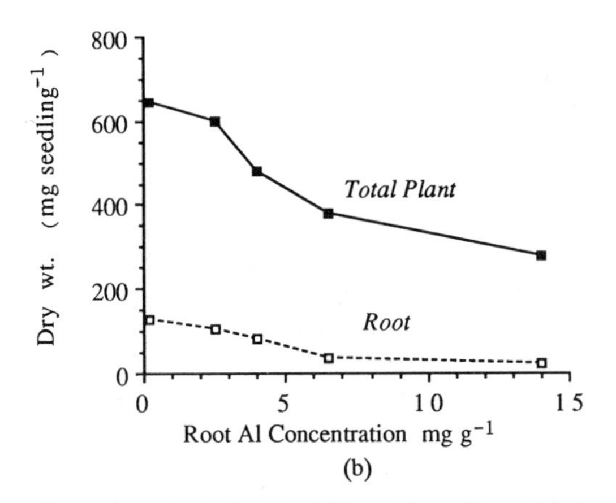

Fig. 2.16. Effect of increasing (*a*) soluble Al and (*b*) root tissue Al on total plant and root biomass in 3–4 month old red spruce seedlings (Cronan et al., 1989). Data are from the hydroponic studies of Thornton et al. (1987), which included a minimum Al treatment of 250 μmol L^{-1}, a pH of 4.0, and a Ca concentration of 50 μmol L^{-1}.

TABLE 2.4. Foliage and Root Chemistry for Both Red and Norway Spruce Trees at Five ALBIOS Sites in North America and Northern Europe[a]

Type	Al	Ca	Ca/Al Molar Ratio
Current foliage	30–65	1700–3015	22–62
O Horizon fine roots	710–1245	4255–6710	2.3–4.1
B Horizon fine roots	4375–9905	3300–4765	0.3–0.5

[a] Sampled by Joslin et al. (1988). Concentration ranges are shown in milligrams per kilogram (mg/kg).

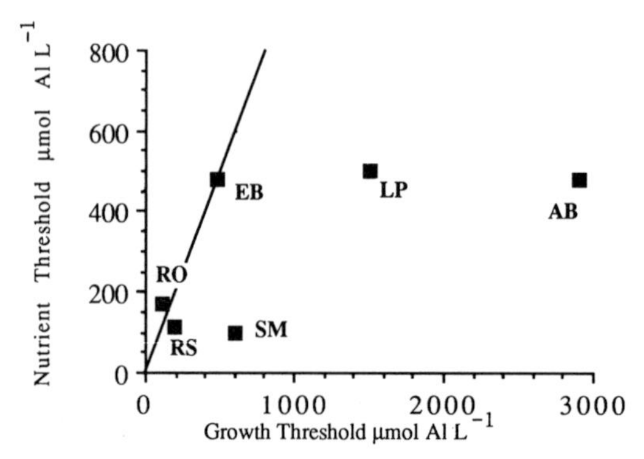

Fig. 2.17. Comparison of Al sensitivity among tree species grown as seedlings under both soil and solution culture conditions. For each species, the point represents the solution Al concentrations where significant decreases in biomass and tissue nutrient concentrations were first observed in the ALBIOS studies summarized by Cronan et al. (1989), Kelly et al. (1990), and Raynal et al. (1990). Species abbreviations include: red spruce (RS), red oak (RO), sugar maple (SM), European beech (EB), loblolly pine (LP), and American beech (AB). Note that there was considerable variability in the thresholds measured for each species among the different experiments (Cronan et al., 1989). The line illustrates the 1:1 relationship for growth and nutrient thresholds; most species plotted to the right of the line, indicating higher thresholds for growth reductions.

basis of these growth responses, the trees could be separated into moderately sensitive species (red spruce, European beech, and sugar maple) and relatively insensitive species (American beech and loblolly pine). For red oak, some of the seedling groups were moderately sensitive to Al (Kelly et al., 1990; DeWald et al., 1990), while others were not (Thornton et al., 1989).

Nutritional effects from Al exposure generally occurred in the ALBIOS plant response studies at lower soluble Al concentrations than those associated with growth reductions (Fig. 2.17). Red spruce, sugar maple, honeylocust, and red oak exhibited significant decreases in tissue Ca and/or Mg at soluble Al concentrations ranging from 100 to 300 µmol L^{-1} (Thornton et al., 1986a,b, 1987; Joslin and Wolfe 1988; Ohno et al., 1988; DeWald et al., 1990). For the other species, reductions in tissue concentrations of Ca and/or Mg occurred at soluble Al concentrations of 600 µmol L^{-1} or less (Thornton et al., 1989; Raynal et al., 1990). Thus, tissue mineral concentration was typically more sensitive to Al than the various growth indexes.

Although the authors of the original ALBIOS plant studies did not examine Ca/Al and Mg/Al ratios extensively in their analyses of plant responses, the data from those studies provide important insights regarding element ratios. For example, Thornton et al., (1987) presented data indicating that tissue Ca/Al molar ratios in red spruce were strongly correlated with Ca/Al molar ratios in the soil solution (Fig. 2.18). As solution Ca/Al ratios decreased below 1.0 in that study, Ca/Al ratios in red spruce needles approached 1.0 and fine root ratios approached 0.1 (Fig. 2.18).

In a companion field study at five of the ALBIOS watersheds, Joslin et al. (1988)

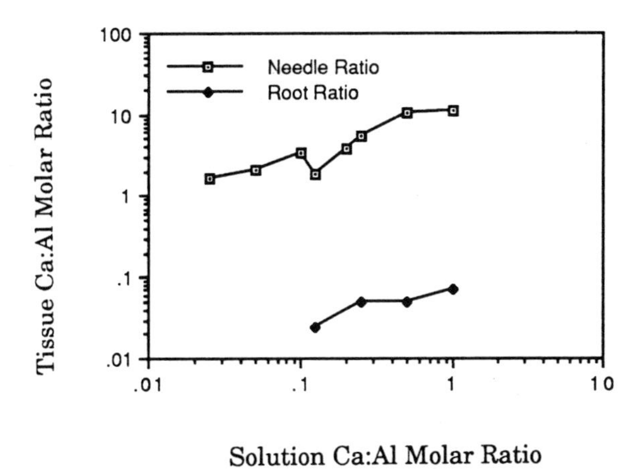

Fig. 2.18. The Ca/Al molar ratios in red spruce needles and fine roots as a function of the Ca/Al molar ratio of the rooting solution. Data are from Thornton et al. (1987).

reported that sites with higher levels of plant-available soil Al supported spruce trees with correspondingly lower foliar levels of calcium and magnesium. The authors interpreted those findings as circumstantial evidence that Al interferes with the uptake and transport of these divalent nutrient cations in some field populations of red and Norway spruce.

Several additional studies of Norway and red spruce have provided evidence supporting the antagonism of Al for Ca and Mg adsorption and uptake (Hüttermann 1983; Van Praag et al., 1985; Shortle and Smith 1988; Asp et al., 1988; Godbold et al., 1988; Cronan 1991). Rost-Siebert (1983) and Abrahamsen (1984) also reported that the Ca/Al ratio is a better predictor of Norway spruce growth than simply Al concentration.

Overall, red spruce was the tree species whose growth and nutrition were most sensitive to soluble Al, with statistically significant biomass and tissue nutrient reductions first occurring at Al concentrations of approximately 200–250 μmol L^{-1} and calculated Al^{3+} ion activities of approximately 100–130 μmol L^{-1} (Thornton et al., 1987; Joslin and Wolfe, 1988). This threshold corresponds to a solution Ca/Al molar concentration ratio of about 1.0. These findings are consistent with the survey results of Raynal et al. (1990) who found that of all the coniferous species examined thus far, North American spruce species appear to be the most sensitive to Al.

In summarizing the multiple ALBIOS studies of red spruce, Raynal et al. (1990) concluded that fine root production was more sensitive to Al than was needle production. The authors also reviewed the relationship of media Ca and Al concentrations to red spruce growth, and reported that significant biomass reductions were generally observed at low Ca/Al ratios (0.03–1.0 in solution), although growth reductions were seen at Ca/Al ratios as high as 37.8 in soil culture. Taken as a whole, the ALBIOS solution culture studies with red spruce demonstrated strong negative correlations between root or foliar biomass and both the concentration of labile Al and the activity of trivalent Al in the soil solution (Raynal et al., 1990).

For some species included in the ALBIOS studies, there were major variations in Al sensitivity that appeared to be the result of genetic variation. For example, loblolly pine seedlings in one hydroponic experiment tolerated up to 3000 $\mu mol\ L^{-1}$ of soluble Al without significant growth reductions (Raynal et al., 1990). However, in a second experiment with the same conditions except for seed source, the loblolly pine exhibited significant reductions in root growth at 1500 μmol of Al L^{-1}. Similar seed source effects on Al sensitivity thresholds have been reported by Williams (1982) for several other southern pines.

The ALBIOS experiments also showed that tree sensitivity to Al may be strongly influenced by the ionic strength or background chemistry of the growth medium (Cronan et al., 1989; Raynal et al., 1990). For example, Thornton et al. (1987) found that when an experiment with red spruce was repeated with the same soluble Al concentration, but with one fifth the background salt concentration, the activity of trivalent Al increased 30% and the severity of the Al toxicity increased. Thus, at the same Al concentration, seedlings in the one fifth strength medium had much lower tissue Mg concentrations and less shoot elongation than seedlings in the full-strength medium. These differences may have been related to changes in the activity of trivalent Al and/or to changes in the Ca/Al or Mg/Al ratios in solution. Similarly, studies with European beech have shown that Ca availability and solution ionic strength play a major role in determining the solution Al threshold at which changes in plant morphology and biomass occur (Kelly et al., 1990).

Another question of concern in the ALBIOS investigation was the influence of experimental methods on the quantification of Al sensitivity in a given species (Cronan et al., 1989). When results were compared from solution and soil culture experiments with the same seed sources, there were some instances where there was remarkably close agreement and other cases where there were substantial inconsistencies between experiments with the same species. As shown by Joslin and Wolfe (1988, Fig. 2.4) and Raynal et al. (1990), most of the solution and soil culture experiments demonstrated uniformly that significant growth reductions in red spruce seedlings occurred at soluble Al concentrations between 200 and 300 $\mu mol\ L^{-1}$. By contrast, solution culture studies gave notably different estimates of Al toxicity thresholds for red oak than did soil culture studies (Kelly et al., 1990).

5. CONCLUSIONS

Based upon a survey of 14 watersheds, important interregional differences were found in the concentrations of aqueous Al in soil waters and surface waters of forested ecosystems in eastern North America and northern Europe. These variations in labile Al were a function of several interacting factors, including soil base saturation, solution pH, anion chemistry, pH-dependent solubility and adsorption relationships, and soil hydrology. The highest concentrations of labile Al were found in watersheds characterized by soils with less than 10–15% base saturation and elevated solution concentrations of strong acid anions. These kinds of environmental conditions were observed at the following ALBIOS sites: Plastic Lake, Ontario; Big Moose, NY; Huntington Forest, NY; Birkenes, Norway; Lake Gårdsjon, Sweden; Lange Bramke, Germany; and Solling, Germany. There are also reports of similar conditions in some high elevation watersheds in the southern Appalachian Mountains (Johnson et al., 1991), in oak woodlands in the Netherlands (van Breemen et al., 1989), and in forest stands in Denmark (Rasmussen, 1986). For such sites, there is a strong positive relationship between acidic deposition, as a source of mobile anions, and Al mobilization. Future increases or decreases in mobile anions at watersheds such as these are likely to produce corresponding changes in both the concentrations of labile Al and the potential for Al toxicity to organisms. The magnitude of the response will be influenced by atmospheric inputs and net ecosystem retention of strong acid anions.

Forest health is influenced by a complex array of interacting biotic and abiotic factors. Most of the documented cases of tree dieback and forest decline are probably the result of multiple stress factors that have combined to decrease the vigor or fitness of a tree population or community. The role of Al toxicity as a stress factor affecting forest health has not been established unequivocally in field studies with natural forest communities. However, based upon the body of biogeochemical and ecotoxicological evidence from previous laboratory and field studies, it is clear to this author that Al toxicity is an important stress factor affecting the mineral nutrition, growth, and vitality of trees at sensitive sites such as those described earlier.

What are the linkages and assumptions in this chain of evidence? (a) We can identify "sensitive" acidic soils with high concentrations of Al and low Ca/Al and Mg/Al molar ratios in the soil solution. (b) Theoretical calculations and empirical data show a strong positive relationship between labile Al and strong acid anion concentrations in these "sensitive" soils. (c) Because of the high affinity of Al for plant roots, there is a tendency for root cell wall exchange sites to become progressively saturated with Al under the prevailing conditions of these "sensitive" acidic forest soils. This process can effectively concentrate Al near the region of active growth and uptake in fine root tips. (d) Once it enters the rhizosphere and interacts with the root system, Al is capable of interfering with plant growth and nutrient balance through a number of specific mechanisms. (e) Finally, atmospheric deposition is a major source of protons and strong acid anions and, as such, can serve to increase Al mobilization and Ca loss in "sensitive" forest soils, thereby promoting the chronic or acute symptoms of Al toxicity and antagonism in forest tree species. As a whole, this chain of evidence and inference supports the hypothesis that inputs of acidic deposition to acid "sensitive" forest soils can increase the cycling of Al and enhance the risk of Al stress for tree species such as red spruce, Norway spruce, and other intolerant forest tree species.

ACKNOWLEDGMENTS

This study would not have been possible without the following ALBIOS co-investigators: R. April, R. Bartlett, P. Bloom, C. Driscoll, S. Gherini, R. Goldstein, G. Henderson, J.D. Joslin, J.M. Kelly, R. Newton, H. Patterson, D. Raynal,

M. Schaedle, C. Schofield, E. Sucoff, H. Tepper, and F. Thornton. The assistance of the following site collaborators is also gratefully acknowledged: P. Dillon, B. LaZerte, D. Schindler, J. Bockheim, S. Ursic, W. Swank, O. Bricker, N. Christophersen, H.M. Seip, F. Andersson, H. Hultberg, B. Ulrich, E. Matzner, and M. Hauhs. The overall study was inspired by Professor Ellis B. Cowling. Funding was provided by the Electric Power Research Institute and the Maine Agricultural Experiment Station. This is MAES publication 1730.

REFERENCES

Abrahamsen G (1984): Effects of acidic deposition on forest soil and vegetation. Philos Trans R Soc London 305:369–382.

Asp H, Bengtsson B, Jensen P (1988): Growth and cation uptake in spruce (*Picea abies Karst.*) grown in sand culture with various aluminum contents. Plant Soil 111:127–133.

Baker JP and Schofield CL (1982): Aluminum toxicity to fish in acidic waters. Water, Air, Soil Pollut 18:289–309.

Bredemeier M, Matzner E, Ulrich B (1990): Internal and external proton load to forest soils in northern Germany. J Environ Qual 19:469–477.

Budd WW, Johnson AH, Huss JB, Turner RS (1981): Aluminum in precipitation, streams, and shallow groundwater in the New Jersey Pine Barrens. Water Resour Res 17:1179–1183.

Campbell PGC., Thomassin D, Tessier A (1986): Aluminum speciation in running waters on the Canadian pre-Cambrian shield: kinetic aspects. Water, Air, Soil Pollut 30:1023–1032.

Cozzarelli IM, Herman JS, Parnell RA (1987): The mobilization of aluminum in a natural soil system: effects of hydrologic pathways. Water Resour Res 23:859–874.

Cronan CS (1980): Solution chemistry of a New Hampshire subalpine ecosystem: a biogeochemical analysis. Oikos 34:272–281.

Cronan CS (1985): Biogeochemical influence of vegetation and soils in the ILWAS watersheds. Water, Air, Soil Pollut 26:355–371.

Cronan CS (1991): Differential adsorption of Al, Ca, and Mg by roots of red spruce (*Picea rubens* Sarg.). Tree Physiol 8:227–237.

Cronan CS, April R, Bartlett R, Bloom P, Driscoll C, Gherini S, Henderson G, Joslin JD, Kelly JM, Newton R, Parnell R, Patterson H, Raynal D, Schaedle M, Schofield C, Sucoff E, Tepper H, Thornton F (1989): Aluminum toxicity in forests exposed to acidic deposition: the ALBIOS results. Water, Air, Soil Pollut 48:181–192.

Cronan CS, Driscoll CT, Newton RM, Kelly JM, Schofield CL, Bartlett RJ, April R (1990): A comparative analysis of aluminum biogeochemistry in a northeastern and a southeastern forested watershed. Water Resour Res 26:1413–1430.

Cronan CS, Goldstein RA (1989): ALBIOS: a comparison of aluminum biogeochemistry in forested watersheds exposed to acidic deposition. In Adriano DC, Havas M (eds): Advances in Environmental Science Acid Precipitation Series. New York: Springer-Verlag, pp 113–135.

Cronan CS, Schofield CL (1979): Aluminum leaching response to acid precipitation: effects on high-elevation watersheds in the Northeast. Science 204:304–306.

Cronan CS, Schofield CL (1990): Relationships between aqueous aluminum and acidic deposition in forested watersheds of North America and northern Europe. Environ Sci Tech 24:1100–1105.

Cronan CS, Walker WJ, Bloom PR (1986): Predicting aqueous aluminum concentrations in natural waters. Nature (London) 324:140–143.

Dahlgren RA, Ugolini FC (1989): Aluminum fractionation of soil solutions from unperturbed and tephra-treated Spodosols, Cascade Range, Washington, USA. Soil Sci Soc Am J 5:559–566.

David MB, Driscoll CT (1984): Aluminum speciation and equilibria in soil solutions of a Haplorthod in the Adirondack Mountains (New York, USA). Geoderma 33:297–318.

DeWald LE, Sucoff EI, Ohno T, Buschena C (1990): Response of northern red oak (*Quercus rubra L.*) seedlings to soil solution aluminum. Can J For Res 20:331–336.

Driscoll CT (1984): A procedure for the fractionation of aqueous aluminum in dilute acidic waters. Int J Environ Anal Chem 16:267–283.

Driscoll CT, Baker JP, Bisogni JJ, Schofield CL (1980): Effect of aluminum speciation on fish in dilute acidified waters. Nature (London) 284:161–164.

Driscoll CT, van Breemen N, Mulder J (1985): Aluminum chemistry in a forested Spodosol. Soil Sci Soc Am J 49:437–443.

Driscoll CT, Wyskowski BJ, Cosentini CC, Smith ME (1987): Processes regulating temporal and longitudinal variations in the chemistry of a low-order woodland stream in the Adirondack region of New York. Biogeochemistry 3:225–241.

Driscoll CT, Wyskowski BJ, DeStaffan P, Newton RM (1989): Chemistry and transfer of aluminum in a forested watershed in the Adirondack region of New York, USA. In Lewis T (ed): Environmental Chemistry and Toxicology of Aluminum. Chelsea, MI: Lewis Publishers, Inc., pp 83–105.

Godbold DL, Fritz E, Huttermann A (1988): Aluminum toxicity and forest decline. Proc Natl Acad Sci USA 85:3888–3892.

Graustein WC (1981): The effects of forest vegetation on solute acquisition and chemical weathering: a study of the Tesuque Watersheds near Santa Fe, New Mexico. Ph.D. thesis, Yale University, New Haven, CT.

Graustein WC, Cromack K, Sollins P (1977): Calcium oxalate: occurrence in soils and effect on nutrient and geochemical cycles. Science 198:1252–1254.

Haug A (1984): Molecular aspects of aluminum toxicity. CRC Crit Rev Plant Sci 1:345–373.

Hauhs M (1985): Wasser- und stoffhaushalt im einzugsgebiet der Langen Bramke (Harz). Berichte des Forschungszentrums Waldokosysteme/Waldsterben, Bd. 17.

Hauhs M (1989): Lange Bramke: an ecosystem study of a forested catchment. In Adriano DC, Havas M (eds): Acidic Precipitation Volume 1 Case Studies. New York: Springer-Verlag, pp 275–305.

Hecht-Bucholz C, Jorns CA, Keil P (1987): Effects of excess aluminum and manganese on Norway spruce seedlings as related to magnesium nutrition. J Plant Nutr 10:1103–1110.

Hem JD Roberson CE (1967): Form and stability of aluminum hydroxide complexes in dilute solution. Geol Survey Water-Supply Paper 1827-A.

Hendershot WH, Dufresne A, Lalande H, Courchesne F (1986): Temporal variation in aluminum speciation and concentration during snowmelt. Water, Air, Soil Pollut 31:231–237.

Hutchinson TC, Bozic L, Munoz-Vega G (1986): Responses of five species of conifer seedlings to aluminum stress. Water, Air, Soil Pollut 31:283–294.

Hüttermann A (1983): Auswirkungen "sauer Depositionen" auf die Physiologie des Wurzelraumes von Waldokosystemen. Allg Forst Holzwirtsch Ztg 38:663–664.

Johnson DW, Van Miegroet H, Lindberg SE, Todd DE, Harrison RB (1991): Nutrient cycling in red spruce forests of the Great Smoky Mountains. Can J For Res 21:769–787.

Johnson NM (1979): Acid rain: neutralization within the Hubbard Brook ecosystem and regional implications. Science 204:497–499.

Johnson NM, Driscoll CT, Eaton JS, Likens GE, McDowell WH (1981): Acid rain, dissolved aluminum, and chemical weathering at the Hubbard Brook Experimental Forest, N.H. Geochim Cosmochim Acta 45:1421–1437.

Joslin JD, Kelly JM, Wolfe MH, Rustad LE (1988): Elemental patterns in roots and foliage of mature spruce across a gradient of soil aluminum. Water, Air, Soil Pollut 40:375–390.

Joslin JD, Mays PA, Wolfe MH, Kelly JM, Garbe RW, Brewer PF (1987): Chemistry of tension lysimeter water and lateral flow in spruce and hardwood stands. J Environ Qual 16:152–160.

Joslin JD Wolfe MH (1988): Responses of red spruce seedlings to changes in soil aluminum in six amended forest soil horizons. Can J For Res 18:1614–1623.

Joslin JD Wolfe MH (1989): Aluminum effects on northern red oak seedling growth in six forest soil horizons. Soil Sci Soc Am J 53:274–281.

Joslin JD Wolfe MH (1992): Red spruce soil solution chemistry and root distribution across a cloudwater deposition gradient. Can J For Res 22:893–904.

Kelly JM, Schaedle M, Thornton FC, Joslin JD (1990): Sensitivity of tree seedlings to aluminum: II. Red oak, sugar maple, and European beech. J Environ Qual 19:172–179.

Likens GE, Bormann FH, Pierce RS, Eaton JS, Johnson NM (1977): Biogeochemistry of a Forested Ecosystem. New York: Sringer-Verlag, 146 pp.

Litaor IM (1987): Aluminum chemistry: fractionation, speciation, and mineral equilibria of soil interstitial waters of an alpine watershed, Front Range, Colorado. Geochim Cosmochim Acta 51:1285–1295.

Manley EP, Chesworth W, Evans LJ (1987): The solution chemistry of podzolic soils from the eastern Canadian shield: a thermodynamic interpretation of the mineral phases controlling soluble Al^{3+} and H_4SiO_4. J Soil Sci 38:39–51.

Matzner E (1989): Acidic precipitation: case study Solling. In Adriano DC, Havas M (eds): Acidic Precipitation. Volume 1: Case Studies. New York: Springer-Verlag, pp 39–84.

McCormick, LH Steiner KC (1978): Variation in aluminum tolerance among six genera of trees. For Sci 24:565–568.

Muniz IP, Leivestad H (1980): Acidification—effects on freshwater fish. In Ecological Impact of Acid Precipitation. Oslo: SNSF project, pp 84–92.

Nilsson SI Bergkvist B (1983): Aluminum chemistry and acidification processes in a shallow podzol on the Swedish west coast. Water, Air, Soil Pollut 20:311–329.

Ohno T, Sucoff EI, Erich MS, Bloom PR, Buschena CA, Dixon RK (1988): Growth and nutrient content of red spruce seedlings in soil amended with aluminum. J Environ Qual 17:666–672.

Paganelli DJ, Seiler JR, Feret PP (1987): Root regeneration as an indicator of aluminum toxicity in loblolly pine. Plant Soil 102:115–118.

Pillai-Nair VD (1978): Aluminum species in soil solutions. Gottinger Bodenkundliche Ber. 52:1–122.

Plankey BJ, Patterson HH, Cronan CS (1986): Kinetics of aluminum fluoride complexation in acidic waters. Environ Sci Tech 20:160–165.

Rasmussen L (1986): Potential leaching of elements in three Danish spruce forest soils. Water, Air, Soil Pollut 31:377–383.

Raynal DJ, Joslin JD, Thornton FC, Schaedle M, Henderson GS (1990): Sensitivity of tree seedlings to aluminum: III. Red spruce and loblolly pine. J Environ Qual 19:180–187.

Reuss JO (1983): Implications of the Ca-Al exchange system for the effect of acid precipitation on soils. J Environ Qual 12:591–595.

Reuss JO, Johnson, DW (1986): Acid Deposition and the Acidification of Soils and Waters. In Ecological Studies 59, New York: Springer-Verlag.

Rost-Siebert K (1983): Aluminum-Toxizitat und Toleranz an Keimpflanzen von Fichte (*Picea abies* Karst.) and Buche (*Fagus sylvatica* L.) Allg Forst Holzwirtsch Ztg 26/27:686–689.

Rustad LE (1988): The biogeochemistry of aluminum in a red spruce (*Picea rubens* Sarg.) forest floor in Maine. Ph.D. Thesis, University of Maine, Orono.

Rustad LE, Cronan CS (1988): Element loss and retention during litter decay in a red spruce stand in Maine. Can J For Res 18:947–953.

Rustad LE, Cronan CS (1989): Cycling of aluminum and nutrients in literfall of a red spruce *Picea rubens* Sarg. stand in Maine. Can J For Res 19:18-23.

Schier GA (1985): Response of red spruce and balsam fir seedlings to aluminum toxicity in nutrient solution. Can J For Res 15:29–33.

Shortle WC Smith KT (1988): Aluminum-induced calcium deficiency syndrome in declining red spruce. Science 240:1017–1018.

Sucoff EI, Thornton FC, Joslin JD (1990): Sensitivity of tree seedlings to aluminum: I. Honeylocust. J Environ Qual 19:163–171.

Sullivan TJ, Christophersen N, Muniz IP, Seip HM, Sullivan PD (1986): Aqueous aluminum chemistry response to episodic increases in discharge. Nature (London) 323:324–326.

Tepper HB, Yang CS, Schaedle M (1989): Effect of aluminum on growth of root tips of honeylocust and loblolly pine. Environ Exp Bot 29:165–173.

Thornton FC, Schaedle M, Raynal DJ, Zipperer C (1986a): Effects of aluminum on honeylocust seedlings in solution culture. J Exp Bot 37:775–785.

Thornton FC, Schaedle M, Raynal DJ (1986b): Effect of aluminum on the growth of sugar maple in solution culture. Can J For Res 16:892–896.

Thornton FC, Schaedle M, Raynal DJ (1987): Effects of aluminum on red spruce seedlings in solution culture. Environ Exp Bot 27:489–498.

Thornton FC, Schaedle M, Raynal DJ (1989): Tolerance of red oak, and American and European beech seedlings to aluminum. J Environ Qual 18:541–544.

Turner RS, Johnson AH, Wang D (1985): Biogeochemistry of aluminum in McDonalds Branch watershed, New Jersey Pine Barrens. J Environ Qual 14:314–323.

Ugolini FC, Minden R, Dawson H, Zachara J (1977): An example of soil processes in the *Abies amabilis* zone of central Cascades, Washington. Soil Sci 124:291–302.

Ulrich B, Mayer R, Khanna PK (1980): Chemical changes due to acid precipitation in a loess-derived soil in central Europe. Soil Sci 130:193–199.

van Breemen N, Burrough PA, Velthorst EJ, van Dobben HF, de Wit T, Ridder TB, Reijinders HFR (1982): Soil acidification from atmospheric ammonium sulfate in forest canopy throughfall. Nature (London) 229:548–550.

van Breemen N, Boderie PMA, Booltink HWG (1989): Influence of airborne ammonium sulfate on soils of an oak woodland ecosystem in the Netherlands: seasonal dynamics of solute fluxes. In Adriano DC, Havas M (eds): Acidic Precipitation Volume 1 Case Studies. New York: Springer-Verlag, pp 209–236.

Van Grinsven JJM, van Breemen N, Mulder J (1987): Impacts of acid atmospheric deposition on woodland soils in the Netherlands: calculation of hydrologic and chemical budgets. Soil Sci Soc Am J 51:1629–1634.

Van Praag HJ, Weissen F, Sougnez-Remy S, Carletti G (1985): Aluminum effects on spruce and beech seedlings. II. statistical analysis of sand culture experiments. Plant Soil 83:339–356.

Vogt KA, Dahlgren R, Ugolini F, Zabowski D, Moore EE, Zasoski R (1987): Aluminum, Fe, Ca, Mg, K, Mn, Cu, Zn, and P in above- and belowground biomass. II. pools and circulation in a subalpine *Abies amabilis* stand. Biogeochemistry 4:295–311.

Williams KA (1982): Tolerances of four species of southern pine to aluminum in solution cultures. MS Thesis. University of Florida, Gainesville, FL.

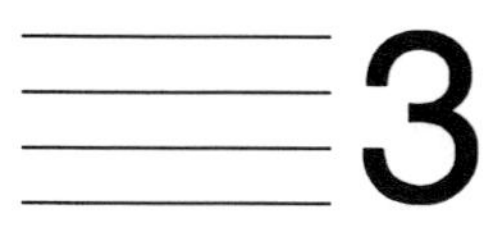

EFFECTS OF ACID DEPOSITION ON SOIL ORGANISMS AND DECOMPOSITION PROCESSES

VOLKMAR WOLTERS

Institut für Zoologie, Universität Mainz, Saarstr. 21, D-55099 Mainz, Germany

MATTHIAS SCHAEFER

Zoologisches Institut, Universität Göttingen, Berlinerstr. 28, D-37073 Göttingen, Germany

1. INTRODUCTION

Considerable attention has been paid to increased anthropogenic emissions of sulfur and nitrogen that can be deposited into forest ecosystems as strong acids (Likens, 1989). Much research has been focused on the possible consequences of these deposits for forest soils (Berdén et al., 1987 for Scandinavia; Binkley et al., 1989 for southeastern United States; Kauppi et al., 1990 for Finland; Olson et al., 1992 for western United Stades; Schulze et al., 1989 for Germany). This chapter deals with the effect of increased input rates of protons on the enormously diverse community of heterotrophic forest soil organisms. Soil biota are part of the soil subsystem, which significantly affects both the structure and function of terrestrial ecosystems. Soil processes affected by edaphic organisms include humification and mineralization of the soils organic matter, as well as soil mixing and turnover. Consequently, soil organisms can affect almost any physicochemical property of soils (e.g., soil chemical and nutritional properties or soil structure and texture). These effects may feed up to the level of the ecosystem (e.g., by altering the growth conditions of the primary producers). It thus seems reasonable to assume that acid

Effects of Acid Rain on Forest Processes, pages 83–127

rain effects on soil biota and on the processes that they control may considerably affect the functioning of the forest ecosystems as a whole. After more than two decades of intense research, however, our view of the causal relationship among "recent forest decline," acid rain induced alterations of soil conditions, and soil biota is still more than scattered. In fact, almost every soil biological parameter measured has been demonstrated to be either reduced, stimulated, or remains unaffected by acid rain (Will et al., 1986). There are several reasons for this somewhat frustrating result:

- Only little is known about the ecology of edaphic organisms due to their cryptic life. We can hardly expect to find general patterns for acid rain effects on soil biota as long as we do not even understand the factors organizing the below-ground community in undisturbed systems.
- Potential effects of acid precipitation on soil chemistry and nutrient availability include decreased base saturation, reduced availability of cations, and increased solubility of aluminum (Al), iron (Fe), and heavy metals (Aber et al., 1982). Increased nitrogen (N) input, which very often accompanies increased H^+ concentration of rain, may additionally contribute to an alteration of soil conditions. The unpredictability of soil organisms in their reaction to acid rain may thus partly occur because soil biota have to respond to a complex and multidirectional alteration of their environment.
- Different disciplines of soil ecology very often use parameters that are not fully compatible. For example, for methodological reasons, it is not generally possible to treat soil microorganisms with the same detailed taxonomic resolution as invertebrates. Microbiological studies concerning the effects of acid emissions into forest soil thus mainly focus on bulk parameters, such as microbial biomass or activity, while the majority of zoological analyses deal with changes in the qualitative and quantitative composition of the animal community. The almost complete absence of integrative studies on functional groups or biotic interrelationships across taxonomic categories may partly be responsible for the lack of a consistent concept about acid rain effects on soil organisms.

It seemed an almost impossible task to write a short and comprehensive review, keeping track of every aspect of acid rain effects on forest soil organisms. We therefore decided to confine ourselves to effects that have been clearly attributed to increased input rates of protons (including both wet and dry deposition). Though actually inseparable from acid rain induced soil acidification, secondary aspects, such as Al toxicity or bioavailability of heavy metals, are generally not explicitly taken into account. In addition, results from related fields of research (e.g., effects of liming, fertilization, or increased N input on soil organisms) have only been included when essentially needed for an evaluation of the data. After a brief review of the approaches used in acid rain research (Section 2), we will try to set the scene for judging acid rain effects on soil biota by giving an overview of the most important groups of soil biota (Section 3.1) and by presenting a scheme on acid rain effects on the below ground community (Section 3.2). The buffering capacity of soil is of overwhelming importance for judging acid rain effects on soil organisms. Therefore, we will proceed with a comparison between biotic activity in base-rich and acid forest soils (Section 4). To give evidence for possible effects of acid rain on ecosystem function we will then present results concerning acid rain effects on decomposition processes (Section 5). Section 6 focuses on acid rain effects on the soil microflora and on soil invertebrates (including functional effects, Section 7). Section 8 presents the conclusions that can be drawn from this chapter and proposes some ideas for further research.

2. EXPERIMENTAL APPROACHES

Five different approaches dominate the studies on the effects of acid rain on the biota of forest soils (Schaefer, 1989): (1) monitoring of long-term changes, (2) comparison of differently polluted sites, (3) gradient analyses within one site, (4) application of simulated acid rain both in the field and in the laboratory, and (5) the exchange of soil cores between different sites. The first three approaches are more or less descriptive, the last two are experimental. As Hagvar (1987a) pointed out, one problem holds for almost every approach: They do not lead to a real understanding of the mechanisms involved. The prognostic capacity of the data sets currently available is thus seriously limited.

A short overview of the various approaches used in acid rain research might help to evaluate the data obtained by them. As a methodological guideline, we will refer to the numbers in front of the following paragraphs at different parts of our paper.

1. The monitoring of long-term changes in acid polluted forest soils would without doubt be the best and most direct way to quantify the effects of acid rain on soil biota. For several practical reasons, however, the time scales that soil biologists can survey are much shorter than those available to soil scientists. Therefore, no real long-term data of biological changes in acidifying forest soils are available.
2. To overcome the problems associated with long-term investigations, a large number of studies focused on the comparison of stands with different rates of acid deposition (e.g., Hogervorst et al., 1993; Rejsek, 1991; van Straalen et al., 1988) or at different stages of soil acidification (e.g., Wolters and Joergensen, 1991). The basic assumption is that the soil biotic conditions at the different sites were initially similar and that most of the differences are attributable to acid induced changes in soil conditions. The most serious methodological problem of this approach is that it is very unlikely, if not impossible, to find at least two forest sites for which this assumption really holds, due to the complex effect of climate, topography, vegetational cover, soil biotic activity, and time on soil state (Jenny, 1941). It is thus difficult to decide whether the observed differences are attributable to acid rain.
3. Investigations along a pollution gradient within one site at least partly avoid this problem. Major shortcomings are that most of the factors listed above

may again become very relevant on a microscale (e.g., when comparing the windward and leeward sides of a mountain range). In addition, the acid rain effects may be overshadowed by microhabitat specific features, which may hinder an extrapolation to larger areas (e.g., when areas near the tree trunk base are compared with areas below canopy; cf. Koenies, 1985; Schäfer, 1988).

4. Application of simulated acid rain in the field or in the laboratory is the most widespread experimental approach used in acid rain research. The serious shortcomings of this approach are that most of the experiments are carried out over a relatively short period of time and with a rather unnatural way of increasing the level of proton input. A very important aspect of this approach is that temporal and spatial variations in the proton input, which inevitably occur in the field, are almost never taken into account. These effects have been demonstrated to significantly alter the effect of acid rain on decomposition processes (Wolters, 1991a). In addition, laboratory experiments may provide an environment so different from field conditions that extrapolation is problematical (van Loon, 1984). This situation is especially true for experiments with pure cultures, though this may be the only way to overcome the "black-box" problem mentioned above.
5. A seldom used approach is the exchange of soil cores between sites of different exposure to acid rain (Wainwright, 1980; Killham and Wainwright, 1981). This approach overcomes some of the problems associated with the simulation experiments, though the comparatively short exposure period as well as large edge effects caused by the small size of the soil cores still set some serious limitations.

3. SETTING THE SCENE

3.1. The Biota of Forest Soils

The contribution of heterotrophic soil organisms to the total ecosystem metabolism varies between 30 and 60% (Reichle, 1977). The community of soil biota is composed of soil microorganisms and soil invertebrates. The components of the soil microflora are bacteria, fungi, and algae. The dominant invertebrate groups in soil belong to the Protozoa, Nematoda, Oligochaeta, and Arthropoda (Table 3.1). The most convenient way of classifying soil invertebrates relies on their body width (Wolters, 1991c; cf. Table 3.3). Soil biota in forests can be characterized by structural parameters, such as species composition, species diversity, population abundance, biomass, spatial distribution, and phenology (cf. Kjoller and Struwe, 1982; Swift, 1987; Paul and Clark, 1989 for microfungi and bacteria; Faber and Verhoef, 1991; Petersen and Luxton, 1982 and Usher, 1985 for animals). As the populations exhibit spatio-temporal dynamics, these parameters change in a stochastic and/or deterministic pattern. The composition of microbio- and zoocoenoses can be related to soil factors, vegetation types, and to ecosystem processes. One example is the series mull–moder–mor (raw humus) (Wallwork, 1976; Petersen and Luxton, 1982; Ellenberg et al., 1986; Schaefer and Schauermann, 1990) (cf. Section 3.2).

Organismic activity in the forest ecosystem manifests itself in the flow of matter, energy, and information. The following rough generalizations about the role of the microflora and fauna in forest soils can be made:

- The microflora contains the largest part of the decomposer biomass. Though soil bacteria are by far the most numerous soil organisms, their contribution to the microbial biomass is small when compared to the fungi. The overwhelming importance of the microflora in ecosystem function is indicated by the fact that it contributes 80–99% of nonroot carbon dioxide (CO_2) losses from soil. Important soil processes, such as the mobilization, immobilization, and transformation of nutrients, are governed by microbial activities.
- With a contribution of 1–20% of nonroot CO_2 losses from soil, the direct effect of soil invertebrates on ecosystem function is comparatively small. It is well established, however, that edaphic invertebrates significantly affect soil processes indirectly by altering the growth and performance of the microflora (Anderson, 1988; Wolters, 1991c). Important faunal effects are transport, modification and mixing of substrates, grazing on microbial populations, and inoculation with microbial propagules.

3.2. A Scheme for Acid Rain Effects on Soil Biota

Three principal pathways of acid rain effects have to be taken into account (Fig. 3.1):

1. Direct effects (every aspect of proton toxicity causing an immediate response of the organisms).
2. Indirect effects via modifications of the abiotic environment [including both chemical stress associated with acidification (e.g., decreased base saturation; reduced availability of cations; increased solubility of Al, Fe, and heavy metals; presence of nonacidic SO_4^{2-},) and alterations of structural characteristics of the soil habitat (e.g., accumulation of litter, modifications in the chemical content of leaf litter, and compaction of pore space)] (cf. Novick et al., 1984).
3. Indirect effects via alterations of the biotic environment (e.g., modifications in competitive relations and/or predation pressure).

In some cases it is impossible to make a clear-cut separation between the two indirect pathways. The nutritional conditions of an invertebrate species may be modified by an alteration of the litter quality (abiotic environment) and/or by an alteration of the microflora (biotic environment). In addition, the action of the two indirect pathways may be significantly interrelated. On one hand, acid rain induced alterations of the chemical environment may affect certain soil invertebrate species

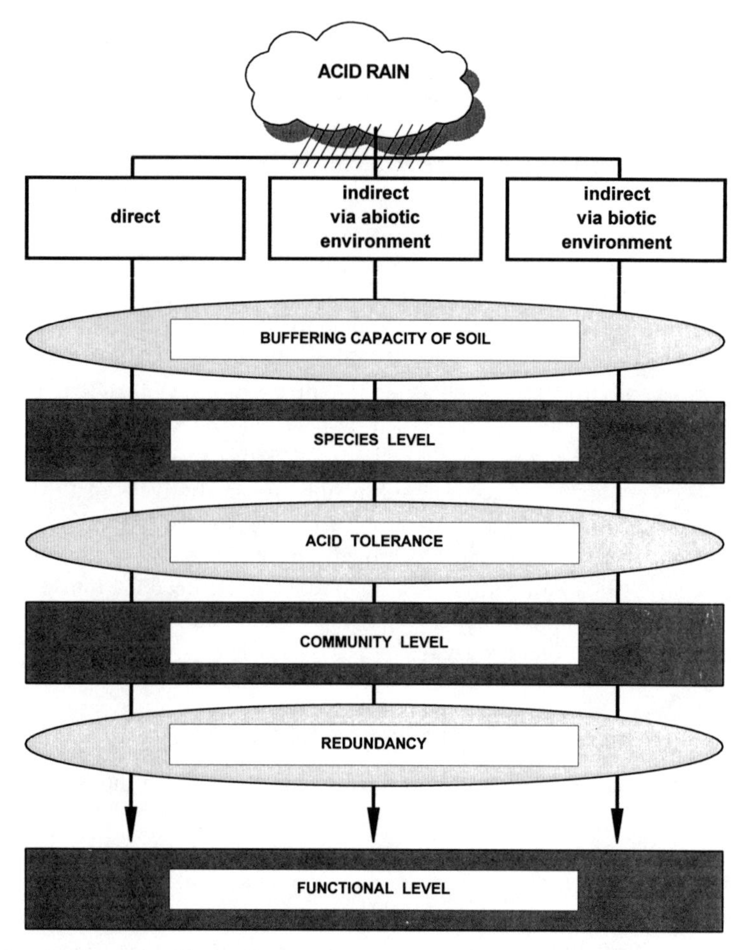

Fig. 3.1. A scheme for acid rain effects on the structure and function of the below-ground community.

via a modification of predation pressure. On the other hand, depression of the microbial biomass in the litter layer may lead to a reduced retention capacity for nutrients, such as calcium (Ca), which in turn may alter the nutritional value of the litter to the macroinvertebrates.

In most cases, many of the direct and indirect effects act together in a synergistic way (Wellburn, 1988). For example, Shriner (1977) concluded that acid rain may have reduced the nematode-induced disease of kidney beans directly by effects of short-time acidification of soil water on free-living larval stages, and indirectly by affecting physiological alterations in the roots of the host plant, which inhibited parasitism and reproduction of adult females. In addition, growth and reproduction of plant parasitic nematodes are inhibited below pH 4.0.

A response to acid rain can be observed at the species, at the community, or at the functional level (Fig. 3.1). In the context of acid rain research, however, the strict differentiation between the species and the community level is pragmatic and thus very often arbitrary. For example, it is often impossible to decide if the acid induced reduction of bulk parameters, such as the microbial biomass, is due to an alteration of the community composition or to a reduction of the cell size (cf. Baath et al., 1984; Section 6.2). Despite the ambiguous nature between the species and the community level, acid rain effects on interspecific interactions will be treated here as indirect effects acting on the species level (modification of the biotic environment). At the species level acid rain may alter physiological, behavioral, and morphological features (Sections 6.1 and 7.1). By modifying important life-history traits, such as growth, fertility, and longevity, this may significantly alter population dynamics. Garden and Davies (1988) showed that acid induced variations in the type and numbers of microbes present on leaf litter result in differences in the nutritional value of the litter to the detritivorous larvae of *Tipula commiscibilis* living in running waters. They pointed out that there is an evident need for a quantification of the effects of these changes on detritivore growth, and possible ramifications in other life history traits. This analogously applies to forest ecosystems. Unfortunately, no consistent investigations have been carried out in this field of research.

At the community level acid rain may alter parameters such as species composition, diversity, and community organization (e.g., food chains or functional groups) (Sections 6.2 and 7.2). At the functional level various alterations of processes may occur (e.g., alterations of carbon and nitrogen mineralization; Section 5). Acid rain research in forest ecosystems largely centers on the functional level (microbial activity, nutrient cycling, and soil properties). Investigations at the community and species level, in contrast, are comparatively sparse.

The extent of acid rain effects depends on the buffering capacity of soil (Section 4.1), the tolerance of the organisms (Sections 6.1 and 7.1), and the redundancy of the community (Sections 6.2 and 7.2) (Bormann, 1987; Belotti, 1989). The organization of these buffers is hierarchical. As long as the increased input of protons is buffered by the soil, acid tolerance of the organisms remains unimportant. Similarly, redundancy of the community (i.e., the ability to replace one species by the other without altering processes) plays no role in buffering acid rain effects, as long as the alterations of soil conditions remain within the tolerance range of the organisms. It should be kept in mind, however, that because a species is still present due to its ability to survive unfavorable environmental conditions in no way means that it will be able to fulfill its ecological function. Buffering of environmental change by redundancy is a topic heavily under discussion at the moment (Lawton and Brown, 1993). An essential point emerging from this debate is that an understanding of the functional implications of redundancy will not be possible until we have a better understanding of "keystone species." Without ignoring the different perceptions of what is "key" (Mills et al., 1993), we pragmatically refer to "keystone species" as soil invertebrate species that are more important for ecosystem structure and function than others.

To quantify the outcome of acid rain effects under field conditions a scale component has to be incorporated into the scheme shown in Figure 3.1. At the spatial scale the heterogeneous distribution of the proton input, soil conditions, resources, and organisms may significantly alter acid rain effects at microsites. At the temporal scale it is necessary at least to distinguish between acute short-term effects and the effects of long-term exposure (Lettl, 1985).

4. COMPARISON BETWEEN BASE-RICH AND ACID FOREST SOILS

The reaction of soil biota to acid rain depends on the buffering capacity of soil for two reasons. First, it simply determines the extent to which increased input rates of protons can affect the soil environment. Second, below-ground communities living in different forest soils adapt differently to soil acidity. Thus, the buffering range of the soil also determines the ability of the soil biota to resist acid rain effects by tolerance and redundancy. The influence of acid rain on functional relationships in soil are most probably also highly soil specific (Ulrich, 1987; Schaefer, 1990). Soil-independent generalizations about acid rain effects on soil organisms may not therefore be possible as a matter of principle.

The acidification of forest soils depends both on the quantity of H^+ deposited from the atmosphere (and other sources of H^+), and the processes that may neutralize H^+ (Binkley, 1992). General assumptions concerning soil specificity of acid rain induced acidification were listed by van Loon (1984):

- No rapid deleterious changes would be expected on well-buffered calcareous soils. Some factors might even enhance productivity at least in the short term.
- Poorly buffered soils of moderate acidity might be most susceptible to current levels of acid rain.
- Intrinsically acidic podzols and other northern forest soils with low base saturation would require relatively large inputs of acidity to further degrade them.

There is considerable doubt that these generalizations also apply for acid rain effects on soil biotic processes, because they strongly rely on the buffering capacity of soil and disregard the soil specific adaptational patterns of the organisms. For example, it has been hypothesized that the effects of acid rain on essential soil processes in base-rich soils may be stronger than expected, because organisms living in these soils are not adapted to an acid environment and may thus be extremely sensitive to acidification (Ulrich, 1987; Schaefer, 1990). This result could have a considerable impact on the functional level.

Ulrich (1987) presented a concept characterizing the stepwise shift of forests from base-rich to acid soil and its functional implications. We will use this as a framework for discussing some of the most important factors affecting soil specificity in acid rain effects on soil biotic processes.

4.1. Forest Soils at Different Stages of Acidification

4.1.1. Base-Rich Soils

For the soil biota living in soils of the carbonate or silicate buffer range (pH 8.6–5) acid tolerance plays no role in competition. This condition offers an explanation for the great diversity of the below-ground community of base-rich forest soils (Ulrich, 1987). The soil is deeply rooted and the roots are homogeneously distributed. The biomass of the soil fauna is dominated by the macrofauna, especially earthworms (Table 3.1) and its element turnover may reach or even exceed annual element input via leaf litter (Anderson and Ineson, 1984). By mixing the leaf litter into the mineral soil (bioturbation) and by producing porous soil crumbs, the macrofauna guarantees that root uptake and organic matter mineralization occur in the same soil compartment. In this respect soil-burrowing animals have a very important function in the ecosystem to approach steady state.

Three hypotheses for the effect of acid rain on soil biotic processes in base-rich forest soils might be drawn:

1. Only a few key macrofauna species are responsible for litter fragmentation and bioturbation (i.e., the formation of mull; some epigeic, saprophagous macroarthropods and some endogeic and anecic earthworms; cf. Schaefer, 1990). The adaptation of these species to acidification is low (cf. Section 7.1). The ability to compensate acid rain effects by acid tolerance is thus minimal. Despite the comparatively high diversity of the invertebrate communities (Section 4.2), buffering by redundancy is also low because of the limited number of species involved in quantitatively important processes.
2. While the buffering capacity of soil is high, that of the thin litter layer on the soils surface is very limited. An acid induced deterioration of the environmental conditions for macroarthropods living in the litter layer may have a significant impact on the mixing of soil and litter by earthworms, because this process strongly depends on litter fragmentation. Accumulation of undecomposed litter on the soil surface may be accelerated by the susceptibility of bacteria dominating in base-rich soils.
3. Soil invertebrates play an important role in buffering acid rain effects in the sensitive surface layer by mediating the contact between mineral soil and litter.

We will return to this scenario in Section 7.3.

4.1.2. Humus Disintegration

The initial phase of the shift from base-rich to acid soils is characterized by a period of humus disintegration (Ulrich, 1987). Nitrate, which was formed from organic bound nitrogen in the mineral soil, is leached. The equivalent proton release causes a change in chemical soil state. A long-lasting decoupling of the ion cycle may lead to a loss of organic nitrogen and organic matter. Humus disintegration usually starts in warm years when the microbial activity in the deeper rooting zone is favored. Soil zoological studies indicate that the biomass of endogeic earthworms may be even higher than in base-rich soils, but that a shift occurs in species composition (e.g., from a *Octolasion* ssp. dominated community to a community dominated by *Aporrectodea caliginosa;* Bonkowski, 1991). Though the nutritional value of the mineral

TABLE 3.1. Synopsis of Species Richness (S), Mean Annual Population Density (N, ind m^{-2}), and Mean Annual Biomass (B, mg dry mass m^{-2}) in the Beech Forests Göttinger Wald (with Mull Soil) and Solling (with Moder Soil)[a]

	Göttinger Wald[b]			Solling[b]		
Animal Group	S	N	B	S	N	B
Microfauna						
Flagellata	—	2.7×10^9	54	—	—	—
Amoebina	—	3.5×10^9	1133	—	—	—
Testacea	65	84×10^6	343	51	57×10^6	256
Turbellaria	3	859	8	3	1,882	4
Nematoda	65	732,000	146	—	—	—
Rotatoria	13	4,893	5	—	—	—
Tardigrada	4	4,207	4	—	41[c]	9[c]
Harpacticoida	—	3,873	2	1	3,300[c]	0.6[c]
Saprophagous and Microphytophagous Mesofauna						
Enchytraeidae	36	22,300	600	15	108,000	1,640
Cryptostigmata	61	25,900	180	72	101,810	195
Uropodina	11	3,390	26	4	1,525	—
Symphyla	2	57	—	1	—	—
Diplura	—	161	—	>1	277	—
Protura	—	2,481	—	>1	278	—
Collembola	48	37,835	153	>11	63,000	246

Zoophagous Mesofauna						
Gamasina	67	2,620	45	13	10,800	397
Saprophagous Macrofauna						
Gastropoda	30	120	430	4	0	0
Lumbricidae	11	205	10,700	4	19	168
Isopoda	6	286	93	0	0	0
Diplopoda	6	55	618	1	0	0
Elateridae larvae	11[d]	37	104	4[d]	332	706
Diptera larvae	37 (>245 spp.)	2,843	161	40	7,415	628
Zoophagous Macrofauna						
Araneida	102	140	47	93	462	173
Pseudoscorpionida	3	35	16	2	89	10
Opilionida	8	19	11	4	20[c]	6[c]
Chilopoda	10	187	265	7	74	155
Carabidae	24	5	144	26	7	93
Staphylinidae	85	103	76	117	314	180

[a] From Schaefer and Schauermann, 1990.
[b] The symbol — = not studied; 0 = not present.
[c] Single measurement.
[d] Families.

soil is reduced due to the loss of soil organic C, the high C input into the soil by the roots of nitrophilous plants helps to maintain a high earthworm biomass.

The function of soil biota during this transient phase is not clear. A naive concept supposes that during the process of acidification, species with low acid tolerance are eliminated and may be partly replaced by species with high acid tolerance. Though this must inevitably occur at the end of the process, things appear to be much more complicated at early stages. For example, it was concluded from a comparative study carried out with six different soils at different stages of acidification that the microflora actively participates in the loss of soil organic C from the mineral soil during the process of humus disintegration (Wolters and Joergensen, 1991; approach 2 in Section 2). This result was indicated because soil acidification promoted a relatively small microflora living on a comparatively inefficient level of C use. The resulting microbially induced loss of C from the mineral soil may be accelerated because the microflora in the more acid soils had a reduced ability to replace the metabolized C with C from fresh leaf litter. Additional treatments were designed to obtain information about the role of soil invertebrates during this process (Wolters and Joergensen, 1992). The results indicate a rather ambiguous nature of earthworm effects. The burrowing activity of *A. caliginosa* almost lost its significance to the microflora in acidified soils. Endogeic earthworms may even accelerate C loss during the process of humus disintegration by favoring the inefficient use of C from freshly fallen litter for metabolic demands. The strong mixing of fragmented beech leaf litter and mineral soil in all *A. caliginosa* treatments, as well as the fact that the incorporation of litter into the microbial biomass was not affected by factors indicating soil acidification, may counteract this process. In the field, however, the positive effect of endogeic earthworms may be diminished, because earthworms tend to colonize deeper layers of the mineral soil (Bonkowski, 1991).

At least in these poorly buffered soils acid rain may simply accelerate the shift from base-rich to acid soils occurring as a result of natural soil acidification. The period of humus disintegration may be very helpful for studying the alterations of soil biotic processes induced by acid deposition under natural conditions. We hypothesize that acid rain effects on the composition of the herb layer and on the behavior of the macroarthropods are of critical importance at this step of ecosystem development.

4.1.3. Acid Soils

At later stages of soil acidification, not only the roots, but also the decomposers tend to minimize the contact with the acid mineral soil. Bacterial activity in the soil is limited and most of the macrofauna species disappear. The loss of soil-burrowing animals leads to an accumulation of a top organic layer, functioning as a decomposer refuge after the mineral soil has become toxic because of the presence of cation acids. Under these conditions microarthropods dominate the sapro- and microphytophagous soil fauna (Table 3.1). According to Ulrich (1987), the disintegration of soil and organic matter and the dominant role of biotic interaction in regulating the nutrient transfer to primary producers reduces the ability of forest ecosystems to maximize the utilization of energy and to minimize the dissipation of matter, which is achieved by combining a high level of energy utilization with a high level of internal structure and a high degree of internal cycling of matter.

Several hypotheses about acid rain effects on soil biotic processes in acid soils can be drawn.

- As most of the buffering capacity of soil is lost, the acid tolerance of the biota living in strongly acidified forest soils, as well as the redundancy of the decomposer community, are the most important buffering mechanisms hindering a strong alteration of soil processes by acid rain.
- Because functional relationships between soil organisms become very important, acid rain effects may very often be indirect, that is, via a modification of the biotic environment.
- Soil biota and a good deal of tree roots are compressed in the very small compartment of the organic layer. Under these conditions active transport of biologically fixed nutrients into the litter layer against soil water flow by invertebrates and fungi may play an important role in recovering from acid rain induced stress.

4.2. Case Study: The Fauna in Mull and Moder Soils

As an example of the fundamental differences between the decomposer system in base-rich and acidified soils, a recently published comparison between the invertebrate communities of a beech forest on mull (Göttinger Wald) and a beech forest on moder (Solling) is summarized in Table 3.1. The following conclusions can be drawn from these and other data summarized by Schaefer and Schauermann (1990):

- Mull, with its base-rich conditions, is dominated by macrofauna species, whereas moder, as an acidic milieu, is dominated by mesofauna species.
- Because mull soils constitute a less extreme environment in comparison to moder soils, species diversity indexes (richness and evenness) are higher in mull soils.
- In moder soils, which are clearly stratified, animal life is concentrated in the organic horizons and a thin upper layer of the mineral soil; in mull soil the fauna penetrates into deeper soil horizons.
- In the mull forest, with its less continuous supply of resources, more sharply defined phenophases occur, whereas more continuous phenological events characterize moder soil conditions.
- In the mull forest, with its diverse herb layer, food utilization patterns are spatially and temporally more diverse, while moder conditions lead to more uniform patterns of food intake.
- Animals of the mull and moder forest are of comparable importance in energy flow. However, the high proportion of primary decomposers under mull conditions is responsible for higher decomposition rates in this forest type.

- Because of the presence of high macrofauna biomass, mineralization processes are more continuous in the mull soil. Hence, mineral cycling is more susceptible to stress in moder soils than under mull humus conditions.

5. EFFECTS OF ACID DEPOSITION ON DECOMPOSITION PROCESSES

In this section we focus on the effect of acid stress on decomposition processes to provide evidence for acid rain effects on ecosystem function. For the sake of brevity we confined ourselves to C and N turnover, without implying that acid rain effects on the turnover of other nutrients are of minor importance.

5.1. Carbon Mineralization and Decomposition

When measuring acid rain induced changes in the C cycle, different parameters are concerned: the mass loss of litter, soil respiration, incorporation of organic matter into the soil, the degradation of definite types of substrates, and the activity of enzymes (Eisenbeis, 1993). At a given area, the course of decomposition depends on litter quality (deciduous versus coniferous leaf litter, woody litter, and belowground litter), soil biota, and soil conditions.

5.1.1. Soil Respiration

Results on the influence of acid precipitation on soil respiration are conflicting (Bitton and Boylan, 1985): In core experiments soil respiration may remain unaffected, increase, or decrease according to the type of soil under consideration. Hovland (1981), Wainwright (1980), and Wood and Rippon (1980) could not find any effect of acid precipitation on soil respiration. However, in most of the studies a decrease in soil respiration induced by acid stress was found (e.g., Lohm et al., 1984). Carbon dioxide production decreased in pH-adjusted acid soils of an oak–pine forest (Francis, 1982). Baath et al. (1984) found lower soil respiration rates in a Scots pine forest even after experimental acidification at low concentrations. In vitro studies for coniferous litter showed that acidic conditions reduced microbial CO_2 evolution (Moloney et al., 1983). In the study carried out by Bewley and Stotzky (1983), a decrease in soil respiration occurred only at pH 2.

5.1.2. Decomposition Rates

Many studies demonstrate a decrease in decomposition rates as a result of acidification (Rechcigl and Sparks, 1985; Tamm, 1976; Wolters, 1991a,b). Litterbag experiments in damaged spruce stands showed decreased decay rates in comparison to undamaged plots (Hartmann et al., 1989). In a beech forest on calcareous soil, simulated acid rain strongly reduced the decomposition rate of ^{14}C-labeled beech leaf litter and hyphae (Fig. 3.2; Wolters, 1991a). It was concluded that acid deposition may cause a very strong accumulation of primary and secondary C compounds

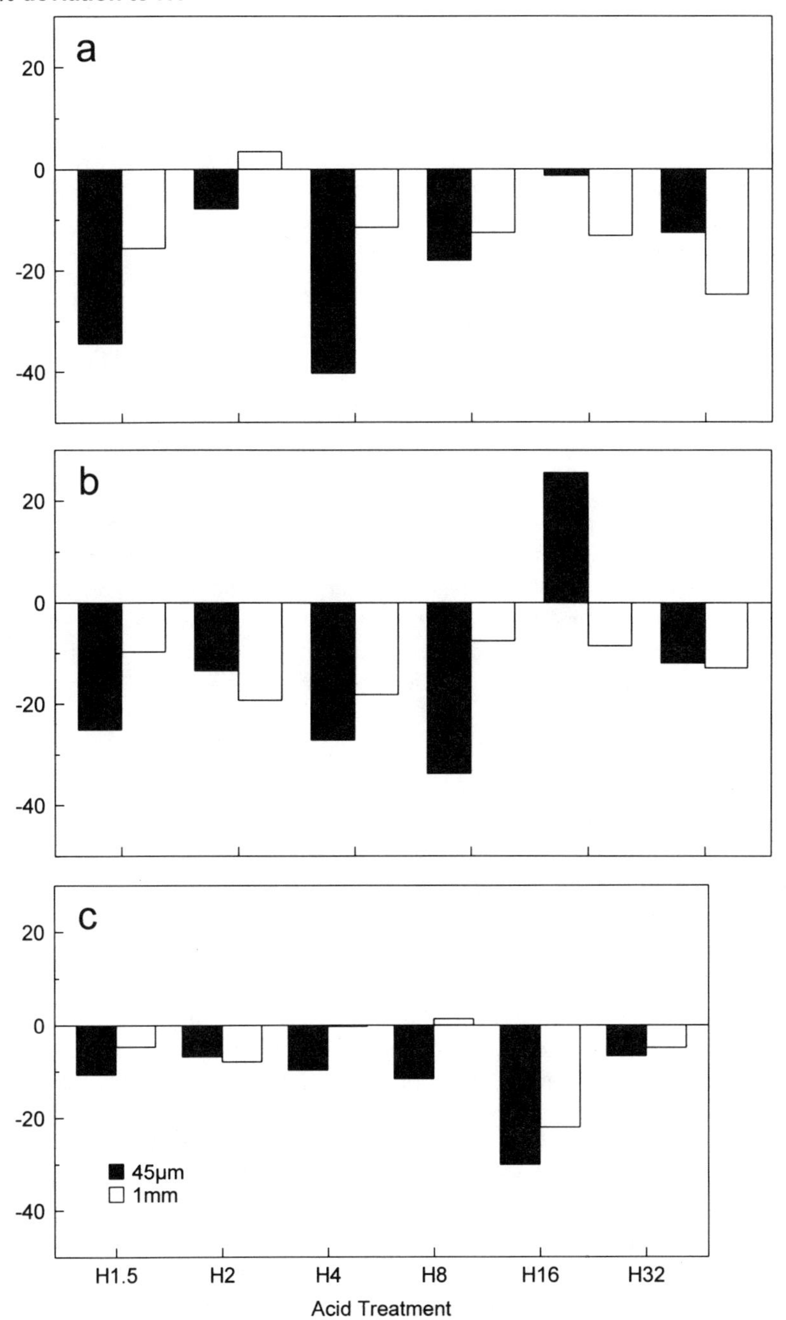

Fig. 3.2. Effect of acid rain on cumulative Co_2–C liberation from labelled C compounds added to beech leaf litter that had been exposed for one year in 45 μm and 1 mm litterbags in the litter layer of a beech forest on calcareous soil (Göttinger Wald, West Germany). (*a*) Co_2–C production from one-year-old litter, (*b*) $^{14}CO_2$–C production from labelled beech leaves, (*c*) $^{14}CO_2$–C production from labelled hyphae. The H^+ imput was increased to 1, 1.5, 2, 4, 8, 16, and 32 times that of the total acid deposition in the Göttinger Wald area (0.013 g h^+ m^{-2} $month^{-1}$). Treatments are referred to as H1 (control), H1.5, H2, H4, H8, H16, and H32 (modified from Wolters, 1991a).

in the litter layer of base-rich soils, even at moderately increased proton inputs. The decomposition rate is strongly related to microbial activity. This result might explain why decomposition of organic matter in acidic coniferous forest soils, supporting a microflora well adapted to low pH values, is apparently only slightly affected by simulated acid rain (Abrahamsen et al., 1980; Section 6.1).

5.1.3. Decomposition of Specific Carbon Compounds

A number of studies have focused on the effects of acid rain on the decomposition of selected C compounds. For example, Grant et al. (1979) and Strayer and Alexander (1981) observed a lower mineralization rate of glucose under acidic conditions. Plants contain 5–30% lignin, 15–60% cellulose, 10–30% hemicellulose, and 2–15% protein (Paul and Clark, 1989). There is evidence that the microbial decomposition of these polymeric compounds is affected by acid rain:

- Cellulose decomposition is known to be a pH-sensitive process (White et al., 1949; Ruschmeyer and Schmidt, 1958). Cellulose degradation in the top 10 cm of the soil near a coking plant was marginally inhibited by acid pollution (Killham and Wainwright, 1981). Bewley and Parkinson (1986) found a reduced C mineralization rate in acidified, cellulose-amended mineral soil. The association of cellulose with more recalcitrant lignin materials in plant litter may alter the susceptibility of the overall decomposition to acid deposition (Bewley and Parkinson, 1986).
- Hovland et al. (1980) found some indication of reduced lignin decomposition in acidified soils.
- The phase of rapid CO_2 evolution in protein hydrolysate-amended soil was delayed by acidification (Grant et al., 1979).

A serious lack of knowledge results from the fact that too little attention has been paid to the breakdown of organic complexes of elements other than N that may be important to plant growth (e.g, P and S) (Alexander, 1980b).

It seems reasonable to assume that substrate specific effects of acid rain on litter decomposition leads to an alteration of the composition and of the state of soil organic matter. This hypothesis was tested in a microcosm experiment (Wolters, unpublished). The ^{14}C labeled beech leaf litter (treated according to the procedure devised by Dickson, 1979) was placed for 132 days on top of the organic layer sampled in a beech forest on moder soil (Solling). The control was watered with simulated rain adjusted to a pH corresponding to that of the normal rain in the research area (pH 4.15, with sulfuric acid [H_2SO_4)], while the acid rain treatment was watered with simulated rain adjusted to pH 2.95 (see Wolters, 1991b for details of the experiment). The effect of increased acid loading on the decomposition of different C components of the labeled litter was investigated at the end of the experiment by means of a sequential extraction of the labeled litter (Joergensen, 1987; cf. Table 3.2). According to Joergensen (1987) the different C fractions can be interpreted as follows: The chloroform extractable fraction contains substances such

TABLE 3.2. Different C Fractions Extracted from ^{14}C Labeled Beech Leaf Litter Placed on Top of the Organic Layer of a Beech Forest on Moder Soil[a]

Extractant (%)	Initial Substatum	Control	Simulated Acid Rain	$LSD_{0.05}$[c]
Chloroform	1.42	1.65	1.55	0.13
Water	5.62	2.80	4.07[b]	0.20
KOH	30.41	28.46	34.39[b]	1.69
H_2SO_4	21.39	23.82	26.68[b]	1.59
Remains	41.18	20.00	19.88	2.22
CO_2 (released within 132 days)		23.27	13.43	

[a]Results of a microcosm experiment (see text for further explanations).
[b]Significantly different to control.
[c]Least significance difference.

as lipids and waxes, the water extractable fraction contains easily soluble C compounds of low molecular weight, the potassium hydroxide (KOH) extractable fraction contains organic compounds such as lignin and proteins, the H_2SO_4 soluble fraction contains mostly cellulose, and the C that cannot be extracted by any of the solvents (remains) contains the overwhelming portion of the large ligno–cellulose complex, typical for beech leaf litter.

The results of the experiment are summarized in Table 3.2. The $^{14}CO_2$–C production was strongly reduced in the acid rain treatment (cf. Wolters, 1991b). This reduction led to a significant accumulation of the ^{14}C compounds extractable by H_2O, KOH, and H_2SO_4. On the other hand, the decomposition of the chloroform extractable fraction and of the insoluble remains was not affected by the application of simulated acid rain. The reduced processing of the water soluble C compounds indicates a general suppression of the litter colonizing microflora and is in accordance with the depression of CO_2. Increased amounts of easily accessible C sources might add to the risk of mineralization flushes. The accumulation of ^{14}C compounds soluble in KOH and H_2SO_4 confirms that acidification strongly hinders the decomposition of organic compounds of high molecular weight (see above). The small effect of acid rain on the chloroform extractable fraction and on insoluble remains points to the specific conditions for the slow biodegradation of recalcitrant C compounds, such as waxes and the ligno–cellulose complex.

5.2. Nitrogen Cycle

The key intermediates of the biogeochemical N cycle are ammonium (NH_4), nitrate (NO_3), and dinitrogen (N_2). Shifts between the different oxidation states of the N atom are commonly mediated by soil microorganisms. The functional groups of microorganisms involved are nitrifying bacteria, denitrifying microorganisms, nitrogen fixing bacteria, nitrate and nitrite ammonifiers, ammonium-assimilating microorganisms, and deaminating microorganisms. Because of the large demand by plants for nitrogen, acid rain effects on different steps of the N cycle have been intensely studied.

5.2.1. *Nitrogen Mineralization*

Investigations by Klein and Alexander (1986) showed that rates of N mineralization were inhibited, stimulated, or not affected and that the use of different rates of additions of artificial rain or different exposure to the simulated precipitation will lead to different conclusions. In summarizing the results of investigations carried out in 12 different forest soils, Stroo and Alexander (1986) concluded that the nature of the effect of acid rain on N mineralization is determined by the soil's content and the C/N ratio of the organic matter. These authors suggest that the soil organic matter neutralizes part of the incoming acidity, and, thus, any negative effect would be observed first in soils with low organic matter content. Persson et al. (1989) confirmed the overwhelming importance of the C/N ratio. In general, N mineralization seems to be less sensitive to acid rain than C mineralization.

It was found in short- and in long-term experiments that ammonification increases rather than decreases in acidified soils (e.g., Tamm, 1976; Haynes and Swift, 1986; Persson et al., 1989). Tamm et al. (1977) attributed this effect to a partial mortality of the microflora followed by mineralization of biomass N, while Aber et al. (1982) regarded it as an artifact caused by unnaturally high doses of acid. Persson et al. (1989) found that mineralization of biomass N in killed soil organisms could only explain 25–30% of the increase in net N mineralization. They concluded that most of the effects found on N mineralization after acid treatment seem to depend on the fact that acidification reduces the availability of C and N to the microorganisms. In addition, the data indicate that acidification reduces the incorporation of N compounds from dead organisms into the soil organic matter and, thereby, results in increased availability of substances rich in N to the microorganisms and increased net N mineralization (Persson et al., 1989).

5.2.2. *Nitrification*

Nitrification is the weakest link in the nitrogen fixation/denitrification/nitrification—ammonification triangle (Francis, 1986). Nitrification in soil is predominantly carried out by chemolithoautotrophic bacteria, which grow optimally at a pH 7.5–8. Though nitrifying bacteria can also be isolated from soils with a pH 4, there is evidence that autotrophic nitrification is drastically reduced at low pH values. However, because of their preference for base-rich forest soils and/or for well-buffered microsites in acid forest soils (e.g., mesofauna faeces), autotrophic nitrifiers may be less affected by acid rain under natural conditions than expected from laboratory experiments.

In acid forest soils, a large part of the nitrate is produced by heterotrophic nitrification (Runge, 1974; Paul and Clark, 1989). Heterotrophic nitrification is less sensitive to acid rain than autotrophic nitrification because it is carried out by a wide range of microorganisms differently adapted to the acidity of their environment

(Alexander, 1980a,b). Suppression of heterotrophic nitrate production by acid rain thus indicates general damage to the microflora rather than to certain microorganisms (Wolters, 1991b).

It should be emphasized that the inhibition of nitrification is not necessarily disadvantageous for the forest ecosystem because forest trees are not essential nitrate plants and because less of the available N is turned into readily leachable nitrate, provided that enough available N is released during decomposition processes (see below).

5.2.3. *Other Steps of the Nitrogen Cycle*

Denitrification is brought about by bacteria that use NO_3^- as an electron acceptor. Characteristically, these bacteria are sensitive to H^+ ions. The rate of denitrification is thus rather slow in acidified soils, and at greater acidities N_2O instead of N_2 (dinitrogen) is the predominant end product. The formation of N_2O in acid soils indicates, however, that denitrifying bacteria are present in low numbers. These bacteria are most probably active in microsites of the soils where the pH is conducive to bacterial activity (Francis, 1982). In general, an inverse relationship exists between soil acidity and the number of denitrifying bacteria (Alexander, 1980a).

Asymbiotic nitrogen fixing algae and bacteria appear to be especially susceptible to acid precipitation (Chang and Alexander, 1983). For example, application of water at pH 5 markedly reduced the rate of acetylene reduction in litter samples (Alexander, 1980a). Thus nitrogen fixation is reduced not only in soils but also in forest litter. At least in acid forests, however, this effect is of minor importance, because N gains by asymbiotic bacterial and fungal fixation seem to be insignificant.

5.2.4. *Nitrogen Leaching*

The NO_3^- ion is more readily leached from the soil than NH_4^+ ion. As ammonification is stimulated and nitrification is inhibited by soil acidification, one would expect reduced N leaching with increasing soil acidity. In fact, some studies indicate that substantial amounts of N might be accumulated in terrestrial ecosystems exposed to acid precipitation. On the other hand, N leaching can be significantly stimulated by increased proton loading rates from rain. For example, application of simulated acid rain to soil cores sampled in the organic layer of a beech forest on moder soil reduced nitrate but increased NH_4 leaching, thereby stimulating the leaching of mineral N corresponding to an annual loss of 17 kg of N ha^{-1} $year^{-1}$ (Wolters, 1991b). Similar results were obtained by other authors (e.g., Kelly and Strickland, 1987; Persson et al., 1989). The following mechanisms may explain these findings (Wolters, 1991b):

- The acid induced decrease in microbial biomass in the F horizon reduced the storage capacity of the microflora, and thus diminished the microbial immobilization of nitrogen.
- Depression of the microflora led to a mobilization of N rich compounds from the microorganisms killed.

- A stress-induced increase in metabolic activity (determined by measuring the metabolic quotient) shifted the microflora in the F layer to a state of less economic use of N.
- The strong suppression of the litter colonizing microflora may have hindered both the transfer into and the storage of N in the carbon-rich litter layer. The most important role of the microflora in the upward transport of nutrients into the L layer has been demonstrated several times (e.g., Brückmann and Wolters, in press).

5.2.5. *Nitrogen Transfer to Plants*

It was argued that acid induced changes in processes such as C and N mineralization could bring about a disruption in nutrient flow to growing plants in a forest ecosystem (McFee and Cronan, 1982). In fact, this was the central justification for carrying out soil biotic studies in the context of acid rain research. Several scenarios were summarized by Kelly and Strickland (1984):

- If decomposition rates are reduced, availability of plant nutrients will be reduced.
- If decomposition increases, two possible scenarios can be formulated: increased nutrient availability could lead to increased productivity of nutrient limited stands, or essential nutrients may be leached from the soil if mineralization occurs faster than nutrient uptake.
- Even if mineralization rates do not increase on an annual basis, an alteration of the temporal pattern can result in increased nutrient leaching.

Dambrine et al. (1992) found a correlation between high nutrient demand of spruce trees, yellowing of needles, soil acidification, and a soil microbial factor. However, experimental data supporting a close relationship between changes of soil biotic activity caused by acid rain and forest vitality are very sparse. The most probable explanation is that several factors could delay or counteract the appearance of growth responses of trees (Aber et al., 1982):

- Acid precipitation may not cause acidification of soils.
- Alteration of soil pH may not cause changes in total N mineralization.
- Increases in N inputs may counteract decreases in N mineralization.
- Drawing on an internal pool of N in the plant may buffer short-term changes in N availability.
- The vegetation could be tolerant to low N availability.

6. EFFECTS OF ACID DEPOSITION ON THE SOIL MICROFLORA

Acid deposition effects on the microflora were reviewed several times (Alexander, 1980a,b; Francis, 1986; Myrold and Nason, 1992). Some general trends for a response to acid rain are

- Acid rain effects are mainly confined to microorganisms living in the organic horizons and near-surface soil layers (Killham et al., 1983).
- The proton content of simulated rain and its effect on microbial processes are not linearly correlated. For example, effects of simulated rain in acid soils were often confined to high input levels of protons (cf. Lettl, 1985) or were especially strong in intermediate acid treatments (Lee and Weber, 1983).
- As a tendency, only very low pH values of simulated acid rain lead to measurable effects. According to Hyvönen and Persson (1990), a reduction in soil pH by 0.3 units or more in soils of pH 4.0–4.5 seems to be required to cause a reduction in microbial respiration of 10% or more.

Microbial reactions to soil acidification strongly depend on environmental conditions. For example, in hardwood forest but not in softwood forest the populations of bacteria, actinomycetes, and fungi closely paralleled changes in soil acidity (Wang et al., 1980). In a microcosm experiment, the reaction of the litter colonizing microflora to the termination of acid rain in a natural beech forest soil was strongly different from that in a limed soil (Wolters, 1991b). It is thus not very surprising that studies on the response of microorganisms to acid stress demonstrate contrasting results. In many acidification experiments microbial activity was not even substantially impaired (Wainwright, 1980).

We will refer to the well-established concept of a differentiation between the species level (individuals and populations) and the community level in the following sections. This differentiation makes it easier to arrange the various studies concerning acid rain effects on the soil microflora currently available. We are nevertheless aware that the application of this concept to microbial assemblages is very often problematical (Cooke and Rayner, 1984).

6.1. The Species Level

6.1.1. *General*

It was already mentioned that soil microbiological field studies at the species level are very sparse. One important reason is that no appropriate methods are available for carrying out quantitative investigations focusing on certain species in a medium such as soil. Some information on acid rain effects at the species level comes from studies carried out with pure cultures in the laboratory. Alexander (1980a) listed some of the limitations of this approach:

- Stresses that are studied in this way are rarely the ones that have an influence on the microbial populations in nature.
- It is extremely rare that one knows which microorganism groups are really involved in a specific process.

Additional limitations were already mentioned in different sections of this chapter. Consequently, most of what is known about acid rain effects on certain species of the microflora is thus based on casual observations or on incidental results of experiments carried out under insufficient conditions.

There are good reasons to believe that specialized species may be particularly sensitive to acid deposition. Evidence exists, for example, that the populations of highly specialized microorganisms involved in decomposition of soil organic N will be affected by acid rain (Haynes and Swift, 1986). For the sake of consistency, this aspect was already discussed in the section on N cycle (Section 5.2). Similarly, the specialized populations in the rooted zone may react very sensitively to acid rain (Alexander, 1980a). Several hypothetical pathways for acid rain effects on the rhizosphere microflora can be derived from the data summarized by Bruck and Shafer (1984):

- Direct acid toxicity may occur due to the excretion of protons by the root tissue.
- Effects of low-pH rain on roots near the soil surface could lead to the presence of senescent tissue. This finding could significantly alter the nutritional conditions for the microflora.
- Changes in root exudates from exposure to acid rain would alter microorganism interrelationships involving potential pathogens in the rhizosphere.
- Interactions with soil invertebrates could change. For example, many plant–parasitic nematodes frequently occur in slightly acidic soil (pH 5–6). Growth and reproduction of many species are inhibited below pH 4 (cf. Section 3). This finding could lead to reduced infections by nematodes and may thus alter biotic interactions in the rhizosphere.

6.1.2. Adaptations

Microbial adaptations to acid deposition at the species level include genotypic and phenotypic changes. Though no direct evidence exists, genotypic adaptations may prevail and are the reason for the amazing capacity of microorganisms to adapt to environmental changes (Myrold and Nason, 1992). An example of physiological adaptation (i.e., phenotypic) is given by Padan (1984). Although the principles of pH homeostasis appear identical among all bacteria, the pH controlling machinery is poisoned at a different intracellular pH: 6.5 (acidophiles), 7.8 (neutralophiles), and 9.8 (alkalophiles). An important phenotypic preadaptation that enables resistance to acid rain is the ability to survive adverse environmental conditions by increased sporulation (Baath et al., 1980, 1984). In Swedish forests gram-negative bacteria were reduced by the application of acid rain, while spore forming bacteria seemed to predominate (Baath et al., 1980).

Some generalizations concerning the tolerance of the soil microflora to soil acidity are:

- Most bacteria are less acid tolerant than fungi.
- Bacteria nevertheless exhibit a wide range of mechanisms that are adaptable to an alteration of external pH.
- Fungi dwelling in an acid soil are well adapted to low pH values. Small changes in the soil pH will therefore affect these fungi only to a very small degree (Baath et al., 1984).

The causal mechanisms behind these patterns are not sufficiently understood. Padan (1984) provides some explanations for the high susceptibility of bacteria to acidity:

- From both the intracellular and extracellular milieu the procaryotic cell of the bacteria, more than the eucaryotic cells of fungi, is exposed to fluctuations in proton concentrations.
- The H^+ concentration of the extracellular environment is in direct contact with the cytoplasmatic membrane of the bacterial cell.
- The proton, more than any other ion, is involved in many physicochemical as well as biochemical reactions, solvolysis, ionization, and oxidoreduction. The high proton concentration of acidic environments leads to high concentrations of dissolved metals that can reach toxic levels.

6.2. The Community Level

6.2.1. *Bulk Parameters*

Many studies have demonstrated that acid rain may significantly alter bulk parameters such as biomass, total microbial activity (sometimes separately for bacteria and fungi), FDA-active fungal biomass, ATP content, and heat production. In fact, it has been stated that species composition of the microbial community may be less sensitive to the acid treatment than bulk and functional parameters. For example, soil respiration and FDA-active fungal biomass were significantly different from control treatments at all levels of acidification investigated by Baath et al. (1984), while no general response could be detected on the species level. Though these results have to be handled with great care because of the methodological problems mentioned in Section 6.1, its implications have to be tested in the future. As a very rough generalization most bulk parameters are depressed by a strongly increased input of protons:

- In a Scots pine forest, for example, FDA-active fungal mass decreased after acidification (Baath et al., 1984).
- In the F layer of a beech forest on moder soil the biomass of the microflora was significantly depressed by simulated acid rain (Wolters, 1991b).
- Acid precipitation reduced the microbial biomass N in the uppermost fermentation layer of a spruce forest soil from 4.8% of total N to 1.8% and that in the lower fermentation layers from 2 to 0.5% (von Lützow et al., 1992).

6.2.2. *Structural Parameters*

Alterations of the community structure of the soil microflora induced by acid rain were observed in the field (e.g., Baath et al., 1984; Lettl, 1985). In Adirondack forests, for example, the species diversity of bacteria, actinomycetes, and fungi was lower under acidic conditions (Wang et al., 1980). In studies carried out in the alpine tundra, increased soil loading rates of acid (and N) were significantly associated with decreases in bacterial diversity (Mancinelli, 1986). Baath et al. (1984) investigated the long-term impact of simulated acid rain on the fungal community in a *Pinus sylvestris* stand on Podzol. Four years after acid rain application they found that fungal species composition was still altered by treatments of 100 and 150 kg of H_2SO_4 ha^{-1}, but not by treatments of 50 kg of H_2SO_4 ha^{-1}. Above all, the abundance of *Penicillium spinulosum* and *Oidiodendron* (cf. *echinulatum* II) increased with increasing rates of acid application, while only small changes occurred in other isolated fungal taxa. In general, the acidification appeared to decrease the species diversity only slightly.

Structural alterations may feed up to the process level. This conclusion can be drawn from both the measurement of process parameters (which are largely dominated by the microflora: Sections 3.1 and 5) and because different functional groups are affected differently by acid rain (Mancinelli, 1986):

- Application of simulated acid rain to an alpine tundra soil reduced both populations of nitrifiers and populations expressing lipolytic activity, while populations expressing proteolytic activity were stimulated.
- The percentage of the bacterial population capable of degrading starch, cellulose, or pectin was not significantly affected by increased acid input.

However, the results of shifts of particular functional groups vary greatly among studies (Myrold and Nason, 1992).

Because of the sensitivity of the microorganisms involved, the rhizosphere seems to be a good candidate for investigating community reactions of the soil microflora to increased input rates of protons. According to the data given in Section 6.1, the various stresses or application of chemicals, such as strong acids, to plant foliage alter rhizosphere characteristics, and may therefore alter the microbial community. These effects may be amplified by the fact that the root system in declining forests often exhibits features that occur after N fertilization (Meyer, 1985). Unfortunately, most of the studies on the response of microbial populations in the rhizosphere to acid rain were carried out with crop plants, while little is known about forest plants. According to Shafer (1988), increased acid content of simulated rain stimulated the numbers of total bacteria but suppressed populations of amylolytic bacteria in the rhizosphere of *Sorghum*. Trends of increased fungal populations were associated with acidity.

6.3. Redundancy of the Microbial Community

Most of the microbial transformations in soil are brought about by several populations. Redundancy therefore plays a dominant role in the microbial reaction to acid rain at the community level (cf. Fig. 3.1). Evidence for this was already given in Section 5 (acid rain effects on C mineralization and ammonification).

In a litterbag experiment carried out in a beech forest on mull soil, the amount of protons contained in precipitation was not correlated to its effect on the litter colonizing microflora (Fig. 3.2; approach 4, Section 2; Wolters, 1991a). Depression was very strong when the input of protons was 1.5 times greater than normal acid deposition, but comparatively low at a 32 times greater input. A hypothetical explanation for this phenomenon is that the application of different amounts of protons to the experimental plots over a period of 42 months has favored adaptation of decomposer organisms to acid stress in a treatment-specific way. The comparatively low effect of acid rain in the treatments with 2-fold and 32-fold increased input levels of protons may thus be an expression of the redundancy of the litter colonizing microflora.

7. EFFECTS OF ACID DEPOSITION ON SOIL INVERTEBRATES

Field experiments showed that some invertebrate species are depressed, while some increase in numbers by the application of simulated acid rain (approach 4, Section 2). A differential response can be observed even in groups of closely related species. As a very general tendency, species from the following groups (mainly belonging to the mesofauna) are favored: Thecamoebae (Protozoa), Enchytraeidae, Oribatida, Uropodina, Collembola, Elateridae, some Diptera (e.g., Sciaridae) (Table 3.3). In contrast, species of the following groups (mainly of the macrofauna and microfauna) decrease: Protozoa (except Thecamoebae), Nematoda, Lumbricidae, Gastropoda, Isopoda, Myriopoda, Protura, other macroarthropods (e.g., Coleoptera), some Diptera.

Many of the effects observed depend on environmental conditions. For example, Abrahamsen et al. (1976) compared the effects of different acid treatments on the enchytraeid species *Cognettia sphagnetorum* in a podzol to that in a podzol-brown earth. The results from the podzol experiment indicate an increase of *C. sphagnetorum* when the amount of H^+ ions applied increases from 0.5 to 55 me. Further application reduced the population density. In the podzol-brown earth experiment, in contrast, no effect was observed. The mode of acid rain application may also be important. In the experiment carried out by Persson et al., (1989), for example, the reaction of different feeding groups of nematodes to simulated acid rain was significantly different in the treatment, when the acid was applied with one single dose, from that in the treatment, when low doses of acid were continuously applied.

7.1. The Species Level

7.1.1. General

The soil fauna reacts sensitively to pH gradients (Hagvar and Abrahamsen, 1980). Several studies demonstrated a species specific reaction of soil invertebrates to acid induced alterations of their environment. For example, by comparing different *Pinus sylvestris* stands in Holland (approach 2, Section 2), van Straalen et al. (1988) found no clear correlation between total microarthropod abundance (Oribatida and Collembola) and vitality of stands. On the other hand, the number of the oribatid species *Platynothrus peltifer* as a percentage of Oribatida correlated positively with vitality. They concluded that *P. peltifer,* because of its sensitivity to acidity and its high needs for micronutrients such as Mn, is on the edge of its

TABLE 3.3. Overview of the Most Important Soil Invertebrate Taxa, the Features Characteristizing the Micro-, Meso-, and Macrofauna, and Some Often-Cited Publications Concerning Acid Rain Effects on Soil Invertebrates

Size Class and Ecological Features	Taxonomic Groups	Selected Literature on Acid Rain Effects	Numerical Response to Soil Acidification and/or Acid Rain (Tentative)
Microfauna: (body width <100 μm) Limited to the water film around substrate surfaces; utilize organic compounds of low molecular weight by bacteria and fungi; unable to break through barriers of soil; generation times that roughly correspond to those of bacteria and fungi	Protozoa	Alexander, 1980a; Baath et al., 1984; Hagvar, 1987a; Stachurska-Hagen, 1980	– (except Thecamoeba)
	Nematoda	Focht, 1990; Hagvar, 1987a; Huhta et al., 1983, 1986; Leetham et al., 1980; Persson et al., 1989; Hyvönen and Persson, 1990; Stachurska-Hagen, 1980; Timans, 1986	–
	Rotatoria	Hagvar, 1987a; Leetham et al., 1980; Stachurska-Hagen, 1980	–
	Tardigrada	Leetham et al., 1980	–
Mesofauna: (body width <2 mm) Live in the pore system of the soil; preferentially feed on fungi, but also ingest decomposed plant material and mineral particles; can move through soil, but are only able to break through structural barriers to a limited extent; the generation times are in the range of weeks and months	Enchytraeidae	Abrahamsen et al., 1976; Baath et al., 1980; Eisenbeis and Feldmann, 1991; Graefe, 1988; Hagvar and Abrahamsen, 1977, 1980; Hagvar et al., 1980; Hagvar, 1984a; Heungens, 1984; Huhta, 1984; Huhta et al., 1983, 1986; Persson et al., 1989	+
	Acarina (mainly Oribatida, Gamasina and Uropodina)	Baath et al., 1980; Eisenbeis and Feldmann, 1991; Hagvar and Abrahamsen, 1977, 1980; Hagvar and Amundsen, 1981; Hagvar and Kjöndal, 1981a–c; Hagvar, 1987a and b; Huhta et al., 1983, 1986; Koskenniemi and Huhta, 1986; Larkin and Kelley, 1987; van Straalen et al., 1987 and 1988	+

	Collembola	Baath et al., 1980; Hagvar, 1984a and b, 1987a and b; Hagvar and Abrahamsen, 1977, 1980, Hagvar and Kjöndal, 1981a; Huang, 1985; Huhta et al., 1983, 1986; Larkin and Kelley, 1987; van Straalen et al., 1987, 1988; Vilkamaa and Huhta, 1986; Weber and Makeschin, 1991; Wolters, 1989a, 1991b	+/−
	Protura	Funke et al., 1985; Funke, 1986; Hagvar, 1984b; Nosek, 1982; Wolters, 1989a	−
	Diplura	Wolters, 1989a	−
Macrofauna: (body width >2 mm) Large enough to break through physical barriers of soil; nutrition is based on substrate use as well as on the utilization of associated microflora and fauna; epigeic species that live on the soil surface are mainly involved in litter comminuting and the reworking of relatively undecomposed organic matter, endogeic species that live in the upper layers of the mineral soil influence porosity, aggregate structure, and clay humus complexing of the soil	Gastropoda	Dzieszkowski, 1988; Gärdenfors, 1987	−
	Lumbricidae	Hagvar et al., 1980, Hagvar, 1987a; Ma, 1984; Makeschin, 1991; Scheu and Wolters, 1991b	− (except *D. octaedra*)
	Arthropoda (general)	Craft and Webb, 1984	+/−
	Araneida	Hagvar et al., 1980; Gunnarsson and Johnson, 1989	−
	Isopoda	Scheu, 1989	−
	Myriopoda	Poser and Scheu, 1989; Scheu and Wolters, 1991a and b	−
	Insecta (general)	Rutherford, 1984	(+)/−
	Diptera (larvae)	Garden and Davies, 1988	(+)/−
	Lepidoptera	Jones and Hendershot, 1988	(+)/−
	Coleoptera	Hagvar et al., 1980; Jones and Hendershot, 1988; Makeschin and Habereder, 1991; Schauermann, 1987; Wolters, 1989b	+/−

ecological amplitude in the research area, and is decreasing in abundance when the forest soil is critically depleted of nutrients. It seems that the sites provide habitats that support different densities of microarthropods, for reasons not always related to the vitality of the stand.

In summarizing field and laboratory experiments carried out in the context of the Norwegian research program on acid rain effects, Hagvar (1987a,b) concluded that a great number of microarthropod species (springtails and mites) are affected by strong alterations of soil pH caused by acid rain and liming (approach 4, Section 2). According to their numerical response, the roughly 50 taxa investigated could be sorted into four categories: (1) increased abundance on acidification and/or reduced abundance on liming (acidophilic species); (2) reduced abundance on acidification and/or increased abundance on liming (calciophilic species); (3) reduced abundance recorded both on acidification and liming (stenoecious species); and (4) various other reactions. Similar categorizations can be made for every other soil invertebrate taxon.

Various hypotheses that might explain the different reactions of certain collembolan and mite species to increased input levels of protons were stated by Hagvar (1984a). These hypotheses can readily be related to the pathways of acid rain effects outlined in Section 3.1 (Fig. 3.1):

1. Direct effects (toxicity): Abundance changes in the liming and acidification experiments were directly due to the lime or H_2SO_4 applied, while the soil pH changes were of minor importance.
2. Indirect effects via the abiotic environment: Relations between pH and density of certain microarthropods are caused by factors correlated to soil pH (N content, loss on ignition, humus type, and soil profile).
3. Indirect effects via the biotic environment:
    - Reduced predation pressure in acidified samples allowed certain prey populations to increase.
    - pH dependent alterations of food conditions caused abundance changes of several microarthropod species. Fungal hyphae, for example, are found in the gut contents of many species and are generally considered to be an important food item for microarthropods. Accelerated growth of fungi in acidified soils (Section 6) may thus favor certain microarthropod species.
    - At different soil pH levels, different species are favored in the competition process.
    - Acid-induced alterations in the field were indirectly due to the marked reduction of the ground vegetation (mainly the mosses).

While refuting most of these hypotheses in light of the available data, Hagvar (1987b) strongly relied on competition as a key factor. The most important argument is that while species increasing in acidified soil showed considerable differences in ecology, they had one important feature in common; they all belong to the most dominant microarthropod species in the raw humus soils used in the experiments. Acid rain may thus have shifted the competition capacity towards species that are adapted to an acid soil environment. This conclusion is consistent with our hypothesis that acid rain effects on soil invertebrates in acid soils are very often mediated by the biotic environment (Section 4.1). Several investigations support our hypothesis. For example, the negative response of the enchytraeid species *C. sphagnetorum* to acid stress has been largely attributed to depression of the microflora (Huhta, 1984; approach 4, Section 2).

7.1.2. Adaptation

Several adaptational features to overcome direct proton toxicity have evolved in soil invertebrates. These features range from behavioral reactions to morphological structures and physiological mechanisms. Most of these features cross both taxonomic classifications and ecological categories. The response of the soil fauna to acid rain may thus be as diverse as the soil fauna itself. As an example of major factors determining the sensitivity to acidification we will focus on the exposure of soil invertebrates to the soil solution. Because the mode of adaptation to the acidification of the soil solution shows some parallelisms with the micro-, meso-, macrofauna concept (Table 3.3) it will also help to find some regularities in the morass of adaptational patterns. Of course, we do not suggest that adaptation to soil acidity or reaction to acid rain can be explained by a few simple factors.

Potentially, the greatest physiological problem of terrestrial animals is dehydration (Barnes et al., 1993). As an intermediate medium between aquatic and terrestrial habitats, the soil provides an environment with reduced risk of drought. This condition made it possible for the soil invertebrates to evolve a most interesting dual nature, featuring characteristics of both aquatic and terrestrial animals. The main physiological similarities between soil and water animals are the integumental permeability of soil dwellers and skin respiration (Ghilarov, 1983). These features also increase the permeability for ions and makes soil invertebrates susceptible to an alteration of the composition of the soil solution.

The microfauna is physiologically aquatic and is thus strongly limited to the water film around resource surfaces, soil particles, and roots (Table 3.3). The overwhelmingly negative response of this group to acid rain may be largely due to the fact that it is forced to stay in close contact with the soil solution without being provided with an efficient protection. Probably, the Thecamoebae (Protozoa) are less sensitive to soil acidity because they evolved the ability to secrete a testa. This finding indicates that morphological structures hindering a direct contact between the body surface and the soil solution are a successful mechanism to reduce the risk of acid toxicity. In this context, the ability of many protozoan species to form cysts may be an important feature providing shelter to acid stress.

It was concluded from several investigations that within the mesofauna sensitivity to acid rain decreases in the order enchytraeids > Collembola > mites (Stuanes et al., 1992). Though enchytraeids have many features in common with earthworms, several species, notably *C. sphagnetorum* and several *Achaeta* species, are clearly acidophilic or acid tolerant (Didden, 1993). Consequently, these species reach high densities in acid soils (cf. Table 3.1). However, enchytraeids still remain in close contact with the soil solution, and may therefore be sensitive to acid rain. On the other hand, microarthropods (Collembola and mites) have a water-repellent body surface and direct physiological effects of acid rain are thus reduced (Hagvar, 1987a,b). Collembola, however, efficiently absorb water with their ventral tube to avoid desiccation, thereby increasing their susceptibility to an acidification of the soil water (Jaeger and Eisenbeis, 1984).

The macrofauna followed two different pathways to overcome the barriers of soil. The correlating morphological features affect their acid tolerance differently. Earthworms reduce the risk of dehydration by remaining in close contact with the soil solution. Consequently, earthworms had to evolve strong muscles to burrow through the soil. However, they still possess the physiological features of semiaquatic invertebrates and most of them are very sensitive to proton toxicity. For example, acidity of the soil solution may directly affect the neurons of earthworms due to the high tegumental permeability (Laverack, 1963; see Lee, 1985 for other aspects of earthworm adaptation to low pH values). On the other hand, the evolution of a thick cuticula and a tracheal system enabled the macroarthropods to reduce the direct contact with the soil solution. In principle, this strongly reduces the susceptibility to low pH values. Nevertheless, the large Ca demand caused by the development of the cuticula does not allow most macroarthropods to survive in an acidified and nutrient-poor environment. In addition, at no step of their evolution did they have the chance to adapt to large amounts of protons coming from above.

7.2. The Community Level

The differential reaction at the species level reported in Section 7.1 suggests that acid rain may cause an alteration of the structure of the invertebrate community. An example is given in Fig. 3.3. Simulated acid rain containing different amounts of protons were added to an acid beech forest soil in the Solling area for 4 years (the method of acid application was identical to the one described in Wolters, 1991a). Different reactions of Oribatida and Mesostigmata lead to treatment specific alterations of the mite community structure (Heiligenstadt, 1988). At low and intermediate input levels of protons this was largely caused by the sensitive reaction of Oribatida. In contrast, at high input levels of protons the Mesostigmata significantly declined, while the Oribatida slightly recovered due to a shift in species composition. Among the studies concerning the effect of proton input on soil invertebrate communities, recent and more comprehensive analyses were presented by van Straalen et al. (1988) for a Dutch pine forest, by Funke (1986) and Hartmann et al. (1989) for south German spruce stands, and by Hagvar (1984a, 1987a, 1990) for the Norwegian research sites. The results are conflicting. The following general conclusions can be drawn from the response of the soil fauna to acidic inputs at the community level:

- The structure of the community changes.
- Species numbers and species diversity values tend to decrease.
- Very high concentrations of protons tend to be generally inhibitory.

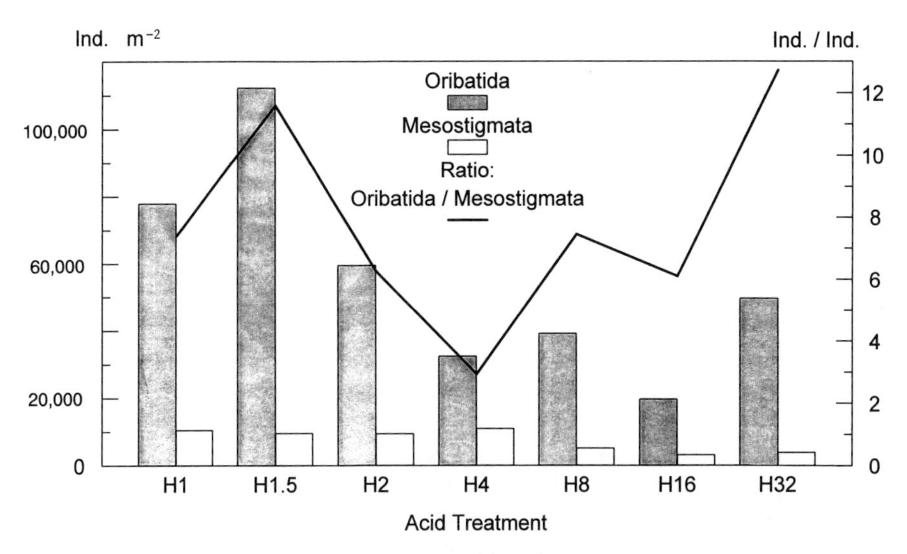

Fig. 3.3. Effect of the application of simulated acid rain containing different amounts of protons to an acid beech forest soil (Solling, West Germany) on Oribatida and Mesostigmata (see Fig. 3.2 for abbreviations of acid treatments) (modified from Heiligenstadt, 1988).

A great number of data supporting these conclusions can be found in the literature cited in Table 3.3.

Alterations of the community organization by acid rain could significantly affect the process level. For example, it was found that the application of simulated acid rain to the soil of a Norway spruce stand affected different functional groups of nematodes in a different way (Hyvönen and Persson, 1990; Persson et al., 1989; approach 4, Section 2): the number of bacterial feeders, omnivores, and predators decreased, while the number of root/fungal feeders remained unchanged or even increased. Though this may primarily be an expression of altered environmental conditions (e.g., microbial populations), the question of to what extent such shifts between different functional groups may cause secondary effects, such as altered turnover of fungal hyphae and/or fine roots, still has to be answered. A serious shortcoming in our present state of knowledge in this field is that most of the studies concerned with acid rain effects on invertebrate communities were carried out at sites with acid soils. The high acid tolerance of the species living in these soils may have overshadowed the effect of increased input rates of protons on the invertebrate community. It is thus not surprising that Hyvönen and Persson (1990) observed a complete recovery of the nematode populations several years after acid rain application.

7.3. Redundancy of the Soil Invertebrate Community

It was stated that acid rain effects at the community level strongly depend on the adaptation of the community to the buffer range of the soil (Section 4.1). Most of the studies carried out on acid soils indicate that the acid induced alterations of soil invertebrate communities were generally not correlated with alterations at the process level. While this may partly be due to the comparatively small contribution of invertebrates to essential soil processes in acid soils, in may also confirm our hypothesis of the importance of acid tolerance and redundancy of the decomposer community for the buffering of acid rain effects in these soils (Section 4.1). However, the acid rain induced alterations of functional effects of soil invertebrates discussed in the following section indicates that these buffer mechanisms may be limited, at least under conditions of acute stress (cf. Section 7.4).

This section focuses on one of the rare examples of a loss of redundancy as a possible cause of strong acid rain effects in base-rich soils and its implications for soil processes under field conditions: The area close around the tree trunk (approach 3, Section 2). Stem flow leads to an increased input of protons (and other air pollutants) in this area compared to throughfall (cf. Ulrich, 1989). As a consequence, soil acidity increases, litter accumulates, and small islands develop exhibiting moder-like conditions (Schäfer, 1988). Because the formation of mull is largely due to the activity of a few macroinvertebrates this suggests a serious alteration in the composition and the performance of the invertebrate community. Soil zoological studies carried out in a beech forest on calcareous soil with the humus form mull confirmed this hypothesis. It was shown that the density of endogeic earthworms was depressed, while epigeic species clearly dominated near tree trunks (Scheu, 1989). On the other hand saprophagous macroarthropods (Diplopoda or Isopoda), enchytraeids, Collembola, Protura, dipterous larvae, predatory macroarthropods (Araneida, Chilopoda, or Staphylinidae) reached higher densities (Scheu, 1989; Wolters 1989a) than in plots remote from trunks. In addition, indications for a depression of the soil microbial biomass was found at the highest polluted sides of the trees (Wolters, unpublished).

The correlation between the reduction of the biomass of endogeic earthworms and a strong accumulation of litter on the soil surface confirms that there is a low ability of the diverse invertebrate community liveing in the base-rich soils to buffer acid rain effects by redundancy (Section 4.1). This result points to the importance of *keystone species* for quantifying shifts in the community structure at the functional level. The significant increase of macroarthropods, in contrast, is not consistent with our hypotheses. This finding is most surprising, because additional experiments showed that the attractivity of beech litter as a food source for one of the dominating macroarthropod species (*Glomeris marginata*) significantly decreased after a leaching of the litter with acidified water (Fig. 3.4). Generally, the preference of surface dwelling species for the thick organic layer close to tree trunks was largely due to the high water content of this substratum. Acid stress may have simply been avoided by moving to less polluted areas at period of strong proton input. By doing so, the animals were also able to compensate for the reduced availability of Ca in the litter close to the tree trunks. This compensation points to the limitation of investigations along a pollution gradient within one site (Section 2).

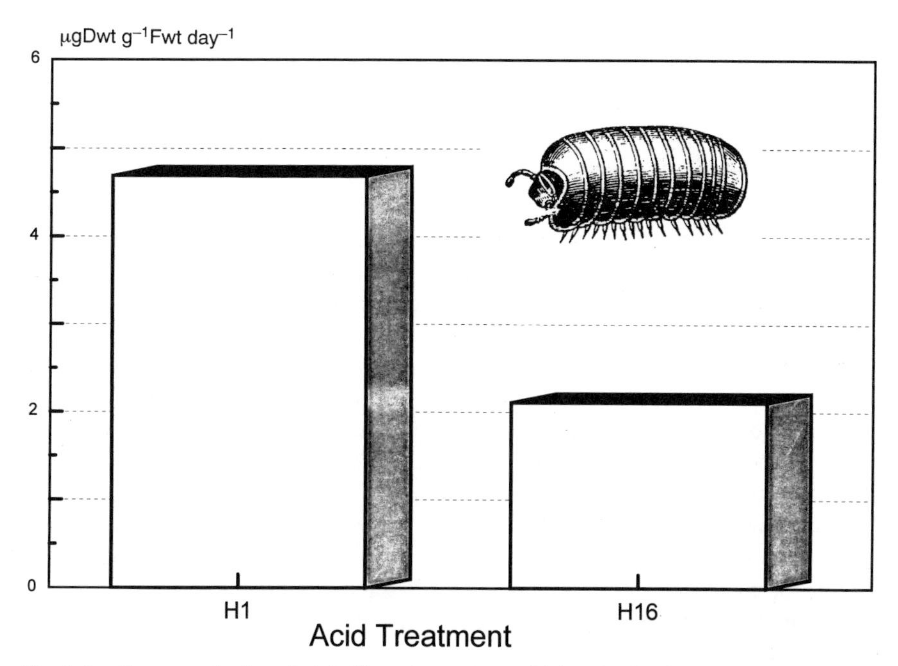

Fig. 3.4. Faeces production of adult *Glomeris marginata* (Diplopoda) feeding on beech leaf litter that had been leached for one week with tap water (H1) or with acidified water (H16; the amount of protons was identical to the H16 treatment described in Fig. 3.2) (Wolters, unpublished).

7.4. Functional Effects of Soil Invertebrates

It has already been stated that soil invertebrates gain most of their ecological importance by functional effects, that is, by the alteration of the microbial growth conditions (Section 3). The quantitative and qualitative outcome of these effects depends on the size and the composition of the invertebrate community, on specific features of the microflora, and on environmental conditions (including climate, resource quality, and soil conditions). There are good reasons to assume that all of these factors are strongly affected by increased acid loading rates of rain. It is surprising that acid rain effects on functional relationships between soil fauna and microflora are a rather neglected field of research.

7.4.1. Acid Soils

Hagvar (1988) carried out microcosm experiments using different microarthropod treatments (*Schwieba* cf. *nova, Isotomiella minor, Mesaphorura yosii,* and a microarthropod mixture). He showed that the presence of one small microarthropod species may increase the dry mass loss of coniferous raw humus and that a simultaneous rise in pH tends to occur. The animal population may have an important stimulating effect on the microflora. While the reason for the pH increase remains unclear, it will most probably have a significant impact on acid rain effects in microsites. We have already pointed to this in the section on nitrification (Section 5.2). There is increasing evidence that heterogeneity resulting from species specific behavioral patterns and interspecific interactions at microsites may contribute significantly to the functioning of soil processes (Anderson, 1987).

It was shown that the collembolan species *Isotoma tigrina* added to soil cores sampled in a beech forest on moder soil reduced C mineralization of the litter colonizing microflora, diminished the biomass and the metabolic quotient of the microflora in the F layer, and accelerated the leaching of mineral N (Wolters, 1991b). Many of these effects were significantly altered by the application of simulated acid rain (pH 2.95). In the acid rain treatments, the negative effect of *I. tigrina* on the mineralization of freshly fallen leaf litter was less pronounced than in the control, while the reduction of the microbial biomass in the F layer was stronger than in the control. In addition, *I. tigrina* increased the metabolic quotient of the microflora and reduced the leaching of mineral N. Most interestingly, for a limited period of time and in contrast to effects otherwise observed, *I. tigrina* stimulated CO_2 production from freshly fallen leaf litter after acid rain had been terminated (Fig. 3.5). This effect was undoubtedly due to the mesofauna mediated transport of microbial propagules into the litter layer and to the deposition of faeces containing Ca-rich particles of hyphae and large amounts of NH_4. This finding suggests that the mesofauna forms an important part of the biological buffering system in the organic layer of acid soils.

Evidence of acid rain effects on functional relationships in acid soils also comes from long-term studies in the field. One example is the study carried out in the beech forests on moder of the Solling area (Ellenberg et al., 1986). The weight of the organic layer of these sites has almost doubled during the last 20 years. There is

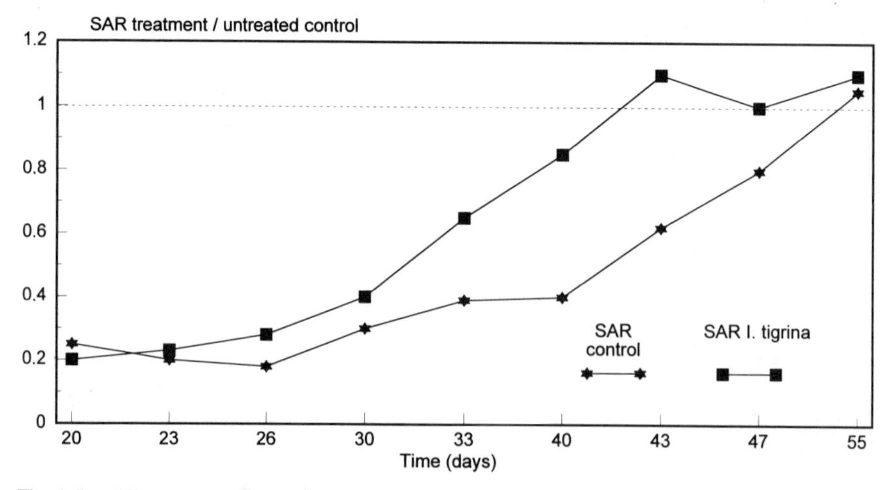

Fig. 3.5. Effect of the collembolan species *I. tigrina* on the recovery of the litter colonizing fauna in a beech forest on moder soil after the termination of simulated acid rain (SAR) (modified from Wolters, 1991).

much evidence that this accumulation of organic material indicates an alteration of decomposition processes caused by acid loading of precipitation and associated trace metal toxicity (Matzner, 1988). Surprisingly, soil zoological investigations revealed that the doubling of the organic layer was not accompanied by significant modification in the structure of the decomposer fauna (Schauermann, 1987). This finding suggests that acid rain altered the time course of litter decomposition indirectly by affecting functional relationships in soil rather than directly by affecting soil organisms.

7.4.2. *Base-Rich Soils*

It was already stated that the macrofauna in base-rich soils plays an important role in buffering acid rain effects on base-rich forest soils by mixing soil and litter (Section 4.1). Experimental evidence for this conclusion will be reported here. The modification of acid rain effects on the decomposition of ^{14}C-labeled beech leaf litter by the macrofauna was studied with microcosms containing calcareous soil (Scheu and Wolters, 1991a,b). Litter fragmentation by the millipede *G. marginata* reduced the acid induced depression of litter mineralization by almost 50%. The burrowing activity of the earthworms *Lumbricus castaneus* and *Octolasion lacteum* led to contact between the intact leaf litter and the mineral soil, which buffered the effect of the acid precipitation on litter decomposition by 82 and 65%, respectively. The epigeic earthworm species *L. castaneus,* by feeding on and removing the fecal pellets of *G. marginata* from the soil surface, almost totally buffered the effect of the acid rain on pellet decomposition. The activity of the endogeic earthworm species *O. lacteum* in feeding on *Glomeris* fecal pellets and mixing them with mineral soil also buffered the effect of the acid rain. However, this effect appeared to be restricted to the first 5 weeks.

Litter bag experiments carried out in a beech forest on mull soil showed that the mesofauna significantly diminished the negative effect of acid rain on C mineralization in a beech forest (Fig. 3.2; Wolters, 1991a). An increased ash content in the litter bags with free access to the mesofauna suggests that this was largely due to a transport of base-rich mineral soil into the litter. As in the acid soil, the mesofauna of base-rich soils may thus also form an important part of the biological buffering system. It should be noted, however, that this effect acts on a completely different level (see above), because it depends on mediating the contact between the well-buffered mineral soil and the litter. Because the reduction in acid rain effects was eliminated when proton input was very strong suggests that the capacity of this buffering system is limited.

8. CONCLUSIONS

It has become evident that acid rain (and the many factors related to it) significantly affects below-ground communities of forest soils. A great number of studies have shown that these alterations may feed up to the process level. Just to name a few:

- The different (sometimes opposite) effects of acid rain on C and N mineralization may significantly alter the nutrient turnover in forest soils.
- The substrate specific effect of acid rain on the decomposition of different C sources can modify both the resource availability to the decomposer organisms and the physicochemical properties of soil organic matter.

The long-term consequences for ecosystem function, however, are still hard to assess. Most probably, the use of different combinations of the approaches listed in Section 2 in interdisciplinary studies is the most promising way of solving this problem. The establishment and/or continuation of long-term monitoring sites is an indispensable prerequisite for an evaluation of the data sets. We also urgently need investigations carried out at pristine sites, which could serve as *zero points* for an evaluation of the data obtained from investigations carried out at more or less polluted sites. Funding organizations should no longer shy away from this seemingly expensive approach. The final pay off will be much bigger than the initial costs.

Buffer ranges of soil and soil specificity in acid rain effects were very important points, which were addressed in this chapter. We stated that soil-independent generalizations about acid rain effects on soil biota will not be possible for principal reasons. Though this seems to be a rather trivial statement, it has almost never been taken into account in the various research programes. We conclude:

- That acid rain may seriously affect seemingly well-buffered soils, because of the low redundancy of the macrofauna community, which is largely responsible for most important characteristics of these soils.
- That the impact of acid rain on poorly buffered soils can significantly be amplified by the destabilization and the restructuring of the decomposer community.
- That acid rain effects on acid soils are most probably much stronger than expected, because essential functional relationships may be altered.

These different effects may reduce forest vitality in a soil specific way, because the acid induced alterations of soil conditions take place at a time scale, which does not allow the trees to adapt. However, much more data are needed to support this hypothesis. The long-term monitoring sites proposed above should be linked to form an international network covering a major gradient of soil conditions, climate, and management practices. As in many fields of ecological research, this comparative approach will enable us to systematically relate acid rain effects on soil biotic processes to the dominant environmental variables (including soil conditions, species specific adaptation patterns, and redundancy of the community). The keystone species concept seems to be an especially fruitful tool in this context. It can help to concentrate the research effort on a comparatively small number of very meaningful questions. The data obtained by this approach will also fit into the models currently developed in the context of other international research programs.

A conceptual framework of future research on acid rain has to take into account three levels of analysis with increasing complexity: structure of the community, role of the biota in ecosystem processes, and properties of the whole system. One major point we made in this chapter is that too little attention has been paid to the different biotic levels affected by acid rain and the linkages between them. The various different parameters used under incompatible conditions made it almost impossible to analyze the causal mechanisms underlying the effects observed. The missing causal analysis hindered the development of a consistent theory for judging the effects of acid stress (and other stress) on soil biotic processes. Though it is inevitably in this case, we know almost nothing about the relationship between the community structure and alterations of functional parameters. The development and testing of soil microbiological methods allowing quantitative studies at the species level is essential for future, more analytical research (this also holds for many invertebrate taxa, especially for those belonging to the microfauna). Though sometimes very artificial, studies on the effect of soil acidification on parameters such as biotic interactions, longevity, fertility, feeding strategies, or resource investment will greatly help us to understand the ecological mechanisms involved. With increasing complexity, these experiments could be related to acid rain effects on the community structure in functionally important compartments of the decomposer subsystem (e.g., the rhizosphere; cf. Section 6.2).

Most of the available data suggest that acid rain effects on soil have shifted many forests away from the intermediate state of ecosystem conditions that most of us generally assume in our predictions of future environmental development. "Acid rain" was one of the main actors on the stage of the environmental research theater during the last decades but has nevertheless been pushed off-scene at the moment. Acid rain has been replaced by "Global change". On a global scale, however, the consequences of climate and land-use change cannot be viewed without considering other anthropogenic stresses such as deforestation or acid deposition (Harte et al., 1992). The acid rain problem may thus return faster into the spotlight than we wish—independent of the emphasis of funding. Similarly to acid rain, for example, the so-called Global change will without doubt change important characteristics of soil organic matter. Both the enormous amount of experience and the large data base gathered during this period of intense acid rain research should therefore be seen as a very useful and versatile basis for future research activities.

ACKNOWLEDGMENTS

We would like to thank Dr. K. Ekschmitt (Mainz) for offering many stimulating suggestions for this paper. We are also grateful to our colleagues Dr. G. Eisenbeis (Mainz), Dr. R.G. Joergensen, and G. Scholle (both Göttingen) for their useful discussion and technical support. Miss A. Klein was very helpful in carrying out the literature survey for Table 3.3.

REFERENCES

Aber JD, Hendry GR, Botkin DB, Francis AJ, Mellilo JM (1982): Potential effects of acid precipitation on soil nitrogen and productivity of forest ecosystems. Water, Air, Soil Pollut 18:405–412.

Abrahamsen G, Bjor K, Horntvedt R, Tveite B (1976): Effects of acid precipitation on coniferous forest ecosystems. In Brakke FH (ed): Impact of Acid Precipitation on Forest and Freshwater Ecosystems in Norway. Oslo: Summary report on the research results from the phase 1 (1972–1975) of the SNSF-Project, pp 37–63.

Abrahamsen G, Hovland J, Hagvar S (1980): Effects of artificial acid rain and liming on soil organisms and the decomposition of organic matter. In Hutchinson TC, Havas M (eds): Effects of Acid Precipitation on Terrestrial Ecosystems. New York: Plenum Press, pp 341–362.

Alexander M (1980a): Effects of acidity on micro-organisms and microbial processes in soil. In Hutchinson TC, Havas M (eds): Effects of Acid Precipitation on Terrestrial Ecosystems. New York: Plenum Press, pp 363–373.

Alexander M (1980b): Effects of acid precipitation on biochemical activities in soil. In Drablos D, Tollan A (eds): International Conference on Ecological Impact of Acid Precipitation. Oslo: SNSF Project, pp 47–52.

Anderson JM (1987): Interactions between invertebrates and microorganisms—noise or necessity for soil processes? In Fletcher M, Gray TRG, Jones JG (eds): Ecology of Microbial Communities. Cambridge: Cambridge University Press, pp 125–145.

Anderson JM (1988): Spatiotemporal effects of invertebrates on soil processes. Biol Fertil Soils 6:216–227.

Anderson JM, Ineson P. (1984): Interactions between microorganisms and soil invertebrates in nutrient flux pathways of forest eco-systems. In Anderson JM, Rayner ADM, Walton DWH (eds): Invertebrate—Microbial Interactions. Cambridge: Cambridge University Press, pp 59–88.

Baath E, Berg B, Lohm U, Lundgren B, Lundquist H, Rosswall T, Söderström B, Wirén A (1980): Effects of experimental acidification and liming on soil organisms and decomposition in a Scots pine forest. Pedobiologia 20: 85–100.

Baath E, Lundgren B, Söderström B (1984): Fungal populations in podzolic soil experimentally acidified to simulate acid rain. Microb Ecol 10:197–204.

Barnes RSK, Calow P, Olive PJW (1993): The Invertebrates–A New Synthesis. Oxford: Blackwell Scientific Publications, 488 pp.

Belotti E (1989): Untersuchungen zur Variabilität und Stabilität von Humusprofilen in Wäldern (unter besonderer Berücksichtigung des anthropogenen Säureeintrags). Ph.D. Thesis, University of Hohenheim, Germany, 183 pp.

Berdén, Nilsson SI, Rosén K, Germund T (1987): Soil Acidification: Extent, Causes, and Consequences. Solna: National Swedish Environment Protection Board Report 3292.

Bewley RJF, Parkinson D (1986): Sensitivity of certain microbial processes to acid deposition. Pedobiologia 29: 73–84.

Bewley RJF, Stotzky G (1983): Simulated acid rain, sulfuric-acid and microbial activity in soil. Soil Biol Biochem 15: 425–430.

Binkley D (1992): Sensitivity of forest soils in the western U.S. to acidic deposition. In Olson RK, Binkley D, Böhm M (eds): The Response of Western Forests to Air Pollution. Ecological Studies 97, Berlin: Springer-Verlag, pp 153–181.

Binkley D, Driscoll C, Allen HL, Schoenberger P, McAvoy D (1989): Acidic Deposition and Forest Soils: Context and Case Studies in the Southeastern United States. Ecological Studies 72, Berlin: Springer-Verlag, 150 pp.

Bitton G, Boylan RA (1985): Effect of acid precipitation on soil microbial activity. 1. Soil core studies. J Environ Qual 14:66–69.

Bonkowski M (1991): Untersuchungen zum Verteilungsmuster der Regenwürmer (Lumbricidae) in einem Buchenwald auf Basdalt und Kalk. Diplomarbeit, University of Göttingen, Germany.

Bormann FH (1987): Landscape ecology and air pollution. In Turner MG (ed): Landscape heterogeneity and disturbance. Berlin: Springer-Verlag, pp 37–57.

Bruck RI, Shafer SR (1984): Effects of acid precipitation on plant diseases. In Linthurst RA (ed): Direct and Indirect Effects of Acidic Deposition on Vegetation. Acid precipitation series, 5 (3), pp 51–63.

Brückmann A, Wolters V (1994): ^{137}Cs in the microbial biomass of forest ecosystems: immobilization and recycling in the organic layer. J Tot Environ, in press.

Chang F-H, Alexander M (1983): Effect of simulated acid precipitation on algal fixation of nitrogen and carbon dioxide in forest soils. Environ Sci Technol 17: 11–13.

Cooke RC, Rayner ADM (1984): Ecology of Saprotrophic Fungi. London and New York: Longman, 415 pp.

Craft CB, Webb JW (1984): Effects of acidic and neutral sulfate salt solutions on forest floor arthropods. J Environ Qual, 13, 436–440.

Dambrine E, Ranger J, Pollier B, Bonneau M (1992): Influence of various stresses on Ca and Mg nutrition of a spruce stand developed on acidic soil. In Teller A, Mathy P, Jeffers JNR (eds): Responses of Forest Ecosystems to Environmental Changes. London, New York: Elsevier Applied Science, pp 465–480.

Dickson RE (1979): Analytical procedures for the sequential extraction of ^{14}C-labelled constituents from leaves, bark and wood of cottonwood plants. Plant 45: 480–488.

Didden WAM (1993): Ecology of Terrestrial Enchytraeidae. Pedobiologia 37: 2–29.

Dzieszkowski A (1988): The gastropod communities of the forest association in Poland and ecological study. Poznan Tow Przyj Nauk Wydz Mat-Przyr Pr Kom Biol 68, 1–117.

Eisenbeis G (1993): Zersetzung im Boden. Inf Natursch Landschaftspfl 6: 53–76.

Eisenbeis G, Feldmann R (1992): Zoologische Untersuchungen zum Status der Bodenfauna im Lennebergwald bei Mainz. Pollichia 23: 521–683.

Ellenberg H, Mayer R, Schauermann J (1986): Ökosystemforschung - Ergebnisse des Sollingprojektes 1966–1986. Stuttgart: Ulmer.

Faber J, Verhoef HA (1991): Functional differences between closely-related soil arthropods with respect to decomposition processes in the presence or absence of pine tree roots. Soil Biol Biochem 23: 15–23.

Focht U (1990): Acidification-induced changes in aerobic metabolism of soil complex. Ekol Pol 38: 257–265.

Francis AJ (1982): Effects of acidic precipitation and acidity on soil microbial processes. Water, Air, Soil Pollut 18: 375–394.

Francis AJ (1986): The ecological effects of acid deposition, part 2. Acid rain effects on soil and aquatic microbial processes. Experientia 42: 455–465.

Funke W, Stumpp J, Roth-Holzapfel M (1985): Soil animals as indicators of forest damage. Verh. Ges. Ökolog. 15: 309–320.

Funke W (1986): Tiergesellschaften im Ökosystem "Fichtenforst" (Protozoa, Metazoa-Invertebrata): - Indikatoren von Veränderungen in Waldökosystemen. Res. Rep. KFK - PEF 9: 1–150.

Garden A, Davies RW (1988): The effects of a simulated acid precipitation on leaf litter quality and the growth of a detritivore in a buffered lotic system. Environ Pollut 52: 303–314.

Gärdenfors U (1987): Impact of airborne pollution on terrestrial invertebrates, with particular reference to molluscs. National Swedish Protection Board, Report 3362: 115 pp.

Ghilarov MS (1983): Some essential physiological characters of soil invertebrates. In Lebrun HM, André A, De Medts C, Grégoire-Wibo C, Wauthy G (eds): New Trends in Soil Biology. Louvain-La-Neuve: Dieu-Brichart, Ottignies, pp 315–320.

Graefe U (1988): Der Einfluss von sauren Niederschlägen und Bestandeskalkungen auf die Enchytraeidenfauna in Waldböden. Manuscript.

Grant IF, Bancroft K, Alexander M (1979): Effect of SO_2 and bisulfite on heterotrophic activity in an acid soil. Appl Environ Microbiol 38: 78–83.

Gunnarsson B, Johnson J (1989): Effects of simulated acid rain on growth rate in a spruce living spider. Environ Pollut 56: 311–317.

Hagvar S (1984a): Ecological studies of microarthropods in forest soils, with emphasis on relation to soil acidity. Ph.D. Thesis, University of Oslo, Norway.

Hagvar S (1984b): Effects of liming and artificial acid rain on Collembola and Protura in coniferous forest. Pedobiologia 27: 341–354.

Hagvar S (1987a): What is the importance of soil acidity for the soil fauna? Fauna (Oslo) 40: 64-72.

Hagvar S (1987b): Why do collemboles and mites react to changes in soil acidity? Ent Medd 55: 115–120.

Hagvar S (1988): Decomposition studies in an easily-constructed microcosm: Effects of microarthropods and varying soil pH. Pedobiologia 31: 293–303.

Hagvar S (1990): Reaction to soil acidification in microarthropods: Is competition a key factor? Biol Fertil Soils 9: 178–181.

Hagvar S, Abrahamsen G (1977): Effect of artificial acid rain on Enchytraeidae, Collembola and Acarina in coniferous forest soil, and on Enchytraeidae in *Sphagnum bog*—preliminary results. Ecol Bull (Stockholm) 25:568–570.

Hgavar S, Abrahamsen G (1980): Colonisation by Enchytraeidae, Collembola and Acari in sterile soil samples with adjusted pH levels. Oikos 34: 245–258.

Hagvar S, Fjellberg A, Klausen FE, Kvamme T, Refseth D (1980): Eksperimentelle forsuringsforsok i skog. 7. Virkning av syrebehandling pa insekter og edderkopper i skogbunnern. - SNSF-projekt IR68/80, Oslos-As, p. 24.

Hagvar S, Amundsen TC (1981): Effects of liming and artificial acid rain on the mite (Acari) fauna in coniferous forest. Oikos 37: 7–20.

Hagvar S, Kjondal BR (1981a): Effects of artificial acid rain on the microarthropod fauna in decomposing birch leaves. Pedobiologia 22: 409–422.

Hagvar S, Kjondal BR (1981b): Decomposition of birch leaves: dry weight loss, chemical changes, and effects of artificial acid rain. Pedobiologia 22:232–245.

Hagvar S, Kjondal BR (1981c): Succession, diversity and feeding habits of microarthropods in decomposing birch leaves. Pedobiologia 22:385–408.

Harte J, Torn M, Jensen D (1992): The nature and consequences of indirect linkages between climate change and biological diversity. In Peters RL, Lovejoy TE (eds): Global Warming and Biological Diversity. New Haven, CT: Yale University Press, 386pp.

Hartmann P, Scheidler M, Fischer R (1989): Soil fauna comparisons in healthy and declining spruce forests. In Schulze E-D, Lange OL, Oren (eds): Forest Decline and Air Pollution. Berlin: Springer-Verlag.

Haynes RJ, Swift RS (1986): Effects of soil acidification and subsequent leaching on levels of extractable nutrients in soil. Plant Soil 95: 327–336.

Heiligenstadt B (1988): Zur Ökologie der Hornmilben (Acari: Oribatida) in einem Moder-Buchenwald: Phänologie und räumliches Verteilungsmuster. Diplomarbeit, University of Göttingen, Germany.

Heungens A (1984): The influence of some acids, bases and salts on the enchytraeid population of a pine litter substrate. Pedobiologia 26: 137–141.

Hogervorst RF, Verhoef HA, van Straalen NM (1993): Five-year trends in soil arthropod densities in pine forests with various levels of vitality. Biol Fert Soils 15:189–195.

Hovland J (1981): The effect of artificial acid rain on respiration and cellulase activity in Norway spruce needle litter. Soil Biol Biochem 13: 23–26.

Hovland J, Abrahamsen G, Ogner G (1980): Effects of artificial acid rain on decomposition of spruce needles and on leaching and on mibilization of elements. Plant Soil 56, 365–378.

Huang P (1985): Belastung und Belastbarkeit streuzersetzender Tiere durch Deposition von Luftverunreinigungen in Waldökosystemen (Labortests). Querschnittsseminar "Bioindikation", 28./29. November, Berlin, UBA.

Huhta V (1984): Response of *Cognettia sphagnetorum* (Enchytraeidae) to manipulation of pH and nutrient status in coniferous forest soil. Pedobiologia 27: 245–260.

Huhta V, Hyvönen R, Koskenniemi A, Vilkamaa P (1983): Role of pH in the effect of fertilization on Nematoda, Oligochaeta and microarthropods. In Lebrun HM, André A, de Medts C, Grégoire-Wibo C, Wauthy G (eds): New Trends in Soil Biology. Louvain-La-Neuve: Dieu-Brichart, Ottignies, pp 61–73.

Huhta V, Hyvönen R, Koskenniemi A, Vilkamaa P, Kaasalainen P, Sulander M (1986): Response of soil fauna to fertilization and manipulation of pH in coniferous forests. Acta For Fenn 195: 1–30.

Hyvönen R, Persson T (1990): Effects of acidification and liming on feeding groups of nematodes in coniferous forest soils. Biol Fertil Soils 9: 205–210.

Jaeger G, Eisenbeis G (1984): pH dependent absorption of solutions by the ventral tube of *Tomocerus flavescens* (Insecta, Collembola). Rev Ecol Biol Sol 21: 519–532.

Jenny H (1941): Factors of Soil Formation. New York: McGraw-Hill, 281 pp.

Joergensen RG (1987): Flüsse, Umsatz und Haushalt der postmortalen organischen Substanz und ihrer Stoffgruppen in Streudecke und Bodenkörper eines Buchenwald-Ökosystems auf Kalkgestein. Göttinger Bodenkundliche Berichte 91, pp 1–409.

Jones ARC, Hendershot W (1988): Dieback of maples in Canada. Its development and some corrective measures. Rev For Fr (Nancy), 40: 20–27.

Kauppi P, Antilla P, Kenträmies K (1990): Acidification in Finland. Berlin: Springer-Verlag, 1237 pp.

Kelly JM, Strickland RC (1984): CO_2 efflux from deciduous forest litter and soil response to acid rain treatment. Water, Air, Soil Pollut 23: 431–440.

Kelly JM, Strickland RC (1987): Soil nutrient leaching in response to simulated acid rain treatment. Water, Air, Soil Pollut 34: 167–181.

Killham K, Wainwright M (1981): Deciduous leaf litter and cellulose decomposition in soil exposed to heavy atmospheric pollution. Environ Pollut (A) 26: 79–85.

Killham K, Firestone MK, McColl JG (1983): Acid rain and soil microbial activity: effects and their mechanisms. J Environ Qual 12: 133–137.

Kjoller A, Struwe S (1982): Microfungi in ecosystems: fungal occurrence and activity in litter and soil. Oikos 39: 389–422.

Klein TM, Alexander M (1986): Effect of the quantity and duration of application of simulated acid precipitation on nitrogen mineralization and nitrification in a forest soil. Water, Air, Soil Pollut 28: 309–318.

Koenies H (1985): Über die Eigenart der Mikrostandorte im Fussbereich der Altbuchen unter besonderer Berücksichtigung der Schwermetallgehalte in der organischen Auflage und im Oberboden. Berichte des Forschungszentrums Waldökosysteme/Waldsterben 9: 288.

Koskenniemi A, Huhta V (1986): Effects of fertilization and manipulation of pH on mite (Acari) populations of coniferous forest soil. Rev Ecol Biol Sol 23: 271–286.

Larkin RP, Kelly JM (1987): Influence of elevated ecosystem sulfur levels on litter decomposition and mineralization. Water, Air, Soil Pollut 34: 425–428.

Laverack MS (1963): The physiology of earthworms. London: Pergamon Press, 206pp.

Lawton JH, Brown VK (1993): Redundancy in ecosystems. In Schulze E-D, Mooney HA (eds): Biodiversity and Ecosystem Function Berlin: Springer-Verlag, pp 255–270.

Lee JJ, Weber DE (1983): Effects of sulfuric acid rain on decomposition rate and chemical element content of hardwood leaf litter. Can J Bot 61: 872–879.

Lee KE (1985): Earthworms. Sydney Academic Press, Australia.

Leetham JW, McNary TJ, Dodd JL, Lauenroth WK (1980): Response of soil nematodes, rotifers and tradigrades to three levels of season-long sulfur dioxide exposures. Water, Air, Soil Pollut 17: 343–356.

Lettl A (1985): SO_2 pollution 1. Influence of acidification on bacterial communities of forest soils. Ekologia (CSSR) 4: 19–31.

Likens GE (1989): Some aspects of air pollution effects on terrestrial ecosystems and prospects for the future. Ambio 18: 172–178.

Lohm U, Larsson K, Nömmik H (1984): Acidification and liming of coniferous forest soil: long-term effects on turn-over rates of carbon and nitrogen during an incubation experiment. Soil Biol Biochem 16: 343–364.

Ma WC (1984): Sublethal toxic effects of copper on growth, reproduction and litter breakdown activity in the earthworm *Lumbricus rubellus,* with observations on the influence of temperature and soil pH. Environ Pollut (A) 33: 207–219.

Makeschin F (1991): Auswirkungen von saurer Beregnung und Kalkung auf die Regenwurmfauna (Lumbricidae: Oligochaeta) im Fichtenaltbestand Höglwald. In Kreutzer K, Göttlein A (eds): Ökosystemforschung Höglwald. Hamburg, Berlin: Parey, pp 117–127.

Makeschin F, Habereder U (1991): Einfluss von saurer Beregnung und Kalkung auf die Laufkäfer (Carabidae: Coleoptera) im Fichtenaltbestand Höglwald. In Kreutzer K, Göttlein A (eds): Ökosystemforschung Höglwald. Hamburg, Berlin: Paul Parey, pp 128–134.

Mancinelli RL (1986): Alpine tundra soil bacterial responses to increased soil loading rates of acid precipitation nitrate and sulfate front range Colorado USA. Arct Alp Res 18: 269–276.

Matzner E (1988): Der Stoffumsatz zweier Waldökosysteme im Solling. Ber Forschungszentrums Waldökosysteme (A) 40: 1–217.

McFee WW, Cronan CS (1982): The action of wet and dry deposition components of acid precipitation on litter and soil. In D'Itri (ed): Acid Precipitation—Effects on Ecological Systems. Michigan: Ann Arbor Science Publications, pp 435–451.

Meyer FH (1985): Einfluss des Stickstoff-Faktors auf den Mykorrhizabesatz von Fichtensämlingen im Humus einer Waldschadensfläche. Allg Forstz 40: 208–219.

Mills LS, Soulé E, Doak DF (1993): The keystone-species concept in ecology and conservation. Bioscience 43: 219-224.

Moloney KA, Stratton LJ, Klein RM (1983): Effects of simulated acidic, metal-containing precipitation on coniferous litter decomposition. Can J Bot 61: 3337–3342.

Myrold DD, Nason GE (1992): Effect of acid rain on soil microbial processes. Environmental Microbiology, 1992. Wiley–Liss, Inc. pp 59–81.

Nosek J (1982): Indikationsbedeutung der Proturen. Pedobiologia 21: 249–253.

Novick NJ, Klein TM, Alexander M (1984): Effect of simulated acid precipitation on nitrogen mineralization and nitrification in forest soils. Water, Air, Soil Pollut 23: 317–330.

Olson RK, Binkley D, Böhm M (1992): The Response of Western Forests to Air Pollution. Ecological Studies 97, Berlin: Springer-Verlag, 532 pp.

Padan E (1984): Adaptation of bacteria to external pH. In Klug MJ, Reddy CA (eds): Current Perspectives in Microbial Ecology. Washington, DC: American Society for Microbiology, pp 49–55.

Paul EA, Clark FE (1989): Soil Microbiology and Biochemistry. San Diego: Academic Press.

Persson T, Lundquist H, Wirén A, Hyvönen R, Wessén B (1989): Effects of acidification and liming on carbon and nitrogen mineralization and soil organisms in mor humus. Water, Air, Soil Pollut 45: 77–96.

Peterson H, Luxton M (1982): A comparative analysis of soil fauna populations and their role in decomposition processes. Oikos 39: 287–388.

Rechcigl JE, Sparks DL (1985): Effect of acid rain on the soil environment: a review. Commun Soil Sci Plant Anal 16: 653–680.

Reichle DE (1977): The role of soil invertebrates in nutrient cycling. In Lohm U, Persson T (eds): Soil Organisms as components of Ecosystems. Ecol Bull (Stockholm) 25: 145–156.

Rejsek K (1991): Acid phosphomonoesterase activity of ectomycorrhizal roots in Norway spruce pure stands exposed to pollution. Soil Biol Biochem 23: 667–671.

Runge M (1974): Die Stickstoff-Mineralisation im Boden eines Sauerhumus-Buchenwaldes II. Die Nitratproduktion. Oecol Plant 9: 219–230.

Ruschmeyer OR, Schmidt EL (1958): Cellulose decomposition in soil burial beds, II. Cellulolytic activity as influenced by alteration of soil properties. Appl Microbiol 6: 115–120.

Rutherford GK (1984): Toxic effects of acid rain on aquatic and terrestrial ecosystems. Can J Physiol Pharmacol 62: 986–990.

Schäfer H (1988): Auswirkungen der Deposition von Luftschadstoffen auf die Streuzersetzung in Waldökosystemen. Berichte des Forschungszentrums Waldökosysteme (A) 37, 243 pp.

Schaefer M (1989): Effect of acid deposition on soil animals and microorganisms: influence on structures and processes. In Ulrich B (ed): International Congress of Forest Decline Research: State of Knowledge and Perspectives. Karlsruhe: Kernforschungszentrum, pp 415–430.

Schaefer M (1990): The soil fauna of a beech forest on limestone: Trophic structure and energy budget. Oecologia 82: 128–136.

Schaefer M, Schauermann J (1990): The soil fauna of beech forests: comparison between a mull and a moder soil. Pedobiologia 34: 299–314.

Schauermann J (1987): Tiergesellschaften der Wälder im Solling unter dem Einfluss von Luftschadstoffen und künstlichem Säure- und Düngereintrag. Gesellschaft für Ökologie, Verhandlungen Bd 16: 53–62.

Scheu S (1989): Die saprophage Makrofauna (Diplopoda, Isopoda und Lumbricidae) in Lebensräumen auf Kalkgestein: Sukzession und Stoffumsatz. Ph.D. Thesis, University of Göttingen, Germany.

Scheu S, Wolters V (1991a): Influence of fragmentation and bioturbation on the decomposition of C-14-labelled beech leaf litter. Soil Biol Biochem 23: 1029–1034.

Scheu S, Wolters V (1991b): Buffering of the effect of acid rain on decomposition of C-14-labelled beech leaf litter by saprophagous invertebrates. Biol Fertil Soils 11:285–289.

Schulze E-D, Lange OL, Oren R (1989): Forest Decline and Air Pollution. Berlin: Springer-Verlag.

Shafer SR (1988): Influence of ozone and simulated acidic rain on microorganisms in the rhizosphere of *Sorghum*. Environ Pollut 51: 131–152.

Shriner DS (1977): Effects of simulated acid rain on host-parasite interactions in plant diseases. Phytopathology 68: 213–218.

Stachurska-Hagen T (1980): Effects of acidification and liming on some soil animals: Protoza, Rotifers and Nematoda. SNSF project IR 74/80, Norwegian Institute for Forest Research, As-NLH.

Strayer RF, Alexander M (1981): Effects of simulated acid rain on glucose mineralization and some physicochemical properties of forest soils. J Environ Qual 10: 460–465.

Strayer RF, Lin C-J, Alexander M (1981): Effect of simulated acid rain on nitrification and nitrogen mineralization in forest soils. J Environ Qual 10: 547–551.

Stroo HF, Alexander M (1986): Role of soil organic matter in the effect of acid rain on nitrogen mineralization. Soil Sci Soc Am J 50: 1219–1223.

Stuanes AO, van Miegroet H, Cole DW, Abrahamsen G (1992): Recovery from acidification. In Johnson DW, Lindberg SE (eds): Atmospheric Deposition and Forest Nutrient Cycling. Ecological Studies 91, Berlin: Springer-Verlag, pp 467–494.

Swift MJ (1987): Organization of assemblages of decomposer fungi in space and time. In Gee JHR, Giller PS (eds): Organization of communities, past and present. Oxford: Blackwell Scientific Publications, pp 229–253.

Tamm CO (1976): Acid precipitation: biological effects in soil and on forest vegetation. Ambio 5: 235–238.

Tamm CO, Wiklander G, Popovic B (1977): Effects of application of sulfuric acid on poor pine forests. Water Air Soil Pollut 8: 75–87.

Timans, U (1986): Einfluss der sauren Beregnung und Kalkung auf die Nematodenfauna. Forstwiss Central 105: 335–337.

Ulrich B (1987): Stability, elasticity, and resilience of terrestrial ecosystems with respect to matter balance. In Schulze E-D, Zwölfer H (eds): Potentials and Limitations of Ecosystem Analysis, Berlin: Springer-Verlag. pp 50–67.

Ulrich B (1989): Effects of acid deposition on forest ecosystems in Europe. Adv Environ Sci 4.

Usher MB (1985): Population and community dynamics in the soil ecosystem. In Fitter AH (ed): Ecological interactions in soil. Oxford, London: Blackwell Scientific Publications, pp 243–265.

van Loon GW (1984): Acid rain and soil. Can J Physiol Pharmacol 62: 991–997.

van Straalen NM, Geurs M, van der Linden JM (1987): Abundance, pH-preference and mineral content of Oribatida and Collembola in relation to vitality of pine forests in the Netherlands. In Perry R, Harrison RM, Bell JNB, Lester JN (eds): Acid Rain Scientific and Technical Advances. London: Publications Division, Selper Ltd.

van Straalen NM, Kraak MHS, Dennemann CAJ (1988): Soil microarthropods as indicators of soil acidification and forest decline in the Veluwe area, the Netherlands. Pedobiologia 32: 47–55.

Vilkamaa P, Huhta V (1986): Effects of fertilization and pH on communities of Collembola in pine forest soil. Ann Zool Fennici 23: 167–174.

von Lützow M, Zelles L, Scheunert I, Ottow JCG (1992): Seasonal effects of liming, irrigation, and acid precipitation on microbial biomass N in a spruce (*Picea abies* L.) forest soil. Biol Fertil Soils 13: 130–134.

Wainwright M (1980): Effect of exposure to atmospheric pollution on microbial activity in soil. Plant Soil 55: 199–204.

Wallwork JA (1976): The Distribution and Diversity of Soil Fauna. London, New York: Academic Press.

Wang CJK, Ziobro RJ, Setliff DL (1980): Effects of acid precipitation on microbial populations of forest litter and soil. In Actual and potential effects of acid precipitation on a forest ecosystem in the Adirondack Mountains. ERDA-80, pp 5-1–5-60. Albany, New York: New York State Energy and Development Authority.

Wellburn A (1988): Air Pollution and Acid Rain. New York: Wiley.

Weber R, Makeschin F (1991): Einfluss von saurere Beregnung und Kalkung auf die oberflächenaktiven Collembolen. In Kreutzer K, Göttlein A (eds), Ökosystemforschung Höglwald. Hamburg, Berlin: Parey, pp 134–143.

White JW, Holbech FJ, Jeffries CD (1949): Cellulose decomposing power in relation to reaction of soils. Soil Sci 68: 229–235.

Will ME, Graetz DA, Roof BS (1986): Effect of simulated acid precipitation on soil microbial activity in a typic quartzipsamment. J Environ Qual 15: 399–403.

Wolters V (1989a): Die Wirkung der Bodenversauerung auf Protura, Diplura und Collembola (Insecta, Apterygota)—Untersuchungen im Stammfussbereich von Buchen. Jber Naturwiss Ver Wuppertal 42: 45–50.

Wolters V (1989b): The influence of omnivorous elaterid larvae on the microbial carbon cycle in different forest soils. Oecologia 80: 405–413.

Wolters V (1991a): Effects of acid rain on leaf litter decomposition in a beech forest on calcareous soil. Biol Fertil Soils 11: 151–156.

Wolters V (1991b): Biological processes in two beech forest soils treated with simulated acid rain—a laboratory experiment with *Isotoma tigrina* (Insecta, Collembola). Soil Biol Biochem 23: 381–390.

Wolters V (1991c): Soil invertebrates—effects on nutrient turnover and soil structure. Z Pflanzenernähr Bodenkd 154: 389–402.

Wolters V, Joergensen RG (1991): Microbial carbon turnover in beech forest soils at different stages of acidification. Soil Biol Biochem 23: 897–902.

Wolters V, Joergensen RG (1992): Microbial carbon turnover in beech forest soils worked by *Aporrectodea caliginosa* (Savigny) (Oligochaeta, Lumbricidae). Soil Biol Biochem 24: 171–177.

Wood MJ, Rippon JE (1980): Effect of acid precipitation on soil microbial processes: pot experiments. In Second Int Symp on Microbial Ecology, University of Warwick, Coventry, England.

HUMUS DISINTEGRATION AND NITROGEN MINERALIZATION

JOHANNES EICHHORN

Abt. Waldschäden, Hessische Forstliche Versuchsanstalt, D-34346 Hann. Münden, Germany

ALOYS HÜTTERMANN

Abt. Technische Mykologie, Forstbotanisches Institut, Universität Göttingen, Büsgenweg 2, D-37077 Göttingen, Germany

1. INTRODUCTION

Nitrogen (N) holds an exceptional position among plant nutrients. On one hand, nitrogen is the element for which plants have the highest demand. On the other hand, it exhibits the special feature that the nitrogen content of the bedrock minerals is neglegible, and thus of no importance for plant nutrition. Because of this lack of nitrogen stores in minerals, all the nitrogen present in a forest ecosystem must come from the atmosphere. The input of nitrogen was the limiting factor for plant growth in most ecosystems until the beginning of the industrial revolution (Kimmins, 1987). In most forest ecosystems all over the world nitrogen is the limiting factor for biomass productivity (e.g., Krzak, 1981; Tamm, 1991). Nitrogen comes from two sources: atmospheric input, which in unperturbed forests is in the range of about 1–5 kg of N ha^{-1} $year^{-1}$ (Kimmins, 1987), and nitrogen fixation, which in the case of free-living microorganisms amounts to 2–5 kg of N ha^{-1} $year^{-1}$ (Paul and Clark, 1989). Forest ecosystems with an appreciably high degree of symbiotic nitrogen fixation can accumulate much higher amounts of nitrogen [e.g., the values for Alder are 26–300 kg ha^{-1} $year^{-1}$ (Paul and Clark, 1989)]. These ecosystems, however, are only exceptions in temperate forests of the northern hemisphere and will not be discussed in this chapter.

Because of this limited nitrogen budget, terrestrial ecosystems, and especially

Effects of Acid Rain on Forest Processes, pages 129–162

forest ecosystems, developed into systems where a pronounced nitrogen limitation leads to a buildup of organic matter in the soil and a concomitant storage of nitrogen there. During evolution, plants evolved into structures that have an exceptionally high C/N ratio (> 400 in the case of wood). This high C/N ratio allows them to built as much plant material per unit of nitrogen as possible.

The seasonal litter fall acts as an interface between the two major compartments of forest ecosystems: The trees (mostly above ground) and the soil (below ground). The falling leaves and dying fine roots contain very little nitrogen and have a high C/N ratio.

Microorganisms, on the other hand, have rather low C/N ratios (between 10 and 20) in their biomass and a limited efficiency to completely metabolize substrates without a concomitant biomass production. Thus the optimal C/N ratio for microbial substrates is in the range of 20–30. The high C/N ratio of plant litter therefore cannot support a steady and optimal growth of microorganisms. Thus it is detrimental for a rapid degradation of the plant material by them, even in the case of readily degradable carbon sources, and will lead to nitrogen deficiency. This fact alone leads to a retarded degradation of the plant material, and, together with the unavoidable autolysis cycles (Deacon, 1980), results into an immobilization of nitrogen. The latter phenomenon is an inevitable result of the liberation of intracellular radical-forming peroxidases set free by the autolysis. Because of these basic mechanisms of litter mineralization accompanied by autolysis cycles, which are discussed below in more details, most of the nitrogen stores in soils are present in the form of humic substances, which in turn are the result of biological processes accompanying the mineralization of plant material (cf. Schachtschabel et al., 1989).

Any nitrogen that is free and available will inevitably be subject of competition between the above-ground (plant) and below-ground (soil-flora) compartments of a forest ecosystem. Both plant and soil organisms are in dire need of nitrogen and limited in their growth by the available nitrogen supply. Consequently, any ammonium or nitrate present or formed in the soil solution will be taken up immediately by microorganisms (Alexander, 1977) which in turn leads to a high rate of carbon mineralization. The C/N ratio in the soil and the amount of available nitrogen thus determine both the rate of litter mineralization and the availability of nitrogen for plants. It is a well-known fact that the addition of straw (C/N=50 - 100) to soil with a good nitrogen supply will immediately lead to an immobilization of the nitrogen in the microbial population and result in nitrogen deficiency in the plants (Schachtschabel et al., 1989). Melillo et al.(1982) showed that the rate of decomposition of hardwood leaf litter is strictly limited by the amount of available nitrogen.

Thus the fate of nitrogen in a forest ecosystem is closely dependent on its carbon cycling. During the long succession leading to forest ecosystems, starting from virgin soils poor in or free from organic carbon, an enormous amount of organic matter is accumulated in the forest floor, which can amount up to 320 tons of carbon per hectare (Schlesinger, 1977) in temperate and boreal coniferous forests. Concomitantly, nitrogen is stored in the form of "immobile N", that is, nitrogen bound to the humic substances that can amount up to some 12,000 kg of N accumulated and stored in the forest floor. Taking into account the yearly litter fall and turnover of fine roots in the soil, it becomes obvious that the dynamics and fate of the humus are the most important factors controlling both the carbon and the nitrogen regime in forest soils.

In German forests there have been strong indications during the last few decades that the nitrogen balance in forest ecosystems is severely impeded and out of equilibrium (Ulrich, 1981). An example of this phenomenon is shown in Figure 4.1*a*., which shows the nitrate and chloride concentrations of the source Friedrichsaue, that is used by the town Zierenberg, in northern Hessen, for its water supply. From the hydrology of the watershed area it is clear that this aquifer is dominated in its composition by the Zierenberg Stadtwald, a forest ecosystem that definitely would be considered to be well supplied with nutrients and that is in a relatively healthy condition (Eichhorn, 1991; for detailed information see Section 5.3.1 in this chapter). This forest is one of the permanent experimental sites of the Forschungszentrum Waldökosysteme of the University of Göttingen and managed by the Hessische Forstliche Versuchsanstalt. The water from this source has been analyzed continuously since 1965. Since the aquifer is characterized by a very high constancy of the dissolved minerals (Lischeid et al., 1991), the data can be taken as indicators of long-term trends. The steady increase in nitrate outflow together with a decrease in chloride indicate that the forest ecosystem that feeds the aquifer is indeed in a stage of humus disintegration and subsequent N mineralization.

Probably, for the first time in history, a forest ecosystem, which for millenia was adapted to consider N as the limiting element and developed numerous strategies to cope with severe N limitations, is now no longer able to keep and utilize its N stores.

In this chapter, the interrelation between humus degradation and N mineraliza-

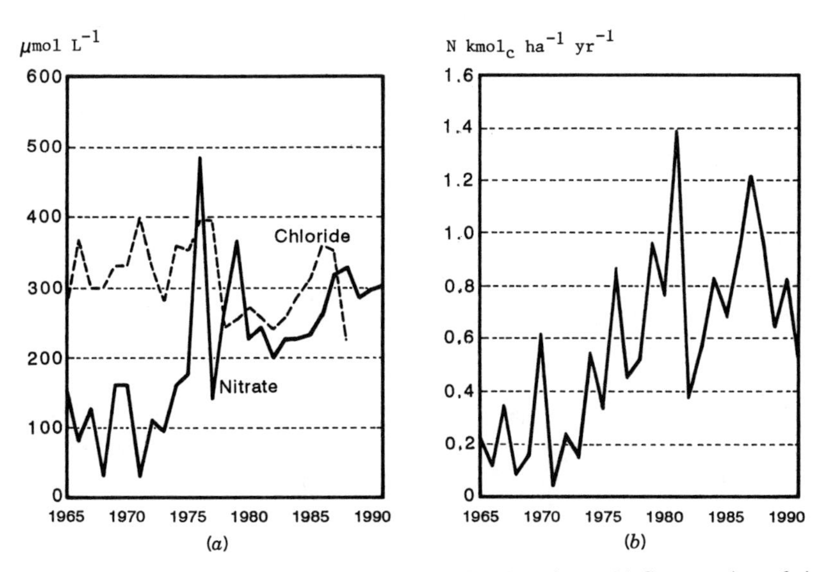

Fig. 4.1. Nitrate output at Friedrichsaue spring in the Zierenberg forest. (*a*) Concentrations of nitrate and chloride over the last 25 years. (*b*) Nitrate flux leaching at Friedrichsaue spring.

tion will be discussed together with their implications for the stability of forest ecosystems.

2. MECHANISMS OF FORMATION OF SOIL ORGANIC MATTER AND ITS DYNAMICS

2.1. Facts To Be Considered

Because of the complexity of humus formation, scientists are still far from being able to understand the reactions that lead to the formation, and finally the degradation, of soil organic matter (for recent reviews see: Haider et al., 1975; Stevenson, 1982; Paul and Clark, 1989). Any model that wants to explain the mechanisms of its dynamics has to consider the following facts:

- Soil organic matter is a highly complex and little defined array of covalently bound molecules of an aromatic and aliphatic nature (Schachtschabel et al., 1989; Paul and Clark, 1989).
- The majority of N present in soils is bound to the humus complex in a form that protects it from rapid microbiological degradation (Stevenson, 1982).
- The different organic compounds present in litter and root debris exhibit extremely different decay characteristics. Whereas under laboratory conditions the first-order decay constant of glucose is about 1 day, the constant of aromatic compounds are considerably longer. The first-order decay constant for lignin under the same conditions was 1 year (cf. Paul and Clark, 1989). The labile macromolecules are degraded and lost during the microbial degradation of the litter, while refractory compounds or biopolymers, such as lignin, cutin, suberins, N-containing heterocyclic, or aliphatic compounds are enriched and transformed to humin (Hatcher and Spike, 1988).
- In soils, even very reactive aromatic compounds of low molecular weight are incorporated fairly quickly into the humus fraction instead of being metabolized (Martin and Haider, 1976, 1977; for a more recent review see Hüttermann et al., 1989b).
- During its residence time in soils, the organic matter is subject to constant metabolic changes that lead to both carbon dioxide (CO_2) mineralization and the transformation to compounds that are even more resistant to microbial attack (cf. Coleman et al., 1983).
- Although the organic matter of soils is rather recalcitrant against microbial degradation and may have long residence times of more than 1000 years even in aerated soils, the bulk of carbon finally is metabolized and mineralized to CO_2 (cf. Ertel, 1988; Anon , 1989). The humic material is in a dynamic state of equilibrium, its synthesis being compensated for by the gradual mineralization of existing material (Stevenson, 1982). If this would not be the case, the carbon cycle of the biosphere would have ceased to function long ago. A

theoretical analysis of the long-term dynamics of C and N in soils, together with the analysis of the situation in Scots pine forest soil was made by Agren and Bosatta (1987).

2.2. Mechanisms of Humus Formation

Although controversial discussions may still be found in the literature, the relevant textbooks and major review articles agree with the following general view of humus formation:

The first step is the breakdown of the organic polymers of the plant material to monomeric constituents, then these are subsequently polymerized by radical mediated processes. During this process, the more readily degradable aliphatic, (i.e., carbohydrate) moieties of the litter are preferentially degraded, whereas the aromatic constituents of the plant material are enriched. This result can at best be seen by analysis of undisturbed soils or composts by cross-polarization–magic angle spinning–nuclear magnetic resonance (CP–MAS–NMR) spectrometry (Fründ and Lüdemann, 1989; Almendros et al., 1991). Concomitantly, the degradation of the aromatic moiety (i.e., lignin) also takes place.

The oxidation of the degradation products of the aromatic plant constituents, the polyphenols to the corresponding quinones, is believed to be a key step in the formation of humic matter in soil (Flaig et al., 1975). This reaction may be a spontaneous chemical autoxidation, or an oxidation catalyzed by microbial enzymes, such as laccases, polyphenoloxidases, and peroxidases. The aromatic rings that serve as building blocks for the humic acid core may originate either from lignin degradation (Christman and Oglesby, 1971), transformation, or from the metabolism of soil microorganisms (e.g., Haider et al., 1975; Atlas and Bartha, 1987; Flaig, 1988; Hedges, 1988; Pfaender, 1988; Ziechmann, 1988; Paul and Clark, 1989).

It is evident from the literature that at least two pathways exist by which phenolic substances can be transformed to humic acids:

1. The abiotic oxidation catalyzed by minerals or metal salts.
2. The enzymatic oxidation by laccase and other oxidative enzymes.

The abiotic polymerization of phenolic compounds was discovered by Huang and his group (e.g., Wang et al., 1978; Wang and Huang, 1987, 1989). In addition, this group was able to demonstrate the catalytic effects of certain oxides of Mn, Fe, Al, and Si to polymerize phenols to polymers (Shindo and Huang, 1984)

Because of the pioneering work of Haider and Martin, ample evidence was accumulated for the notion that phenoloxidases [i.e., peroxidases or polyphenoloxidases (laccases)] have an important function in the structural assembly of the humic acid complex (Haider and Martin, 1967, 1970; Martin et al., 1967, 1972; Martin and Haider, 1969). In their earlier works, the investigators developed the concept that the aromatic structures of the humic acid complex may be derived predominantly from substances secreted and subsequently polymerized by soil organisms, especially soil fungi. In later contributions, they showed that phenolic compounds of nonfungal origin could serve as humic acid precursor as well. Martin et al. (1979), for example, described the decomposition of ^{14}C-labeled catechol and its incorporation into the stable organic fraction of the soil and could relate both phenomena to the action of phenole-oxidizing enzymes. Since then the cross-coupling of phenolic compounds with regard to its importance to humus formation has been studied extensively by several laboratories using both peroxidases (e.g., Berry and Boyd, 1984, 1985; Schnitzer et al., 1984; Simmons et al., 1988) and laccase-type phenoloxidases (Martin and Haider, 1980; Bollag et al., 1980, 1982; for a review see Filip and Preusse, 1985). Suflita and Bollag (1981) showed that the transformation of phenolics to humic acid type substance also could be achieved in vitro by enzymes extracted from soils. Leonowicz and Bollag (1987) extracted and biochemically characterized the laccase activities from soils.

It was demonstrated for the enzyme laccase from white-rot fungi, that, in this special study, during the polymerization of polyphenols, with lignin used as a model phenol, the strength of the binding is extremely high (Haars et al., 1989; Hüttermann et al., 1989b). These investigators were able to use this reaction for the production of wood composites, such as particle boards.

The vigor and chemical reactivity of the phenolic radicals can at best be demonstrated by the fact that even amino acids and sugars like glycine, lysine, cysteine, and glucosamine, are incorporated into the humic acid type substances and, after incorporation, are recalcitrant to rapid microbial attack in soils (Martin and Haider, 1980). Recent results from our group demonstrated that even macromolecules like lignin will incorporate molecules like glucose when laccase and oxygen are present as the catalyst and reactant (Milstein and Hüttermann, 1993).

From the evidence accumulated so far, it is reasonable to assume that the unspecific polymerization of phenols and lignins is the basic mechanism by which the core of the humic acid type soil substances are formed. These reactions are catalyzed either by radical donating enzymes like peroxidase or laccase, ions, or oxides that readily exhibit charge transfers like Mn or Fe, or even oxides from Al or Si and minerals present in the soil. The origin of the aromates may either be the aromatic moiety of the litter and root organic substances, or metabolites from litter decomposing microorganisms. From the chemical background of this reaction it is obvious that a free phenolic group is needed for this polymerization, a fact that can be extracted from virtually all the studies quoted above.

2.3. Mechanisms of Humus Degradation

Although there is little doubt that all the organic substances present in soils eventually are degraded and underly a constant turnover (Ertel, 1988; Schachtschabel et al., 1989), the organisms and enzymes responsible for this process are still more or less obscure.

From the structural models of humic acids (Fig. 4.2) it is obvious that the degradation of this compound is bound to be catalyzed by enzymes that must have similar activities to those described for the lignin degrading white-rot fungi (cf. Kirk and Farrell, 1987). It is a well-accepted fact, however, that these fungi, which are the most vigorous lignin degraders in nature, are not common soil organisms and are unable to compete with the endogenous soil flora (cf. Hüttermann and Haars, 1987). On the other hand, so far no lignolytic activity has been unequivocably demonstrated for soil bacteria (cf. Kirk and Farrell, 1987).

The question thus remains: Which organisms are responsible for the degradation of lignin or humus in soils? One group of microorganisms that are well known and prominent soil organisms, and for which the possibility of being humus degraders has been discussed for almost 50 years, are the mycorrhizal fungi. Norkrans (1950), Falck and Falck (1954), Lyr (1963), and Coleman et al. (1983) discussed the possibility that, under various conditions, mycorrhizal fungi might get their carbohydrates not only from the trees but in addition, might possess the ability to degrade the macromolecules of the litter as well. These researchers came to the conclusion that this feature, if it would play a role in soil metabolism, would change our whole picture of soil dynamics and could account for humus degradation. Trojanowski et al. (1984), using ^{14}C-labeled lignin and lignocellulose, were for the first time able to demonstrate that ectomycorrhizal fungi indeed were able to mineralize even ring-labeled DHP–lignin to CO_2 at a rate comparable to the degradation of this compound by typical white-rot fungi. These findings were confirmed by Oelbe-Farivar (1985) and later by other groups as well (Haselwandter et al., 1990). Until other organisms are known to degrade lignin or humus at an appreciable rate, it appears reasonable to consider mycorrhizal fungi at least as a major group of soil organisms who could contribute to humus degradation.

2.4. Dissolved Organic Carbon in Soil Solutions

Soluble low molecular weight substances in the soil solution are very important factors for the dynamics of humus formation and degradation. Although, judging from our experience with lignin polymerization (cf. Milstein et al., 1989; Kharazipour et al., 1991), some of the reactions leading to humus polymerization can be expected to occur at the surface of the soil matrix evolving water-insoluble partners as well. This dissolved organic carbon (DOC) in the soil solution is an important factor during pedogenesis (Scheffer and Ulrich, 1960; Schachtschabel et al., 1989). A large spectrum of different organic compounds are summarized under the term DOC. By the use of temperature-resolved pyrolysis-field ionization mass spectrometry (MS), Hempfling and Schulten (1990) found the following classes of substances in the DOC fraction from leachates of moder profiles under mixed pine and oak forests (mentioned in the order of their quantitative importance):

- Monomeric and dimeric subunits of lignin.
- Benzene derivatives: phenols and their methoxyl, carboxyl, and alkyl derivates.
- Amino acids, peptides, proteins, *N*-acetylated amino sugars, and other nitrogen-containing compounds.
- Carbohydrates.

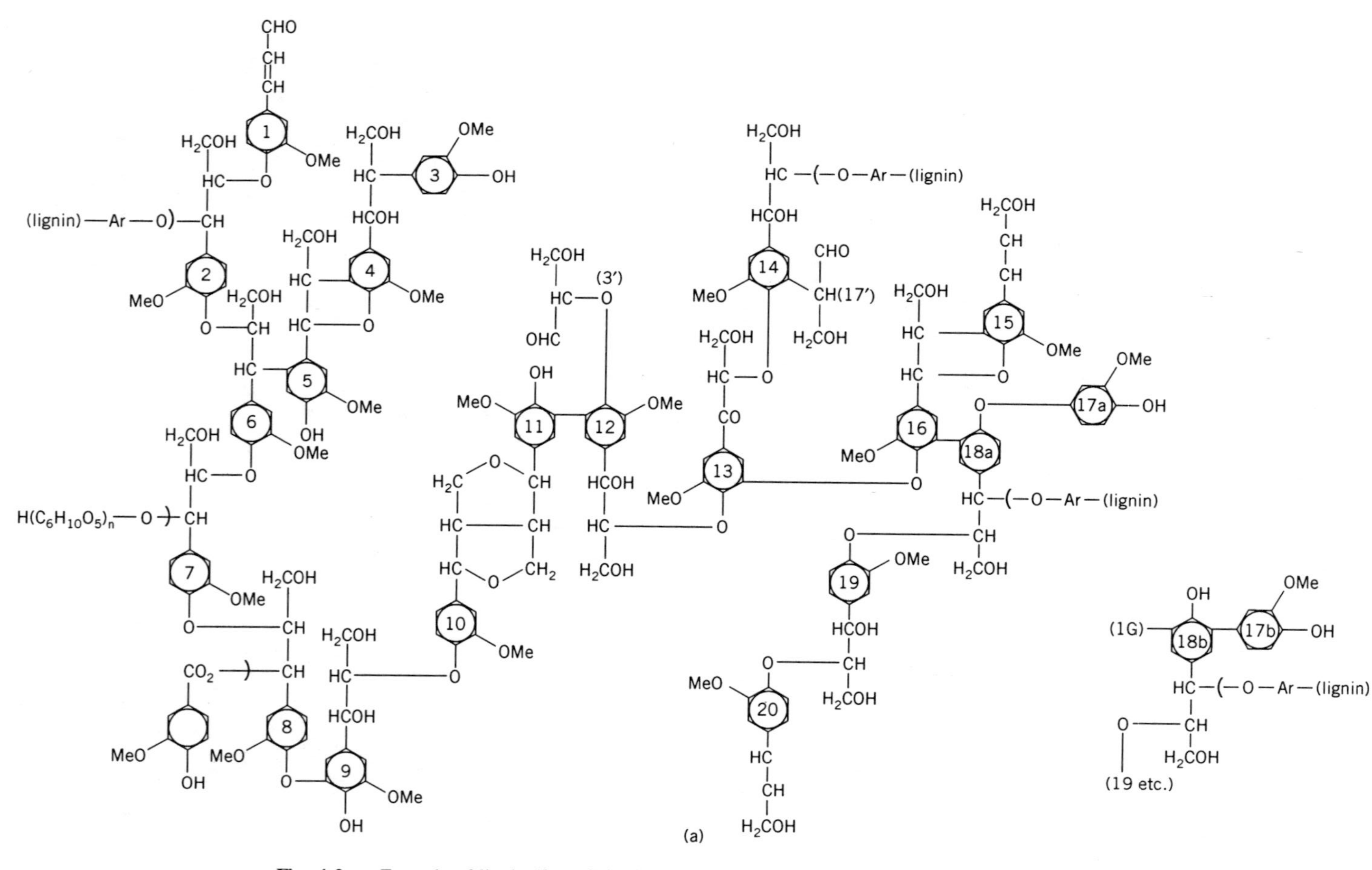

Fig. 4.2a. Formula of lignin (from Schachtschabel et al., 1989, with permission of Enke Verlag).

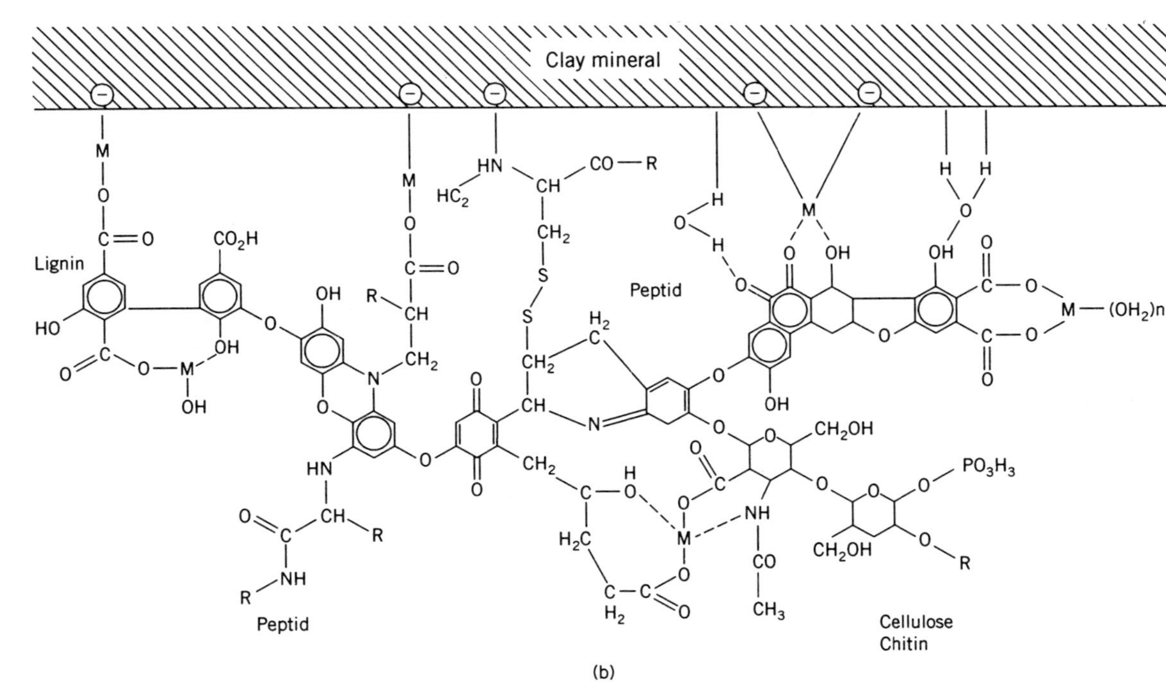

Fig. 4.2b. Formula of humic substance (from Schachtschabel et al., 1989, with permission of Enke Verlag).

These compounds were found together with low amounts of sterols, steranes and tocopheroles, fatty acids, alkanes, and alkenes.

Thus the whole metabolism of the soil biota (animals, plants, and microorganisms) is reflected in the composition of the DOC, the majority of the chemicals detectable being of aromatic nature.

Göttlein (1988) followed a different approach in his thesis. He divided the total organic fraction into two general groups, "polar" and "less polar", which were different according to their molecular weight in high-performance liquid chromatography (HPLC), with the polar compounds having a much smaller size. With this experimental approach, he was able to derive a semiquantitative classification of the DOC compounds, without being able to indicate their structure. The polarity, namely, acidity of the DOC molecules, is governed by the amount of phenolic and carboxyl groups present. According to Cronan (1985) the fraction of polar compounds is higher in more acidic soils. Such analyses are extremely difficult, because of the variation of compounds in different extracts. On the other hand, the extraction method itself is determined to a great extent by the measured composition of the soil solution via the rate of recovery of individual compounds.

All authors working on the composition of the DOC in forest soils, however, agree that sugars and phenols are the main components. In acid soils, the sugar content of the soil solution varies considerably from season to season. This content can, in autumn, amount to two thirds of the total DOC (Göttlein et al., 1991 a,b). These values are considerably higher than those reported by Greenland and Oades (1975) for various forest ecosystems. The variation in composition of the DOC, as measured by Göttlein et al. (1991 a,b), gives some hints concerning the dynamics of the balance between humus formation and disintegration.

2.5. A Model for Formation and Breakdown of Humic Acid-Type Substances

By considering the facts and relevant theories about humus formation and degradation, we can suggest the following model, which is similar to the actual sequence leading to the degradation of lignin by white-rot fungi:

- The construction of the structural backbone of humic acids is catalyzed by enzymes or chemicals which, via oxygen radicals, polymerize phenolic substances. The chemical *conditio sine qua non* for this reaction is a free phenolic group.
- The degradation and subsequent mineralization of the stable organic substance in the soil is performed by enzymes that have properties similar to the ligninolytic system of the white-rot fungi. One group of soil organisms for which such ligninolytic ability was unequivocably shown are ecto- and ericoid mycorrhizae.

The idea that the mechanisms of humus degradation should be similar to the metabolic regulation of lignin degradation is supported by the early and well-established findings that, like lignin degradation, humus degradation is also enhanced by the presence of an easy metabolizable cosubstrate, such as glucose or proteins (for a summary of the early literature see Scheffer and Ulrich, 1960).

In Figure 4.3 a simple model is given for the interrelations of the different polymerization and deploymerization steps involved in humus dynamics. This model was derived from similar considerations concerning the degradation of lignin (Hüttermann, 1989).

2.6. Explanation of the Phenomenon of Humus Disintegration

Based on these considerations and the literature data available at that time for lignin dynamics as outlined above, Ulrich (1981) proposed a hypothesis that could explain the nitrate leaching presently observed in German forests on the basis of the change of the chemical climate in the soils because of acid precipitation (for details see Chapters 1 and 10). His model has the following basic assumptions:

1. The continuous partial degradation and repolymerization of the soils organic matter, in other words, humus dynamics is based upon two different enzymatic reactions:

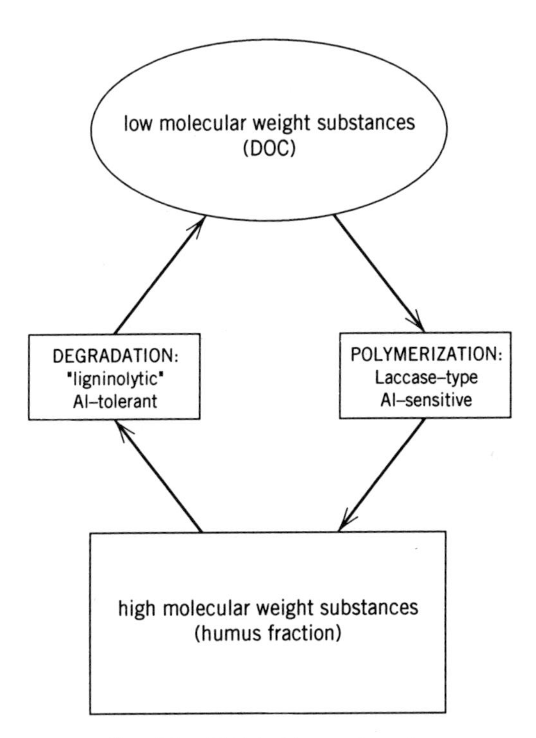

Fig. 4.3. Model of humus dynamics.

(a) A polymerization by laccase-type enzymes.

(b) An enzymatic degradation by enzymes similar to the known lignolytic systems of white-rot fungi.

2. The polymerizing enzymes have properties similar to fungal laccase, having a pH-optimum at about pH 5 and requiring a free phenolic OH group. Thus, it is sensitive both to low pH and/or OH complexing ions like Al and Fe (III).
3. The humus degrading enzyme system has properties similar to the ligninolytic system of white-rot fungi (i.e., it has a pH optimum at pH 3). Because of the pK_a values of the phenols in question, there should be much less sensitivity of the enzyme towards phenole-complexing ions.

In the course of acid deposition, two major changes take place in forest soils. First, a decrease of the pH and concomitantly a liberation of Al and Fe cations that complex the free phenolics. This result means that the polymerization of the phenolics is inhibited, whereas the degradation of the humic acids is not influenced by these early events of acid deposition. Ultimately, this will lead to a disturbance in the balance between humus degradation and repolymerization, shifting the equilibrium in the direction of humus degradation.

Indicators of this phenomenon, which were coined by Ulrich as humus disintegration, are two very important changes in the chemistry of the soil solution:

- A release of high amounts of soluble organic carbon (DOC).
- A release of nitrogen, in the form of nitrate, into the soil solution.

2.7. Experimental Evidence in Favor of the Theory

2.7.1. *Sensitivity of the Reaction of Laccase to Aluminium Ions*

Jentschke (1984) in his masters thesis, studied the influence of the presence of Al^{3+} ions on the oxidation of phenols catalyzed by laccase. In the first part of his study, he investigated the formation of an Al complex by a variety of phenolic substances. Using differential spectroscopic techniques, he was able to determine the degree of complexation of a given compound by the Al ions. In the second part, he found that only in the case of phenolic compounds, that form complexes with Al ions is the laccase-catalyzed oxidation inhibited by the presence of Al ions. Figure 4.4 shows the relationship between the rate of enzymatic oxidation and its inhibition by Al ions. This finding demonstrates that Al concentrations, which are present in solutions of acidified soils, readily inhibit the enzymatically catalyzed polymerization of phenols bound to it.

It is well known that Al (or Fe) complexes of phenolic compounds are present in soil solutions (Behmel and Ziechmann, 1987) or acid waters (Tipping and Backes, 1988; Tipping et al., 1991). Spiteller (1985) measured the molecular weight distribution of the dissolved organic carbon in the equilibrium soil solution taken from forest sites in the Harz and found that for all the measured molecular weights Al was present at a concentration that indicated a complete stochiometric binding of this ion

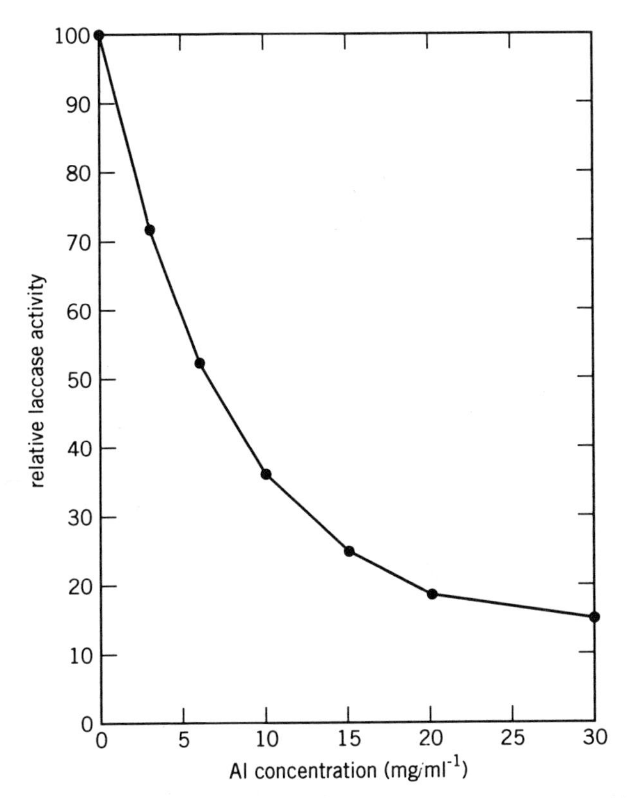

Fig. 4.4. Inhibition of laccase activity by Al ions (taken from Jentschke, 1984, with permission).

to all free phenolic groups. Therefore, it appears reasonable to assume that the presence of Al or Fe ions in acid soils, together with a lower pH, will inhibit the action of laccase, thus blocking the repolymerization of the humic substances.

2.7.2. *Influence of Aluminium Ions on the Degradation of Lignin by Ectomycorrhizal Fungi*

The influence of the following chemical parameters on lignin degradation by different species of ectomycorrhizal fungi was studied by Oelbe-Farivar (1985). The following fungi were studied: *Cenococcum geophilum, Rhizopogon luteolus, Suillus variegatus,* and *Tricholoma aurantiacum.* As shown in Figures 4.5 and 4.6, for a typical fungus (*C. geophilum*) a lowering of the pH value of the solution from 5 to 3 resulted in a marked increase of the lignolytic activity (Fig. 4.5) and Al concentrations of up to 40 mg L^{-1} had no significant inhibitory effect on the lignolytic activity (Fig. 4.6). No significant influence of nitrogen supply on the lignolytic activities of all tested fungi was detectable (Oelbe-Farivar, 1985). Apparently, the mycorrhizal fungi studied in this thesis, unlike the situation reported for the well-studied white-rot fungus *Phanaerochaete chrysosporium* (Kirk et al., 1978), exhibit

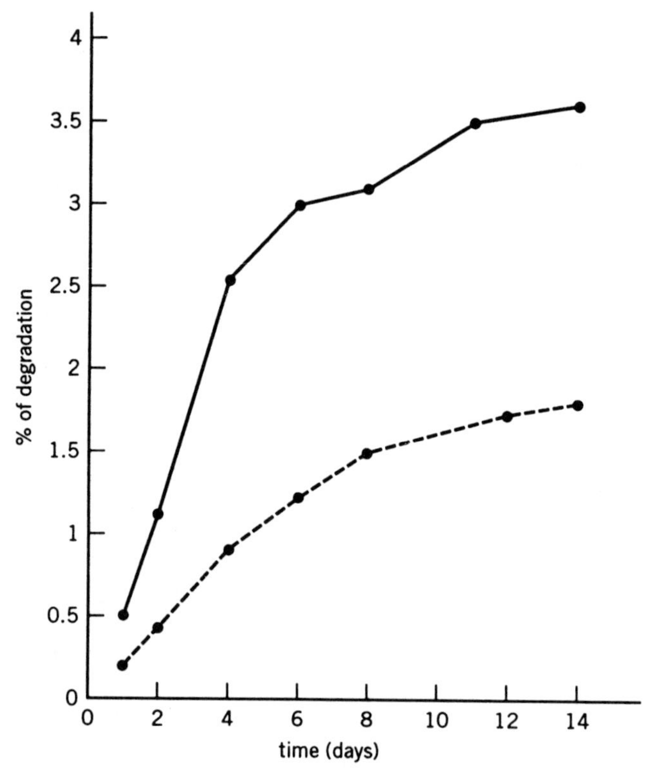

Fig. 4.5. Influence of pH on ligin degradation by *C. geophilum*. Solid line, pH 5.5; dashed line, pH 3.0 (from Oelbe-Farivar, with permission).

the same regulation of lignolytic enzymes with regard to N as was observed earlier for the forest pathogen *Heterobasidion annosum* (Hüttermann et al., 1984).

The conclusion to be drawn after summarizing the experimental evidence available is that the hypothesis of Ulrich (1981) concerning the mechanism of humus disintegration to date has not been disproven. Hopefully, this phenomenon, which appears to play a very important role in the forest ecosystems of the northern hemisphere, will in the future attract the attention of experimental soil scientists. We expect that when more data become available the mechanism of humus disintegration will become clearer.

3. SHORT-TERM ACIDIFICATION PUSHES

Northern and central Europe are characterized by a cool and wet climate, where temperature is the factor that limits microbial activity and mineralization in soils. A dry and warm year influences an ecosystem not only by drought, but also by increased mineralization. If the amount of nitrate that is formed by nitrogen mineralization cannot be taken up by plants, the result will be a temporary accumulation

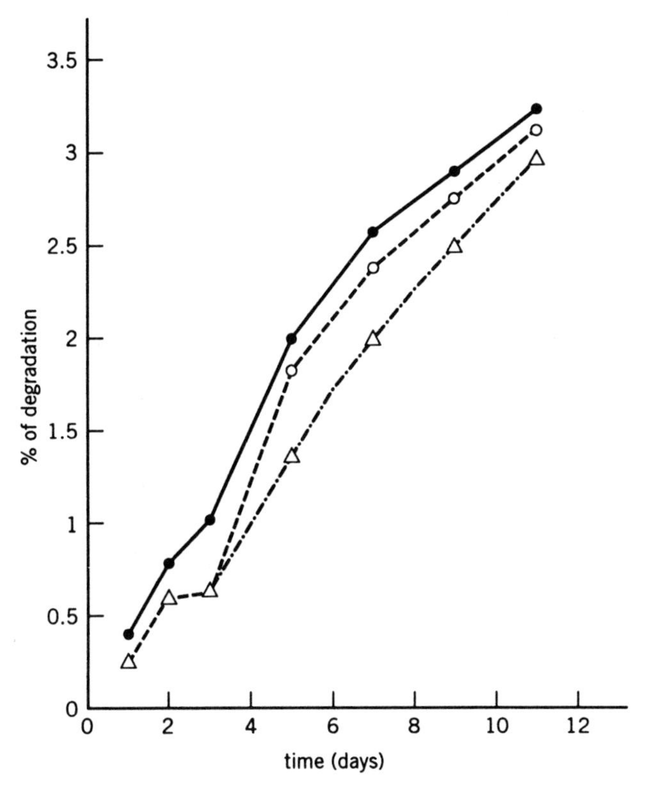

Fig. 4.6. Influence of Al ions on lignin degradation by *C. geophilum*. ●, Control; ○, 0.37mm Al; △, 1.48 mmAl (from Oelbe-Farivar, 1985, with permission).

of nitric acid. These climatically induced changes of soil acidity as a result of a temporary lack of equilibrium between the metabolism of the soil microorganisms and plant roots, were first postulated by Ulrich (1980). The initial data on this phenomenon were collected during two successive years in the experimental site in the Solling mountains near Göttingen: 1981 when the summer was cool and wet and 1982 with a warm and dry summer (Murach, 1983; for a review see Hüttermann and Ulrich, 1984). A summary of these results is given in Figure 4.7.

The data indicate that, unlike the results found in 1981, during the warm and dry summer of 1982, a steady increase in nitrate was observed in the soil solution, accompanied by a steady decrease of pH. This acidification push resulted in a significant change in the relationship of living and dead fine roots. Whereas during the summer of 1981, the mass of living fine roots always exceeded the root necromass, this relation was changed considerably during the next vegetation period. The amount of dead roots continuously increased, and finally exceeded the mass of living roots. Murach (1983) was able to disprove all other possible explanations for this phenomenon and postulated a causal relationship between the acid push and the observed root dieback.

Note that this type of internal acidification push is operating in an ecosystem that

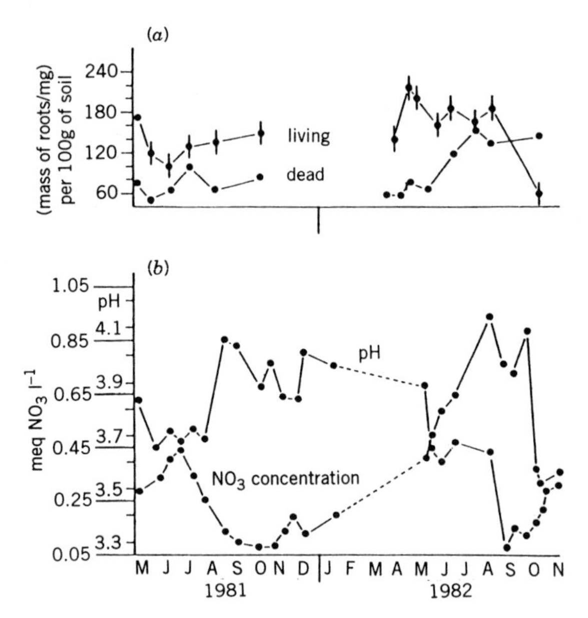

Fig. 4.7. Acidification push during autumn 1982 in a spruce forest (Solling). (*a*) Fine roots, living and dead. (*b*) pH (inner scale) and NO³ concentration (outer scale) in the soil solution (from Hüttermann and Ulrich, 1984, with permission of the Royal Society).

is otherwise undisturbed and has to be separated from nitrate losses after clear cutting and felling of a whole stand, as described by Vitousek et al. (1979). However, a seasonal pattern of nitrogen mineralization, although seen after the clearing of a stand, was also observed by Vitousek and Matson (1985). Since the events described in Figure 4.7 are triggered by frequent and subtile changes in the climate, they have to be considered as an indication of a nonstable equilibrium in the nutrient cycle of such forests.

Meanwhile, such acidification pushes were also observed on other sites. In the case studies Oberwarmensteinbach and Wülfersreuth (Fichtelgebirge) Guggenberger and Beudert (1989) noted that the DOC concentrations below the forest floor undergo high variations with regard to the seasonal pattern. The highest concentrations of DOC were always found when periods of drought were followed by heavy rains. Besides N mineralization, the subsequent mineralization push caused a pronounced decrease in pH and an increase in DOC.

4. FOREST ECOSYSTEMS IN THE PHASE OF PERMANENT HUMUS DISINTEGRATION

4.1. Definition of Nitrogen Saturation in Forest Ecosystems

A permanent enrichment of nutrients can lead to saturation phenomena in ecosystems. In the literature, nutritional saturation is defined in different ways. By taking the dose-response curves established in mineral nutrition studies, Nilsson (1986) defines N saturation as the stage in which the primary production of an ecosystem does not respond to and cannot be increased by additional supplements of N fertilizers. This definition *per constructionem* obviously implies otherwise well-balanced optimal nutritional conditions, which at present cannot be expected per se in forest ecosystems under the influence of human activities. Since the cation balance in most German forests is severely impeded (e.g., Forschungsbeirat Waldschäden/Luftverunreinigungen, 1989), this definition therefore does not appear to be operationally applicable to our current situation. In addition, the growth of a forest stand is not as easily to determine as the state of plants in experimental trials, and it is impossible to extrapolate the present growth data over the whole production time, which is in the order of centuries. Note, however, that, in spite of the undisputable observed overall decline of German forests, the present increment values by far exceed the expected growth according to the Zuwachstabellen (Forschungsbeirat Waldschäden/Luftverunreinigungen, 1989). Because of the very scarce supply of base cations, this increased growth rate might be taken as an indicator for a general overfertilizing of our forests with nitrogen.

A more suitable definition for forest ecosystems appears to be that of Skeffington and Wilson (1987) who define an ecosystem as saturated with N when the physiological N demand of the primary producers is satisfied and at the same time a measurable export of N has taken place. The problems with this definition, however, reside in the difficulties of experimentally describing the borderline of a fulfilled biological demand of the primary producer. This problem is avoided by the definition of N saturation by Agren and Bosatta (1987), which simply state that N saturation is reached when in long-term studies the N export surpasses the N input. This definition is based on a better understanding of the anthropogenic manipulations of forests. The problems with this definition are (a) a complete water balance is difficult to quantify and (b) dry deposition of N, mist, and fog are not easy to measure in the forests. In addition, it disregards the fact that the storage capacity of forest ecosystems for N may not be constant but can vary both with regard to long-term development of a stand over many years and during the seasons of an individual year.

In view of these problems and for the sake of simplicity, Cole (1992) and van Miegrot et al. (1992) define a ecosystem to be in the state of N saturation when atmospheric input and N mineralization together exceed the retention capacity of a given ecosystem and leads to a permanent leaching of nitrate. An advantage to this definition in that it is easy to monitor. In addition, it relates to the biologically logical concept of the critical load. According to these authors, for each given ecosystem there exists a defined level of maximal tolerable N input. If the N input is below this limit, no excess nitrate will be exported (Nilsson and Grennfelt, 1988).

4.2. Inventory Data from Case Studies

A summary of all 28 German permanent forest ecosystem study sites with regard to their N balance (Fig. 4.8*a*) reveals a variation in the input rates between less than 1

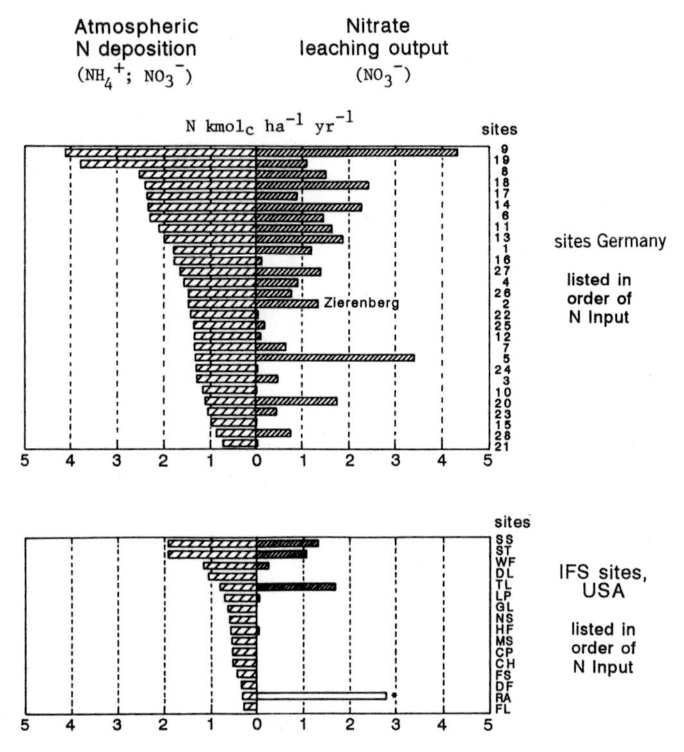

Fig. 4.8. Nitrogen input and output fluxes in Germany and the United States.

and more than 4 kmol of N eq $ha^{-1}a^{-1}$ The majority of the sites have values between 1.3 and 2.5 kmol of N eq $ha^{-1}a^{-1}$, the mean being 1.7 kmol.

The data on N export are rather different with regard to the different sites studied. In the Hils mountains, a site in the state of Niedersachsen, a N export of almost 5 kmol of N was measured, the mean export rates varied between 1.5 and 2.5 kmol. In four sites, the export of N is higher than the input, in 19 others the difference between input and leaching is less than 1.5 kmol. It must be taken into account that the average N incorporation into the annual increment of a mature stand amounts to about 1.5 kmol of N. This result means that the overwhelming majority (82%) of the permanent study sites in Germany are in the stage of humus disintegration and N mineralization. Only five sites do not leach nitrate into the seapage. According to the currently discussed definitions of N saturation (see Section 4.1), these are the only sites that can be considered not to be in a state of N saturation.

A grouping of the data according to the content of exchangeable Ca and Mg in the soils (data not shown) indicates that there is no direct correlation between the base saturation of the soil and the export of N and its subsequent leaching into the ground water.

A comparison of these data with the results of the integrated forest study (IFS) in the eastern United States (Fig. 4.8*b*) reveals that in this case 63% (10 out of 16) of the sites do not show significant leaching of N. In one case study, marked in the figure with an asterix, the high nitrate exports can be attributed to the nitrogen fixation of the red alders planted at this site. Although such an interpretation has to be made with great caution, the comparison of the German and US data shows that a permanent nitrate efflux is not the normal reaction of a forest ecosystem.

4.3. Zierenberg Forest, A Typical Site in Which Humus Disintegration Takes Place

4.3.1. *Site Characteristics*

According to the results of the IFS, humus disintegration occurs mainly in forest ecosystems, which are characterized by high C and N contents in the forest floor, low C/N ratios, and a high base saturation in the soil, which indicates a very good nutritional status (Van Miegrot et al., 1992). The site that to date was most intensively studied for the possibility of humus disintegration is the Zierenberg forest, which is one of the several case study sites of the Forschungszentrum Waldökosysteme der Universität Göttingen. For 3 years the N budget in this stand was monitored at a very high intensity, together with special studies pertaining to humus disintegration. For a detailed description of this site and the first summary of results from the studies of the participating 11 different research groups that cover a wide spectrum of forest ecosystems research, see Eichhorn (1991). The site that has an enormously high N store of about 11 tons of N ha^{-1} in its forest floor is located in northern Hessen, on the hillside Kleine Gudenberg, close to the town of Zierenberg, northwest of Kassel. The most important data characterizing the site are given in Table 4.1. Originally a Melico-Fagetum (Roloff and Hubeny, 1991) forest, it now carries a 151-year-old beech forest with a few ash trees in between. The trees have a mean height of 35 m and belong to productivity class I.7 (Ertragsklasse I.7 nach Schober). The ground vegetation is dominated at present by stinging nettles (*Urtica dioca*). From local records it is clear that this site was continuously covered with a

TABLE 4.1. Data Characterizing the Zierenberg Site

Natural regional area	Northwest Hessian mountains
Forest district	
Elevation	450 m a.s.l.
Substrate	Basalt over lime with loess
Soil type	Mainly brown soils
Moisture	Moist
Nutrient condition	Eutrophic
Tree species	*Fagus sylvatica*
Age	151 years
Height	34 m
Natural association	*Melico fagetum Circea lutetiana*
Annual temperature	7.0°C
Annual precipitation	650 mm

forest since the early settlement of this area, even during the time between 1640 and 1850, when forest coverage in Germany was at a minimum. This fact is probably due to its rather steep slopes. Therefore, it can be stated that in spite of forest management, the site still can be characterized as a stand that is close to the characteristics of a natural stand. The stand is situated on a basalt cone over shell lime and Röt. Below the basalt cone there is a zone that is dominated by basalt rubble, followed by a zone of limestone. Pedogenesis of the soil included residues of sandy and loamy sediments from the Untermiozän. Drifting soils consisting of a variety of materials resulted in the rather high heterogeneity of the soil status on the site (Jochheim, 1991).

In the transient zone between shell lime and Röt, springs the source, the Friedrichsaue. Although the hydrological situation is not optimal, it was shown, however, that the ion content at this source reflects the integrated export of ions from the site rather well (Lischeid et al., 1991). The nitrate and chloride concentrations of this source over the 25-year period 1965–1990 were shown in Figure 4.1. These data were converted to actual mass flows on the basis of a climatological water balance using the precipitation data from nearby weather stations Zierenberg and Escheberg and the models developed for similar calculations in the Solling by Matzner (1988). The data shown in Figure 4.1*b* indicate that the increase in actual export of nitrate from the site over this time span is considerably higher than was expected based of concentration data only. From these data it is clear that the Zierenberg forest is indeed in the stage of humus disintegration and N mineralization.

As mentioned previously, a large amount of N is stored in the soil, more than 10 tons ha^{-1}. It is of course impossible to reconstruct how this store was built, considering the mean annual imput of N in pristine forests, which is on the order of 2–5 kg, it should have taken about 2000–5000 years. This time span definitely is much shorter than the time that elapsed since the end of the last ice age (~10,000 years ago), which was the time when the reforestation of this area began along with pedogenesis.

In the history of the stand, there have been two main periods during which anthropogenic factors influenced the N budget of the stand. The first was the time from about 1100–1300 AD, when the top of the hill was occupied by a small castle and the area was deforestated. This period definitely lead to a partial deterioration of the soil and concomitant loss of N stores. We do not know, however, what amount of human and animal faeces can be assumed to have been deposited during that time in the Zierenberg forest by the inhabitants of the castle. On the other hand, it is unlikely that their occupancy was accompanied by a significant builtup of humus and N in the soil, which would have made up the earlier losses in soil organic matter. After the castle was removed and the township of Zierenberg established, the area was neglected, because of its steep slopes, and gradually taken over by the forest. Since the eighteenth century, the area was subject to planned (i.e., sustained) yield managed forestry. From the first complete and modern record taken in the year 1840, we know that the area was fully aforrested. The second time span during which anthropogenic N input has to be considered is this century, especially the second one half of the century with its N emissions from industry, agriculture, and traffic. Obviously, the second time period caused the overloading of N, which lead to the observed loss of N to the aquifer.

4.3.2. *Spatial Variability of Nitrogen and Sulfur Mineralization in the Forest Floor*

The data on total N stores in the soil with regard to the soil depths, together with the actual concentrations of nitrate and ammonium, are given in Figure 4.9. For a comparison, the mean values of the same data of 23 sites in Hessen are also given. Figure 4.9 indicates that in Zierenberg the overall stores of nitrogen are somewhat higher than the average of the Hessian sites: the differences being in the order of some 30%. Figures 4.9*b* and *c* show that the concentrations of soil nitrate and ammonium, however, are much higher in the top soil layers, five times in the case of nitrate and twofold with regard to ammonium. These data again indicate the exceptionally high degree of humus disintegration occurring at this site.

As mentioned above, a characteristic feature of the stand is its high variability of bedrock minerals in the soil. Therefore it is, like a mosaic, composed of small patches of areas with different chemical soil conditions. According to the hypothetical mechanism of humus disintegration outlined above, in such a stand the influence of variable climate and weather should result in a high variability of nitrate formation. Figure 4.10 shows that this is indeed the case in this stand, especially in the topsoil stratum. In addition, these data reveal that most of the time the highest nitrate concentrations were found in the deepest layer of the soil. Such an increase of nitrate load with increasing soil depths was found by Hildebrand (1992) for the ARINUS site at the Schluchsee, in the Blackforest. This N enrichment could be taken as an indicator that the N releasing process is an active one, which takes place over the whole soil profile. This increase in concentration with regard to soil depth was also observed in the same samples for sulfate (Fig. 4.10).

Probably the same factors (i.e. metabolic activities of microorganisms and enzymes) are responsible for the release of both nitrate and sulfate. The same concentration gradient with increasing soil depths was observed for both anions. However, whereas in the top layer of the soil nitrate is taken up by the soil microorganisms and/or plants during summer, the sulfate is more or less unaffected by the seasons.

4.3.3. *Dissolved Organic Carbon and Ions*

The concentrations of DOC in the different soil depths are given in Figure 4.10. The concentrations reveal a similar variability over time as was observed for nitrate, especially in the uppermost layer of the soil. It is of interest to compare these values with the aluminum present in the same solutions (Fig. 4.11). The highest amounts of Al are found in the upper layer of the profile, where the influence of acid deposition is the highest. The methods employed for ion analysis do not allow us to distinguish between free and bound Al. Because of the high Al content of the solution in the upper layer, and the affinity of this ion for organic ligands (Schierl, 1989; Baur et al., 1988; Göttlein, 1988) it appears to be safe to assume that in those

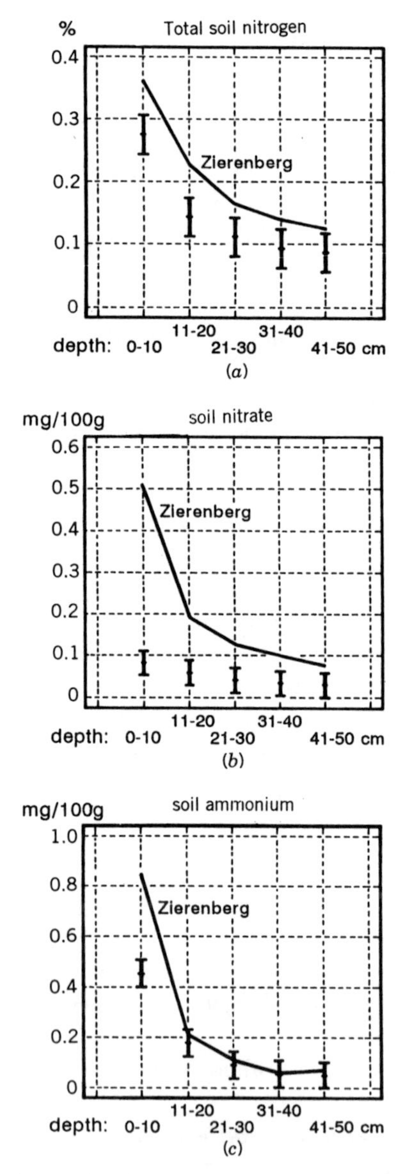

Fig. 4.9. Concentrations of total soil nitrogen (*a*), soil nitrate (*b*), and soil ammonium (*c*) at the Zierenberg site (line drawing), in comparison to the mean values of the 23 case studies in Hessen.

soil solutions most of the possible binding sites are occupied by Al ions, as was reported for soil solutions from the Harz (Spiteller, 1985). The loss of Al during the passage to deeper soil depths (Fig. 4.12) is difficult to interprete at present. It may be that the higher pH values of the deeper soil profiles influence the Al binding.

It is important to note that the highest concentration of DOC is found in the upper

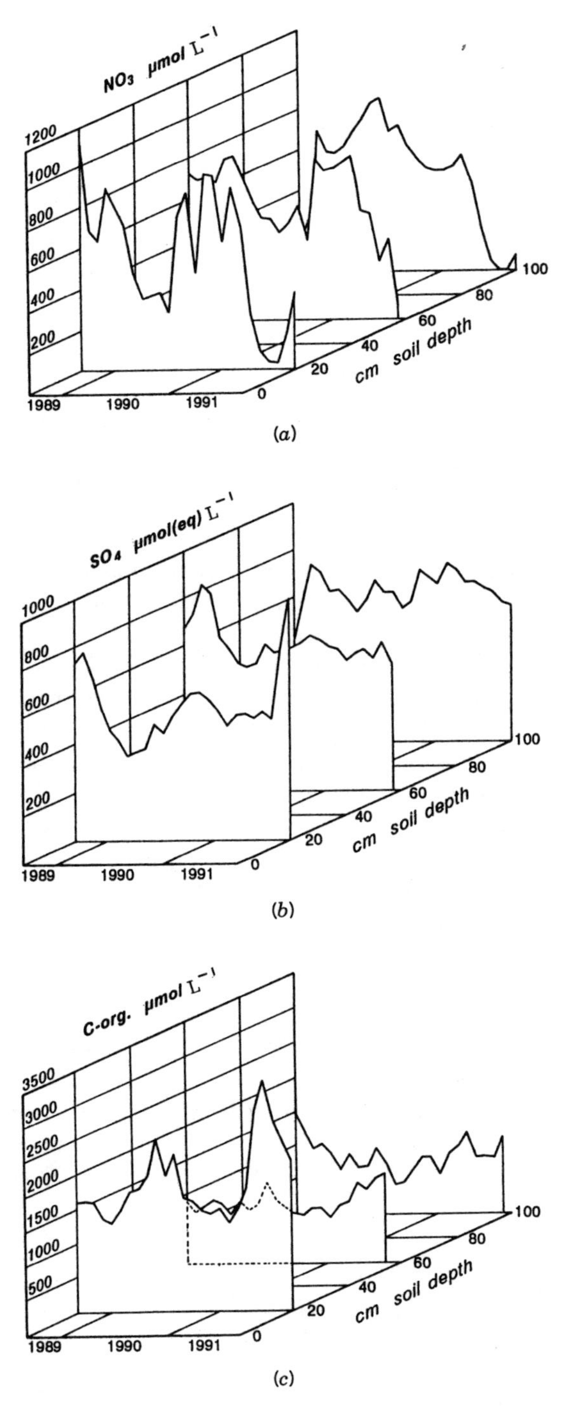

Fig. 4.10. Temporal and spatial variations of nitrate (*a*), sulfate (*b*), and DOC (*c*) at the Zierenberg site.

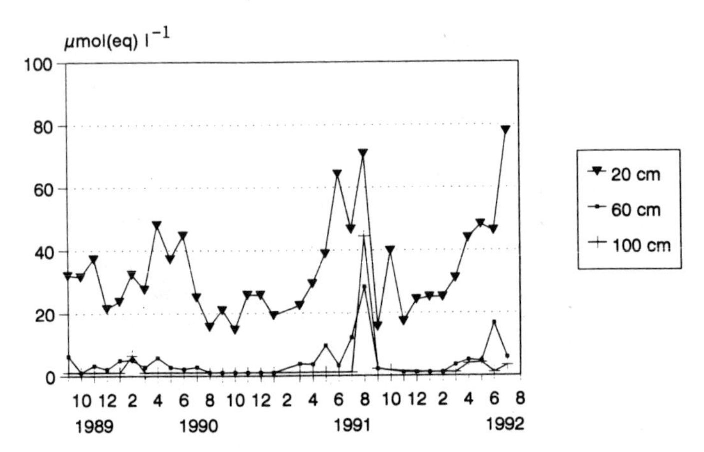

Fig. 4.11. Temporal and spatial variations of Al at the Zierenberg site.

layer of the profile, with decreasing concentration at increasing depths. This result shows that the highest liberation of DOC occurs in the top layer of the soil profile that has the highest humus content. During the passage to deeper horizons, the water obviously loses some of its DOC freight. This finding may result from precipitation with inorganic colloidial molecules or uptake by either microorganisms or mycorrhizal fungi. Tamm and Hallbäcken (1988) found that microbial degradation of the carbohydrate moiety of the DOC can occur in rather deep soil horizons. The view that at least part of the DOC are readily usable nutrients for microorganisms is supported by a comparison of the nitrate and DOC concentrations over the seasons (Fig. 4.12). Whereas during most of the year the concentrations of both compounds exhibit similar trends, the nitrate values show a minimum during the warm summer months. Since the liberation of both DOC and nitrate is regulated by the same metabolic events, this disappearance of nitrate from the upper soil horizons can only be interpreted as uptake by metabolically active microorganisms.

The distribution pattern of DOC over the soil profile was different in the much more acidic soil of the Schluchsee site (Hildebrand, 1992). In this site, the highest DOC freight was observed in the third soil horizon (A_{he2}). Hildebrand interpreted this as an indication for a currently occurring disintegration of those humus moities that were accumulated by the beech trees that had grown on the stand before the presently growing spruce were planted (for the history of this stand see Feger et al., 1988).

4.3.4. Physiological Reaction of the Stand

The data obtained in the Zierenberg case study offer some information about the physiological reaction of the stand with regard to N oversaturation.

The first indication of the mechanism by which trees handle a high amount of N can be gained from the N concentrations in the leaves (Fig. 4.13). It shows that the N concentration in the leaves of the Zierenberg site are almost exactly the mean of the Hessen sites. These data indicate that the N content of the leaves does not

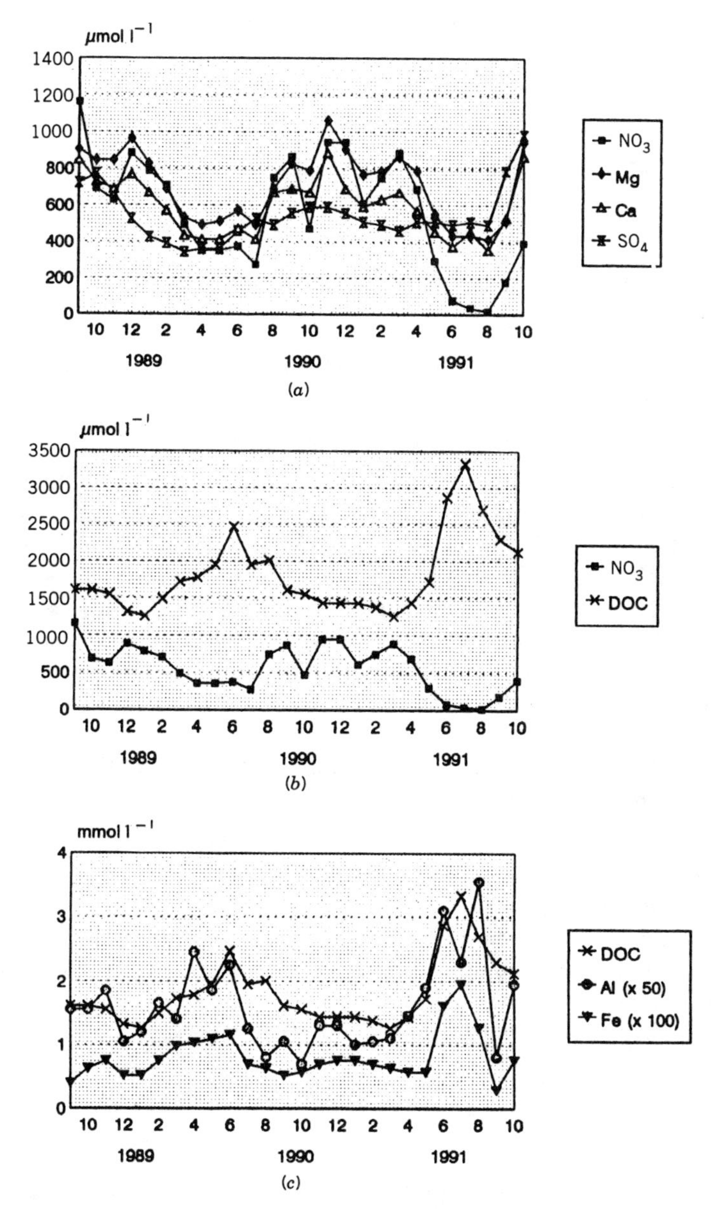

Fig. 4.12. Comparison of ions (*a*), nitrate (*b*), and aluminum and iron (*c*) with the DOC concentration in the seepage at 20 cm soil depth at Zierenberg.

increase above a certain level with an increasing N load of the ecosystem. This finding shows that above a certain level of N supply, the demand of the leaves is satisfied and additional N does not lead to an additional uptake. For those ecosystems, the definition of N saturation according to Skeffington and Wilson (1988) certainly is fulfilled.

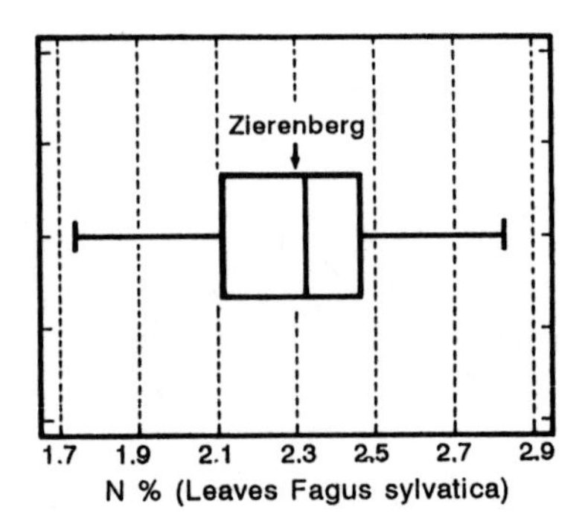

Fig. 4.13. Nitrogen content in the leaves of beech trees of the 23 Hessian case study sites.

Analysis of the nitrate input data reveals that over the 3 years of study the N input in the stand (the sum of nitrate throughfall and stemflow) always exceeded the open field precipitation by a factor of more than two (Fig. 4.14). The only possible interpretation for this highly reproducible and consistent set of data is that the trees leach nitrate through their leaves. This result means that the roots take up more N than the trees can handle in their N metabolism and dispose of the excess N through the leaves. This stand obviously is so heavily oversaturated with N that the trees use the nitrate as counterions for the leached cations. Such a phenomenon has not yet been observed in North American forests.

4.3.5. Changes of the Species Diversity and Composition of the Ground Vegetation

The ground vegetation in forests serves as a biological indicator for changes in the nutrient balance in the ecosystem. This result can best be seen in sites that are monitored for several decades. Roloff (1989) reinvestigated in 1988 32 sites that were surveyed for the years 1971–1977. In spite of the relatively small time span between the first and the second study, Roloff was able to demonstrate a very dramatic change in the sociology of the ground vegetation. In 60% of the sites the composition of the ground vegetation was significantly changed, with a higher

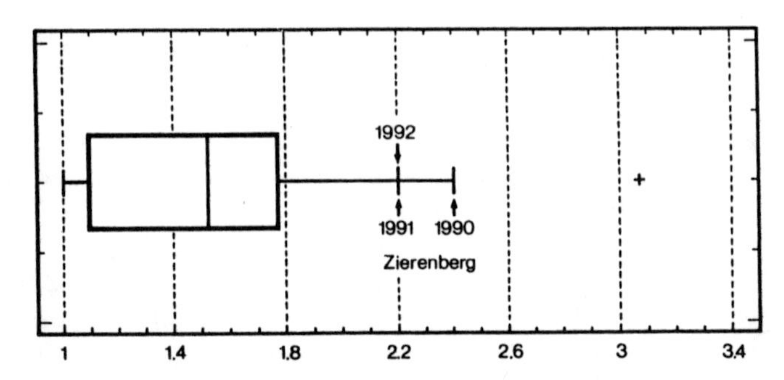

Fig. 4.14. Relation of nitrate throughfall plus stemflow to open field precipitation at Zierenberg compared to the 23 Hessian case study sites.

portion of nitrophilic plants. The indicator values for N (Ellenberg, 1979) at those sites was shifted to significantly higher values. Roloff reports an increase in many sites of nitrophilic species like *Sambucus nigra, Galeopsis trehalit, Solanum dulcamara, Moehringa trinerva,* and especially *Urtica dioica* (stinging nettle). The dominance of such nitrophilic species can amount to mass colonization of stands where they did not occur a few decades ago. A similar increase in nitrophilic species because of high depositions of N was also reported for southern Sweden (Falkengren-Grerup, 1990). At present, this change to more nitrophilic species still leads to a higher number of species in the ground vegetation of forests.

Because of the vast amount of data available on the integrated study at Zierenberg, it was at last possible to compare the degree of ground coverage of the stinging nettle to the nitrate content of the soil. Figure 4.15 shows that a good correlation exists between these two sets of data, confirming the assumption that the increase in distribution of this species during the last decades is a direct result of the N input into our ecosystems (cf. Reif et al., 1985)

However, this development is highly detrimental to the beech forests where it takes place. Because of its development of a perennial, widespread, and intensively branched root system, the stinging nettle apparently competes with and suppresses the young beech seedlings. As soon as this species achieves a ground coverage on the forest floor higher than 30%, the natural regeneration of the beech is completely inhibited (Wenzlik, 1993).

The overloading of N together with a decrease in base saturation has to be considered as a novel and very important selective factor in our forests. At present, some plants are still able to cope with these changing conditions (Falkengren-Grerup, 1990) and we can, in some forests, still observe an increase in species numbers and diversity. However, if this trend continues, in the long run a decrease in species has to be expected (Roloff, 1989), which eventually will lead to a different type of forest.

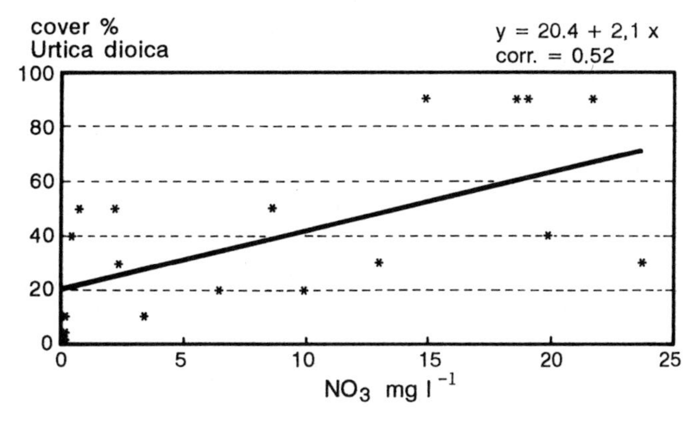

Fig. 4.15. Correlation between nitrate concentration in the soil solution and the degree of coverage of stinging nettle (*U. dioica*) (from Roloff and Hubeny, 1991, with permission).

5. CONCLUSIONS

The importance of the humus disintegration for the overall balance of N in a typical German forest can be seen from the N stores and fluxes in the Zierenberg forest for the years 1990–1992 (Fig. 4.16; data taken from Eichhorn, 1993). The data reveal that the amount of N processed annually in this ecosystem amounts to 8.71 kmol of N. Of this N, 7.18 kmol come from the soil and 1.53 kmol comes from the atmosphere. This number amounts to about 1% of the total N store in the soil. Of the N processed in the above ground vegetation, 1.57 kmol are transferred to and stored in the standing timber, 7.16 kmol go back to the soil. About 0.12% of the total N store (0.93 kmol of N) leave the ecosystem via leaching and appear in the aquifer.

Although the actual amount of leachate appears to be rather small with regard to the N that is internally processed, this phenomenon has to be taken seriously. As outlined above, humus disintegration is the result of severe changes in the soil chemistry and microflora activity. In addition, the saturation of the soil solution with N may have severe detrimental effects on soil biology as a whole. Since only a few studies were used to interpret this phenomenon, the known set of data is still rather scarce. The few studies dealing with the influence of soil nitrate on soil biology and biochemistry definitely show that high concentrations of nitrate have a

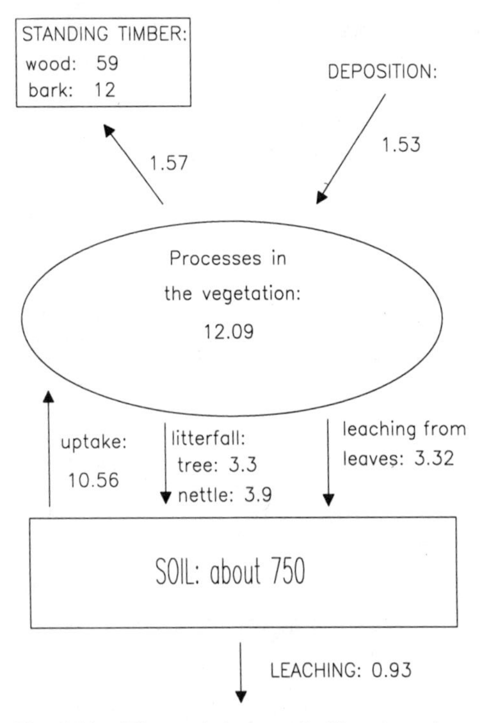

Fig. 4.16. Nitrogen balance at the Zierenberg site.

significant influence on the mycorrhizal fungi (Rastin et al., 1990 a,b) and general soil biochemistry (Rastin, 1993).

To date, the most alarming report on the effect of N oversaturation is the observation that the stinging nettle, a species that is highly favored by a change in N supply competes with and suppresses the natural regeneration of the beech (Wenzlik, 1993), the dominant natural tree in our forests. If this phenomenon is expected to occur in a wide range of forests, it would mean that we have to expect a shift in species composition of our forests and a discontinuation of our presently established practice of forest management. Three hundred years of forestry, during which the foresters of central Europe and Scandinavia developed the concept of sustained yield management of forests, which basically influenced forest practice all over the world (cf. Schenck, 1955; Hüttermann, 1987), are endangered by our very civilization.

ACKNOWLEDGMENTS

The Zierenberg study was funded by the German Bundesministerium für Forschung und Technologie within the program Stabilitäsbedingungen von Waldökosystemen coordinated by the Forschungszentrum Waldökosysteme der Universität Göttingen.

REFERENCES

Agren GL, Bosatta E (1987): Theoretical analysis of the long-term dynamics of carbon and nitrogen in soils. Ecology 68:1181–1189.

Alexander M (1977): Introduction to Soil Microbiology 2nd ed. New York: Wiley 467 pp.

Almendros G, Fründ R, Gonzales-Vila FJ, Haider KM, Knicker H, Lüdemann H-D (1991): Analysis of ^{13}C and ^{15}N CPMAS NMR-spectra of soil organic matter and composts. FEBS Lett 282:119–121.

Anonymous (1989): Protecting the Earth's Atmosphere: An International Challenge. Interim Report of the Study Commission of the 11th German Bundestag. German Bundestag, Publications Section. Bonn, 595 pp.

Atlas RM, Bartha R (1987): Microbial Ecology: Fundamentals and Applications. New York: Benjamin/Cummings, 532 pp.

Baur S, Feger HK, Brahmer G (1988): Mobilität und Bindungsformen von Aluminium in Wassereinzugsgebieten des Scharzwaldes. Mitt Dtsch Bodenkd Ges 57:141–146.

Behmal P, Ziechmann W (1987): The determination of phenolic constituents of soils, their ability for humification and chelate formation with Al^{3+}-Ions and $H[AlCl_4]$. Chem Erde 47:243–253.

Berry DF, Boyd SA (1984): Oxidative coupling of phenols and anilines by peroxidase:Structure–activity relationships. Soil Sci Soc Am J 48:565–569.

Berry DF, Boyd SA (1985): Reaction rates of phenolic humus constituents and anilines during cross-coupling. Soil Biol Biochem 17:631–636.

Bollag J-M, Liu Shu-Yen, Minard RD (1980): Cross coupling of phenolic humus constituents and 2,4-dichlorophenol. Soil Sci Soc Am J 44:52–56.

Bollag J-M, Liu Shu-Yen, Minard RD (1982): Enzymatic oligomerization of vanillic acid. Soil Biol Biochem 14:157–163.

Christman RF, Oglesby RT (1971): Microbial degradation and formation of humus. In Lignins, Occurence, Formation, Structure and Reactions. Sarkanen KV, Ludwig CH (eds): New York: Wiley–Interscience, pp 769–795.

Cole DW (1992): Nitrogen chemistry, deposition, and cycling in forests—overview—. In Atmospheric Deposition and Forest Nutrient cycling. A synthesis of the Integrated Forest Study. Ecological Studies Vol. 91. Johnson DW, Lindberg SE (eds): Berlin: Springer-Verlag, pp 150–152.

Coleman DC, Reid CPP, Cole CV. (1983): Biological strategies of nutrient cycling in soils systems. In Macfayden A, Ford ED (eds): Advances in Ecological Research, Vol. 13, New York: Academic Press, pp 1–55.

Cronan SC (1985): Comparative effects of precipitation acidity on three forest soils: carbon cycling responses. Plant Soil 88:101–112.

Deacaon JW (1980): Introduction to Modern Mycology. London: Blackwell Scientific Publications, 197 pp.

Eichhorn J (ed) (1991): Fallstudie Zierenberg: Stress in einem Buchenwaldökosystem in der Phase einer Stickstoffübersättigung. Forschungsberichte des Hess. Minist. f. Landesentwi., Wohnen, Landwirt, Forsten und Naturschutz, Vol. 13, Wiesbaden, 117 pp.

Eichhorn J (1993): Gesamtbericht Koordinationseinheit P 6.3.2.: Stress in einem Buchenwaldökosystem in der Phase des Stickstoff-Vorratsabbaus. In Stabilitätsbedingungen von Waldökosystemen L. Berichte des Forschungszentrums Waldökosysteme, Göttingen. 6 pp.

Ellenberg H (1979): Zeigerwerte der Gefässpflanzen Mitteleuropas. Scr Geobotan 9:1–122.

Ertel JR (1988): Genesis. Group Report. In Frimmel FH, Christman RF (eds): Humic substances and their Role in the Environment. Report of the Dahlem Workshop. Chichester: Wiley, pp 105–112.

Falck R, Falck M (1954): Die Bedeutung der Fadenpilze als Symbionten der Pflanzen für die Waldkultur. Frankfurt: Sauerländer's.

Falkengren-Grerup U (1990): Distribution of field layer species in Swedish deciduous forests in 1929–54 and 1979–88 as related to soil pH. Vegetatio 86:143–150.

Feger K-H, Zöttl HW, Brahmer G (1988): Projekt ARINUS II. Einrichtung der Messstellen und Vorlaufphase. In Horsch F, Filby WEG, Fund N, Gross S, Hanisch B, Pilz E, Seidel A (eds): 4. Statuskolloquium des Projekts Europäisches Forschungszentrum für Massnahmen zur Lufreinhaltung, KFK-PEF 35, Vol. 1, Kernforschungszentrum Karlsruhe, pp 27–38.

Filip Z, Preusse T (1985): Phenoloxidierende Enzyme—ihre Eigenschaften und Wirkungen im Boden. Pedobiologia 28:133–142

Flaig, WH (1988): Generation of model chemical precursors. In Frimmel FH, Christman RF (eds): Humic substances and their Role in the Environment. Report of the Dahlem Workshop. Chichester: Wiley, pp 75–92.

Flaig W, Beutelspacher H, Rietz E (1975): Chemical composition and physical properties of humic substances. In Gieseking JE (ed): Soil Components. Vol. 1. Organic Components. pp 1–211 New York: Springer-Verlag

Forschungsbeirat Waldschäden/Luftverunreinigungen (1989). Third Report, ISSN 0931-7805, Kernforschungszentrum Karlsruhe, 611 pp.

Fründ R, Lüdemann H-D (1989): The quantitative analysis of solution and CPMAS-C-13-NMR spectra of humic material. Sci Tot Environ 81/82: 157–168

Göttlein A (1988): Einfluss von sauer Beregnung und Kalkung auf wasserlösliche organische Stoffe eines Waldbodens unter Fichte. Ph.D. Thesis University of München, Munchen, Germany, 178 pp.

Göttlein A, Kreutzer K, Sschierl R (1991a): Beiträge zur Charakterisierung organischer Stoffe in wässrigen Bodenextrakten unter dem Einfluß von sauer Beregnung und Kalkung; In Kreutzer K, Göttlein A (eds): Ökosystemforschung Höglwald: Beiträge zur Auswirkung von sauer Beregnung und Kalkung in einem Fichtenbestand. Hamburg: Verlag Parey, pp 212–221.

Göttlein A, Pruschka H (1991b): Statistische Auswertung des Einflusses von sauer Beregnung und Kalkung auf die Wasserlöslichkeit organischer Bodeninhaltsstoffe; In Kreutzer K, Göttlein A (eds):

Ökosystemforschung Höglwald: Beiträge zur Auswirkung von saurer Beregnung und Kalkung in einem Fichtenbestand. Hamburg: Verlag P. Parey, pp 221–227.

Greenland DJ, Oades JM (1975): Saccharides. In Gieseking JE (ed): Soil Components Vol. I. Organic Components. Berlin: Springer-Verlag, pp 213–261.

Guggenberger G, Beudert G (1989): Zur Dynamik des gelösten organischen Kohlenstoffs (DOC) in unterschiedlich immissionsbelasteten Waldstandorten. Mitt Dts Bodenkd Ges 59:367–372.

Haars A, Kharazipour A, Zanker H, Hüttermann A (1989): Room-temperature curing adhesives based on lignin and phenoloxidases. In Hemingway RW, Conner, AH, Branham SF (eds): Adhesives from Renewable Resources: ACS Symposium Series No. 385, American Chemical Society, Washington DC, pp 126–134.

Haider K, Martin JP (1967): Synthesis and transfromation of phenolic compounds by *Epicoccum nigrum* in relation to humic acid formation. Soil Sci Soc Am Proc 31:766–771.

Haider K, Martin JP (1970): Humic acid-type phenolic polymers from *Aspergillus sydowi* culture medium, *Stachybotrys* ssp. cells and autoxidized phenol mixtures. Soil Biol Biochem 2:145–156.

Haider K, Martin JP, Filip Z (1975): Humus biochemistry. In Paul EA, McLaren AD, (eds): Soil Biochemistry Vol. 4., New York: Dekker, pp 195–244.

Hatcher PG, Spike EC (1988): Selective Degradation of Plant Biomolecules. In Frimmel FH, Christman RF (ed): Humic substances and their Role in the Environment. Report of the Dahlem Workshop. Chichester: Wiley, pp 59–74.

Haselwandter K, Bobleter O, Read DJ (1990): Degradation of 14-C-labelled lignin and dehydropolymer of coniferyl alcohol by ericoid and ectomycorrhizal fungi. Arch Microbiol 153:352–354.

Hedges JI (1988): Polymerization of humic substances in natural environments. In Frimmel FH, Christman RF, (eds): Humic substances and their Role in the Environment. Report of the Dahlem Workshop. Chichester: Wiley pp 45–58.

Hempfling R, Schulten H-R (1990): Direct chemical characterization of dissolved organic matter in water by pyrolysis-field ionization mass spectrometry. Int J Environ Anal Chem 43:55–62.

Hildebrand EE (1992). Die chemische Untersuchung ungestört gelagerter Waldbodenproben—Methoden und Informationsgewinn—Göttingen: Habilitationsschrift 180 pp.

Hüttermann A (1987): History of Forest Botany (Forstbotanik) In Germany. Ber Deutsch Bot Ges 100:107–141.

Hüttermann A (1989): Verwendung von Weissfäulepilzen in der Biotechnologie. GIT Fachz Lab 10/89:943–950.

Hüttermann A, Haars A (1987): Biochemical control of forest pathogen inside the tree. In Chet I (ed): Innovative approaches to plant disease control. New York: Wiley, pp 275–296.

Hüttermann A, Haars A, Trojanowski J, Milstein O, Kharazipour A (1989a): Enzymatic modification of lignin for its technical use-strategies and results. In Glasser WG, Sarkanen S (eds): New Polymeric Materials from Lignin. ACS Symposium Series No. 397, American Chemical Society, Washington DC, pp 361–371.

Hüttermann A, Loske D, Majcherczyk A (1989b): Biologischer Abau von Organochlor-Verbindungen in Böden, im Abwasser und in der Luft. In VDI-Berichte 745: Halogenierte organische Verbindungen in der Umwelt. Vol. II. Düsseldorf: VDI-Verlag, pp 911–926.

Hüttermann A, Trojanowski J, Haars A, Böttcher U (1984): Physiology and mechanisms of lignin degradation by *Heterobasidion annosum*. In Kile GA (ed): Proceedings of the Sixth International Conference on Root and Butt Rots of Forest Trees. Melbourne: CSIRO, pp 257–268.

Hüttermann A, Ulrich B (1984): Solid phase-solution-root interactions in sols subjected to acid deposition. Philos Trans R Soc London B 305:353–368

Jentschke G (1984): Untersuchungen zum biochemischen Mechanismus der Humusdisintegration - Versuch eine Falsifizierung. Berichte des Forschungszentrums Waldökosysteme/Waldsterben. Vol. 5. Göttingen: pp 175–256.

Jochheim H (1991): Chemische Bodeneigenschaften der Fest- und Lösungsphase in einem Buchenwald-

Ökosystem in der Phase der Humusdisimntegration. In Eichhorn J. (ed): Fallstudie Zierenberg: Stress in einem Buchenwaldökosystem in der Phase einer Stickstoffübersättigung. Forschungsberichte des Hess. Minist. f. Landesentwi., Wohnen, Landwirt., Forsten und Naturschutz, Vol. 13, Wiesbaden: pp 20–25.

Kharazipour A, Haars A, Shekholeslami M, Hüttermann A (1991): Enzymgebundene Holzwerkstoffe auf der Basis von lignin und Phenoloxidasen. Adhasion 5:30–36.

Kimmins JP (1987): Forest Ecology. New York: Macmillan, 531 pp.

Kirk TK, Schultze E, Connors WG, Lorentz LF, Zeikus JG (1978): Influence of culture parameters on lignin metabolism by *Phanaerochaete chrysosporium.* Arch Microbiol 117: 277–285.

Kirk TK, Farrell RL (1987): Enzymatic "combustion": The microbial degradation of lignin. Annu Rev Microbiol 41:465–505.

Krzak J (1981): A model of forest nitrogen cycling to assess the effects of management intensity on long-term productivity in Douglas-fir forests of the Pacific Northwest. Ph.D. Thesis, Oregon State University, Corvallis, OR, 232 pp.

Leonowicz A, Bollag J-M (1987): Laccases in soils and the feasability of their extraction. Soil Biol Biochem 19:237–242.

Lischeid G, Schmitt J, Hauhs M (1991): Wasserhaushalt. In Eichhorn, J (ed): Fallstudie Zierenberg: Stress in einem Buchenwaldökosystem in der Phase einer Stickstoffübersättigung. Forschungsberichte des Hess. Minist. f, Landesentwi., Wohnen, Landwirt., Forsten und Naturschutz, Vol. 13, Wiesbaden: pp 26–30.

Lyr H (1963): Zur Frage des Streuabbaus durch ektotrophe *Mykorrhiza* In Mykorrhiza Internationales Mykorrhizasymposium 25–30. April in Weimar. Jena: Fischer Verlag.

Martin JP, Haider K (1969): Phenolic polymers of *Stachybotrys atra, Stachybotrys chartarum* and *Epicoccum nigrum* in relation to hunic acid formation. Soil Sci 107: 260–270.

Martin JP, Haider K (1976): Decomposition of specifically carbon-14-labeled ferulic acid: Free and linked into model humic acid-type polymers. Soil Sci Soc Am J 40:377–380.

Martin JP, Haider K (1977): Decomposition in soil of specifically carbon-14-labeled DHP and corn stalk Lignins and coniferyl alcohols. In Soil Organic Matter Studies, Vol. II. International Atomic Energy Agency, Vienna.

Martin JP, Haider K (1980): A comparison of the use of phenolase and peroxidase for the synthesis of model humic acid-type polymers. Soil Sci Soc Am J 44:983–988.

Martin JP, Haider K, Linares LF (1979): Decomposition and stabilization of ring-^{14}C-labeled catechol in soil. Soil Sci Soc Am J 43:100–104.

Martin JP, Haider K, Wolf D (1972): Synthesis of phenols and phenolic polymers by *Hendersonula toruloidea* in relation to humic acid formation. Soil Sci Soc Am Proc 36: 311–315.

Martin JP, Richards SR, Haider K (1967): Properties and decomposition and binding action in soil of "humic acid" synthezised by *Epicoccum nigrum.* Soil Sci Soc Am Proc 31: 657–662.

Matzner E (1988): Der Stoffumsatz zweier Waldökosysteme in Solling. Berichte der Forschungszentrums Waldökosysteme der Universität Göttingen, Vol. A 40, 217 pp.

Melillo JM, Amber JD, Muratore JF (1982): Nitrogen and lignin control of hardwood leaf litter decomposition dynamics. Ecology 63: 621–626.

Milstein O, Nicklas B, Hüttermann A (1989): Oxidation of aromatic compounds in organic solvents with laccase from *Trametes versicolor.* Appl Microbiol Biotechnol 29: 70–74.

Milstein O, Hüttermann A (1993): Enzymatic coupling of glucose to lignin by the action of laccase, submitted.

Murach D (1983): Die Reaktionen von Fichtenfeinwurzeln auf zunehmende Bodenversauerung. Allg Forstz 38: 683–686.

Nilsson J (ed) (1986): Critical loads for nitrogen and sulphur. Miljörapport 1986:II. Copenhagen: 223 pp.

Nilsson J, Grennefelt P (1988): Critical loads for sulphur and nitrogen. Report from the Nordic Working Group. Nord, 11 pp.

Norkrans B (1950): Studies on growth and cellulolytic enzymes of *Tricholoma*. Symb Bot Upsal 11, 1–126.

Oelbe-Farivar M (1985): Physiologische Reaktionen von Mykorrhizapilzen auf simulierte saure Bodenbedingungen. Ph.D. Thesis, Universität Göttingen, Göttingen, Germany, 68 pp.

Paul EA, Clark FE (1989): Soil Microbiology and Biochemistry, San Diego: Academic Press, 275 pp.

Pfaender FK (1988): Generation in controlled model ecosystems. In Frimmel FH, Christman RF (eds): Humic substances and their Role in the Environment. Report of the Dahlem Workshop. Chichester: Wiley pp 92–104.

Rastin N (1993): Biochemischer und mikrobiologischer Zustand verschiedener Waldböden. Habilitationsschrift D.Sc. Thesis, Forstwissenschaftlicher Fachbereich der Universität Göttingen, Göttingen, Germany, 134 pp.

Rastin N, Schlechte G, Hüttermann A (1990a): Soil macrofungi and some biological, biochemical and chemical investigations on the upper and lower slope of a spruce forest. Soil Biol Biochem 22: 1039–1047.

Rastin N, Schlechte A, Hüttermann A, Rosenplänter K (1990b): Seasonal fluctuation of some biological and biochemical soil factors and their dependance on certain soil factors on the upper and lower slope of a spruce forest. Soil Biol Biochem 22: 1049–1061.

Reif A, Teckelmann M, Schultze E-D (1985): Die Standortamplitude der Grossen Brennessel (*Urtica dioica* L.)—eine Auswertung vegetationskundlicher Aufnahmen auf der Grundlage der Ellenbergschen Zeigerwerte. Flora 176: 365–382.

Roloff A (1989): Pflanzen als Bioindikatoren für Umweltbelastungen. I. Prinzipien der Bioindikation und Beispiel Waldbodenvegetation. Forstarchiv 60: 184–187.

Roloff A, Hubeny C (1991): Vegetationskundliche Fragestellungen auf der Fläche Zierenberg. In Eichhorn J. (ed): Fallstudie Zierenberg: Stress in einem Buchenwaldökosystem in der Phase einer Stickstoffübersättigung. Forschungsberichte des Hess. Minist. f. Landesentwi., Wohnen, Landwirt., Forsten und Naturschutz, Vol. 13, Wiesbaden: pp 87–91.

Schachtschabel P, Blume H-P, Brümmer G, Hartge K-H, Schwertmann U (1989): Lehrbuch der Bodenkunde. Stuttgart: Enke Verlag, 491 pp.

Scheffer F, Ulrich B (1960): Humus und Humusdüngung—Morphologie, Biologie, Chemie und Dynamik des Humus. Stuttgart: Enke Verlag, 266 pp.

Schenck CA (1955): The Biltmore Story. Recollections of the Beginning of Forestry in the United States. American Forest History Foundation, Minnesota, 224 pp.

Schierl R (1989): Bestimmung des Komplexierungsgrades von Al, Fe und Mn in Bodenlösungen durch Kationenaustausch. Vom Wasser 73: 161–165.

Schlesinger WH (1977): Carbon balance in terrestrial detritus. Annu Rev Ecol Syst 8:51–81.

Schnitzer M, Barr M, Hartenstein R (1984): Kinetics and properties of humic acids produced from simple phenols. Soil Biol Biochem 16: 371–376.

Shindo H, Huang PM (1984): Catalytic effects of Manganese (IV), Iron (III), Aluminium, and Silicon oxides on the formation of phenolic polymers. Soil Sci Soc Am J 48:927–934.

Simmons KE, Minard RD, Bollag J-M (1988): Oxidative coupling and polymerization of guaiacol, a lignin derivative. Soil Sci Soc Am J 52: 1356–1360.

Skeffington RD, Wilson EJ (1988): Excess nitrogen deposition: issues for consideration. Environ Pollut 54: 159–184.

Spiteller M (1985): Untersuchungen zum Stickstoff-Voratsabbau zweier Waldökosysteme im Harz und Mont-Morency-Forest, Kanada. Mitt. Dtsch Bodenkd Ges 43: 477–482.

Stevenson FJ (1982): Humus chemistry, Genesis, composition, reactions. New York: Wiley.

Suflita JM, Bollag J-M (1981): Polymerization of phenolic compounds by a soil-enzyme complex. Soil Sci Soc Am J 45: 297–302.

Tamm CO (1991): Nitrogen in Terrestrial Ecosystems. Questions of Productivity, Vegetational Changes, and Ecosystem Stability. In Ecological Studies Vol. 81, Heidelberg: Springer-Verlag, 116 pp.

Tamm CO, Hallbäcken L (1988): Changes in soil acidity in two forest areas with different acid deposition: 1920s to 1980s. Ambio 17: 56–61.

Tipping E, Backes CA (1988): Organic complexing of Al in acid waters: model-testing by titration of a streamwater sdample. Water Res 22: 593–595.

Tipping E, Woolf C, Hurley MA (1991): Humic substances in acid surface waters; modelling aluminium biding, contriobution to ionic charge balance, and control of pH. Water Res 25: 425–435.

Trojanowski J, Haider K, Hüttermann A (1984): Decomposition of 14-C-labelled lignin, holocellulose, and lignocellulose by mycorrhiza fungi. Arch Microbiol 139: 202–206.

Ulrich B (1980): Die Wälder in Mitteleuropa: Messergebnisse ihrer Umweltbelastung, Theorie ihrer Gefährdung, Prognose ihrer Entwicklung. Allg Forstz 35: 1198–1202.

Ulrich B (1981): Theoretische Betrachtungen des Ionenkreislaufs in Waldökosystmen. Z Pflanzenernähr Bodenk 144: 647–659.

Van Miegrot H, Cole DW, Foster NW (1992): Nitrogen Distribution and Cycling. In Johnson DW, Lindberg SE (eds): Atmospheric Deposition and Forest Nutrient Cycling—A synthesis of the Integrated Forest Study Ecological Studies Vol. 91. Heidelberg: Springer-Verlag, pp 178–195.

Vitousek PM, Gosz JR, Grier CC, Melillo JM, Reiners WA (1979): Nitrate losses from disturbed ecosystems. Science 204: 469–474.

Vitousek, PM, Matson PA (1985): Disturbance, nitrogen availability, and nitrogen losses in an intensively managed Loblolly Pine plantation. Ecology 66: 1360–1376.

Wang MC, Huang PM (1987): Polycondensation of pyrogallol and glycine and the associated reactions as catalysed by birnessite. Sci Tot Environ 62: 435–442.

Wang MC, Huang PM (1989): Abiotic cleavage of pyrogallol and the associated reactions as catalysed by a natural soil. Sci Tot Environ. 81/82: 501–510.

Wang TSC, Li Song Wu, Huang PM (1978): Catalytic polymerisation of phenolic compounds by a latosol. Soil Sci 126: 81–86.

Wenzlik A (1993): Vorkommen stickstoffliebender Pflanzen in nordhessischen Buchenwäldern auf Basalt. Hildesheim/Holzminden: Diplomarbeit Fachhochschule, 57 pp.

Ziechmann W (1988): Evolution of structural models from consideration of physical and chemical properties. In Frimmel FH, Christman RF (eds): Humic substances and their Role in the Environment. Report of the Dahlem Workshop. Chichester: Wiley, pp 113–132.

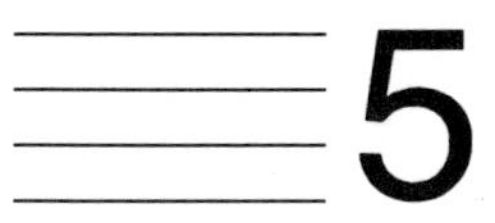

RELATIONSHIPS OF ACID DEPOSITION AND SULFUR DIOXIDE WITH FOREST DISEASES

JAMES J. WORRALL

College of Environmental Science and Forestry, State University of New York, Syracuse, New York 13210

1. INTRODUCTION

Because of the importance of diseases in natural and managed forests, they must be considered in any thorough evaluation of a pollutant's impact on a forest. A strong interest in the effects of air pollution on plant disease is evidenced by the frequency of literature reviews (Bruck and Shafer, 1984; Darley and Middleton, 1966; Heagle, 1973; Horn, 1985; Laurence, 1981; Manning, 1975; Shriner and Cowling, 1980; Treshow, 1975). This chapter is concerned with the interactions of acid deposition and the related pollutant sulfur dioxide (SO_2) with forest diseases. Because disease caused by acid deposition is considered elsewhere in this volume, the term "disease" in this chapter will exclude damage by air pollutants and nutrient imbalances related to acid deposition.

Although the primary focus of this volume is on acid deposition, most information on disease interactions is based on field observations, where it is not always possible to distinguish between acid deposition and direct effects of SO_2. Saunders (1966) concluded, as others had, that SO_2 dissolved in water as undissociated sulfurous acid (H_2SO_3) is the agent lethal to microorganisms. "Acid deposition" usually refers to deposition of sulfuric acid (H_2SO_4) and nitric acid (HNO_3), which are derived from SO_2 and other pollutants by complex reactions in the atmosphere.

Effects of Acid Rain on Forest Processes, pages 163–182

Because of the difficulty in distinguishing these effects in the field, all comprehensive reports dealing with acid deposition or SO_2 are considered here.

Another problem with field studies is that it is difficult, if not impossible, to find replicated stands that are alike in all respects except for the pollutant, and data are not often quantitative. Thus, natural variation in abundance of the diseases due to local site factors may confound the data. Those studies that do provide such comparisons involve a localized effect of pollutants from a point source, which may be different from the regional effects that are attributed to pollutants more often today.

Fundamentally, the relationships between air pollution and forest diseases may be considered in two categories. First, the disease may influence the tree's susceptibility to air pollutant damage. There has been relatively little work in this area, but experiments have demonstrated that potted *Betula pendula* Roth, with cherry leaf roll virus suffered severe symptoms and growth loss from combined exposure to SO_2 and ozone (O_3); uninfected plants were apparently unaffected by the fumigation (Kontzog et al., 1990). A similar experiment (but with different gas concentrations) showed that virus infection caused an increase of some pollutant symptoms but apparently protected against others (Hamacher and Giersiepen, 1989). Second, air pollutants may influence the severity or incidence of the disease. This interaction is the one that is of greatest concern and the primary subject of this chapter. In practice, a third relationship between pollutants and disease must also be considered: One may be confused with the other.

2. CONFUSION INVOLVING POLLUTANT DAMAGE AND OTHER DISEASES

2.1. Mistaking One for the Other

Because diseases are difficult to detect and diagnose, and because their external symptoms are often similar to those described for declines where pollutants are thought to be involved, it may be difficult to reliably distinguish diseases from such declines. Evaluation of pollutant damage by those unfamiliar with forest diseases makes it likely that diseases go unrecognized. For instance, Braun and Sauter (1983) stated that symptoms of needle diseases caused by *Lophodermium* and *Chrysomyxa* spp. are frequently confused with those of pollutant damage. The symptoms on pine of SO_2 injury and needle blight caused by *Rhizosphaera kalkhoffii* Bub. are also quite similar (Tanaka, 1980).

However, the situation is probably worse with root diseases, which are more difficult to diagnose. Kandler (1990, 1992) concluded that diseases and suboptimal forest conditions that have been known for years have been incorrectly regarded as novel forest damage and attributed to air pollution. He noted cases where trees in higher damage rating classes had more root and stem decays than those in lower classes. Schüler (cited in Courtois, 1983) found that four fir trees representing four stages of "Tannensterben" (fir death or fir decline) had roots decayed in direct proportion to the degree of needle loss.

Other recent studies suggest that more work is needed to distinguish damage by natural agents from pollutant-related declines. In Slovenia, a preliminary study attributed 30% of damage in a polluted forest to well known, easily recognized biotic and abiotic factors other than pollution (Solar and Jurc, 1989). In the northeastern United States, although certain internal US Forest Service memoranda and various committee reports may suggest that diseases play no primary role in the observed symptoms and mortality of *Picea rubens* Sarg. (Johnson et al., 1992), there is evidence to the contrary. For instance, Weiss et al. (1985) reviewed the many natural agents, including insects and biotic and abiotic diseases, that have been historically important in spruce-fir forests. Chronic wind stress, with its attendant crown and root damage, is an overlooked but important cause of crown dieback and growth decline in subalpine spruce-fir forests (Harrington, 1986; Rizzo and Harrington, 1988). In addition to chronic wind stress, a wide variety of biotic diseases and the spruce beetle could account for most of the mortality in Appalachian spruce-fir forests (Worrall and Harrington, 1988).

Rickettsia-like bacteria, mycoplasma-like organisms, and viruses were found in many of the tree species of interest (Nienhaus and Castello, 1989; Seemüller, 1989). Tomato mosaic virus was recently isolated from *P. rubens* in northeastern North America and reinoculated into seedlings of *P. rubens* and other conifers (Jacobi et al., 1992; Jacobi and Castello, 1992). Its influence on growth and decline are unknown.

2.2. Diseases, Declines, and Stress

The disease triangle (Fig. 5.1) is a simple but sturdy conceptual model of plant disease that has been used for years. It shifts focus away from the pathogen and onto the interaction of the pathogen, the host, and their abiotic and biotic environment. The length of each side depends on the favorability of that component for disease and partially determines the amount of disease (the area of the triangle).

Pathogens often infect and cause disease under "normal" conditions that are otherwise favorable to plant growth. In other cases, disease is favored by less common conditions that, for example, eliminate antagonists, permit germination and growth of pathogen propagules, or stress the host and compromise defense mechanisms. Particularly if host stress is thought to be involved, such pathogens may be called secondary pathogens or weak parasites.

For such cases, especially with respect to trees, models more detailed than the

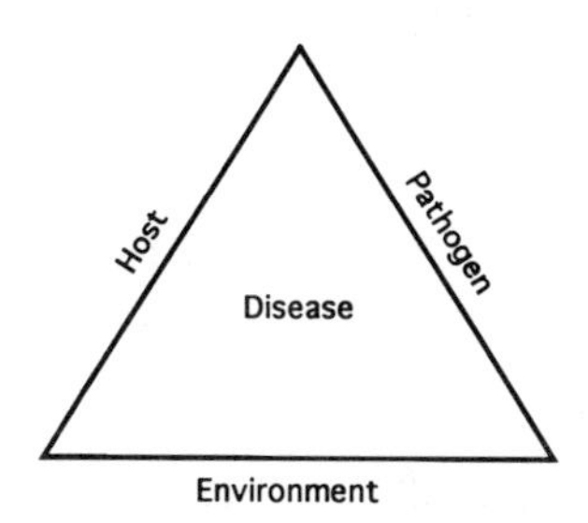

Fig. 5.1. The disease triangle.

disease triangle have been developed. Hüttermann (1984) summarized the concept of "Kettenkrankheiten" (literally, chain diseases) developed by Falck in 1918. Houston (1981) envisioned a cumulative effect of environmental stress from which trees may recover unless secondary organisms attack, causing irreversible decline. He used the term "dieback-decline diseases." Manion (1991) recognized three kinds of factors influencing decline diseases: (1) long-term predisposing factors; (2) short-term inciting factors; and (3) contributing factors, which are usually biotic agents that lead to death. The models share their emphasis on the point that environmental stress reduces the tree's ability to defend itself and allows otherwise innocuous agents to attack and kill it.

Such models are useful if one keeps in mind that they refer to a specific kind of disease and in no way represent all diseases. In recent analyses of forest problems and tree mortality, confusion arises when only decline diseases are recognized and pathogens are interpreted as evidence of stress. As will be discussed below, many, if not most, diseases actually *decrease* in stands stressed by air pollutants. In these cases, abundance of diseases could just as well be interpreted as evidence of lack of pollutant impact.

Johnson and Siccama (1983) suggested that *Climacocystis (Polyporus) borealis* (Fr.) Koutl. & Pouz., *Phellinus (Fomes) pini* (Thore:Fr.) A. Ames and *Armillaria mellea* (Vahl:Fr.) Kumm. found in "dieback" areas of Vermont and New Hampshire are "normally present but inconsequential in vigorous stands. Moreover, their invasion is triggered by stress in dieback-decline diseases." In fact, the few studies of C. borealis to date have not suggested that invasion is triggered by stress. Atkinson (1901) found that it infects basal wounds such as caused by falling trees, grows into the roots and up the stem in the heartwood, encroaches on the cambium, and may kill the tree. It can induce resinosis, a sign of a vigorous host response in resinous conifers. It has since been observed to play a role in the decline and death of mature *P. rubens* (Worrall and Harrington, 1988). The latter two pathogens are certainly consequential in many otherwise vigorous forests. Even *Armillaria,* for which host predisposition is a factor in many cases (Wargo and Harrington, 1991), is also known to attack vigorous young trees (Hartig, 1874; Livingston, 1990). Thus, the assumption cannot be made that pathogens are indicators of prior stress without case-by-case study.

In this connection it may be advisable to avoid confusion caused by using the word "dieback" as a general reference to decline diseases. Dieback is a specific symptom expressed by individual trees. Although the symptom is common in many decline syndromes (Manion, 1991), it is better not to confuse a symptom with a disease.

3. INFLUENCE OF ACID DEPOSITION AND SULFUR DIOXIDE ON DISEASES

Even in the early part of this century, differences were noted in abundance of certain diseases in polluted and unpolluted areas (see Donaubauer, 1966). It later became clear that, although some diseases increased in frequency in polluted areas, others clearly decreased.

3.1. Increased Disease Incidence

3.1.1. *Abiotic Diseases*

Sunscald may be a common but little recognized secondary effect of air pollutants. When trees especially susceptible to air pollutants die in a closed canopy, the sudden solar exposure of other, thin-barked species on flat sites or those on a southern aspect may contribute to the breakup of stands (Donaubauer, 1966).

Frost damage and winter injury have been associated with pollutant impacts, especially with conifers. Frost damage results from freezing of tissues by early or late frosts before hardening in the fall or, more commonly, after bud break in the spring. Winter injury, death of tissues during winter (although symptoms may not become apparent until spring), is often associated with unusual cycles of springlike weather in the winter followed by severe cold (Rehfuess and Rodenkirchen, 1984; Schwerdtfeger, 1981).

Widespread injury to conifers in Germany in 1955–1956 was associated with an extended cold period in February, with a monthly average of −8°C and lows to −36°C (Huber, 1956; Wentzel, 1956). Immediately beforehand there had been a springlike period with temperatures above 10°C. Damage was much more severe in areas receiving high amounts of pollutants, with many trees expected to die. A clear pattern of winter injury associated with a local source of SO_2 was also observed. Similar damage following a similar weather pattern was observed in Bavaria (Rehfuess and Rodenkirchen, 1984).

There is growing evidence that winter injury is a significant cause of damage to *P. rubens* in northeastern North America. There have been fairly frequent, though mostly unpublished, reports of significant winter injury, sometimes leading to mortality, in the last 50 years (Barnard et al., 1990). Most of the reports suggest that the damage is more severe at high elevations and, in contrast to experience in Europe (Zieger et al., 1958), on younger needles. Cold tolerance of current-year needles in December and January is often barely sufficient to protect from temperatures recorded rather frequently (DeHayes et al., 1990). *Picea rubens* is more susceptible to such injury than other northern conifers in the region. There is limited evidence that acid deposition may increase susceptibility to winter injury. In experiments on mature trees, branches exposed to lower than ambient ozone levels and protected from cloud acidity had less winter injury, but differences were not significant (Barnard et al., 1990). Shoots receiving those treatments also tolerated lower temperatures in freezing experiments. Additional evidence for a role of winter injury and its modification by airborne pollutants was summarized recently (Barnard et al., 1990; Johnson et al., 1992).

Autumn frost hardiness does not appear to play a significant role in the reported spruce-fir decline of North America (Sheppard et al., 1989). However, experimental treatment of conifer seedlings with simulated acid rain can increase susceptibility to frost damage (Flückiger et al., 1988; Fowler et al., 1989). In the latter study, autumn frost hardiness of seedlings was directly related to the pH of mist treatments in the range 2.5–5.0, and the effects were judged to be biologically significant in view of ambient deposition and autumn temperatures in the region of interest. The relationship between frost hardiness and susceptibility to winter injury is unclear because the mechanisms of winter injury are uncertain. To the extent that freezing damage is involved in winter injury, these studies also support the hypothesis that acid deposition may increase winter injury.

3.1.2. Foliage Diseases

Some *Lophodermium* spp. are thought to live in needles as symptomless endophytes until stress or senescence triggers their further development and fruiting (Osorio and Stephan, 1991). In young stands of *Picea abies* affected by air pollution in Czechoslovakia, heavy attacks of needlecast (*Lophodermium abietis* Rostr.) were observed (Kudela and Nováková, 1962). *Lophodermium piceae* (Fuckel) Höhn on the same host was more frequent in declining than in healthier stands (Suske et al., 1989). However, it caused asymptomatic infections and was not detected in most reddened needles locally characteristic of decline. Others associated the fungus with such symptoms, but winter injury and/or SO_2 were thought to also play a role (Rehfuess and Rodenkirchen, 1984).

An unprecedented outbreak of needle blight of *Pinus densiflora* Sieb. et Zucc. caused by *Rhizosphaera kalkhoffii* was noted in Japan in the mid-1960s when SO_2 pollution was highest (Chiba and Tanaka, 1968; Tanaka, 1980). When seedlings were inoculated and placed in unpolluted and polluted environments, only the latter showed symptoms. Similarly, plants receiving carbon-filtered air suffered much less severe needle blight than those receiving unfiltered air (Tanaka, 1980). Although *R. kalkhoffii* can be a serious pathogen of trees planted outside their natural range (Nicholls et al., 1974), it may function strictly as a saprobe in some situations (Diamandis, 1978). In Europe, the disease is said to be a typical indicator of pollutant damage on needles (Rack and Butin, 1984). On the other hand, Dotzler (1991) found *R. kalkhoffii* caused little or no infection and that SO_2, O_3, and/or H_2O stress had no effect. However, because of evidence that spores did not germinate on needle surfaces, it is possible that conditions for germination and infection after inoculation may not have been adequate.

Coniothyrium fuckelii Sacc., which can cause needle blight of *Pinus sylvestris* L., was isolated from a greater percentage of lesions (39%) in industrial areas than in an area relatively free of pollution (1.8%) (Kowalski, 1981). The relative frequency of lesions with respect to pollution was not reported.

Anthracnose caused by *Discula destructiva* Redlin suddenly appeared on *Cornus florida* L. in the eastern United States in the mid-1970s. The pathogen may have been introduced to North America (Hibben and Daughtrey, 1988; Redlin, 1991), which alone could explain the outbreak. However, it was suggested that acid deposition may contribute to the epidemic. This hypothesis was supported by findings that inoculations of leaves failed unless preceded by an acid mist treatment (Anderson et al., 1989). Subsequent, more sophisticated experiments with simulated rain (pH 2.5–5.5) demonstrated a negative relationship between pH and infection (R.L. Anderson, personal communication). In areas where the disease is epidemic, rain and cloud pH often falls to levels that were sufficient to increase susceptibility in the experiments. Although these data are consistent with the hypothesis that the epidemic was facilitated by acid deposition, further study is necessary to determine the role of anthropogenically derived acids in disease severity in the field.

3.1.3. Root and Stem Decays

Armillaria is a commonly cited example of a secondary pathogen invading trees weakened by stresses such as air pollution (Wargo and Harrington, 1991). Numerous workers, beginning as early as 1907, documented an increased incidence of the root pathogen in areas obviously affected by air pollution (Grzywacz and Wazny, 1973; Kudela and Nováková, 1962; Scheffer and Hedgcock, 1955; and earlier references cited in Donaubauer, 1966).

Several studies suggest that lower soil pH may promote Armillaria root rot. In a comparison of seedling infection in two soils, higher infection rates were obtained in the soil with lower pH and less Ca and Mg (Singh, 1983). Low soil pH was associated with field occurrence of Armillaria root rot in *Pseudotsuga menziesii* (Mirb.) Franco but not in *Abies grandis* Lindl. (Shields and Hobbs, 1979). The area studied was free of significant acidic deposition. However, a preference for acidic soils could conceivably contribute to observed increases of *Armillaria* in areas subject to acidic deposition.

In spruce-fir forests of northeastern United States, *Armillaria* is associated with a reported decline and death of trees, but primarily at low elevations (Carey et al., 1984; Worrall and Harrington, 1988). The higher elevation forests are considered to be more polluted, but a comparison based on pollution alone is not reasonable because the forests of differing elevation differ in so many other respects.

Conclusions regarding *Armillaria* are complicated by two caveats. First, it sometimes decreases in polluted areas, as discussed below. Second, *Armillaria* can, under certain circumstances, attack and kill vigorous and otherwise healthy trees, and was known as a major pathogen for many years (Hartig, 1874). The situation is complicated by the breakup of *A. mellea sensu lato* into a number of closely related species that differ in several ways, including aggressiveness and requirement for stress (Blodgett and Worrall, 1992a,b; Wargo and Harrington, 1991).

A list of wood-decay fungi found in damaged forests of the Erzgebirge in Czechoslovakia is available (Jancarik, 1961). Some were found only in an area of slight damage and others only in a more severely damaged area. Many of these fungi are saprobes or have extensive saprobic phases, so any increase might be explained by increased availability of wood killed directly by pollution.

3.1.4. Feeder Root Diseases

Cylindrocarpon destuctans (Zinss.) Schol. and *Mycelium radicis-atrovirens* Melin are known as seedling pathogens but have been isolated from fine roots of declining conifers (Kowalski, 1982; Schönhar, 1984). It was suggested that they may be able to attack mature trees that are weakened by air pollutants. Liss et al. (1984) suggested that, under the indirect influence of pollutants, feeder root pathogens contributed to loss of fine roots in *Picea abies* (L.) Karst. More root damage and higher frequency of putative pathogens were found in a forest with acid soil than in one with less acid to neutral soil (Schönhar, 1989). Their role in the reported decline was not assessed.

An apparent effect of pH on susceptibility to root diseases was demonstrated by a study in which seedlings of *Acer saccharum* Marsh. were treated at pH 2.4, 3.0, and 3.4 before planting (Roman and Raynal, 1980). When planted and given tapwater, survival was 7, 53, and 93%, respectively. An unspecified fungicide added after planting prevented the mortality, suggesting that fungal pathogenesis was facilitated by the prior acidic treatment.

3.2. No Effect or Decreased Disease Incidence

3.2.1. *Rusts*

Many diseases clearly decrease in the presence of pollutants. Diseases caused by obligate parasites, such as rusts and powdery mildews, appear to be particularly sensitive. Around two smelters in western North America, the forests naturally supported a rich flora of rust fungi of both leaves and stems in the genera *Coleosporium, Cronartium, Gymnosporangium, Melampsora, Peridermium, Phragmidium, Puccinia,* and *Pucciniastrum* (Scheffer and Hedgcock, 1955). All were either absent in the area of direct tree damage or present at lower levels than in surrounding forests. Similar effects were seen in another study (Grzywacz and Wazny, 1973). In four zones around smelters in Ontario, there was no white pine blister rust in the inner zones and progressively higher levels up to 10% of trees in the relatively unpolluted zone (Linzon, 1971). An exception, in which *Melampsora pinitorqua* (Braun) Rostr. was epidemic in a polluted stand, was explained by the especially conducive stand composition and dense stand structure that protected the understory from pollutants (Domanski, cited in Krzan and Siwecki, 1980).

In agreement with the observations above, experiments demonstrated a negative effect on rust of simulated acid rain treatments. Frequency of fusiform rust of *Pinus taeda* L. caused by *Cronartium quercuum* (Berk.) Miyabe ex Shirai f. sp. *fusiforme* dropped from 2.6 to 1.6 galls per 20 trees in treatments with pH 5.2 and 4.0, respectively, and lower pH treatments continued the trend (Bruck et al., 1981). Similarly, 86% inhibition of telia production on *Quercus* sp. was associated with pH 3.2 compared with 6.0 (Shriner, 1977).

3.2.2. *Foliage Diseases*

Köck (1935) observed that powdery mildew of oak (*Microsphaera alni* (Wallr.:Fr.) G. Wint. var. *quercina*), which was very common elsewhere in Austria, was completely absent from a region influenced by a paper mill that discharged sulfuric acid into the atmosphere. Later studies in Poland confirmed that oak mildew (as *M.*

alphitoides Griff. & Maubl.) was relatively sparse in industrial areas (Grzywacz and Wazny, 1973). Similarly, lilac mildew (*Microsphaera alni*) was conspicuously absent in polluted areas (Hibben and Taylor, 1975). Even trees near a roadside had lower levels of mildew than trees 200 m from the road, though higher leaf temperatures near the road may have been involved (Flückiger and Oerteli, 1978). In contrast, Kowalski (1983) reported one and cited another case of abundant oak mildew in industrial areas of Poland, but other factors may have been involved, and no comparison was made with unpolluted areas.

In contrast to earlier cited examples, some studies suggested no effect or a decrease of needlecasts in response to SO_2 and related pollutants. The percentage of forest area occupied by *Lophodermium pinastri* (Schrad.:Fr.) Chev. on *P. sylvestris* did not clearly differ between industrial and less industrial regions of Poland (Grzywacz and Wazny, 1973). Percent of colonized needles in litter was very low near a point source of SO_2, rose to 70% somewhat further away, and then declined again, albeit less steeply, with distance. Kowalski (1981) found a trend of decreasing frequency of isolation of *Lophodermium seditiosum* Minter, St. & Mil. in zones with increasing pollutant damage. Needlecasts caused by fungi in the genera *Lophodermium, Hypoderma,* and *Hypodermella* near the smelter at Trail, British Columbia, were less frequent in polluted areas (Scheffer and Hedgcock, 1955). For example, *L. pinastri* was absent in the zone with greatest SO_2 impact, sparse where there was moderate injury, and more abundant where injury was least. Frequencies of most other needle and twig pathogens were rather low and about the same in and outside the areas of injury.

Tar spot of maple, caused by *Rhytisma acerinum* (Pers.:Fr.) Fr., was absent in areas with high SO_2 levels, and fewer lesions occurred in areas of moderate SO_2 than in areas with low SO_2 (Brevan and Greenhalgh, 1976; Greenhalgh and Bevan, 1978). Black spot of rose (*Diplocarpon rosae* Wolf) is also reduced or absent where SO_2 exceeds 100 $\mu g\ m^{-3}$ (Saunders, 1966).

3.2.3. Root Diseases

Although Armillaria root and butt rot was mentioned as a classic example of a disease that increases in stands stressed by pollution, there are some contradictory data. Grzywacz and Wazny (1973) reported a trend near a point source of SO_2 in which four zones ranging from severely damaged to undamaged had 0.2, 0.6, 1.2, and 1.0% of *P. sylvestris* seedlings with symptoms caused by *Armillaria*. Only in the "slightly damaged" zone was *Armillaria* more frequent than in the least polluted zone, and the increase was slight. Using the same zone classes, Domanski (1978) found a progressive decrease in *Armillaria* with pollution; the fungus was not found in the strongly damaged zone.

Annosum root and butt rot, caused by *Heterobasidion annosum* (Fr.) Bref., occupied a lower proportion of land area (0.87%) in 40 polluted forests in Poland than in all Polish forests considered together (1.37%) (Grzywacz and Wazny, 1973). Also, an effect of distance from source was observed in one more intensively studied area. This effect was more frequent in the moderate damage zone than in zones more or less impacted by SO_2. Similar results were reported by Domanski (1978). Because annosum root rot is often most severe on alkaline soils, a decrease in disease might be expected if soil acidification occurred.

Natural populations of *Phytophthora cinnamomi* Rands, a pathogen in some forest systems, decreased in the upper 4 cm of soil when treated with simulated acid rain at pH < 4 (Shafer et al., 1982). The magnitude of the effect was not reported. Simulated acid rain treatments decreased frequency of infection of lupine and numbers of sporangia formed on uniformly infected lupine radicles (Shafer et al., 1985). However, plants ultimately became infected in all acid treatments, and the investigators concluded that acid rain probably has little effect on Phytophthora root rot, except over decades and in poorly buffered soils.

3.2.4. Stem Diseases

Other than for rusts, there are no clear examples of substantial decrease of stem diseases by SO_2 or acid deposition. In Scheffer and Hedgcock's (1955) study of forests around smelters, frequencies of stem decays were unaffected by the pollution. Near the Anaconda, Montana smelter and in the surrounding, unaffected forest about 7% of *Pinus contorta* Dougl. had stem decay and 72% of *Pseudotsuga menziesii* Franco had either *Phaeolus schweinitzii* (Fr.) Pat. or *Phellinus pini* (Thore:Fr.) A. Ames. However, Linzon (1966) reported that stem decays were less frequent near the Sudbury, Ontario smelters than in less polluted forests. A study in polluted stands of *Larix decidua Mill.* (Kowalski, 1982) compared effects of several harvesting systems on frost damage and a wide variety of fungi, but no comparison with stands less affected by pollutants was made. There is some evidence that beech bark disease may be somewhat decreased by sulfur pollution (Decourt et al., 1980).

Although there was speculation in the late 1970s and early 1980s that acid rain predisposed trees both in Scandinavia and the northeastern United States to Scleroderris canker, caused by *Ascocalyx (Gremmeniella) abietina* (Lag.) Schl., evidence has not convincingly demonstrated any relationship. Field experiments in the United States showed slightly, but not significantly, less infection of seedlings in more acid treatments (pH 3.5 vs. 5.6) (Bragg and Manion, 1984; Setliff and Manion, 1980). However, natural disease increase rates were somewhat higher on more acid soils. The overall conclusion of these studies was that acid rain was not a significant factor for the disease. Donaubauer (1984) found no correlation between acid deposition or other pollutants and occurrence of Scleroderris canker in Europe. Fumigation with SO_2 in the field did not affect the infection rate of inoculated pine (Laurence et al., 1984). In laboratory experiments, however, acid mist treatments (pH 3.5 with a weekly pulse of pH 2.8) did lead to increased infection of *Picea abies* cuttings (Barklund and Unestam, 1988).

A regional decline of *Quercus* spp. in northern Germany reportedly involves cambial necrosis in the lower stem (Balder and Dujesiefken, 1989; Hartmann et al., 1989). Early reports suggested that industrial emissions may be involved (Anonymous, 1983), but drought stress now appears to be more important. Several *Ophiostoma* spp. may be involved and *O. piceae* (Münch) Syd. & P. Syd. causes symptoms like those in the field, but only on trees that were subjected to drought stress (Eisenhauer, 1991). *Ophiostoma* spp. were not found in a study of fungi attacking oaks in an industrial area where trees were presumably weakened by immissions (Kowalski, 1983). Similarly, *Pezicula cinnamomea* (DC.) Sacc. causes cankers in such trees, but evidence suggests a predisposition by drought stress (Kehr, 1991). Fungi known as pathogens of vigorous oaks were infrequently associated with oak decline in Poland (Kowalski, 1991).

3.3. Mechanisms

Acid deposition may influence forest diseases via the following mechanisms (based in part on Shriner and Cowling, 1980):

- Direct contact with and effect on pathogens by acidic moisture.
- Effect on microbes antagonistic (or protagonistic) to pathogens.
- Effect on host susceptibility (predisposition), including:
  - Increased rate of cuticle weathering, rendering plants more susceptible to certain pathogens
  - Interference with stomatal function
  - Plant injury, creating infection court
  - Leaching of germination inhibitors from host
  - Leaching of nutrients from host
  - Other physiological disturbances, such as soil nutrient effects,
  - Synergistic interaction with pollutants, light and temperature.

3.3.1. Direct Effects on Pathogens

In the case of powdery mildews disappearing in regions of H_2SO_4 deposition, Köck (1935) noted that the effect was probably similar to the control of mildews in agriculture with finely ground elemental sulfur. Sulfur dioxide fumigation of lilac also suggested direct inhibition, because fumigation before inoculation had no effect (Hibben and Taylor, 1975).

An experiment with tarspot of maple suggests that the pathogen is vulnerable only prior to infection. Although polluted air apparently prevents infection, as discussed earlier, no inhibition of disease was observed in plants exposed to polluted air after infection (Greenhalgh and Bevan, 1978).

Rusts and wood-decay fungi are said to be unusually sensitive to SO_2 (Heagle, 1973; Weinstein et al., 1975). This sensitivity may explain the clear decrease in rust diseases near smelters noted above. However, in the case of wood-decay fungi, fumigation with SO_2 resulted in stimulation of growth at concentrations as high as 1 mg m^{-3} and less than 50% inhibition at 1000 mg m^{-3} (Grzywacz, 1973), suggesting quite high tolerance compared to plants. Also, wood-decay fungi, such as those causing stem decays and root and butt rots, are probably less directly exposed to the pollutants, regardless of their form, and consequently less vulnerable.

3.3.2. *Effects on Antagonists*

Microbial antagonists function in the natural biological control of plant diseases. Phylloplane microbes may influence the prepenetration development of foliage pathogens. Simulated acid rain partially reduced total colony forming units of fungi on birch leaves at some sampling dates (Helander and Rantio-Lehtimäki, 1990). Bacteria, also important members of phylloplane communities, are generally more sensitive to low pH than are fungi, and their sensitivity to simulated acid rain on leaf surfaces was confirmed (Lacy et al., 1981; Shriner, 1977).

As mentioned above, acid mist treatments in one study favored the development of *A. abietina* on Norway spruce (Barklund and Unestam, 1988). The mist treatments were found to also greatly reduce the epiphytic and endophytic flora of the shoots (Barklund et al., 1984; Barklund and Unestam, 1988). The pathogen normally has a period of latent infection before cold temperatures permit further invasion and symptom development. The authors suggested that epiphytes regulate establishment of latent infections and endophytes play a role in controlling further invasion. Thus, acidification may increase frequency and severity of infections by inhibiting the antagonistic microflora.

3.3.3. *Effects on Host Susceptibility*

Ulrich's (1981) hypothesis to account for forest decline involves damage of fine roots by Al and Mn released by acidified soil, followed possibly by entry of weak pathogens. Aerial plant parts could become susceptible because of a water deficit induced by root damage (Schütt, 1981). These particular mechanisms have not been demonstrated with respect to diseases, but increases in susceptibility were shown in a number of experimental systems.

Treatment of plants with pollutants before introducing the pathogen were used to demonstrate effects on susceptibility in several systems. Although the pH values used may be unrealistic, the study cited above by Roman and Raynal (1980) suggested an increase in host susceptibility that was not mediated by soil cations. The acid treatment was suspended before seedlings were placed in the soil, where they were exposed to the pathogen(s). The increase of dogwood anthracnose by acid rain treatments discussed earlier (R.L. Anderson, personal communication) also suggests an experimental effect on host susceptibility. Similarly, a preinoculation treatment with SO_2 caused increased susceptibility of *Pinus densiflora* to *Rhizosphaera kalkhoffii* (Chiba and Tanaka, 1968). Inoculation alone led to no obvious symptoms. Fumigation (5.2 mg m^{-3} for 1–4 h) alone caused leaf damage and inoculation after fumigation caused somewhat more symptoms and led to pycnidial production. However, fumigation after inoculation led to even greater damage and more fruiting, suggesting that the pathogen was less sensitive to the treatment than the host or that pathogenesis was even stimulated by fumigation.

Sulfur dioxide affects infection of *P. sylvestris* by *Mycosphaerella dearnessii* Barr (= *Scirrhia acicola*), cause of brown spot needle blight of pines (Weidensaul and Darling, 1979). There is no report of this disease increasing in the field due to ambient pollution, and *P. sylvestris* is not a natural host. Seedlings exposed to 533

$\mu g\ SO_2\ m^{-3}$ five days after inoculation had significantly more fungal lesions than unexposed seedlings. The fungus in culture was apparently unaffected by much higher levels (2600 $\mu g\ SO_2\ m^{-3}$) (Ham, cited in Weidensaul and Darling, 1979), suggesting that the effect on disease may be mediated by host resistance. However, when plants were fumigated immediately before inoculation, disease was not significantly different from unfumigated plants.

In one study involving mixed pollutants near a highway, two mechanisms of influence on disease with opposing effects were suggested (Flückiger et al., 1989). Apparently, the pollutants increased mycelial invasion and lesion growth, but decreased conidial germination and infectivity. When branches of potted *Platanus X acerifolia* (Ait.) Willd. were inoculated with mycelium of *Apiognomonia veneta* (Sacc. & Speg.) Höhn. (cause of sycamore anthracnose), disease was more severe in plants exposed to ambient air than in those in filtered air. The mechanism may be related to increases in amino acids in the phloem (Bolsinger and Flückiger, 1989). In contrast, leaves inoculated with conidial suspensions had less infection in ambient than in filtered air. The level of disease with respect to distance from the roadway was apparently not studied, so it is unclear which mechanism had the overriding effect.

Susceptibility to forest pathogens may increase through degradation of leaf surfaces (Barg, 1987; Shriner, 1977; Brown et al., 1991). Erosion of epicuticular waxes, damage to leaf trichomes, and degradation of other surface elements may make them more wettable (supporting epiphytic germination and growth of pathogen propagules) and easier to penetrate. However, it remains to be demonstrated that such effects lead to higher frequency of penetration.

3.3.4. Relationship of Dose Response to Mechanisms

A hypothetical scheme of interactions is offered to facilitate interpretation of disease distributions with respect to air pollution and to aid assessment of the major kinds of mechanisms (Fig. 5.2). At the risk of oversimplification, diseases are grouped into necrotrophic (tissues killed before any invasion) and biotrophic (pathogen feeds largely on living tissues).

Direct inhibition of pathogens would obviously decrease both biotrophic and necrotrophic diseases (Fig. 5.2 *a,b*). Tar spot of maple (biotrophic) probably is inhibited largely by this mechanism. The effect of host stress caused by pollutants depends on the type of disease. A number of necrotrophic diseases are apparently favored by pollutant-related stress (Fig. 5.2*c*). Rhizosphaera needlecast, dogwood anthracnose, winter injury, and in some cases Armillaria root rot and Lophodermium needlecasts may fit this mechanism. Biotrophic diseases, in contrast, often do best on vigorous hosts and may decrease in stressed hosts (Fig. 5.2*d*). Powdery mildews and stem and needle rusts likely fit this mechanism to some extent.

If both host stress and direct inhibition of the pathogen are operative, they would have opposite effects on many necrotrophic diseases. The net effect depends on the relative importance of the two. If the host is stressed at concentrations below those that inhibit the pathogen, an initial increase would be expected before the disease

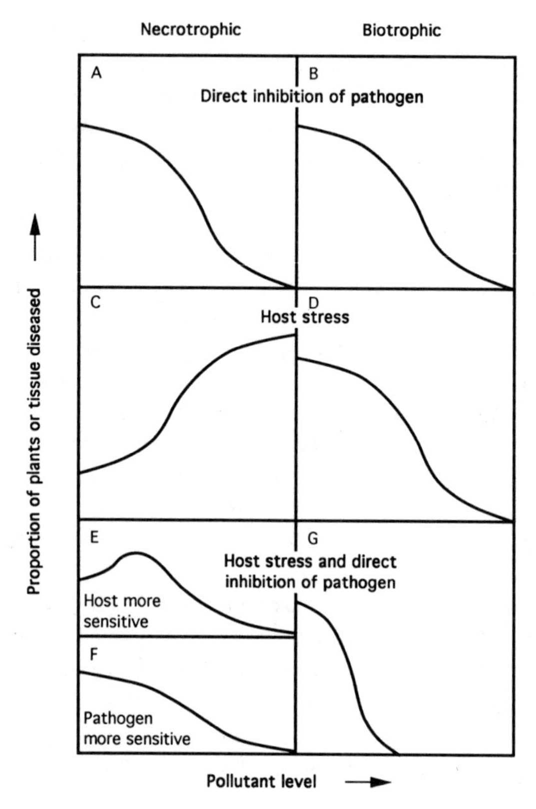

Fig. 5.2. Hypothetical effects of pollutant dose on amount of disease in two kinds of pathosystems with three general mechanisms of pollutant action.

drops to levels below those in unpolluted forests (Fig. 5.2*e*). There is evidence suggesting such a pattern for Armillaria root rot, annosum root rot, and Lophodermium needlecasts in certain cases. This phenomenon may be more common than is now known and could explain some of the apparently contradictory reports. If the pathogen is more sensitive, a steady downward trend in disease would be expected, though it could be mitigated by the effect of host stress (Fig. 5.2*f*). Sycamore anthracnose apparently fits one of the latter two patterns.

For biotrophic diseases, the combined effect would likely be additive or even synergistic, resulting in a steep decline in disease amount (Fig. 5.2*g*). Because biotrophic diseases would be expected to decrease by all of the three mechanisms, it is not possible to deduce the mechanism by fitting one predicted pattern to field observations. With powdery mildews, their largely superficial thalli and known sensitivity to sulfur suggest that direct inhibition is an important mechanism. Further physiological studies are necessary to determine which mechanisms are operative in these cases.

Several additional possibilities are not presented graphically. Cases where a pollutant directly stimulates the pathogen or promotes host vigor are possible, given stimulation of some fungi by certain concentrations of SO_2 and the stimulation of tree growth by nutrient input. An indirect stimulation of root and butt rots could occur if trees killed directly by pollution provide pathogens with greater inoculum potential for attacking residual trees. Effects on antagonists are also not included. A negative effect on antagonists might mimic the effects of host stress on necrotrophic diseases.

4. SUMMARY

The impacts of air pollutants on forests must be considered against the backdrop of the endemic diseases, with which air pollution damage may be confused and which air pollutants may affect. The research reports reviewed here include many examples of both stimulation and inhibition of diseases by acid deposition and SO_2. All reports of leaf and stem rusts, mildews, and several other foliage diseases like tar spot of maple show a clear decrease in polluted air. All reports of abiotic diseases, such as frost damage and winter injury, suggest that pollution leads to an increase in such damage. For Armillaria root rot, Lophodermium needlecasts, and other foliage diseases, such as caused by *Rhizosphaera kalkhoffii,* the preponderance of reports indicate an increase in polluted situations, but contradictory data can be found. In a few cases, a dose-dependent effect has been demonstrated in which disease increases at low doses and decreases at higher doses. The mechanisms for these effects are complex and apparently involve direct effects on pathogens, variable effects of host stress, and other mechanisms.

REFERENCES

Anderson RL, Knighten JL, Dowsett S (1989): Enhancement of *Discula* sp. infection of flowering dogwood (*Cornus florida*) by pretreating leaves with acid mist (Abstract). Plant Dis 73:859.

Anonymous (1983): Zum Erkennen von Immissions-Schäden an Waldbäumen 1983. Allg Forst Z 51/52:1387–1402.

Atkinson GF (1901): Studies of some shade tree and timber destroying fungi. Cornell Univ Agric Expt Sta (Ithaca) Bull. 193:197–235.

Balder H, Dujesiefken D (1989): Neuartige Schadsymptome der Eiche in Norddeutschland. In Internationaler Kongress Waldschadensforschung: Wissenstand und Perspektiven. Poster Kurzfassungen, Band I. Forschungsbeirat Waldschäden/Luftverunreinigung, Germany, pp 59–60.

Barg VS (1987): Plant cuticle as a barrier to acid penetration. In Hutchinson TC, Meema KM (Eds): Effects of Atmospheric Pollutants on Forests, Wetlands and Agricultural Ecosystems. Berlin: Springer-Verlag, pp 145–154.

Barklund P, Axelsson G, Unestam T (1984): *Gremmeniella abietina* in Norway spruce, latent infection, sudden outbreaks, acid rain, predisposition. In Manion PD (ed): Scleroderris Canker of Conifers. The Hague: Martinus Nijhoff/Dr. W. Junk Publisheres, pp. 111–113.

Barklund P, Unestam T (1988): Infection experiments with *Gremmeniella abietina* on seedlings of Norway spruce and Scots pine. Eur J For Pathol 18:409–420.

Barnard JE, Lucier AA, et al. (1990): Changes in Forest Health and Productivity in the United States and Canada. Report 16 of the National Acid Precipitation Assessment Prog. series titled: Acidic Deposition: State of Science and Technology. Washington: US Government Printing Office, 186 pp.

Bevan RJ, Greenhalgh GN (1976): *Rhytisma acerinum* as a biological indicator of pollution. Environ Pollut 10:271–285.

Bolsinger M, Flückinger W (1989): Ambient air pollution induced changes in amino acid pattern of phloem sap in host plants - relevance to aphid infestation. Environ Pollut 56:209–216.

Blodgett JT, Worrall JJ (1992a): Distributions and hosts of *Armillaria* species in New York. Plant Dis 76:166–169.

Blodgett JT, Worrall JJ (1992b): Site relationships of *Armillaria* species in New York. Plant Dis 76:170–174.

Bragg RJ, Manion PD (1984): Evaluation of possible effects of acid rain on Scleroderris canker of red pine in New York. In Manion PD (ed): Schleroderris Canker of Conifers. The Hague: Martinus Nijhoff/Dr. W. Junk Publishers, pp 130–141.

Braun HJ, Sauter JJ (1983): Unterschiedliche Symptome des "Waldsterbens" im Schwarzwald—mögliche Kausalketten und Basis-Ursachen. Allg Forstz 26/27: 656–660.

Brown DA, Windham MT, Graham ET, Trigiano RN, Anderson RL, Berrang P (1991): Effects of acid rain treatment on the flowering dogwood (*Cornus florida* L.) leaf surface: an SEM study (Abstract). Phytopathology 81:1141.

Bruck RI, Shafer SR, Heagle AS (1981): Effects of simulated acid rain on the development of fusiform rust on loblolly pine (Abstract). Phytopathology 71:864.

Bruck RI, Shafer SR (1984): Effects of acid precipitation on plant diseases. In Linthurst RA (ed): Direct and Indirect Effects of Acidic Deposition on Vegetation. Woburn, UK:Butterworths, pp. 19–32.

Carey AC, Miller EA, Geballe GT, Wargo PM, Smith WH, Siccama TG (1984): *Armillaria mellea* and decline of red spruce. Plant Dis 68:794–795.

Chiba O, Tanaka K (1968): The effect of sulfur dioxide on the devlopment of pine needle blight caused by *Rhizosphaera kalkhoffi* Buback (I.) J Japn For Soc 50:135.

Courtois H (1983): Die Pathogenese des Tannensterbens und ihre natürlichen Mechanismen. Allg Forst Jadzg. 154:93–98.

Darley EF, Middleton JT (1966): Problems of air pollution in plant pathology. Annu Rev Phytopathol 4:103–118.

Decourt N, Malphettes CB, Perrin R, Caron D (1980): La pollution soufrée limite-t-elle le développement de la maladie de l'écorce du hêtre? Ann Sci For 37:135–145.

DeHayes DH, Waite CE, Ingle MA, Williams MW (1990): Winter injury susceptibility and cold tolerance of current and year-old needles of red spruce trees from several provenances. For Sci 36:982–994.

Diamandis S (1978): Top-dying of Norway spruce, *Picea abies* (L.) Karst., with special reference to *Rhizosphaera kalkhoffii* Bubák. II. Status of *R. kalkhoffii* in "top-dying" of Norway spruce. Eur J For Pathol 8:345–356.

Domanski S (1978): Fungi occurring in forests injured by air pollutants in the upper Silesia and Cracow industrial regions of Poland. VI. Higher fungi colonizing the roots of trees in converted forest stands. Acta Soc Bot Pol 47:285–295.

Donaubauer E (1966): Durch Industrie-abgase bedingte Sekundärschäden am Wald. Mitt Forst Bundes-Versuchsanst Wien 73:101.

Donaubauer E (1984): Experience with Scleroderris canker on *Pinus cembra* L. in afforestations of high altitude. In Manion PD (ed): Scleroderris Canker of Conifers. Proceedings of the International Symposium. The Hague: Martinus Nijhoff/Dr W. Junk, pp 158–161.

Dotzler M (1991): Infektionsversuche mit *Rhizosphaera kalkhoffii* und *Lophodermium piceae* an unterschiedlich gestressten Jungfichten (*Picea abies* [L.] Karst.). Eur J For Pathol 21:107–123.

Eisenhauer DR (1991): Zur Taxonomie und Pathogenität von *Ophiostoma piceae* (Münch) Syd. in Zusammenhang mit Absterbeerscheinungen in Trauben- und Stieleichenbeständen des mittel- und nordostdeutschen Deluviums. Eur J For Pathol 21:267–278.

Flückiger W, Braun S, Leonardi S (1988): Air pollution, nutrient imbalances and changed resistance to parasites and frost (Abstract). In: Abstracts of Acidic Deposition and Forest Decline, An International Symposium, Faculty of Forestry Miscellaneous Publication No. 20 (ESF88-005). Syracuse: SUNY College of Environmental Science and Forestry, p 8.

Flückiger W, Oerteli JJ (1978): Der Einfluss verkehrsbedingter Luftverunreinigungen auf den Befall der Eiche durch *Microsphaera alphitoides*. Phytopathol Z 93:363–366.

Flückiger, W, v. Sury R, Braun S (1989): Auswirkungen von Verkehrsimmissionen auf den Schadorganismenbefall von Strassenrandgehölzen. Kali-Briefe (Büntehof) 19:767–780.

Fowler D, Cape JN, Deans JD, Leith ID, Murray MB, Smith RI, Sheppard LJ, Unsworth MH (1989): Effects of acid mist on the frost hardiness of red spruce seedlings. New Phytol 113:321–335.

Greenhalgh GN, Bevan RJ (1978): Response of *Rhytisma acerinum* to air pollution. Trans Br Mycol Soc 71:491–494.

Grzywacz A (1973): Sensitivitiy of *Fomes annosus* Fr. Cooke and *Schizophyllum commune* Fr. to air pollution with sulphur dioxide. Acta Soc Bot Pol 42:347–360.

Grzywacz A, Wazny J (1973): The impact of industrial air pollutants on the occurence of several important pathogenic fungi of forest trees in Poland. Eur J For Pathol 3:129–141.

Hamacher J, Giersiepen R (1989): Histologisch-cytologische Veränderungen in Kirschenblattrollvirus-infizierten und durch Luftschadstoffe gestressten *Betula*-Arten. Nachrichtenbl. Dtsch Pflanzenschutzdienst (Braunschweig) 41:124–130.

Harrington TC (1986): Growth decline of wind-exposed red spruce and balsam fir in the White Mountains. Can J For Res 16:232–238.

Hartig R (1874): Wichtige Krankheiten der Waldbäume. Beiträge zur Mykologie und Phytopathologie für Botaniker und Forstmänner. Berlin: Springer. 127 pp.

Hartmann G, Blank R, Lewark S (1989): Das gegenwärtige Eichensterben in Norddeutschland—Verbreitung, Symptome und mögliche Ursachen. In Internationaler Kongress Waldschadensforschung: Wissenstand und Perspektiven. Poster Kurzfassungen, Band I. Forschungsbeirat Waldschäden/ Luftverunreinigung, Germany, pp 61–62.

Heagle AS (1973): Interactions between air pollutants and plant parasites. Annu Rev Phytopathol 11:365–388.

Helander ML, Rantio-Lehtimäki A (1990): Effects of watering and simulated acid rain on quantity of phyllosphere fungi of birch leaves. Microb Ecol 19:119–125.

Hibben CR, Daughtrey ML (1988): Dogwood anthracnose in northeastern United States. Plant Dis 72:199–203.

Hibben CR, Taylor MP (1975): Ozone and sulphur dioxide effects on the lilac powdery mildew fungus. Environ Pollut 9:107–114.

Horn NM (1985): Effects of air pollution and acid rain on fungal and bacterial diseases of trees. Uitvoerig verslag Band 20, nr. 1. Wageningen: Dorschkamp Research Institute for Forestry and Landscape Planning, 69 pp.

Houston DR (1981): Stress triggered tree diseases—the diebacks and declines. United States Department of Agriculture and Forest Service. NE-INF-41-81, 36 pp.

Huber B (1956): Winterfrost 1956 und Rauchschäden. Allg Forstz 11: 609–610.

Hüttermann A (1984): Die Anwendung der Koch'schen Postulate auf die Untersuchungen zur Ursachenforschung des Waldsterbens. Forstarchiv 55(2):45–48.

Jacobi V, Castello JD (1992): Infection of red spruce, black spruce, and balsam fir seedlings with tomato mosaic virus. Can J For Res 22:919–924.

Jacobi V, Castello JD, Flachmann M (1992): Isolation of tomatoc mosaic virus from red spruce. Plant Dis 76:518–522.

Jancarik V (1961): Vyskyt drevokaznych hub v kourem poskozovani oblasti Krusnych hor. Lesnictvi 7:677–692.

Johnson AH, Friedland AJ, Miller EK, Battles JJ, Huntington TG, Vann DR, Grombeck GR (1992): Eastern spruce fir. In Johnson, DW, Lindberg, SE (eds): Atmospheric Deposition and Forest Nutrient Cycling. New York: Springer-Verlag, pp 496–525.

Johnson AH, Siccama TC (1983): Acid deposition and forest decline. Environ Sci Technol 17:294A–305A.

Kandler O (1990): Epidemiological evaluation of the development of Waldsterben in Germany. Plant Dis 74:4–12.

Kandler O (1992): The German forest decline situation: a complex disease or a complex of diseases. In Manion PD, La Chance D (eds): Forest Decline Concepts. St. Paul: APS Press, pp 59–84.

Kehr RD (1991): Pezicula canker of *Quercus rubra* L., caused by *Pezicula cinnamomea* (DC.) Sacc. I. Symptoms and pathogenesis. Eur J For Pathol 21:218–223.

Köck G (1935): Eichenmehltau und Rauchgasschäden. Z Pflanzenkr Pflanzenschultz 45:44–45.

Kontzog H-G, Kleinhempel H, Matschke J (1990): Combined effects of environmental stress and virus infections on the growth of forest trees. Arch Phytopathol Pflanzenschutz 26:359–362.

Kowalski T (1981): Fungi infecting needles of *Pinus sylvestris* in Poland in relation to air pollutuion zone. In Millar, CS (ed): Current Research on Conifer Needle Diseases. Proceedings of the IUFRO Working Party on Needle Diseases, pp 93–98.

Kowalski T (1982): Vorkommen von Pilzen in durch Luftverunreinigung geschädigten Wäldern im Oberschleisischen und Krakauer Industriegebiet. VIII Mykoflora von Larix decidua an einem Standort mit mittlerer Immissionsbelastung. Eur J For Pathol 12:262–272.

Kowalski T (1983): Vorkommen von Pilzen in durch Luftverunreinigung geschädigten Wäldern in Oberschlesischen und Krakauer Industriegebiet. IX. Mykoflora von Quercus robur L. und *Quercus rubra* L. an einem Standort mit mittlerer Immissionsbelastung. Eur J For Pathol 13:46–59.

Kowalski T (1991): Oak decline: I. Fungi associated with various disease symptoms on overground portions of middle-aged and old oak (Quercus robur L.). Eur J For Pathol 21:136–151.

Krzan Z, Siwecki R (1980): Recent studies on *Melampsora larici-populina* and *Melampsora pinitorqua* in Poland. Folia Forestalia 422:14–16.

Kudela M, Novákáva E (1962): Lesní skudci a skody zverí v lesích poskozovanch kourem. Lesnictví 6:493–502.

Lacy GH, Chevone BI, Cannon NP (1981): Effects of simulated acidic precipitation on *Erwinia herbicola* and *Pseudomonas syringae* populations. Phytopathol 71: 888.

Laurence JA (1981): Effects of air pollutants on plant–pathogen interactions. J Plant Dis Protect 88:156–172.

Laurence JA, Reynolds KL, MacLean Jr. DC, Hudler GW, Dochinger LS (1984): Effects of sulfur dioxide on infection of red pine by *Gremmeniella abietina*. In Manion PD (ed): Scleroderris Canker of Conifers. Proc Intl Symp The Hague: Martinus Nijhoff/Dr. W. Junk, pp 122–129.

Linzon SN (1966): Damage to eastern white pine by sulfur dioxide, semimature-tissue needle blight and ozone. J Air Pollut Control Assoc 16:140–144.

Linzon SN (1971): Economic effects of sulfur dioxide on forest growth. J Air Pollut Control Assoc 21:81.

Liss B, Blaschke H, Schütt P (1984): Vergleichende Feinwurzeluntersuchungen an gesunden und erkrankten Altfichten auf zwei Standorten in Bayern—ein Beitrag zur Waldsterbenforschung. Eur J For Pathol 14:90–102.

Livingston WH (1990): *Armillaria ostoyae* in young spruce plantations. Can J For Res 20:1773–1778.

Manion PD (1991): Tree Disease Concepts. Englewood Cliffs: Prentice-Hall. 402 pp.

Manning WJ (1975): Interactions between air pollutants and fungal, bacterial and viral plant pathogens. Environ Pollut 9:87–90.

Nicholls T, Prey AJ, Skilling DD (1974): *Rhizosphaera kalkhoffii* damages blue spruce Christmas tree plantation. Plant Dis Rep 58:1094–1096.

Nienhaus F, Castello JD (1989): Viruses in forest trees. Annu Rev Phytopathol 27:165–186.

Osorio M, Stephan BR (1991): Life cycle of *Lophodermium piceae* in Norway spruce needles. Eur J For Pathol 21:152–163.

Rack, K, Butin H (1984): Experimenteller Nachweis nadelbewohnender Pilze bei Koniferen. I. Fichte (Picea abies). Eur J For Pathol 14:302–310.

Redlin SC (1991): *Discula destructiva* sp. nov., cause of dogwood anthracnose. Mycologia 83:633–642.

Rehfuess KE, Rodenkirchen H (1984): Über die Nadelröte-Erkrankung der Fichte (*Picea abies* Karst.) in Süddeutschland. Forstwis Cbl 103:248–262.

Rizzo DM, Harrington TC (1988): Root movement and root damage of red spruce and balsam fir on subalpine sites in the White Mountains, New Hampshire. Can J For Res 18:991–1001.

Roman JR, Raynal DJ (1980): Effects of acid precipitation on vegetation. In Actual and Potential Effects of Acid Precipitation on a Forest Ecosystem in the Adirondack Mountains. ERDA Report 80-28. National Technical Information Service PB81-20383. Syracuse: SUNY College of Environmental Science and Forestry, pp 4-1–4-63.

Saunders PJ (1966): The toxicity of sulphur dioxide to *Diplocarpon rosae* Wolf causing blackspot of roses. Ann Appl Biol 58:103.

Scheffer TC, Hedgcock GG (1955): Injury to northwestern forest trees by sulfur dioxide from smelters. USDA Technical Bulletin 1117.

Schönhar S (1984): Infektionsversuche an Fichten- und Kiefernkeimlingen mit aus faulen Feinwurzeln von Nadelbäumen häufig isolierten Pilzen. Allg Forst Ju J- Ztg 155:191–192.

Schönhar S (1989): Untersuchungen über Feinwurzelschäden und Pilzbefall in Fichtenbeständen des Nordschwarzwalds und der Schwäbischen Alb In Internationaler Kongress Waldschadensforschung: Wissenstand und Perspektiven. Poster Kurzfassungen, Band I. Forschungsbeirat Waldschäden/ Luftverunreinigung, Germany, pp 265–266.

Schütt P (1981): Erste Ansätze zur experimentellen Klärung des Tannensterbens. Schweiz Z Forstwes 132:443–452.

Schwerdtfeger F (1981): Die Waldkrankheiten. Fourth Edition. Berlin: Verlag Parey. 486 pp.

Seemüller E (1989): Mycoplasmas as the cause of diseases of woody plants in Europe. Forum Mikrobiol 12:144–151.

Setliff EC, Manion PD (1980): Forest pathology and acid rain. In Raynal DJ, Leaf AL, Manion PD, Wang CJK (eds): Effects of Acid Precipitation on a Forest Ecosystem, New York State Energy Research and Development Authority Report 80-28. Albany, Research Foundation of State University of New York, pp. 6-1–6-26.

Shafer SR, Bruck RI, Heagle AS (1982): Survival of *Phytophthora cinnamomi* in soil columns exposed to simulated acid rain (Abstract). Phytopathology 72: 361.

Shafer SR, Bruck RI, Heagle AS (1985): Influence of simulated acid rain on *Phytophthora cinnamomi* and Phytophthora root rot of blue lupine. Phytopathology. 75:996–1003.

Sheppard LJ, Smith RI, Cannell MG (1989): Frost hardiness of *Picea rubens* growing in spruce decline regions of the Appalachians. Tree Physiol 5:25–37.

Shields, WJ, Hobbs SD (1979): Soil nutrient levels and pH associated with *Armillariella mellea* on conifers in northern Idaho. Can J For Res 9:45–48.

Shriner DS (1977): Effects of simulated rain acidified witih sulfuric acid in host–parasite interactions. Water, Air, Soil Pollut. 8:9–14.

Shriner DS, Cowling EB (1980): Effects of rainfall acidification on plant pathogens. In Hutchinson TC, Havas M (eds): Effects of Acid Precipitation on Terrestrial Ecosystems. New York: Plenum Press, pp 435–442.

Singh P (1983): Armillaria root rot: influence of soil nutrients and pH on the susceptibility of conifer species to the disease. Eur J For Pthol 13:92–101.

Solar M, Jurc D (1989): Inventory and evaluation methods of forest die-back and of harmful biotic and abiotic factors in Slovenia (Yugoslavia). In Internationaler Kongress Waldschadensforschung: Wissenstand und Perspektiven. Poster Kurzfassungen, Band I. Forschungsbeirat Waldschäden/ Luftverunreinigung, Germany, pp 39–40.

Suske J, Franz F, Acker G (1989): Latent infection by *Lophodermium piceae* hyphae in green Norway spruce needles: infection frequency, and host–fungus interaction at ultrastructural level. In Internationaler Kongress Waldschadensforschung: Wissenstand und Perspektiven. Poster Kurzfassungen, Band II. Forschungsbeirat Waldschäden/Luftverunreinigung, Germany, pp 625–626.

Tanaka K (1980): Studies on relationship between air pollutants and microorganisms in Japan. In Proceedings of Symposium on Effects of Air Pollutants on Mediterranean and Temperate Forest Ecosystems. U.S. Department of Agriculture Forestry Service General Technology Report PSW-43, pp 110–116.

Treshow M (1975): Interaction of air pollutants and plant diseases. In Mudd, JB, Kozlowski, TT (eds): Responses of Plants to Air Pollution. New York: Academic Press, pp 307–334.

Ulrich B (1981): Eine ökosytemare Hypothese über die Ursachen des Tannensterbens (Abies alba Mill). Forstwiss Centralbe, 100:228–236.

Wargo PM, Harrington TC (1991): Host stress and susceptibility. In Shaw CG, Kile GA (eds): Armillaria Root Disease. USDA Agriculture Hdbk. No. 691. Washington: U.S. Government Printing Office, 233 pp.

Weidensaul TC, Darling SL (1979): Effects of ozone and sulfur dioxide on the host–pathogen relationship of Scotch pine and *Schirrhia acicola*. Phytopathology 69:939–941.

Weinstein LH, McCune DC, Alusio AL, van Leuken P (1975): The effects of sulphur dioxide on the incidence and severity of bean rust and early blight of tomato. Environ Pollution 9:145–155.

Weiss MJ, McCreery LR, Millers I, Miller-Weeks M, Obrien JT (1985): Cooperative Survey of Red Spruce and Balsam Fir Decline and Mortality in New York, Vermont and New Hampshire 1984. USDA Forest Service Northeastern Area NA-TP-11. 53 pp.

Wentzel KF (1956): Winterfrost 1956 und Rauscschäden. Allg Forstz 11:541–543.

Worrall JJ, Harrington TC (1988): Etiology of canopy gaps in spruce-fir forests at Crawford Notch, New Hampshire. Can J For Res 18:1463–1469.

Zieger E, Pelz E, Hornig W (1958): Ergebnisse einer Umfrage über Umfang und Art der Frostschäden des Winters 1955/56 in den staatlichen Forstwirtschaftsbetrieben der DDR. Arch Forstwes 7:316–329.

ACID DEPOSITION AND ECTOMYCORRHIZAL SYMBIOSIS: FIELD INVESTIGATIONS AND CAUSAL RELATIONSHIPS

CHRISTINE RAPP

Institut für Waldbau, Universität Göttingen, Büsgenweg 1, D-37077 Göttingen, Germany

GEORG JENTSCHKE

Forstbotanisches Institut, Universität Göttingen, Büsgenweg 2, D-37077 Göttingen, Germany

1. INTRODUCTION

Only a few years after the alarming reports on forest decline in central Europe and in other parts of Europe and North America, initial observations and investigations indicated a reduction of the mycorrhizal status of forest trees and of the abundance and diversity of ectomycorrhizal fungi. Several research programs were initiated to assess the condition of ectomycorrhizae and to elucidate factors and processes influencing ectomycorrhizal development in forest ecosystems more or less subjected to high deposition rates.

Although mycorrhizae are organlike units consisting of two totally different organisms and reactions to environmental stress cannot easily be attributed to the response of one of the symbionts, this chapter deals with the reactions of the symbiosis to acid deposition with a special emphasis on the fungal partner. Effects on host physiology are comprehensively discussed by Godbold (Chapter 7) including the modifying effects of mycorrhizae on the tolerance of trees to metal stress.

Effects of Acid Rain on Forest Processes, pages 183–230

This chapter aims to give a compilation of available information about the performance of ectomycorrhizae in situations of forest decline and to discuss possible factors and processes influencing mycorrhizal development.

2. OBSERVATIONS AND INVESTIGATIONS OF ECTOMYCORRHIZAL ROOTS IN DECLINING FOREST ECOSYSTEMS

Adverse effects on fine roots and ectomycorrhizae supposedly play an important role in the formation of forest decline symptoms. However, much of the evidence that acid deposition may affect the ectomycorrhizal symbiosis is based on field observations rather than experiments (e.g., Hüttermann et al., 1983; Blaschke, 1981, 1986a; Liss et al., 1984; Meyer, 1984, 1987; Schütt et al., 1983).

Attempts to identify stress indicators for fine roots and the mycorrhizal status are comparative investigations of healthy and declining forest trees and ecosystems. Classification of tree vitality is usually based on the presence of visible symptoms of the above-ground plant parts.

The condition of the ectomycorrhizal symbiosis in the field is characterized in various ways and by use of different parameters: Vitality of mycorrhizal root tips is examined by visual inspection and classified according to morphological and anatomical criteria. In addition, biochemical methods were applied, for example, the determination of adenosine triphosphate (ATP) concentration (Pankow et al., 1988, 1989) or the kinetics of glucose and phosphorus uptake (Niederer and Wieser, 1988). In many studies the intensity of mycorrhizal colonization is calculated by counting mycorrhizal and nonmycorrhizal root tips. The determination of fungus-specific components, such as chitin or ergosterol, has until now found no broad application to field samples. More detailed studies include differentiation in morphotypes or identification of species and species composition. Mycocoenological investigations concentrate on the occurrence of ectomycorrhizal fruitbodies.

2.1. Visible Symptoms of Ectomycorrhizal Roots

Numerous studies on fine roots and ectomycorrhizae of Norway spruce (*Picea abies*) indicate disturbances of the symbiotic relationship at declining forest sites. Observations of striking morphological and anatomical features are interpreted as signs of degradation (e.g., Liss et al., 1984; Blaschke, 1986a; Meyer, 1984, 1987). Some of these studies are summarized in Table 6.1. Photographs and drawings of deteriorated fine roots and ectomycorrhizae were published by several authors [Blaschke, 1981 (Fig. 1a), 1986a (Figs. 3 and 4); Blaschke and Bäumler, 1986 (Fig. 3); Göbl, 1988 (Fig. 2); Liss et al., 1984 (Figs. 2 and 3); Meyer, 1984 (Figs. 11 and 12); Meyer, 1985a (Figs. 7a, b)]. Corresponding studies on the fine roots of European beech (*Fagus sylvatica*) are scarce (Asche and Flückiger, 1987; Brand, 1991; Kumpfer and Heyser, 1986; Vincent, 1989; Zhao et al., 1990) and correlations between ectomycorrhizal status and stages of decline are not clear (Brand, 1991).

TABLE 6.1. Some Characteristics Reported on Fine Roots and Ectomycorrhizae from Declining Forest Stands

Habitus of Fine Root Systems
Reduced ramification[i,j,l]
String shaped long roots[k]
Points of root disturbance[b,d,i,j]
Fine root stumps[b,i,j]
Morphology of Ectomycorrhizae
Dark discoloration[i]
Simple growth forms predominating[c]
Weak ramification[c]
Deformation[c,e]
Root tips penetrating hyphal mantle[c,l]
Beaded appearance[j]
Tapering of tips[j]
Slightly swollen[j]
Anatomy of Ectomycorrhizae
Thinned hyphal mantle[j]
Hypertrophied Hartig net[j]
Intracellular infection of cortical cells[a,c,f,g,j]
Intracellular infection of cortical cells due to pathogens[h]
Increased thickness of cortex cell walls[f,j]
Increased deposition of phenolic compounds in cortical cells[a,c,f,j]

[a]von Alten and Fischer, 1987. [b]Blaschke, 1981. [c]Blaschke, 1986a. [d]Eichhorn, 1987. [e]Göbl, 1988. [f]Haselwandter et al., 1988. [g]Haug, 1987. [h]Haug et al., 1988. [i]Liss et al., 1984. [j]Meyer, 1984. [k]Meyer, 1985a. [l]Meyer, 1987.

Visible manifestation of changing symbiotic relationships is the penetration of root tips through the hyphal mantle. Renewed short root growth without subsequent mantle formation indicates disturbances in the formation of ectomycorrhizae (Blaschke, 1986a). This condition is caused by decreasing activity of the ectomycorrhizal fungus, which no longer inhibits the elongation of short roots (Meyer, 1987). This ability is thought to depend on the hormonal status of the plant and may be reduced by high amounts of nitrogen in the soil (Slankis, 1971, 1974; Meyer, 1985b). The percentage of penetrating root tips without subsequent mantle formation might therefore be a valuable indicator for disturbance, especially in relation to a surplus of nitrogen.

Bead formation, that is, the presence of constrictions at periodic intervals along the root axis, was reported for nonmycorrhizal (James et al., 1978) and ectomycorrhizal roots of field-collected samples (Kessler, 1966; Marks and Foster, 1967; Meyer, 1984; Berg, 1989; Gronbach, 1988). Meyer (1984) observed the beaded appearance of ectomycorrhizae at fine root systems of diseased Norway spruce trees and related it to periodic growth activity. Gronbach (1988) and Berg (1989) investigated ectomycorrhizae of Norway spruce exposed to acid irrigation and liming, and tried to record atypical growth forms or possible signs of pathological change. Both of them found beaded mycorrhizae, but their occurrence was noticed after acid irrigation as well as after liming, and could not be related to the treatments. Other workers supposed relations to soil moisture (Kessler, 1966), concentrations of available aluminum in the soil (James et al., 1978), phosphorus supply (Beslow et al., 1970), or diurnal changes in light and temperature (Thomson et al., 1990). However, the phenomenon of beading was never fully explained. Only recently has the formation of beads been shown to occur even under "constant" environmental conditions in laboratory experiments (Beslow et al., 1970; Downes et al., 1992; Thomson et al., 1990). The beaded appearance is thought to be morphological evidence of discontinuous extension, whose reasons for occurrence are not clear and whose regulation is not yet understood.

Some of the points listed in Table 6.1 are attributed to ectomycorrhizae of affected forest trees as well as to ectomycorrhizae in the later stages of ageing (e.g., Kottke and Oberwinkler, 1986; Downes et al., 1992). There are several reports of ectomycorrhizae from affected Norway spruce stands to show signs of senescence, for example, discoloration (Liss et al., 1984) or deposition of phenolic compounds (Blaschke, 1986a). Other studies report on senescent ectomycorrhizae without giving concrete information about what senescent mycorrhiza look like (Göbl, 1988; Meyer, 1984). Intracellular infection is thought to be a normal process of ageing (Nylund, 1987; Ritter, 1990). However, careful investigation is advisable, because intracellular hyphae might not belong to the symbiotic fungus. Haug et al. (1988) found intracellular infections due to parasitic soil fungi in vascular tissue and meristem of mycorrhizal roots from severely damaged Norway spruce of the Black Forest. They suppose that roots and ectomycorrhizae of damaged trees are more susceptible to infections caused by soil pathogens.

In conclusion, it is doubtful that many of these attributes contribute to the identification of damage to ectomycorrhizae. To date there is little convincing evidence of visible, pathological changes in the morphology and histology of ectomycorrhizae in relation to forest decline. More information is needed about the healthy, normal status of ectomycorrhizae and its natural variation under different conditions in the field. The importance of broadening our knowledge on the natural processes of ontogeny, ageing, and senescence is stressed by several authors (Berg, 1989; Gronbach, 1988; Gronbach and Agerer, 1986; Kottke et al., 1986) and verified by previous investigations (Al Abras et al., 1988; Downes et al., 1992). Nevertheless, repeated reports of high numbers of senescent ectomycorrhizae might indicate an acceleration of the life cycle and the turnover rates under stress conditions. Future investigations should take this into account.

2.2. Quantification of Ectomycorrhizal Status

The object of several studies was to compare the intensity of mycorrhizal colonization of tree roots from healthy forest stands with roots from declining forests. The results indicate decreasing numbers of active mycorrhizae (Blaschke, 1981, 1986a; Liss et al., 1984; Livingston and Blaschke, 1984; Meyer et al., 1988; Schütt et al., 1983) and mycorrhizal frequency (Haselwandter et al., 1988; Meyer et al., 1988; Vincent, 1989) in declining stands (Table 6.2). However, in other studies the mycorrhizal status could not be related to the occurrence of generally assessed aboveground symptoms like needle loss (e.g., Haug et al., 1986). Instead, mycorrhizal colonization could be related to soil conditions (von Alten and Fischer, 1987; von Alten and Rossbach, 1988); to crown transparency, crown losses, twigs, and bark damages (Jansen, 1988, 1991); or to stand age (Jansen, 1988, 1991; Termorshuizen and Schaffers, 1991).

Processing of field samples usually includes sorting of root fragments under the stereomicroscope. Common criteria to recognize ectomycorrhizae are the presence of a hyphal mantle, extramatrical mycelium, and/or rhizomorphs, color, and a swollen look. However, Kottke and Oberwinkler (1986) point out that thin hyphal mantles could easily be misinterpreted as nonmycorrhizal root tips and thus cause underestimation of mycorrhizal colonization. In addition, Al Abras et al. (1988) found that 1- and 2-year-old *Picea excelsa* mycorrhizae had lost their sheath, but a normal Hartig net was still present. To indicate the possible underestimation of their counts, Meyer et al. (1988) referred to ectomycorrhizae as "apparent" ectomycorrhizae. The problems involved with visual inspection of root tips are evident not only with regard to classification of mycorrhizal and nonmycorrhizal roots, but with classification of the vitality of ectomycorrhizae as well (see Section 2.1). Division of ectomycorrhizae into categories like "young–senescent," "active–inactive," or "living–dead" is probably very subjective and represents only rough estimates. As visual differentiation between "damaged" and senescent mycorrhizae is almost impossible, the quantification of the mycorrhizal status is generally based on numbers of "active" or living mycorrhizae and nonmycorrhizal root tips. The "active" condition of mycorrhizae is characterized by the presence of a hyphal mantle, turgidity, and a light-colored apex (e.g., Blasius et al., 1985; Livingston and Blaschke, 1984; Meyer et al., 1988; Vincent, 1989). Brittle, wrinkled, and dark colored fragments are considered "inactive" or dead (e.g., Meyer et al., 1988). However, little is known about the process of ageing and the lifespan of ectomycorrhizae. There exist only a few published data ranging from a few months to a few years (Orlov, 1957; Harley and Smith, 1983; Vogt et al., 1986). Usually, the calculations are based on turgescence as criterion for viability determination (e.g, Vogt et al. 1986), but this method is questionable, as excised mycorrhizae can remain turgid for as long as 8 months (Ferrier and Alexander, 1985). In addition, the rate at which individual mycorrhizae pass through morphological stages of aging varies considerably from tip to tip so that the morphological appearance is not a good indicator of its chronological age (Downes et al., 1992).

Although much information is available, it is very difficult to compare the data presented in Table 6.2. Counts of mycorrhizal tips are related to different parameters, for example, to the number of total root tips (usually called "mycorrhizal frequency"), to units of soil volume, to ground area, or to the dry weight of fine (< 2 mm in diameter) or finest (< 1 mm in diameter) roots (named "relative mycorrhizal

TABLE 6.2. Quantitative Assessments of the Mycorrhizal Status in Declining and Healthy Forest Ecosystems

Tree Species Age (years)	Parameter	Declining	Healthy	References
Picea abies				
5	Relative mycorrhizal frequency[a]	87	102	Blasius et al., 1985
80–100	Mycorrhizal frequency[b]	15–40[c]	10–50[c]	
		50–75[c]	10–70[c]	von Alten and Fischer, 1987
80–100	Mycorrhizal frequency[d]	16	37–40	Haselwandter et al., 1988
80–110	Number of mycorrhizae[e]	Lower	Higher	Liss et al., 1984
90–130	Number of active mycorrhizae per unit volume of soil		1–3-fold	Livingston and Blaschke, 1984
30	Mycorrhizal frequency[b]	27–46	36–68	Meyer et al., 1988
	Relative mycorrhizal frequency[f]	No clear relation		
Pseudotsuga menziesii				
<20, 20–40, >40	Mycorrhizal frequency[b]	9	36	Jansen, 1991
Fagus sylvatica				
60–100	Mycorrhizal frequency[b]	No clear relation		Vincent, 1987, 1989
	Relative mycorrhizal frequency[f]	450[h]	1095[h]	Vincent, 1987
	Living ectomycorrhizae[g]	19[h]	32[h]	Vincent, 1989

[a] Number of ectomycorrhizae per 100 mg of finest roots (smaller 1 mm in diameter).
[b] Percent vital mycorrhizal root tips of total number of root tips.
[c] Soils overlying sandstone or limestone, respectively.
[d] Percent vital mycorrhizal (smooth brown mycorrhiza) root tips to total number of root tips.
[e] No specification of parameter.
[f] Number of ectomycorrhizae per g of fine roots (smaller 2 mm in diameter).
[g] Percent living ectomycorrhizae of total mycorrhizal tips.
[h] Median (calculated by authors) of data from humus-free topsoil. No clear relation in humus (O and A horizon).

frequency" by Kottke and Agerer, 1983). Sometimes this situation is complicated even further by use of terms like "mycorrhization rate" and "degree of mycorrhization" or by unclear definition of the parameters used. Blasius et al. (1985) tried to find adequate parameters for judging the quality of root systems and recommended the use of the "relative mycorrhizal frequency," which is the number of mycorrhizae per 100 mg of dry weight of finest roots (< 1 mm in diameter), for comparative investigations of forest stands.

In conclusion, correlations between the intensity of mycorrhizal colonization and above-ground symptoms were found to be weak. This result may be due to the mode of assessment or to the selection of tree parameters. In addition, many studies are short-term investigations, which do not consider annual variations. The interpretation of quantitative data is further complicated if turnover rates are unknown. There is a need for new methods and for a rapid and reliable assessment of mycorrhizal colonization of root samples from the field. Ergosterol assays might be a suitable tool (Martin et al., 1990; Salmanowicz and Nylund, 1988), but experience in broad application to field-collected samples is still lacking. Furthermore, there should be greater agreement between the terminology used and use of parameters. In accordance with Kottke and Oberwinkler (1986) and Blasius et al. (1985) we suggest that investigations aimed at quantification of mycorrhizal colonization by counting root tips should include the calculation of numbers of ectomycorrhizae in proportion to dry weight of finest roots (1 mm maximum in diameter) and possibly to root length. The former quotient, the "relative mycorrhizal frequency," provides valuable information about the intensity of mycorrhizal colonization, and is therefore a useful parameter for comparative studies in the field.

2.3. Number of Species (Morphotypes) and Composition of the Ectomycorrhizal Community

The response of ectomycorrhizae to the effects of acid deposition is supposed to be species specific. Although progress in identifying ectomycorrhizae is remarkable (e.g., Agerer, 1987, 1991a; Ingleby et al., 1990), identification is still difficult and efforts failed in many cases. Nevertheless, frequencies of morphotypes may indicate changes in species diversity. Although the relative benefits of individual species to the tree are still unknown, these changes may indicate early stress responses. Quantitative investigations fail to reveal such effects, because changes may occur without net change in total numbers of mycorrhizae (Meier, 1991).

After conducting several studies, no correlations between stage of decline and number of ectomycorrhizal species (Brand, 1991) or number of morphotypes (von Alten and Rossbach, 1988; Estivalet et al., 1988; Haug, 1987) could be found. Similar to mycorrhizal colonization (see Section 2.2) the number of species or morphotypes could be related to soil type (von Alten and Rossbach, 1988) or stand age (Jansen, 1991) rather than to tree vitality.

However, the frequency of distinct morphotypes was found to be related to stand condition. Liss et al. (1984) found higher numbers of two filamentous types at healthy Norway spruce trees. Similar results were reported by Markkola and

Ohtonen (1988), who investigated the influence of air pollution due to urban emissions on mycorrhizal status. The species with rhizomorphs seemed to suffer more from pollution, while mycorrhizae of *Cenococcum geophilum* showed a higher abundance as well. In contrast, Eichhorn (1987) found *C.geophilum* mycorrhizae to be more abundant in healthy Norway spruce stands. Investigations in *Pinus sylvestris* stands in The Netherlands showed that mycorrhizae formed by *C.geophilum* were positively correlated with nitrogen deposition (Termorshuizen and Schaffers, 1991). Meier et al. (1989) present results from an experiment with mycorrhizal red spruce (*Picea rubens*) seedlings exposed to simulated acid rain treatments. While the total number of mycorrhizae did not change in response to acid rain, the number of mycorrhizae, which were formed by *C.geophilum*, increased with increasing rain acidity.

Although the results presented are not consistent, there is evidence of decreasing diversity of ectomycorrhizal species under stress conditions. Possible consequences may include the reduction of the potential for adaptation to changes in the environment. However, research in identifying ectomycorrhizae has to be encouraged. We need to know a great deal more about the benefits (or cost) of individual mycorrhizal fungal species to the tree and about species-specific sensitivities of the fungi to stress.

2.4. Fruitbody Production of Ectomycorrhizal Fungi

Air pollution affects both the below-ground structures of the fungal symbionts, and the carpophores of ectomycorrhizal fungi.

Mycocoenological studies in The Netherlands show an impoverishment of the mycoflora due to a decrease in fruitbody production of mycorrhizal fungi (Jansen and van Dobben, 1987; Termorshuizen and Schaffers, 1987, 1991), through mainly those that are associated with coniferous trees (Arnolds, 1988). Similar conditions are reported to exist in Germany (Schlechte, 1986; Winterhoff and Krieglsteiner, 1984) and Czechoslovakia (Fellner, 1988). The impoverishment of the mycoflora is attributed to the influences of air pollution (Arnolds, 1988, 1991).

Decreasing occurrence of mycorrhizal carpophores possibly indicates decreasing mycorrhizal colonization of fine roots, because positive correlations between the above-ground amount of carpophores and the below-ground amount of mycorrhizae were found (Agerer, 1990; Deacon et al., 1983; Jansen, 1988, 1981; Jansen and De Nie, 1988; Laiho, 1970; Termorshuizen and Schaffers, 1991). Jansen (1991) found cumulative abundance of fruitbodies being a reliable index for the number of mycorrhizae.

The amount of fruitbodies produced is [similar to the number of ectomycorrhizae (see Section 2.2)] related to stand age, becoming lower with increasing stand age (Dighton and Mason, 1985; Jansen, 1988, 1991; Termorshuizen and Schaffers, 1987, 1991; Vogt et al., 1983). Young (5–10 year) plots of *Pinus sylvestris* showed no negative correlations with air pollution, but 50–80-year-old stands showed very significant negative correlations between the deposition of nitrogen compounds and both the number of carpophores and the number of fruiting species (Termorshuizen and Schaffers, 1991). These investigations suppose that mycorrhizal development in young plots is favored due to disturbance of the soil before planting. The decline of carpophores in the old plots seemed to precede a decline of ectomycorrhizae (Fellner, 1988; Termorshuizen and Schaffers, 1991). However, these results should be verified by additional studies.

Pauperization of the macromycet flora might be a sensitive signal for the disturbance of the symbiotic relationship. The value of carpophores for bioindication of the amount of air pollution (Fellner, 1988), soil properties (Hansen, 1988), or soil acidification (Tyler, 1985) is pointed out, but further evidence is needed.

3. HOW ARE ECTOMYCORRHIZAE AFFECTED BY AIR POLLUTION?

The decreasing intensity of mycorrhizal colonization and the numbers of ectomycorrhizal carpophores are considered to be related to air pollution. Possible mechanisms are still unknown. Two hypotheses were proposed to explain how ectomycorrhizae may be affected by acid deposition (Arnolds, 1988, 1991; Jansen and Dighton, 1990; Meyer, 1985b; Persson, 1988; Stroo and Alexander, 1985; Fig. 6.1):

1. Changes in the carbon supply. Atmospheric pollutants affect the photosynthetic activity of trees, production, and transport of assimilates to fine roots and ectomycorrhizae. Insufficient nutrient supply causes early senescence and reduced regeneration of mycorrhizae (e.g., Liss et al., 1984; Schütt et al., 1983).
2. Changes in soil chemistry (see Chapters 1 and 2). Acid deposition causes changes in soil chemistry, for example, acidification and nitrogen saturation, which affect ectomycorrhizae (e.g., Arnolds, 1988; Meyer, 1985 a, b; Nihlgård, 1985; Jansen and Dighton, 1990; Jansen, 1991).

3.1. Carbohydrate Supply From the Host Plant

From the gaseous pollutants that may directly affect above-ground tree performance and patterns of carbon allocation, ozone (O_3) is considered to be one of the most important agents (Smith, 1990). There are several reports on the effects of above-ground application of O_3 on ectomycorrhizal colonization. However, the results of these investigations are conflicting. Some authors found that O_3 treatments at concentrations up to 400 $\mu g\ m^{-3}$ impaired ectomycorrhizal development (e.g., Blaschke, 1990; Meier et al., 1990; McQuattie and Schier, 1992), whereas others found no such effect (Reich et al., 1985; Stroo et al., 1988; Wöllmer and Kottke, 1990; Adams and O'Neill, 1991; Gorissen et al., 1991; Edwards and Kelly, 1992; Shaw et al., 1992). This inconsistency may be attributed to differences in the sensitivity of the tree species investigated or differences in experimental conditions and methods applied. More important, most of these studies were conducted with young seedlings and their reactions to gaseous pollutants may differ from those of mature trees. In fact, experimental evidence (Skärby et al., 1987; Koch and Lautenschlager,

1988; Zimmermann et al., 1988; Smidt and Hermann, 1992) suggests that gaseous pollutants do not significantly alter photosynthesis under field conditions. Hampp (1992) comprehensively reviewed the recent literature on direct effects of air pollutants on the carbon metabolism of forest trees and concluded that at realistic concentrations effects of O_3 are rather limited.

In addition, this view is further supported by studies of the carbohydrate contents in fine roots of declining and healthy trees. In these studies, no relationship was found between the starch contents of fine roots and the above-ground symptoms of Norway spruce (Persson, 1988) and European beech (Asche and Flückiger, 1987; Flückiger et al., 1986; Wiebe and Blaschke, 1988). Vital fine roots of damaged Norway spruce had even higher starch contents compared to healthy trees (Eichhorn, 1987). Meyer et al. (1988) found no correlations between foliar carbohydrate levels and mycorrhizal colonization in damaged and healthy Norway spruce stands in Germany.

To date there is no clear evidence of the carbon supply being insufficient for fine roots and ectomycorrhizae in declining stands. Recently, however, Schier et al. (1990) reported on synergistic effects of O_3 and Al on mycorrhizal pine seedlings. Other work (Dubé and Bornmann, 1992; Temple et al., 1993) also suggests that the interaction of above-ground and below-ground stresses may be important for tree performance. These interactions may be particularly significant at high elevation forests where O_3 concentrations are higher compared to sites at low altitude (Lovett and Kinsman, 1990; Puxbaum et al., 1991). Future research should focus on these interactions.

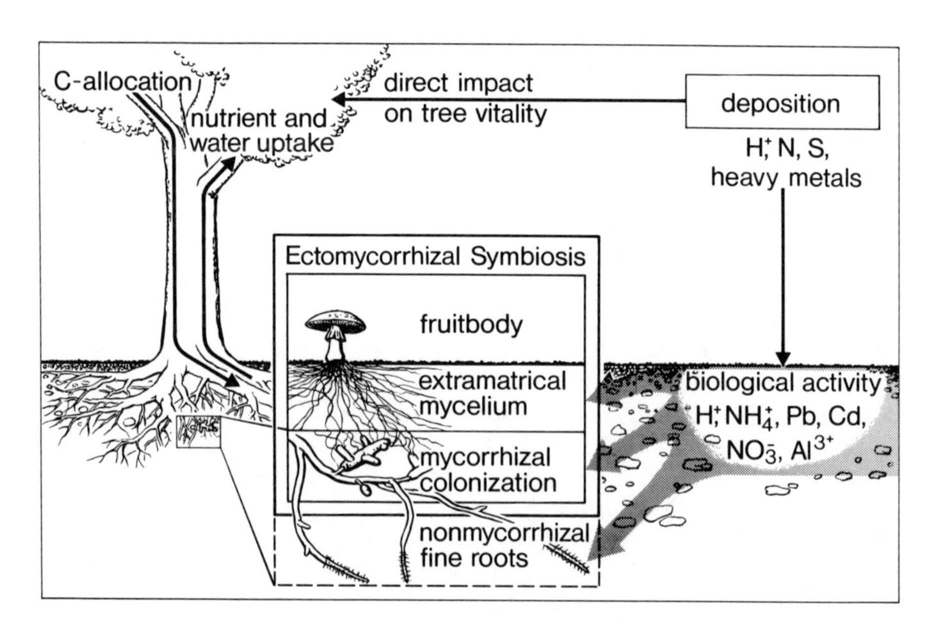

Fig. 6.1. The impact of air pollutants on ectomycorrhizal symbiosis. Tree-mediated versus soil-mediated effects.

3.2. Soil Acidification

The following sections describe driving variables of changes in the ectomycorrhizal status of forest trees affected by acid deposition (Fig. 6.1). However, the interactions between the soils chemical and biotic factors are numerous and highly complex. Acid deposition affects soil biotic processes (Wolters and Schaefer, Chapter 3), which in turn are important factors for the formation of different types of humus. Since mycorrhizae are adapted to certain soil biological conditions [e.g., bacterial populations (Duponnois and Garbaye, 1990)] and since the species composition of mycorrhizal communities (Scherfose, 1990), as well as the abundance of mycorrhizae (Meyer, 1967), depend on humus and soil types, any change in the biotic conditions of a soil may affect the ectomycorrhizal status of the trees. However, these indirect effects of acid deposition on ectomycorrhizae are poorly understood due to the lack of available data. The following discussion will, therefore, concentrate on the "more direct" effects of air pollutants, which were studied in a number of investigations.

Results of Meyer et al. (1988) suggest that the formation of ectomycorrhizae in the mineral soil of a Norway spruce stand in the Fichtelgebirge (Germany) was affected by soil chemical factors reflecting podsolization or soil acidification. The number of apparent mycorrhizae per ground area was positively correlated with the pH and concentrations of Ca^{2+} and Mg^{2+} in the soil solution, as well as with the molar ratio of Ca^{2+} to Al in mineral soil extracts. The number of ectomycorrhizae was negatively correlated to the concentration of ammonium (NH_4^+) in the soil solution. Kumpfer and Heyser (1986) found a correlation between the spatial distribution of ectomycorrhizal morphotypes in the stemflow area of European beech trees and the molar Ca/Al ratio of soil–water extracts. These investigators observed one morphotype (type 1: without extramatrical mycelium) being the most dominant in samples below the stem base. Relations between soil pH and the abundance of species (morphotypes) are reported by Zhao et al. (1990), who found a positive correlation between *Lactarius subdulcis* mycorrhizae and decreasing pH (7.0–3.0) in the mineral topsoil (0–15 cm) of 20 European beech stands in Austria. Von Alten and Rossbach (1988) found three morphotypes being more abundant at pH 6–7 and two morphotypes at pH 3–4 in a severely damaged Norway spruce site in Germany.

Hence, there is increasing evidence that soil acidification is a main factor impairing ectomycorrhizal development. The combination of low pH and high nitrogen content of soils seems to be particularly unfavorable for most ectomycorrhizal fungi (Arnolds, 1988).

3.2.1. Effects of Artificial Acidification and Compensatory Liming in the Field

Several studies on the effects of simulated acid rain on fine roots and ectomycorrhizae were carried out either in a greenhouse or in a laboratory (see Section 3.2.2).

TABLE 6.3. The Höglwald Research Project in Bavaria[a]

A. Experimental Design

Treatment	No Irrigation	Irrigation (pH 2.7–2.8)	Irrigation (pH 5–5.5)
No lime	A1	B1	C1
Lime (4000 kg ha^{-1})	A2	B2	C2

B. Ectomycorrhizal Response to Treatments

Parameter	A1	B1	C1	A2	B2	C2
Mycorrhizal colonization (%)[b]						
(4 years after treatment)						
Humus	42	29	49	34	37	64
Mineral soil	33	51	40	38	47	47
Number of species/morphotypes	12[c]	14[c]	14[c]	13[d]	14[c]	13[d]
Number of species/morphotypes exclusively on plot			1[c]	3[d]	1[c]	2[d]
Morphology/anatomy[c,d]	No clear relations to treatments					
Cumulative relative fruitbody production[e]						
Ectomycorrhizal macrofungi	181	180	440	29	445	48
Litter macrofungi	437	343	525	435	362	255

[a]Kreutzer and Göttlein, 1991. [b]Blaschke, 1988. [c]Gronbach, 1988. [d]Berg, 1989. [e]Agerer, 1989, 1991b.

Little information exists about the effects of acid rain treatments in the field. This section presents results of acid irrigation and of liming experiments in the field.

The experimental design of the Höglwald Research Project (Kreutzer and Göttlein, 1991) in an 83-year-old (1990) spruce (*P. abies*) stand in Bavaria includes plots with acid irrigation (sulfuric acid), a combination of acid irrigation and compensatory liming (dolomite), and corresponding control plots (see Table 6.3.A). The influence of the different treatments on mycorrhizae (see Table 6.3.B) are reported by Blaschke (1986a,b, 1988), Berg (1989), Gronbach (1986, 1988), Gronbach and Agerer (1986), and on ectomycorrhizal fruitbody production by Agerer (1989, 1991b).

After 2 years of treatment Blaschke (1986b) found only slight differences in the number of mycorrhizae per 100 mg of fine root biomass between plot B1, which was irrigated with acidified (pH 2.8) rain, and the control plot without any treatment (Al). However, in the humus layer mycorrhizal colonization decreased after 2 more years of acid irrigation compared to the control (Al). After irrigation with "normal" rain (pH 5.0) mycorrhizal colonization increased. In contrast, mycorrhizal colonization increased in the mineral soil after acid irrigation (see Table 6.3.B), suggesting that the effect of acid irrigation depends on the soil substrate. Eight species (morphotypes) were common on all plots, but the abundance of other types varied depending on the treatment (Gronbach, 1988; Berg, 1989). For example, a striking increase in the abundance of one distinct morphotype named *Piceirhiza gelatinosa* was observed on the acid irrigation plot (Gronbach and Agerer, 1986). The effects of acid irrigation on the production of ectomycorrhizal fruitbodies was investigated 4 years after initiation of treatments (Agerer, 1989). Irrigation with rain (pH 5.0) increased productivity by more than one half. However, the stimulating effect caused by irrigation was eliminated by a low pH on the acid irrigation plot. Litter macrofungi showed positive response to irrigation as well, but the negative effect of acid irrigation was less compared to ectomycorrhizal fungi (Agerer, 1991b).

In the southern United States Esher et al. (1992) applied acidified throughfall (pH 4.3 and 3.6) to throughfall excluded 1-m^2 plots in loblolly pine (*Pinus taeda*) and longleaf pine (*Pinus palustris*) stands. After 2 years of acid treatment they found a significant decrease in percent and number of ectomycorrhizae, whereas saprophytic fungi were not affected. Acid treatment significantly increased nitrate (NO_3^-) levels in the litter leachate and NH_4^+ levels in the fermentation layer. Although there was no detectable long-term change in soil pH it was assumed that there may have been a temporarily lowered soil pH, exposing sensitive root meristems and mycorrhizae to H^+ or Al^{3+} ions causing damage by loss of exchangeable Ca^{2+} ions.

The results of acid irrigation experiments indicate that a decrease in mycorrhizal colonization and changes in species composition may be caused. However, causal interpretation of results from this kind of field experiment is difficult. The addition of acid to forest soils may induce different soil reactions depending on soil conditions. It will make a great difference if additional acid is applied to a base-saturated, well-buffered soil or to a poor soil already showing signs of podsolization. The same amount of acid rain may or may not cause an effect. In addition, short-term and long-term effects may be different. Experimental conditions have to be compared carefully before any generalizations are to be drawn.

Liming is common practice in forestry and was recently recommended to counteract anthropogenic acidification of forest soils due to acid deposition (Ulrich, 1986; Antibus and Linkins, 1992). However, the effects of liming are complex and do not only cause an increase of soil pH, but the microbial activity, mineralization, and nutrient availability, especially of nitrogen, are increased as well. Effects of liming are therefore similar to those caused by nitrogen fertilization (see Section 3.3.1). The effects of liming on different ectomycorrhizal parameters are summarized in Table 6.4.

In the Höglwald Research Project (see Table 6.3) compensatory liming reduced mycorrhizal colonization in the humus layer, but not in the mineral soil. Liming in combination with "normal" irrigation (pH 5.0) increased mycorrhizal colonization in both soil substrates (Blaschke, 1988). The application of 1500 kg of CaO ha^{-1} reduced the numbers of ectomycorrhizae per square meter in a mature European beech (*Fagus sylvatica*) stand in France (Blaise and Garbaye, 1983). However, no change was found in mycorrhizal colonization if the numbers of mycorrhizae were related to fine root dry weight. The effects of monthly application (during growing season) of crushed limestone (6000 kg of Ca ha^{-1} per season) on mycorrhizae were

TABLE 6.4. The Effects of Lime Application to Forest Soils on the Ectomycorrhizal Symbiosis

Tree Species Age (years)	Treatment per Hectare	Parameter	Effect	References
A. Mycorrhizal Colonization				
Picea abies 83	Dolomite 4000 kg	Mycorrhizal frequency	− 8% (humus) + 5% (mineral soil)	Blaschke, 1988
Pinus resinosa 59	Limestone 6000 kg Ca	Ectomycorrhizae/100 cm^3 soil	+ 9% (1986) + 41% (1987)	Antibus and Linkins, 1992
Fagus sylvatica 90	CaO 1500 kg	Ectomycorrhizae/g root dry wt	No effect	Blaise and Garbaye, 1983
Fagus sylvatica 139	Dolomite 30,000 kg	Relative mycorrhizal frequency	+ 117% (humus) + 15% (mineral soil)	Rapp, 1991
B. Number of Species or Morphotypes				
Picea abies 83	Dolomite 4000 kg	No. of morphotypes	+ 3	Berg, 1989
Pinus resinosa 59	Limestone 6000 kg Ca	No. of morphotypes	No effect	Antibus and Linkins, 1992

C. Frequency of Species or Morphotypes				
Pinus resinosa	Limestone	*Piloderma bicolor*	Increase	Antibus and Linkins, 1992
59	6000 kg Ca	Grainy-brown type	Increase	
		Cenococcum geophilum	No effect	
Picea abies		*Cenococcum geophilum*	Decrease	Kottke and Oberwinkler, 1988
Pinus sylvestris	Limestone	*Cenococcum geophilum*	+ 18%	Lehto, 1984
	2000–	With external mycelium	+ 8%	
	6000 kg	A-mycorrhiza[a]	− 20%	
D. Occurrence of Carpophores				
Picea abies	Dolomite	Relative fruitbody production	− 84%	Agerer, 1989
83	4000 kg			
Picea abies	CaO	No. of species	− 10%	Fiedler and Hunger, 1963
70	7300 kg	Grams of dry weight per hectare	− 77%	
	Dolomite	No. of species	− 18%	
	16,000 kg	Grams of dry weight per hectare	− 64%	
Pinus sylvestris	$CaCO_3$	No. of species	Decrease	Kuyper and De Vries, 1990
25	3000–	*Laccaria bicolor*	− 42%	
	18,000 kg	*Lactarius hepaticus*	− 42%	
		Xerocomus badius	− 15%	
		Paxillus involutus	No effect	

[a]Modification of Melin's classification.

studied in a 59-year-old red pine (*Pinus resinosa*) plantation in New England (Antibus and Linkins, 1992). Mean numbers of viable mycorrhizae per 100 cm^3 of soil were higher on the limed plot, but differences were small relative to spatial and seasonal variability over two field seasons.

Lime application in the Höglwald Research Project caused the additional occurrence of species (morphotypes) which were not present on the control plot (Berg, 1989). In other studies, however, liming did not appear to affect the diversity of morphotypes (Antibus and Linkins, 1992; Kottke and Oberwinkler, 1988), but the relative frequency of single species or morphotypes. The abundance of *Piloderma bicolor* and of a grainy-brown morphotype was increased (Antibus and Linkins, 1992), whereas the frequency of *Cenococcum geophilum* mycorrhizae seemed to be lower on limed plots (Kottke and Oberwinkler, 1988). In contrast, the results presented by Lehto (1984) showed an increase of *C.geophilum* mycorrhizae from 12 to 30% after lime application in Scots pine (*Pinus sylvestris*) stands.

The fruitbodies of ectomycorrhizal fungi showed marked reactions to liming in abundance and productivity (Agerer, 1989, 1991b; Fiedler and Hunger, 1963; Hora, 1959; Kuyper and De Vries, 1990). In the Höglwald Research Project liming without and in combination with "normal" irrigation sharply decreased fruitbody production. Litter macrofungi showed indifferent reactions regarding lime application (Agerer, 1989, 1991b). Liming reduced the abundance and production of ectomycorrhizal fruitbodies, but increased the abundance of saprophytic fungi in 70-year-old Norway spruce stands (Fiedler and Hunger, 1963). Formation of fruitbodies of *Paxillus involutus* and *Lactarius rufus* was depressed after liming a Scots pine forest (Hora, 1959). The number of ectomycorrhizal species was reduced after liming in a 25-year-old *P. sylvestris* stand in The Netherlands (Kuyper and De Vries, 1990). The number of fruitbodies of *Lactarius hepaticus* and *Laccaria bicolor* and *Xerocomus badius* decreased significantly, whereas *P. involutus* was unaffected by liming. Kuyper and De Vries suggested that the negative effects of liming might be caused by increased nitrogen availability due to enhanced mineralization.

The number of investigations on this topic is small and the results are not consistent due to the complex properties of soil substrates, to the use of different parameters, and to the natural variations in space and time. The production of carpophores seems to be affected the most. Whereas the number of ectomycorrhizal species (morphotypes) is not changed strongly, the behavior of distinct species (morphotypes) indicates more or less species-specific reactions. As mentioned earlier, causal interpretation of field experiments is difficult, but negative effects of liming on ectomycorrhizae cannot be excluded. Recommendations for application should be given with care.

3.2.2. Effects of pH

Field experiments suggest that acidic inputs into forest ecosystems or compensatory liming affect ectomycorrhizal development and community structures. The mechanisms involved are not clear, however. A number of studies (e.g., Stroo and Alexander, 1985; Shafer et al., 1985; Reich et al., 1985, 1986; Dighton et al., 1986; Göbl, 1986; Dighton and Skeffington, 1987; Entry et al., 1987; Keane and Manning, 1987, 1988; McAfee and Fortin, 1987; Stroo et al., 1988; Danielson and Visser, 1989; Meier et al., 1989; Simmons and Kelly, 1989; Blaschke, 1990; Wöllmer and Kottke, 1990; Adams and O'Neill, 1991; Walker and McLaughlin, 1991; Edwards and Kelly, 1992) were undertaken in the greenhouse or with tree seedlings outplanted to forest sites or open top chambers to test the response of ectomycorrhizae to acidic inputs into soils. The results of these experiments were highly variable (for reviews see Dighton and Jansen, 1991; Andersen and Rygiewicz, 1991) making it difficult to draw general conclusions. The reason for the variability of results in acid irrigation experiments in both the greenhouse and the field may be due to the complex way in which treatments affect soil chemistry and biology, and the dependence of changes in soil chemistry on the original status of the soil. Hence, in field and greenhouse experiments, effects are not ascribable to single factors. In order to elucidate principal pathways of the impact of soil acidification on mycorrhizae, laboratory experiments are required.

However, it should be kept in mind that laboratory experimental results are often difficult to apply to situations found in the field, because growth conditions in laboratory cultures differ widely from natural soil conditions. In addition, a number of ectomycorrhizal fungal species, among those, members of the important genus *Russula,* have in most cases failed to grow under laboratory conditions. Another limitation is that in dual cultures with the host included, the symbiotic relationship can only be studied in the very early part of the tree's life cycle. In spite of these limitations, carefully conducted laboratory experiments may provide valuable information on major processes affecting ectomycorrhizal development.

Since the early research on physiological processes of ectomycorrhizal symbiosis, many investigations (e.g., Melin, 1924; How, 1940; Modess, 1941) on the effect of substrate pH on the growth of ectomycorrhizal fungi in pure culture have been published (see a review by Hung and Trappe, 1983). The recent literature includes work by Hung and Trappe (1983), Dennis (1985), Oelbe-Farivar (1985), Erland et al. (1990), Willenbourg et al. (1990), and Jongbloed and Borst-Pauwels (1990b, 1992). Even though in all of these pure culture studies the growth conditions were varied widely (e.g., sometimes buffered and sometimes unbuffered media were used) most studies confirm that ectomycorrhizal fungi are acidophilic. However, the use of unbuffered media in most cases may have overestimated pH optima due to acidification of media during growth (Jongbloed and Borst-Pauwels, 1990b). On the other hand, buffers in growth media have their own problems. In many studies di- or tri-carboxylic acids like tartric or citric acid were used. Buffer systems of these acids contain increasing amounts of undissociated and, therefore, uncharged acid molecules with decreasing pH. These may enter the plasmalemma and alter membrane properties leading to increased efflux of cations (Jackson and Taylor, 1970). In addition, passive permeation of the plasmalemma by undissociated acids and subsequent dissociation may cause acidification of the cytoplasma and thus result in a significant disturbance of fungal biochemical and physiological processes. Hence, growth inhibition of certain ectomycorrhizal fungi found on

TABLE 6.5. Growth Inhibition of Ectomycorrhizal Fungi by Aluminum in Pure Culture

		Culture Conditions			
Mycobiont	Critical[a] Al conc (mM)	pH	Nitrogen Source	Activity of Metal Ions Potentially Reduced by	References
Amanita muscaria	19	NR[b]	NH_4^+	PO_4, SO_4^{2-}, Chel.[c]	Hintikka, 1988
A. muscaria	2	3.0	NH_4^+	Agar	Zel et al., 1991
A. muscaria	1.5	4.0	NH_4^+	PO_4, *Chel.*	Burt et al., 1986
A. muscaria	0.4	3.0	NO_3^-		Oelbe-Farivar, 1985
Cenococcum geophilum	4	3.0	NH_4^+		Oelbe-Farivar, 1985
C. graniforme	2	3.4	NH_4^+	Agar, PO_4, Chel.	Thompson and Medve, 1984
C. geophilum	0.4	3.0	NO_3^-		Oelbe-Farivar, 1985
Hebeloma crustuliniforme	0.4	3.8	NH_4^+/NO_3^-		Browning and Hutchinson, 1991
Laccaria bicolor	3	4.0	NH_4^+	PO_4, Chel.	Burt et al., 1986
L. laccata	>>1[d]	3.5	NH_4^+		Jongbloed and Borst-Pauwels, 1992
Lactarius hepaticus	0.1	3.5	NH_4^+		Jongbloed and Borst-Pauwels, 1992
L. piperatus	>>20	3.5	NH_4^+	Agar	Zel and Gogala, 1989
L. rufus	19	NR[b]	NH_4^+	PO_4, SO_4^{2-}, Chel.	Hintikka, 1988
L. rufus	0.8	4.0	NH_4^+		Klam and Godbold (unpublished)
L. rufus	0.03	3.5	NH_4^+		Jongbloed and Borst-Pauwels, 1992
Leccinum testaceoscabrum	19	NR[b]	NH_4^+	PO_4, SO_4^{2-}, Chel.	Hintikka, 1988
Pisolithus tinctorius	>>20	3.0	NH_4^+	Agar	Zel et al., 1991
P. tinctorius	>>2	4.0	NH_4^+		Klam and Godbold (unpublished)
P. tinctorius 210	2	3.4	NH_4^+	Agar, PO_4, Chel.	Thompson and Medve, 1984

P. tinctorius 230	2	3.4	NH_4^+	Agar, PO_4, Chel.	Thompson and Medve, 1984
P. tinctorius 250	2	3.4	NH_4^+	Agar, PO_4, Chel.	Thompson and Medve, 1984
Paxillus involutus	150	NR[b]	NH_4^+	PO_4, SO_4^{2-}, Chel.	Hintikka, 1988
P. involutus	1.5	4.0	NH_4^+	PO_4, Chel.	Burt et al., 1986
P. involutus MAI	0.8	4.0	NH_4^+		Klam and Godbold (unpublished)
P. involutus 533	0.2	4.0	NH_4^+		Klam and Godbold (unpublished)
P. involutus NAU	0.2	4.0	NH_4^+		Klam and Godbold (unpublished)
Rhizopogon luteolus	1.5	3.0	NH_4^+		Oelbe-Farivar, 1985
R. roseolus	4	3.0	NH_4^+		Oelbe-Farivar, 1985
R. roseolus	0.4	3.0	NO_3^-		Oelbe-Farivar, 1985
R. rubescens	0.4	3.8	NH_4^+/NO_3^-		Browning and Hutchinson, 1991
Suillus bovinus	460	NR[b]	NH_4^+	PO_4, SO_4^{2-}, Chel.	Hintikka, 1988
S. bovinus	0.4	4.0	NH_4^+	PO_4, Chel.	Burt et al., 1986
S. luteus	150	NR[b]	NH_4^+	PO_4, SO_4^{2-}, Chel.	Hintikka, 1988
S. luteus	19	3.4	NH_4^+	Agar, PO_4, Chel.	Thompson and Medve, 1984
S. tomentosus	0.4	3.8	NH_4^+/NO_3^-		Browning and Hutchinson, 1991
S. variegatus	19	NR[b]	NH_4^+	PO_4, SO_4^{2-}, Chel.	Hintikka, 1988
S. variegatus	19	4.5	NH_4^+	Agar, PO_4, Chel.	Väre and Markkola, 1990
S. variegatus	>>4	4.8	NH_4^+	PO_4, Chel.	Paulus and Bresinsky, 1989
S. variegatus	2	3.5	NH_4^+	Agar	Zel and Gogala, 1989
Thelephora terrestris	2	3.4	NH_4^+	Agar, PO_4, Chel.	Thompson and Medve, 1984

[a] Aluminum concentration decreasing fungal growth by at least 30%.

[b] The pH at low Al concentrations not reported (NR).

[c] Chelation with organic ligands.

[d] No biologically significant inhibition at the maximum concentration tested.

buffered media at low pH (<4), may not only be the result of high proton concentrations per se but also a result of the buffer used. Willenborg et al. (1990) found little difference in growth response to pH changes in the range 2.5–5.5 using unbuffered medium. However, growth of most fungi tested, with the exception of *Hebeloma crustuliniforme,* was poor in the submerged cultures.

Erland et al. (1990) compared the radial growth rates of six ectomycorrhizal strains in pure culture with growth rates of the extramatrical mycelium growing in association with pine seedlings. These investigators found much higher growth rates in the hyphae growing out from mycorrhizal root tips compared to hyphae growing saprophytically on buffered agar medium or autoclaved peat. Moreover, growth depression observed at low pH in the pure culture experiments were not evident in treatments with seedlings. These results show that most ectomycorrhizal fungi are much more acid tolerant than pure culture experiments suggest. Hence, the decrease in soil pH per se caused by acid precipitation is unlikely to explain solely the effects on ectomycorrhizae observed in the field.

3.2.3. Metal Toxicity

Acidification of mineral soils may increase solubilization of Al ions from silicate and clay minerals (Ulrich, Chapter 1; Cronan, Chapter 2). In humus top layers, however, Al ions are less abundant. Here the mobilization of heavy metals (Tyler et al., 1987) may be a major process affecting biotic processes (Tyler et al., 1989). A number of greenhouse studies showed that addition of Al (Entry et al., 1987; McQuattie and Schier, 1992) or heavy metals (Burton et al., 1984; Dixon, 1988; Dixon and Buschema, 1988; Cappelka et al., 1991) to soil or sand substrates impairs ectomycorrhizal colonization. However, from these studies it remains unclear whether these effects were primarily related to the impact of the metals on the growth and physiology of the fungal symbionts or were due to an indirect effect via impaired host performance. In addition, pot studies of this type suffer from several shortcomings (cf. Section 3.3.2). In many cases, these shortcomings include ecologically questionable high doses of added metals and the lack of information on biologically relevant metal concentrations in the solution phase. Hence, results from these pot studies are difficult to transfer to the situation found in forest soils.

Direct effects of Al (Table 6.5) or heavy metals (e.g., Cd and Pb, Table 6.6) on the physiology of ectomycorrhizal fungi were studied using pure culture techniques. Culture conditions varied greatly among these studies with regard to the type of medium (agar or liquid), pH, or concentrations of polyvalent anions (especially phosphate) and organic ligands. Because any of these factors may severely affect the activity of the free metal ions in the solution phase by means of, for example, polymerization, precipitation (pH, phosphate), and chelation reactions, or binding to charged surfaces (agar), and because the toxicity of dissolved metals in most cases correlates with the activity of the free metal ions (Godbold, Chapter 7; Tyler et al., 1989), the great variation in the sensitivity of the fungal symbionts towards these metals found between different studies (Tables 6.5 and 6.6) is not surprising. In addition, metal toxicity was often reduced by increasing concentrations of di-

TABLE 6.6. Critical Cadmium and Lead Concentrations Inhibiting Growth of Ectomycorrhizal Fungi in Pure Culture

Range of Critical[a] Metal Concn (μM)	Number of Fungi Tested	Culture Conditions Activity of Metal Ions Potentially Reduced by	References
Cd			
220–890	15	Agar, PO_4	Willenborg et al., 1990
13–150	9	Agar, PO_4	McCreight and Schroeder, 1982
9–450	11	Agar, PO_4	Colpaert and von Assche, 1992b
0.3	1	Agar[b], PO_4	Darlington and Rauser, 1988
0.1–10	3	Agar[b], PO_4	Jongbloed and Borst-Pauwels, 1990a
Pb			
250–2500	9	Agar, PO_4, Chel[c]	McCreight and Schroeder, 1982
50–500	4	Agar, not reported	Pachlewski and Chrusiak, 1986
10–100	7		Marschner, 1994

[a] Metal concentration decreasing fungal growth by at least 25%.
[b] No difference between agar and liquid cultures.
[c] Chelation with organic ligands.

valent cations like Ca^{2+} and Mg^{2+} (Gadd and Griffith, 1978; Kessels et al., 1985; Taylor, 1988; Marschner, 1991) and modified by the type of nitrogen source (Oelbe-Farivar, 1985; Chapter 7). Only a few investigations addressed these problems using low pH, low nutrient concentrations, and liquid media without chelators. The data from these studies suggest that growth depression of mycorrhizal fungi may occur at a concentration of 30–100 μM of Al, 10 μM of Pb, or 0.1 μM of Cd. This result suggests that toxicity threshold values of Pb are one order of magnitude higher in the fungal symbionts than in *P. abies* seedlings (see Chapter 7). For Al, the data indicate that the tolerance of the mycorrhizal partners is in the same range. For Cd the situation is quite different, since the sensitivity may be much higher in the mycobiont than in the host (*P. abies,* see Chapter 7). However, these are rough estimates and comparisons of other host–fungus combinations may give different results. Future research should provide more realistic data on fungus and tree sensitivity to Al and heavy metals.

Although data obtained in the laboratory cannot be easily transferred to the situation in the field (due to differences in, e.g., metal speciation; Tyler et al., 1989; Godbold, Chapter 7), a comparison of metal concentrations inhibitory to mycorrhizal fungi in pure culture experiments with those found in soil solutions of forest stands in the German Solling area (Al concentrations up to 1 mM, Matzner and Prenzel, 1992) suggests that in acidified mineral soils Al has the potential to directly impair ectomycorrhizal fungal growth. Hence, during soil acidification Al tolerance may be an important feature in the competition and succession of ectomycorrhizal species and genotypes in mineral soil horizons. Cadmium and Pb concentrations found in humus or soil lysimeters of the Solling sites (up to 0.03 and 0.3 μM, respectively;

Mayer et al., 1980; Lamersdorf, 1989) are 3- to 30-fold lower than potentially toxic concentrations estimated in pure culture experiments. This result suggests that Cd and Pb may not directly affect the physiology of ectomycorrhizal fungi in the Solling area. However, experiments with mycorrhizal Norway spruce seedlings show that the extramatrical mycelium reacts much more sensitively to Pb than the reaction patterns of mycorrhizal colonization or hyphal growth in pure culture suggest (Table 6.7). Since the Pb tolerance of the host tree may be lower than that of the fungus, the data show that the fungi were indirectly affected by reduced performance of the host root (see Chapter 7). Hence, in the forest, Pb may indirectly affect fungal performance at much lower Pb concentrations than estimates by pure culture toxicity tests indicate. Future laboratory work should take this into account.

The above comparison of critical Al, Cd, and Pb concentrations found in laboratory experiments and soil solution data is based on the total concentration of metals in soil solutions rather than estimates on the activities of free metal ions (For a discussion of the effect of metal speciation on the toxicity potential the reader is referred to Chapter 7). In acidified mineral soils with low content of organic matter, monomeric Al often forms the dominant Al species (Cronan, Chapter 2). Hence, for these soil horizons total and critical Al concentrations found in laboratory experiments may be compared directly. From the heavy metals considered in this chapter, Pb forms stable complexes with dissolved organic compounds, whereas complexation of Cd in acidic soil solutions is believed to be of minor importance (Tyler et al., 1989; Livens, 1991). In forest soils in Sweden 29–36% of total dissolved Pb was present as cationic species (Tyler et al., 1987) indicating that complexation of Pb by fulvic acids may not be quantitative. Although complexation undoubtedly decreases the toxic effects of Pb, the risk of Pb toxicity is still evident since metal speciation in soil solutions is subject to substantial seasonal variations. At times of high internal production of acidity in the soil (Matzner and Cassens-Sasse, 1984), a decrease in pH may lead to enhanced dissociation of dissolved metal organic complexes (Strnad, 1987; Fischer, 1987). In addition, Al (Matzner and Cassens-Sasse, 1984) and heavy metals (Tyler et al., 1987; Lamersdorf, 1989) bound to soil surfaces may increasingly be mobilized. These processes may account for a considerable increase in the activity of free metal ions and thus in the toxicity potential of the soil solution.

TABLE 6.7. Decrease (%) of Mycorrhizal Colonization, Growth of the Extramatrical Mycelium, and Hyphal Growth in Pure Culture After Exposure to Lead[a]

Fungus	Degree of Mycorrhizal Colonization[b]	Length of Extramatrical Hyphae	Pure Culture Mycelium Biomass
Paxillus involutus	−34	−83	−8
Pisolithus tinctorius	− 3	−48	−7

[a]Mycorrhizal *Picea abies* seedlings were grown in sand culture and exposed to 5 μM $PbCl_2$. Growth reductions in pure culture were assessed after exposure to 10 μM $PbCl_2$ in liquid cultures. Data are relative values expressed as percent decrease of corresponding control values (modified from Marschner, 1994).

[b]Percent of root tips converted to mycorrhizae.

In addition, the risk for heavy metal toxicity may be affected by the metal content in the humus layer. The deposition of heavy metals increases with increasing altitude of the forest stand (Glavac et al., 1987; Lovett and Kinsman, 1990) and this may result in higher contents of heavy metals in the humus layer. A comparison of mycorrhizal spruce seedlings grown in forest soils with low and high Pb content shows that with increasing metal content of the soil, metal transfer into the mycorrhizal organs is enhanced (Table 6.8). This finding means that at high deposition sites at high altitude and with high contents of heavy metals in the humus layers (e.g., in the German Harz mountains, Table 6.8), the risk of heavy metal toxicity may be much higher compared to the Solling site, which is at low altitude.

It should also be considered that metal availability may be significantly different in the rhizosphere compared to the bulk soil (Marschner et al., 1986). Alkalization of the rhizosphere may reduce the solubility of metals and enhance their binding to mucilage and soil particles thus reducing metal stress (Marschner, 1991). In contrast, acidification or a change in redox potential may mobilize metals in the rhizosphere (Marschner et al., 1986). In addition, during water uptake by the roots, metals dissolved in the soil solution are transported by mass flow toward the root surfaces. This process may account for a considerable accumulation of toxic metals in the rhizosphere (see Chapter 7). By depending on water inflow rates and soil properties, the increase in metal concentration at the root surface may reach a factor of 10 compared to the bulk soil (Nietfeld, 1993). From the above discussion it may be concluded that the gap between concentrations found in soil solutions and threshold values of heavy metal toxicity found in laboratory experiments, may be diminished considerably in some circumstances.

In forest soils at locations in southcentral Virginia, which naturally varied in the contents of Cu, Zn, and Pb, differences in the number of ectomycorrhizae per soil volume were attributed to differences in the heavy metal content of the soils (Bell et al., 1988). Although results from field work are difficult to interpret, when no attempt was made to quantify concentrations and speciation of metals in the soil solutions and other soil factors were not conclusively excluded as a cause of the observed changes, the data suggest that heavy metals may be important factors impairing ectomycorrhizal development in forest soils even at soil contents comparable to those found in central European forests (Lamersdorf, 1988).

TABLE 6.8. Contents of Lead in Mycorrhizal Roots of Norway Spruce Seedlings Grown in Forest Humus of Low and High Lead Content

	Pb Content	
Compartment	Solling	Harz
Humus (mmol kg_{DW}^{-1})	2.0	6.0
Whole root system (mmol kg_{DW}^{-1})	0.8	1.6
Cortex cell wall of *Paxillus involutus* mycorrhiza (mmol dm^{-3})[a]	1.2	3.9

[a]Lead contents of tissues were determined by energy-dispersive X-ray microanalysis. Data from P. Marschner and G. Jentschke (unpublished).

The variation of metal tolerance among different ectomycorrhizal species (see Table 6.5 and 6.6) and even within the same species is high (Denny and Wilkins, 1987; Colpaert and Van Assche, 1987, 1992a,b). This finding implies that at low metal concentrations sensitive fungi may be affected, whereas other more tolerant species are not affected. This would result in a shift in species composition due to altered competition potentials of the fungi involved. This general view is supported by the data of Rühling et al. (1984) and Rühling and Söderström (1990) who estimated the frequency of basidiomycete fruitbodies in coniferous forests in Sweden along gradients of heavy metal pollution from smelters. These investigators found that with increasing heavy metal contents in the humus layer, total fruitbody frequencies strongly decreased. However, fruitbody frequencies of the genus *Lactarius* were not affected and those of *Amanita* tended to increase (Rühling and Söderström, 1990). The better performance of *Amanita* species coincides with results from pure culture experiments, where *Amanita muscaria* often showed higher tolerance to Zn (Brown and Wilkins, 1985), Cd (McCreight and Schroeder, 1982), or Ni (Jones and Hutchinson, 1988) compared to other fungi tested. Further evidence for the presence of a selection process due to heavy metal pollution is provided by laboratory work showing that ectomycorrhizal fungi isolated from metal polluted soils often had higher tolerance towards the polluting metals than isolates of the same species from unpolluted soils (Colpaert and van Assche, 1987, 1992a,b).

However, it is difficult to transfer results from sites heavily polluted by emitting sources at short distances to the majority of forest sites affected by acid deposition where heavy metals accumulate at slow rates and soils are concurrently acidified. In addition, realistic toxicity threshold values were only estimated for a few fungal strains. Hence, generalizations at this stage of research underlie considerable uncertainty. Additional information about different species and ecotypes is required. With our present knowledge, the role of heavy metal deposition in the change of the mycorrhizal status of forest trees remains unclear. However, a contribution to some of the observed changes, at least at high deposition sites, is likely and cannot be disproved.

3.3. Nitrogen Supply

Nitrogen availability is a limiting factor in many boreal and northern temperate forests. In this region forest management may include nitrogen fertilization. However, in the present situation of high nitrogen deposition in central Europe, problems originate from nitrogen saturation and related soil acidification rather than from nitrogen deficiency.

3.3.1. Ectomycorrhizae in the Situation of High Nitrogen Supply in the Field

One of the first reports on relationships between air pollution and mycoflora was published by Schlechte (1986). He investigated two spruce stands in Germany exposed to different levels of pollution and found considerable differences in the number of ectomycorrhizal species and biomass production of carpophores. The stand exposed to a high level of pollution with respect to nitrogen ~ 41 kg of N ha^{-1} within 6 months) showed a fivefold reduction in the number of species and had a three times lower biomass production of carpophores. Investigations of Rastin et al. (1990) in the same area showed a possible relationship between reduced biomass production of ectomycorrhizal carpophores and high NO_3^- concentrations in the soil solution.

Parts of The Netherlands are exposed to a large nitrogen input from the atmosphere. In extreme cases the dry and wet deposition of ammonia (NH_3) and ammonium (NH_4^+) in forests amounts to 70 kg ha^{-1} $year^{-1}$ (see Table 6.11). The decline of the ectomycorrhizal fungus *Cantharellus cibarius* was reported to be strongest in parts of The Netherlands with the highest levels of air pollution (Jansen and Van Dobben, 1987). In 25 Douglas fir stands Jansen (1991) found that the number and frequency of mycorrhizae and the number and species of fruitbodies were inversely related to the loads of atmospheric pollutants. Investigations on carpophores and mycorrhizal status in young (5–10 years) and old (50–80 years) *P. sylvestris* stands throughout The Netherlands showed a highly significant negative correlation in the old plots among the estimated pollution by nitrogen compounds and both the number of carpophores and the number of fruiting species (Termorshuizen and Schaffers, 1987, 1991).

It is difficult, however, to relate effects on ectomycorrhizal fungi observed in areas with high loads of atmospheric pollutants to single factors like deposition of nitrogenous compounds. High loads of atmospheric pollutants may include increased input of nitrogen, but at the same time the input of other substances (nutrients, as well as acidifying substances or heavy metals) probably increases as well. The composition of element inputs may vary in different regions and the effects caused may be different depending on stand and soil characteristics.

Nitrogen-fertilization experiments in the field were carried out under different aspects with different nitrogen forms (ammonium, nitrate, organic nitrogen, etc.), quantities, and combinations with other nutrients (see a review by Kuyper, 1989). Nitrogen-fertilizers inhibit growth of ectomycorrhizal fungi in many cases (see Table 6.9), but not necessarily when applied in balanced combination with P and/or K. Menge et al. (1977) studied the effects of N, P, NP, and NPK fertilization in an 11-year-old *Pinus taeda* plantation in North Carolina. Six weeks after the application of ammonium nitrate (NH_4NO_3) (56 and 112 kg of N ha^{-1}) the number of mycorrhizal tips per 100 cm^3 of soil showed a significant 14–20% decrease. No decrease was noticed after the NPK treatment (112 kg of N, 25 kg of P, 50 kg of K ha^{-1}). In the same experiment Menge and Grand (1978) counted the numbers of ectomycorrhizal fruitbodies and found lower numbers compared to the control after fertilization with NH_4NO_3, NP, and NPK, indicating that fruitbody production may be more sensitive to nitrogen addition than mycorrhizal colonization. The application of NPKCa (200 kg of NH_4NO_3 ha^{-1}) in a mature beech forest (*Fagus sylvatica*) in France, however, decreased the number of mycorrhizal tips per gram of root dry weight by 43% (Blaise and Garbaye, 1983). Morphotypes with external mycelium were partly replaced by mycorrhizae with smooth hyphal mantles.

TABLE 6.9. The Effects of Nitrogen Fertilization on the Ectomycorrhizal Symbiosis

Tree Species Age (years)	Treatment per Hectare	Parameter	Effect	References
A. Mycorrhizal Colonization				
Picea abies 45	$(NH_4)_2SO_4$ 700 kg	Relative mycorrhizal frequency	Decrease	Haug and Feger, 1990
P. sitchensis 35	$(NH_4)_2SO_4$ 300 kg N	Mycorrhizal frequency	− 5–19%	Alexander and Fairley, 1983
Pinus taeda 11	NH_4NO_3 56/112 kg N	Ectomycorrhizae/100 cm^3 soil	− 14–20%	Menge et al., 1977
	NP		− 13–15%	
	NPK		No effect	
Fagus sylvatica 90	NH_4NO_3 (PKCa) 200 kg	Ectomycorrhizae/g root dry wt	− 43%	Blaise and Garbaye, 1983
F. sylvatica 139	$(NH_4)_2SO_4$ 660 kg	Relative mycorrhizal frequency	− 87% (humus) No effect (mineral soil)	Rapp, 1991

B. Frequency of Species or Morphotypes				
Pinus taeda 11	NH_4NO_3 56/112 kg N	"Genus" Ja[a]	− 7%	Menge et al., 1977
		Cenococcum geophilum	No effect	
P. sitchensis 35	$(NH_4)_2SO_4$ 300 kg N	*C. geophilum*	Decrease	Alexander and Fairley, 1983
Fagus sylvatica 90	NH_4NO_3 (PKCa)	Type A2[b]	− 43%	Blaise and Garbaye, 1983
C. Occurrence of Carpophores				
Pinus taeda 11	NH_4NO_3 56/112 kg N	Number of fruitbodies	− 67–83%	Menge and Grand, 1978
P. sylvestris 110	300 kg N	Fresh weight per hectare	− 42–66%	Ritter and Tölle, 1978
P. sylvestris	$NH_4Cl/NaNO_3$ 80 kg N	Number of fruitbodies	Decrease	Termorshuizen et al., 1988
Fagus sylvatica 120	NH_4NO_3 60/180 kg N	Number of fruitbodies	− ca. 100%	Rühling and Tyler, 1991

[a]Morphotypes after Dominik, 1959.
[b]Morphotype after Voiry, 1981.

Alexander and Fairley (1983) found a 15% decrease in the production of mycorrhizae in a 35-year-old Sitka spruce (*Picea sitchensis*) plantation after treatment with ammonium sulfate [$(NH_4)_2SO_4$] (300 kg of N ha^{-1}). Mortality, however, was reduced by 30%, which indicated an increase of longevity of ectomycorrhizae after fertilization. *Cenococcum geophilum* mycorrhizae were significantly reduced after fertilization. In other studies $(NH_4)_2SO_4$ was applied in forest ecosystems to simulate situations with high deposition of acidity and subsequent soil acidification (Haug and Feger, 1990; Rapp, 1991). Both experiments showed a decrease in mycorrhizal colonization after treatment (Table 6.9). However, direct impact of the fertilizer on fine roots and ectomycorrhizae was not excluded. The detrimental effects of $(NH_4)_2SO_4$ on ectomycorrhizae are reported by Loub (1963).

Fruitbody formation of mycorrhizal fungi showed marked reactions to additional nitrogen (see Table 6.9). The effects of various inorganic and organic nitrogen fertilizers on the production of *Lactarius rufus* fruitbodies were studied in three pine (*P. sylvestris*) stands differing in age and water regime in Finland (Ohenoja, 1988). The abundance of fruitbodies varied with site type and the age of the forest. On a very dry site (60–80 years) frequency increased just after fertilization but after 4–5 years of treatment fruitbody production declined. The dryish, older forest (80–100 years) seemed to be very sensitive to fertilization. All treatments reduced the production of *L. rufus* and other mycorrhizal fungi. In the mesic forest yields of *L. rufus* increased on urea and nitroform plots, but decreased on the plots receiving other fertilizers. It should be mentioned, however, that some mycorrhizal fungi can be considered nitrophilous with regard to fruitbody production after nitrogen fertilization, for example, *Paxillus involutus* (Hora, 1959; Laiho, 1970; Ohenoja, 1978, 1988; Kuyper, 1989; Kuyper and de Vries, 1990) and *Lactarius rufus* (Hora, 1959; Ohenoja, 1978).

Some of the investigations indicate that the influence of large nitrogen input is stronger on fruitbody formation than on below-ground mycorrhizal colonization (Menge et al., 1977; Menge and Grand, 1978; Kuyper, 1989; Termorshuizen and Schaffers, 1991). However, there are only a few studies including the assessment of above-ground as well as below-ground organs of the ectomycorrhizal fungi. In addition, there is no information about short-term and long-term effects of fertilizer treatments. Causal interpretation of nitrogen fertilization experiments in the field is difficult. The application of nitrogen is not equivalent to simply "adding" nitrogen, because at the same time soil chemical processes (e.g., soil acidification) are induced and microbial populations are able to be changed. The resulting soil reactions may be various and complex and response reactions cannot be related to a single factor.

3.3.2. Regulation of Root and Mycorrhizal Growth

A reduction of the mycorrhizal status of the trees in terms of the number of mycorrhizae per unit of soil or hectare by nitrogen can be the result of several different processes related to (a) root growth, (b) root branching, (c) mycorrhizal colonization, or (d) turnover of mycorrhizal (and if present nonmycorrhizal) roots (see

Section 2.2). Clearly, most of these factors can be expected to interfere with each other or cannot easily be differentiated from one another in the field due to lack of appropriate methods. However, much confusion in the literature could have been avoided by the choice of suitable parameters making it possible to at least approximately distinguish among the different processes (see Section 2.2 and Kottke and Oberwinkler, 1986). Nitrogen is one of the major factors controlling root growth (Marschner, 1986). In agricultural crops increased nutrient availability, especially of nitrogen, affects the shoot/root ratio in favor of the shoots. This shift is seen as an adaptation mechanism to different levels of soil fertility. At low nutrient availability the plant increases root growth, and hence the root surface area, to enhance nutrient acquisition (for reviews see Marschner, 1986; Fitter, 1985, 1987; Hetrick, 1991).

This mechanism was also found in nonmycorrhizal tree seedlings (Ingestadt and Kähr, 1985; Ingestadt et al., 1986; Wallander and Nylund, 1991; Gijsman, 1990a; Olsthoorn et al., 1991). Mycorrhizal seedlings show a similar regulatory pattern to different levels of nitrogen nutrition (Ingestadt et al., 1986; Jentschke et al., 1991; Godbout and Fortin, 1992; Wallander and Nylund, 1991, 1992). Although all of the cited investigations were carried out with young seedlings, the results may also be valid for mature forest stands. Vogt et al. (1983) investigated fine root and mycorrhizal root biomass in Douglas fir stands of different age and site productivity in western Washington, and, after canopy closure, found lower fine and mycorrhizal root biomass in stands with higher productivity. Similar results were obtained in 63–91-year-old *P. abies* forests in southern Germany (Kottke and Agerer, 1983).

In contrast, local application of nitrogen enhanced root growth and branching in agricultural crops (for a review see Marschner, 1986). To our knowledge there is little information on chemical factors controlling root branching in forest tree species. Meyer (1966) grew beech seedlings in split pots filled with mor and mull soil and showed that root growth and the number of root tips was higher on the mull side. Since the mull soil provided more nutrients (especially nitrogen) to the roots, the improved root growth may be due to the locally increased nutrient availability. However, physical, chemical, and biological conditions differ largely between mull and mor soil. Hence, causal interpretation of Meyer's results is difficult.

Godbold et al. (1988) grew nonmycorrhizal seedlings of *P. abies* in hydroponic culture with a range of nitrate concentrations. They found that lateral root proliferation, but not primary root elongation, was enhanced by increasing the nitrate concentration up to 1 m*M*.

The situation is further complicated because the form of nitrogen (NH_4^+ vs. NO_3^-) has a profound influence on root growth and morphology (Leyton, 1952; van den Driessche, 1978; Marschner, 1986; Gijsman, 1990a). Hence, any comparison of root growth under different conditions of nitrogen supply should also consider differences in the chemical form of nitrogen supplied to the roots.

Much of the recent and older literature deals with the causal relationship of nutrient supply, carbohydrate, or hormone levels in roots and the intensity of ectomycorrhizal colonization. This work was reviewed in detail by Slankis (1974), Harley and Smith (1983), Nylund (1988), and by Gogala (1991). The results from many investigations confirmed that nitrogen addition to pot cultures of mycorrhizal seedlings often resulted in an optimum response curve. Nitrogen added at a suboptimal rate often stimulated growth and mycorrhizal formation, while over optimum supply caused a significant reduction of the mycorrhizal status estimated as the percentage of root tips converted to mycorrhizae. However, there are at least two facts that make a comparison with field conditions difficult. (1) In many studies nonmycorrhizal seedlings were planted into forest or nursery soil. The roots were either colonized by indigenous inoculum in the soil or by artificial inoculum. The way of primary infection in these experiments can be expected to differ largely from the infection process in forest soils, where much of the infection may be due to hyphae growing in contact with already established mycorrhizae (Fleming, 1983; Brownlee et al., 1983). Some authors (e.g., Termorshuizen and Ket, 1991 and those cited in the following paragraph) addressed this problem by establishing mycorrhizal colonization prior to the experimental period. This procedure avoids side effects from the different nitrogen regimes on the artificial establishment of mycorrhizae and restricts treatment effects to the recolonization of new roots by older parts of the root system. (2) In most pot studies, no information on nutrient concentrations in the solution phase is given and in many cases these are suspected to be much higher than in forest soil solutions (Zöttl, 1990). Hence, there is a potential risk of overestimating nitrogen effects by pot studies.

With the development of new culture techniques (Kähr and Arveby, 1986; Nylund and Wallander, 1989; Jentschke et al., 1991) ectomycorrhizal seedlings may be grown in relatively defined nutritional conditions. Ingestadt et al. (1986) grew ectomycorrhizal pine seedlings at high growth rates and showed that mycorrhizal formation was not inhibited at high nutrient application rates. However, in this work exact measurements of mycorrhizal colonization were not assessed.

Wallander and Nylund (1991, 1992, see Table 6.10) also used Scots pine seedlings and a similar culture technique and showed that high NH_4^+ and NO_3^- concentrations had less negative effects on mycorrhizal formation as previously thought (cf. Slankis, 1967). No root tip data were presented in these studies. Instead, the fungal biomass was estimated by the ergosterol assay. The data showed that the fungal biomass per unit of root dry weight was affected by high nitrogen supply and that, from the two nitrogen forms applied, NH_4^+ was more effective than NO_3^-. It is likely that the amount of fungal mycelium present in each individual mycorrhizal root tip was affected by the different nitrogen regimes. Hence, reduced fungal biomass per unit root weight does not necessarily mean a similar reduction of the number of mycorrhizal root tips. Estimates of fungal biomass may therefore tend to overestimate nitrogen effects on the structural components of the symbiosis, the mycorrhizae.

In experiments with spruce seedlings inoculated with *Lactarius* species and grown with different concentrations of nitrogen, Jentschke et al. (1989, 1991) measured the number of mycorrhizal and nonmycorrhizal root tips and found a slight-to-moderate reduction in the percentage of root tips converted to mycorrhizae (Table 6.10). Again, NH_4^+ was more effective than NO_3^-.

Average concentrations of nitrate in forest soil solutions at sites with high inputs of nitrogen may not exceed values of 3 m*M* (Table 6.11). In contrast, NH_4^+ binds to

TABLE 6.10. Effects of Nitrogen on Mycorrhizal Colonization

Mycobiont	Nitrogen Conc (m*M*) of High N Treatment		Phosphorous Supply	Effect[a]	References
	NH_4^+	NO_3^-			
Pinus sylvestris (Effects on ergosterol concentration in roots)					
Laccaria bicolor	7		Not limiting	−72%	Wallander and Nylund, 1991
		7	Not limiting	−41%	Wallander and Nylund, 1991
	7	7	Not limiting	−51%	Wallander and Nylund, 1992
	21		Not limiting	−82%	Wallander and Nylund, 1991
		21	Not limiting	−59%	Wallander and Nylund, 1991
	14		Limiting	+ 3%	Wallander and Nylund, 1992
Hebeloma crustuliniforme	7	7	Not limiting	−17%	Wallander and Nylund, 1992
	14		Limiting	+ 7%	Wallander and Nylund, 1992
Suillus bovinus	7		Not limiting	−59%	Wallander and Nylund, 1992
Picea abies (Effects on % mycorrhizal root tips of total number of root tips)					
Lactarius rufus	1.5	1.5	Limiting	− 1%	Jentschke et al. 1991
	2.7	0.3	Limiting	− 7%	Jentschke et al. 1991
	0.3	2.7	Limiting	− 1%	Jentschke et al. 1991
Lactarius theiogalus	1.5	0.3	Limiting	−25%	Jentschke et al. 1989
	0.3	2.7	Limiting	− 1%	Jentschke et al. 1989

[a]Reduction/increase of mycorrhizal colonization compared to low nitrogen controls.

TABLE 6.11. Nitrogen Concentrations in Forest Soil Solutions

Forest Site[a]	N Deposition (throughfall) ($kg\ ha^{-1}\ yr^{-1}$)		Soil Depth (cm)	Concentration[b] in Soil Solution (m*M*)		References
	NH_4^+	NO_3^-		NH_4^+	NO_3^-	
Fagus sylvatica						
Harste (GER)	13	9	80	<0.01	0.4 (0.06–1.0)	Bredemeier et al., 1990
Solling (GER)	13	12	Humus	0.1 (0–1.3)	0.2 (0–1.4)	Matzner et al., 1982
			100	0.01 (0–0.2)	0.02 (0–0.2)	Bredemeier et al., 1990
Picea abies						
Spanbeck (GER)	15	13	10	0.03 (0–0.2)	0.7 (0.1–1.6)	Bredemeier, 1987
			80	0.02 (0–0.1)	1.0 (0.5–1.5)	Bredemeier et al., 1990
Solling (GER)	16	16	Humus	0.2 (0–0.7)	0.3 (0–1.6)	Matzner et al., 1982
			100	0.01 (0–0.1)	0.2 (0.1–0.4)	Bredemeier et al., 1990
Wingst (GER)	41	12	Humus	0.4 (0.1–1.2)	0.4 (0–0.8)	Büttner, 1992
			30	0.1 (0–0.4)	0.7 (0.1–2.8)	Bredemeier et al., 1990; Meyer, 1992

Pinus sylvestris						
Harderwijk (NL)	31	12	Humus	0.2 (0–0.4)	0.4 (0.1–1.7)	PHB de Visser, pers. com.
			20	0 (0–0.1)	0 (0–0.4)	PHB de Visser, pers. com.
Gerritsfles (NL)	57	16	0–10	0.9[c]	1.0	van Breemen and Verstraten, 1991
Tongbersven (NL)	63	9	0–10	0.8[c]	0.4	van Breemen and Verstraten, 1991
Pseudotsuga menziesii						
Kootwijk (NL)	29	10	Humus	0.3 (0.1–0.7)	1.0 (0.4–1.5)	Gijsman and Willigen, 1991; PHB de Visser, pers. com.
			20	0.1 (0–0.2)	0.8 (0.3–1.4)	Gijsman and Willigen, 1991
			90	low	0.7 (0.2–2.0)	Gijsman and Willigen, 1991
Speuld (NL)	31	11	0–10	0.4[c]	0.5	van Breemen and Verstraten, 1991
Zelhem (NL)	67	13	0–10	2.0[c]	4.2	van Breemen and Verstraten, 1991
Quercus robur + *Betula pendula*						
Hackfort Estate (NL)	41	13				
Acid sandy loam soil			10–90	0.1 ± 0.1(SD)	2.3 (0.4–8)	van Breemen et al., 1987
Calcareous soil			10–90	0.02 ± 0.02 (SD)	2.7 (0.4–8)	van Breemen et al., 1987

[a] Germany = GER; The Netherlands = NL.
[b] Average value, concentration range (>95% of data) in parentheses. SD = standard deviation.
[c] Recalculated data.

soil surfaces or may readily be nitrified (Tietema et al., 1992) and its concentration in the soil solution, even in ecosystems with high atmospheric input of NH_4^+, is relatively low (except for sites with extreme inputs of nitrogen) and may not exceed 1.3 m*M* in the organic layers of the German Solling area (Table 6.11). At these concentrations mycorrhizal colonization by *L. rufus* or *L. theiogalus* was not severely affected. No data are available for other species in the concentration range of interest. However, from the data of Wallander and Nylund (1991, 1992) we may expect that, at realistic concentrations of nitrogen, the reduction of the mycorrhizal colonization may be much less than the effects found in their experiments. Moreover, Wallander and Nylund (1992) showed that nitrogen effects are significantly modified by the P supply. Contrary to the results of some fertilization experiments (e.g., Menge et al., 1977; see Section 3.3.1) they showed that nitrogen effects were absent when a restricted supply of P was used. Since P often is a growth limiting factor in forest soils, this effect may be of great ecological significance.

In the same paper Wallander and Nylund (1992) showed that the extramatrical mycelium is affected to a greater extent by a high supply of nitrogen than the mycorrhizae themselves. Arnebrant and Söderström (1992) grew mycorrhizal pine seedlings in rhizotrones filled with peat that were fertilized with different concentrations of nitrogen. They found that the extension of the extramatrical mycelium of one unidentified species was severely inhibited by nitrogen, while other fungi were less sensitive. Since no information was given on the nitrogen concentration in the solution phase of the rhizotrones, it remains unclear whether the effects found by Arnebrant and Söderström (1992) are of ecological significance. However, the above data suggest that the extramatrical mycelium is the most sensitive part of the mycorrhizal symbiosis and this sensitivity is species dependent. The decrease in growth of the extramatrical mycelium may likely affect the potential to colonize new roots. Hence, nitrogen sensitive species would be displaced by other less sensitive species (Arnebrant and Söderström, 1992).

There are several points at which nitrogen may interfere with root growth and mycorrhizal development. Some of them were discussed and it may be concluded that a reduction in the number of mycorrhizae per unit of soil volume may be explained in terms of root growth dynamics and not by the impairment of mycorrhizal development. Reduced fine root biomass related to nitrogen does not necessarily reflect root damage and reduced tree health, but may on the contrary indicate a highly efficient root system. More roots are not necessary and would only be uneffective sinks for carbohydrates, thus they would reduce above-ground tree growth.

3.3.3. Surplus of Nitrogen: NH_4^+ Toxicity

In a situation when nitrogen is no longer growth limiting, an excessive supply of nitrogen may negatively affect plant physiological processes and may contribute to tree damage (Nihlgård, 1985; Schulze, 1989). Experimental evidence suggests that from the two major inorganic nitrogen forms, NH_4^+ may be detrimental to agricultural crops (Puritch and Barker, 1967; Mehrer and Mohr, 1989; Hecht and Mohr, 1990), and, at high concentrations, to forest tree species (Flaig and Mohr, 1992). In laboratory experiments, Flaig and Mohr (1992) found that pine seedlings exposed to high concentrations of NH_4^+ took up much more NH_4^+—N than they could incorporate into proteins. As a result, internal pools of low molecular nitrogen compounds (amino acids and free NH_4^+) increased strongly, indicating the presence of NH_4^+ toxicity (Mehrer and Mohr, 1989; Hecht and Mohr, 1990). In contrast, this accumulation was not detected when the nitrogen source was NO_3^-. High contents of the basic amino acid arginine were found in The Netherlands, in needles of damaged pine trees from stands with high inputs of NH_4^+ (van Dijk and Roelofs, 1988). This finding suggests that the disorder of the nitrogen metabolism observed in young seedlings exposed to high NH_4^+ concentrations may also be significant in the field situation. However, in the field, a considerable proportion of nitrogen may be assimilated directly by canopy uptake of NH_3/NH_4^+ (Roelofs et al., 1985; Schulze, 1989), thus making a comparison with laboratory grown seedlings difficult. Ammonium ion uptake leads to an acidification of the rhizosphere (Nye, 1981; Marschner et al., 1986, 1991; Gijsman, 1990b) and interferes with cation uptake of both mycorrhizal fungi (Jongbloed et al., 1991; Boxman et al., 1986) and higher plants (Marschner, 1986; Boxman and Roelofs, 1988). All these effects may contribute to the overall negative effects of excessive NH_4^+ on plant growth (Gijsman, 1990a; Flaig and Mohr, 1992). It is thus likely that besides a direct impairment of fungal growth by high NH_4^+ (Boxman et al., 1986) mycorrhizal development may also be indirectly affected by reduced performance of the host root.

Rygiewicz et al. (1984) measured NH_4^+ uptake and proton excretion in mycorrhizal and nonmycorrhizal Douglas fir seedlings and found that the mycorrhizal seedlings excreted less H^+ per unit of nitrogen absorbed than nonmycorrhizal seedlings by using short-term experiments. These investigators concluded that mycorrhizae ameliorate negative effects of NH_4^+ uptake. However, as stated by Gijsman (1990a) the seedlings used by Rygiewicz and co-workers were far from steady-state conditions. Hence, long-term ionic balances of mycorrhizal and nonmycorrhizal seedlings may have been different from the results obtained in short-term experiments. There is a need for more conclusive studies on the physiological effects of high but realistic concentrations of NH_4^+ on mycorrhizal and nonmycorrhizal tree seedlings before the significance of NH_4^+ toxicity in forest ecosystems heavily loaded with NH_3/NH_4^+ can be confirmed.

4. CONCLUSIONS: CONSEQUENCES FOR TREE NUTRITION AND WATER SUPPLY

Field and laboratory work suggest that the ectomycorrhizal status of forest trees is affected by acid precipitation. However, in the field it is not always clear whether a reduction of the number of ectomycorrhizae is caused by adverse soil conditions due to soil acidification or by a change in nutrient availability. The latter effect can be regarded as a normal adaptation mechanism of root and mycorrhizal development to improved nutrient availability. Laboratory work suggests that the uptake of the major nutrients N and P is regulated on different levels of root and mycorrhizal development, each of which would affect the active uptake surface of a root system: root growth, root architecture including root branching, and growth of the extramatrical mycelium.

With regard to the acquisition of Ca^{2+} and Mg^{2+}, the situation is quite different. Sufficient concentrations of these cations (especially Ca^{2+}, which is phloem immobile and cannot be retranslocated to growing parts of the root system) are a prerequisite for root growth (Marschner, 1986, 1991) and mycorrhizal development as shown for vesicular arbuscular mycorrhizae (Habte and Aziz, 1991; Gryndler et al., 1992). For ectomycorrhizae it was shown that an eightfold increase in Ca^{2+} and Mg^{2+} concentration increased the number of mycorrhizae per seedling by 25–40% (Jentschke et al., 1989, 1991). A limited supply of Mg^{2+}, contrary to P, did not enhance the development of mycorrhizae or extramatrical hyphae (Wallander and Nylund, 1992). This means that, in a situation in which the active uptake surface decreases, a shortage of Ca^{2+} and Mg^{2+} does not alter root or mycorrhizal growth in order to improve acquisition of these nutrients. Hence, the risk for the development of nutrient deficiencies would increase.

In addition, changes in the biomass of fine roots and mycorrhizal mycelium may affect water uptake and drought sensitivity. Since water stress may be an important feature in the development of damage symptoms, especially in Norway spruce (Ulrich, 1989; Gruber, Chapter 8), reduced mycorrhizal development may also contribute to this damage. However, further research is required to evaluate the significance of these mechanisms.

It is likely that not only quantitative but also qualitative changes of the mycorrhizal status are important factors in tree performance (Meier, 1991). A shift in species composition may occur due to several reasons: tree age (Mason et al., 1982), soil acidification including metal toxicity, change in humus type, or nutrient availability. Several of these points were discussed in detail in this section. Mycorrhizal fungi differ in their ability to improve nutrient (Mitchell et al., 1984; Dixon and Hiol-Hiol, 1992a; Lamhamedi et al., 1992a) and water uptake (Dixon and Hiol-Hiol, 1992b; Lamhamedi et al., 1992b). Amelioration of metal toxicity may also be species dependent (Jones and Hutchinson, 1986; Denny and Wilkins, 1987; Colpaert and van Assche, 1992a; see Chapter 7). Hence, any loss of important fungal symbionts may have severe consequences for tree nutrient and water supply and, thus, may contribute to tree damage. However, there is a great lack of knowledge concerning the ecological role of individual mycorrhizal fungal species and genotypes. Much more information is needed before the significance of both qualitative changes in species composition and quantitative changes in root and mycorrhizal development can be understood on the ecosystem level.

REFERENCES

Adams MB, O'Neill EG (1991): Effects of ozone and acidic deposition on carbon allocation and mycorrhizal colonization of *Pinus taeda* L. seedlings. For Sci 37: 5–16.

Agerer R (1987): Colour Atlas of Ectomycorriza. Schwäbisch-Gmünd: Einhorn-Verlag.

Agerer R (1989): Impacts of artificial acid rain and liming on fruitbody production of ectomycorrhizal fungi. Agr Ecosyst Environ 28:3–8.

Agerer R (1990): Gibt es eine Korrelation zwischen Anzahl der Ektomykorrhizen und Häufigkeit ihrer Fruchtkörper? Z Mykol 56:155–158.

Agerer R (1991a): Characterization of ectomycorrhiza. In Norris JR, Read DJ, Varma AK (eds): Techniques for the study of mycorrhiza. London, San Diego, New York, Boston, Sydney, Tokyo, Toronto: Academic Press, Methods of microbiology, Vol. 23, pp. 25–74.

Agerer R (1991b): Streuzersetzende Grosspilze im Höglwald-Projekt: Reaktionen im vierten Jahr der Behandlung. In Kreutzer K u. Göttlein (eds): Ökosystemforschung Höglwald—Beiträge zur Auswirkung von saurer Beregnung und Kalkung in einem Fichtenaltbestand. Beih Forstwiss Centralbl 39:99–103.

Al Abras K, Bilger I, Martin F, Le Tacon F, Lapeyrie F (1988): Morphological and physiological changes in ectomycorrizas of spruce [*Picea excelsa* (Lam.) Link] associated with ageing. New Phytol 110:535–540.

Alexander IJ, Fairley RI (1983): Effects of N-fertilization on populations of fine roots and mycorrhizas in spruce humus. Plant Soil 71:49–53.

Andersen CP, Rygiewicz PT (1991): Stress interactions and mycorrhizal plant response — understanding carbon allocation priorities. Environ Poll 73:217–244.

Antibus RK, Linkins AE (1992): Effects of liming a red pine forest floor on mycorrhizal numbers and mycorrhizal and soil acid phosphatase activities. Soil Biol Biochem 24:479–487.

Arnebrant K, Söderström B (1992): Effects of nitrogen on the mycelial extension of four different ectomycorrhizal fungi grown in symbiosis with *Pinus silvestris*. In Read DJ, Lewis DH, Fitter AH, Alexander IJ (eds): Mycorrhizas in Ecosystems. Wallingford, UK: CAB International, pp 369–370.

Arnolds E (1988): The changing macromycet flora in The Netherlands. Trans Br Mycol Soc 90:391–406.

Arnolds E (1991): Decline of ectomycorrhizal fungi in Europe. Agric Ecosyst Environ 35:209–244.

Asche N, Flückiger W (1987): Erste Ergebnisse der Wurzeluntersuchungen in zwei Buchen-Beständen in der Nordwest-Schweiz. Allg Forstz 42:758–761.

Bell R, Evans CS, Roberts ER (1988): Decreased incidence of mycorrhizal root tips associated with soil heavy-metal enrichment. Plant Soil 106:143–145.

Berg B (1989): Charakterisierung und Vergleich von Ektomykorrhizen gekalkter Fichtenbestände. Ph.D. Thesis, Universität München, Germany.

Beslow DT, Hackskaylo E, Melhuish JH (1970): Effects of environment on beaded root development. Bull Torrey Bot Club 97:248–252.

Blaise T, Garbaye J (1983): Effects of fertilization on the mycorrhization of roots in a beech forest. Acta Oecol Oecol Plant 4:165–169.

Blaschke H (1981): Veränderungen bei der Feinwurzelentwicklung in Weisstannenbeständen. Forstwiss Centralbl 100:190–195.

Blaschke H (1986a): Vergleichende Untersuchungen über die Entwicklung mykorrhizierter Feinwurzeln von Fichten in Waldschadensgebieten. Forstwiss Centralbl 105:477–487.

Blaschke H (1986b): Einfluss von saurer Beregnung und Kalkung auf die Biomasse und Mykorrhizierung der Feinwurzeln von Fichten. Forstwiss Centralbl 105:324–329.

Blaschke H (1988): Mycorrhizal infection and changes in fine-root development of Norway spruce influenced by acid rain in the field. In Jansen AE, Dighton J, Bresser AHM (eds): Ectomycorrhiza and Acid Rain. Bilthoven, The Netherlands; pp 112–115.

Blaschke H (1990): Mycorrhizal population and fine root development on Norway spruce exposed to controlled doses of gaseous pollutants and simulated acidic rain treatments. Environ Pollut 68:409–418.

Blaschke H, Bäumler W (1986): Über die Rolle der Biogeozönose im Wurzelbereich von Waldbäumen. Forstwiss Centralbl 105:122–130.

Blasius D, Kottke I, Oberwinkler F (1985): Zur Bewertung der Güte von Fichtenwurzeln geschädigter Bestände. Forstwiss Centralbl 104:318–325.

Boxman AW, Roelofs JGM (1988). Some effects of nitrate versus ammonium nutrition on the nutrient fluxes in *Pinus sylvestris* seedlings. Effects of mycorrhizal infection. Can J Bot 66:1091–1097.

Boxman AW, Sinke RJ, Roelofs JGM 1986. Effects of NH_4^+ on the growth and K^+ (^{86}Rb) uptake of various ectomycorrhizal fungi in pure culture. Water, Air, Soil, Pollut 31:517–522.

Brand F (1991): Ektomykorrhizen an *Fagus sylvatica*. Charakterisierung und Identifizierung, ökologische Kennzeichnung und unsterile Kultivierung. Lib Botan 2:1–229.

Bredemeier M (1987): Stoffbilanzen, interne Protonenproduktion und Gesamtsäurebelastung des Bodens in verschiedenen Waldökosytemen Norddeutschlands. Ph.D. Thesis, Universität Göttingen, Germany. Ber Forschungszentrum Waldökosysteme Universität Göttingen A33:1–183.

Bredemeier M, Matzner E, Ulrich B (1990): Internal and external proton load to forest soils in Northern Germany. J Environ Qual 19:469–477.

Brown MT, Wilkins DA (1985): Zinc tolerance of *Amanita* and *Paxillus*. Trans Br Mycol Soc 84:367–370.

Browning MHR, Hutchinson TC (1991): The effects of aluminum and calcium on the growth and nutrition of selected ectomycorrhizal fungi of jack pine. Can J Bot 69:1691–1699.

Brownlee C, Duddridge JA, Malibari A, Read DJ (1983): The structure and function of mycelial systems of ectomycorrhizal roots with special reference to their role in forming inter-plant connections and providing pathways for assimilate and water transport. Plant Soil 71:433–443.

Burt AJ, Hashem AR, Shaw G, Read DJ (1986): Comparative analysis of metal tolerance in ericoid and ectomycorrhizal fungi. In Gianinazzi-Pearson V, Gianinazzi S (eds): Physiological and genetical aspects of mycorrhizae. Paris: INRA, pp 377–382.

Burton KW, Morgan E, Roig A (1984): The influence of heavy metals upon the growth of Sitka-spruce in South Wales forests. II. Greenhouse experiments. Plant Soil 78:271–282.

Büttner G (1992): Stoffeinträge und ihre Auswirkungen in Fichtenökosystemen im nordwestdeutschen Küstenraum. Ph.D. Thesis, Universität Göttingen, Germany. Ber Forschungszentrum Waldökosysteme Universität Göttingen A84:1–165.

Chappelka AH, Kush JS, Runion GB, Meier S (1991): Effects of soil-applied lead on seedling growth and ectomycorrhizal colonization of loblolly pine. Environ Pollut 72:307–316.

Colpaert JV, van Assche JA (1987): Heavy metal tolerance in some ectomycorrhizal fungi. Funct Ecol 1:415–421.

Colpaert JV, van Assche JA (1992a): Zinc toxicity in ectomycorrhizal *Pinus sylvestris*. Plant Soil 143:201–211.

Colpaert JV, van Assche JA (1992b): The effects of cadmium and the cadmium–zinc interaction on the axenic growth of ectomycorrhizal fungi. Plant Soil 145:237–243.

Danielson RM, Visser S (1989): Effects of forest soil acidification on ectomycorrhizal and vesicular-arbiscular mycorrhizal development. New Phytol 112:41–47.

Darlington AB, Rauser WE (1988): Cadmium alters the growth of the ectomycorrhizal fungus *Paxillus involutus:* A new growth model accounts for changes in branching. Can J Bot 66:225–229.

Deacon JW, Donaldson SJ, Last FT (1983): Sequences and interactions of mycorrhizal fungi on birch. Plant Soil 71:257–262.

Dennis JJ (1985): Effect of pH and temperature on in vitro growth of ectomycorrhizal fungi. Information Report Pacific Forestry Centre, Canadian Forestry Service, Victoria, B.C. BC-X-273:3–19.

Denny HJ, Wilkins DA (1987): Zinc tolerance in *Betula* spp. III. Variations in response to zinc among ectomycorrhizal associates. New Phytol 106:545–544.

Dighton J, Jansen AE (1991): Atmospheric pollutants and ectomycorrhizae—More questions than answers. Environ Pollut 73:179–204.

Dighton J, Mason PA (1985): Mycorrhizal dynamics during forest tree development. In Moore D

Casselton LA, Wood DA, Franklund JC (eds): Development biology of higher fungi. Cambridge: Cambridge University Press, pp 117–139.

Dighton J, Skeffington RA (1987): Effects of artificial acid precipitation on the mycorrhizas of Scots pine seedlings. New Phytol 107:191–202.

Dighton J, Skeffington RA, Brown KA (1986): The effects of sulphuric acid (pH 3) on roots and mycorrhizas of *Pinus sylvestris*. In Gianinazzi-Pearson V, Gianinazzi S (eds): Physiological and genetical aspects of mycorrhizae. Paris: INRA, pp 739–743.

Dixon RK (1988): Response of ectomycorrhizal *Quercus rubra* to soil cadmium, nickel and lead. Soil Biol Biochem 20:555–559.

Dixon RK, Buschena CA (1988): Response of ectomycorrhizal *Pinus banksiana* and *Picea glauca* to heavy metals in soil. Plant Soil 105:265–271.

Dixon RK, Hiol-Hiol F (1992a): Mineral nutrition of *Pinus caribaea* and *Ecualyptus camaldulensis* seedlings inoculated with *Pisolithus tinctorius* and *Thelephora terrestris*. Soil Science Plant Anal 23:1387–1396.

Dixon RK, Hiol-Hiol F (1992b): Gas exchange and photosynthesis of *Eucalyptus camaldulensis* seedlings inoculated with different ectomycorrhizal symbionts. Plant Soil 147:143–149.

Dominik T (1959): Synopsis of a new classification of the ectotrophic mycorrhizae established on morphological and anatomical characteristics. Mycopathology 11:359–367.

Downes GM, Alexander IJ, Cairney JWG (1992): A study of ageing of spruce [*Picea sitchensis* (Bong.) Carr.] ectomycorrhizas. I. Morphological and cellular changes in mycorrhizas formed by *Tylospora fibrillosa* (Burt.) Donk and *Paxillus involutus* (Batsch. ex Fr.) Fr. New Phytol 122:141–152.

Dubé SL, Bornman JF (1992): Response of spruce seedlings to simultaneous exposure to ultraviolet-B radiation and cadmium. Plant Physiol Biochem 30:761–767.

Deponnois R, Garbaye J (1990): Some mechanisms involved in growth stimulation of ectomycorrhizal fungi by bacteria. Can J Bot 68:2148–2152.

Edwards GS, Kelly JM (1992): Ectomycorrhizal colonization of loblolly pine seedlings during 3 growing seasons in response to ozone, acidic precipitation, and soil Mg status. Environ Pollut 76:71–77.

Eichhorn J (1987): Vergleichende Untersuchungen von Feinwurzelsystemen bei unterschiedlich geschädigten Altfichten (*Picea abies* Karst.). Ph.D. Thesis, Universität Göttingen, Germany.

Entry JA, Cromack Jr K, Stafford SG, Castellano MA (1987): The effect of pH and aluminum concentration on ectomycorrhizal formation in *Abies balsamea*. Can J For Res 17:865–871.

Erland S, Söderström B, Andersson S (1990): Effects of liming on ectomycorrhizal fungi infecting *Pinus sylvestris* L. 2. Growth rates in pure culture at different pH values compared to growth rates in symbiosis with the host plant. New Phytol 115:683–688.

Esher RJ, Marx DH, Ursic SJ, Baker RL, Brown LR, Coleman DC (1992): Simulated acid rain effects on fine roots, ectomycorrhizae, microorganisms and invertebrates in pine forests of the Southern United States. Water Air Soil Pollut 61:269–278.

Estivalet D, Perrin R, Le Tacon F, Garbaye J (1988): Microbial aspects of forest decline in the Vosges. In Jansen AE, Dighton J, Bresser AHM (eds): "Ectomycorrhiza and Acid Rain." Bilthoven, The Netherlands: pp 79–103.

Fellner R (1988): Effects of acid deposition on the ectotrophic stability of mountain forest ecosystems in central Europe. In Jansen AE, Dighton J, Bresser AHM (eds): Ectomycorrhiza and Acid Rain. Bilthoven, The Netherlands: pp 116–121.

Ferrier RC, Alexander IJ (1985): Persistence under field conditions of excised fine roots and mycorrhizas of spruce. In Fitter AH, Atkinson D, Read DJ, Usher MB (eds): Ecological interactions in soil. Oxford: Blackwell Scientific Publications, pp 175–179.

Fiedler HJ, Hunger W (1963): Über den Einfluss einer Kalkdüngung auf Vorkommen, Wachstum und Nährelementgehalt höherer Pilze im Fichtenbestand. Arch For 12:936–962.

Fischer WR (1987): Complexation of heavy metals by water soluble humic substances. In Lindberg SE,

Hutchinson TC (eds): Heavy metals in the environment. Edinburgh: CEP Consultants, Vol. 1, pp 304–306.

Fitter AH (1985): Functional significance of root morphology and root system architecture. In Fitter AH, Atkinson D, Read DJ, Usher MB (eds): Ecological interactions in soil. Oxford: Blackwell Scientific, pp 87–106.

Fitter AH (1987): An architectural approach to the comparative ecology of plant root systems. New Phytol 106 Suppl:61–77.

Flaig H, Mohr H (1992): Assimilation of nitrate and ammonium by the Scots pine (*Pinus sylvestris*) seedling under conditions of high nitrogen supply. Physiol Plant 84:568–576.

Fleming LV (1983): Succession of mycorrhizal fungi on birch: infection of seedlings planted around mature trees. Plant Soil 71:263–267.

Flückiger W, Braun S, Flückiger-Keller H, Leonardi S, Asche N, Bühler U, Lier M (1986): Untersuchungen über Waldschäden in festen Beobachtungsflächen der Kantone Basel-Landschaft, Basel-Stadt, Aargau, Solothurn, Bern, Zürich and Zug. Schweiz Forstwes 137:917–1010.

Gadd GM, Griffiths AJ (1978): Microorganisms and heavy metal toxicity. Microbiol Ecol 4:303–317.

Gijsman AJ (1990a): Nitrogen nutrition of Douglas-fir (*Pseudotsuga menziesii*) on strongly acid sandy soil. 1. Growth, nutrient uptake and ionic balance. Plant Soil 126:53–61.

Gijsman AJ (1990b): Nitrogen nutrition of Douglas-fir (*Pseudotsuga menziesii*) on strongly acid sandy soil. 2. Proton excretion and rhizosphere pH. Plant Soil 126:63–70.

Gijsman AJ, de Willigen P (1991): Modelling ammonium and nitrate uptake by a mature Douglas-fir stand from a soil with high atmospheric NHx input. Neth J Agric Sci 39:3–20.

Glavac V, Koenies H, Jochheim H, Heimerich R (1987): Relief effects on the deposition of air pollutants in forest stands—lead deposition as an example. In Mathy P (ed): Air pollution and ecosystems. Dordrecht, The Netherlands: Reidel, pp 520–521.

Göbl F (1986): Wirkung simulierter saurer Niederschläge auf Böden und Fichtenjungpflanzen im Gefässversuch. III. Mykorrhizauntersuchungen. Centralbl Forstwes 103:89–107.

Göbl F (1988): Mykorrhiza- und Feinwurzeluntersuchungen im Waldschadensgebiet Gleingraben/ Steiermark. Österr Forstz 6:16–18.

Godbold DL, Dictus K, Hüttermann A (1988): Influence of aluminum and nitrate on root growth and mineral nutrition of Norway spruce (*Picea abies*) seedlings. Can J For Res 18:1167–1171.

Godbout C, Fortin JA (1992): Effects of nitrogen fertilization and photoperiod on basidiome formation of *Laccaria bicolor* associated with container-grown jack pine seedlings. Can J Bot 70:181–185.

Gogala N (1991): Regulation of mycorrhizal infection by hormonal factors produced by hosts and fungi. Experientia 47:331–340.

Gorissen A, Joosten NN, Jansen AE (1991): Effects of ozone and ammonium sulphate on carbon partitioning to mycorrhizal roots of juvenile Douglas fir. New Phytol 119:243–250.

Gronbach E (1986): Influence of acid rain precipitation and calcium on frequency, morphology and anatomy of ectomycorrhizae. In Sylvia DH, Hung LL, Graham JH (eds): Mycorrhizae in the next decade: practical applications and research priorities. Proceedings of the 7th NACOM, p 97.

Gronbach E (1988): Charakterisierung und Identifizierung von Ektomykorrhizen in einem Fichtenbestand mit Untersuchungen zur Merkmalsvariabilität in sauer beregneten Flächen. Ph.D. Thesis, Universität München, Germany.

Gronbach E, Agerer R (1986): Charakterisierung und Inventur der Fichten-Mykorrizen im Höglwald und deren Reaktion auf saure Beregnung. Forstwiss Centralbl 105:329–335.

Gryndler M, Vejsadova H, Vancura V (1992): The effect of magnesium ions on the vesicular-arbuscular mycorrhizal infection of maize roots. New Phytol 122:455–460.

Habte M, Aziz T (1991): Relative importance of Ca, N, and P in enhancing mycorrhizal activity in *Leucaena leucocephala* grown in an oxisol subjected to simulated erosion. J Plant Nut 14:429–442.

Hampp R (1992): Comparative evaluation of the effects of gaseous pollutants, acidic deposition, and mineral deficiencies on the carbohydrate metabolism of trees. Agr Ecosyst Environ 42:333–364.

Hansen PA (1988): Prediction of macrofungal occurrence in Swedish beech forests from soil and litter variable models. Vegetatio 78:31–44.

Harley JL, Smith SE (1983): Mycorrhizal symbiosis. London: Academic Press.

Haselwandter K, Berreck M, Hauser M, Schuler R (1988): Untersuchungen der Ektomykorrhiza sowie der mikrobiellen Biomasse und Basalatmung des Rhizosphärenbodens von unterschiedlich geschädigten Fichtenbeständen auf verschiedenen Bodentypen. In Führer E, Neuhuber F (eds): Waldsterben in Österreich. Theorien, Tendenzen, Therapien. FIW-Symposium 1988. Bundesministerium Wiss. u. Forschung Wien, pp 201–213.

Haug I (1987): Licht- und elektronenmikroskopische Untersuchungen an Mykorrhizen von Fichtenbeständen im Schwarzwald. Ph.D. Thesis, Universität Tübingen, Germany.

Haug I, Feger KH (1990): Effects of fertilization with $MgSO_4$ and $(NH_4)_2SO_4$ on soil solution chemistry, mycorrhiza and nutrient content of fine roots in a Norway spruce stand. Water Air Soil Pollut 54:453–467.

Haug I, Kottke I, Oberwinkler F (1986): Licht- und elektronenmikroskopische Untersuchungen von Mykorrhizen der Fichte (Picea abies (L.) Karst.) in Vertikalprofilen. Z Mykol 52:373–392.

Haug I, Weber G, Oberwinkler F (1988): Intracellular infection by fungi in mycorrhizae of damaged spruce trees. Eur J For Pathol 18:112–120.

Hecht U, Mohr H (1990): Factors controlling nitrate and ammonium accumulation in mustard (*Sinapis alba*) seedlings. Physiol Plant 78:379–387.

Hetrick BAD (1991): Mycorrhizas and root architecture. Experientia 47:355–362.

Hintikka V (1988): High aluminum tolerance among ectomycorrhizal fungi. Karstenia 28:41–44.

Hora FB (1959): Quantative experiments on toadstool production in woods. Trans Br Mycol Soc 42:1–14.

How JE (1940): The mycorrhizal relations of larch. I. A study of *Boletus elegans* Schum. in pure culture. Ann Bot 4:135–150.

Hung LL, Trappe JM (1983): Growth variation between and within species of ectomycorrhizal fungi in response to pH in vitro. Mycologia 75:234–241.

Hüttermann A, Becker A, Gehrmann J, Tischner R (1983): Einfluss von Schadstoffen und Kalkdüngung auf die Morphologie der Wurzeln von *Fagus sylvatica*. In Böhm W, Kutschera L, Lichtenegger E (eds): Wurzelökologie und ihre Nutzanwendung. International Symposium Gumpenstein 1982, Verlag Gumpenstein, pp 637–652.

Ingestad T, Arveby AS, Kähr M (1986): The influence of ectomycorrhiza on nitrogen nutrition and growth of *Pinus sylvestris* seedlings. Physiol Plant 68:575–582.

Ingestad T, Kähr M (1985): Nutrition and growth of coniferous seedlings at varied relative nitrogen addition rate. Physiol Plant 65:109–116.

Ingleby K, Mason PA, Last FT, Fleming LV (1990): Identification of ectomycorrhizas. ITE research publication no. 5, Edinburgh Research Station.

Jackson PC, Taylor JM (1970): Effects of organic acids on ion uptake and retention in Barley roots. Plant Physiol 46:538–542.

James H, Court MN, Macleod DA, Parsons JW (1978): Relationships between growth of Sitka spruce (*Pinus sitchensis*), soil factors and mycorrhizal activity on basaltic soil in Western Scotland. Forestry 51:105–120.

Jansen AE (1988): Relation between mycorrhizas and fruitbodies and the influence of tree vitality in Douglas fir plantations in The Netherlands. In Jansen AE, Dighton J, Bresser AHM (eds): Ectomycorrhiza and Acid Rain. Bilthoven, The Netherlands, pp 68–76.

Jansen AE (1991): The mycorrhizal status of Douglas Fir in The Netherlands: its relation with stand age, regional factors, atmospheric pollutants and tree vitality. Agric Ecosyst Environ 35:191–208.

Jansen AE, de Nie HW (1988): Relations between mycorrhizas and fruitbodies of mycorrhizal fungi in Douglas fir plantations in The Netherlands. Acta Bot Neerl 37:243–249.

Jansen AE, Dighton J (1990): Effects of air pollutants on ectomycorrhizas. A review. Air Pollution Research Report 30, EEC, Brussels.

Jansen AE, van Dobben HF (1987): Is decline of *Cantharellus cibarius* in The Netherlands due to air pollution? Ambio 16:211–213.

Jentschke G, Godbold DL, Hüttermann A (1989): Effects of ammonium and nitrate on mycorrhizal infection of Norway spruce seedlings under controlled conditions. Agric Ecosyst Environ 28:201–206.

Jentschke G, Godbold DL, Hüttermann A (1991): Culture of mycorrhizal tree seedlings under controlled conditions—Effects of nitrogen and aluminium. Physiol Plant 81:408–416.

Jones MD, Hutchinson TC (1986): The effect of mycorrhizal infection on the response of *Betula papyrifera* to nickel and copper. New Phytol 102:429–442.

Jones MD, Hutchinson TC (1988): The effects of nickel and copper on the axenic growth of ectomycorrhizal fungi. Can J Bot 66:119–124.

Jongbloed RH, Borst-Pauwels GWFH (1990a): Differential response of some ectomycorrhizal fungi to cadmium in vitro. Acta Bot Neerl 39:241–246.

Jongbloed RH, Borst-Pauwels GWFH (1990b): Effects of ammonium and pH on growth of some ectomycorrhizal fungi in vitro. Acta Bot Neerl 39:349–358.

Jongbloed RH, Borst-Pauwels GWFH (1992): Effects of aluminum and pH on growth and potassium uptake by three ectomycorrhizal fungi in liquid culture. Plant Soil 140:157–165.

Jongbloed RH, Clement JMAM, Borst-Pauwels GWFH (1991): Kinetics of NH_4^+ and K^+ uptake by ectomycorrhizal fungi — Effect of NH_4^+ on K^+ uptake. Physiol Plant 83:427–432.

Kähr M, Arveby AS (1986): A method for establishing ectomycorrhiza on conifer seedlings in steady-state conditions of nutrition. Physiol Plant 67:333–339.

Keane KD, Manning WJ (1987): Effects of ozone and simulated acid rain and ozone and sulfur dioxide on mycorrhizal formation in paper birch and white pine. In Perry R, Harrison RM, Bell JNB, Lester, JN (eds): Acid Rain: Scientific and Technical Advances. London, pp 608–613.

Keane KD, Manning WJ (1988): Effects of ozone and simulated acid rain on birch seedling growth and formation of ectomycorrhizae. Environ Pollut 52:55–65.

Kessels BGF, Belde PJM, Borst-Pauwels, GWFH (1985): Protection of *Saccharomyces cerevisiae* against Cd^{2+} toxicity by Ca^{2+}. J Gen Microbiol 131:2533–2537.

Kessler KJ (1966): Growth and development of mycorrhizae of sugar maple (*Acer saccharum* Marsh.). Can J Bot 44:1413–1425.

Koch W, Lautenschlager K (1988): Photosynthesis and transpiration in the upper crown of a mature spruce in purified and ambient atmosphere in a natural stand. Trees 2:213–222.

Kottke I, Agerer R (1983): Untersuchungen zur Bedeutung der Mykorrhiza in älteren Laub- und Nadelwaldbeständen des Südwestdeutschen Keuperberglandes. Mitt Ver Forstl Standortskd Forstpfl 30:30–39.

Kottke I, Oberwinkler F (1986): Mycorrhiza of forest trees—structure and function. Trees 1:1–24.

Kottke I, Oberwinkler F (1988): Vergleichende Untersuchungen der Feinstwurzelsysteme und der Anatomie von Mykorrhizen nach Trockenstress und Düngemassnahmen. Kernforschungszentrum Karlruhe Projekt Europ Forschungszentrum 39:1–19.

Kottke I, Rapp C, Oberwinkler F (1986): Zur Anatomie gesunder und "kranker" Feinstwurzeln von Fichten: Meristem und Differenzierungen in Wurzelspitzen und Mykorrhizen. Eur J For Pathol 16:159–171.

Kreutzer K, Göttlein (eds) (1991): Ökosystemforschung Höglwald—Beiträge zur Auswirkung von saurer Beregnung und Kalkung in einem Fichtenaltbestand. Beih Forstwiss Centralbl 39:1–261.

Kumpfer W, Heyser W (1986): Effects of stemflow on the mycorrhiza of beech (*Fagus sylvatica* L.). In Gianinazzi-Pearson V, Gianinazzi S (eds): Physiological and genetical aspects of mycorrhizae. Paris: INRA, pp 745–750.

Kuyper TW (1989): Auswirkungen der Walddüngung auf die Mykoflora. Beiträge zur Kenntnis der Pilze Mitteleuropas 5:5–20.

Kuyper TW, de Vries BWL (1990): Effects of fertilization on the mycoflora of a pine forest. In Oldeman RAA, Schmidt P, Arnolds EJM, Staudt F (eds): Forest ecosystems and their components. Wageningen: Pudoc, pp 102–111.

Laiho O (1970): *Paxillus involutus* as a mycorrhizal symbiont of forest trees. Acta Fenn 106:1–72.

Lamersdorf N (1988): Verteilung und Akkumulation von Spurenstoffen in Waldoekosystemen. Ph.D. Thesis, Universität Göttingen, Germany. Ber Forschungszentrum Waldökosysteme Universität Göttingen A36:1–263.

Lamersdorf NP (1989): The behavior of lead and cadmium in the intensive rooting zone of acid spruce forest soils. Toxicol Environ Chem 18:239–247.

Lamhamedi MS, Bernier PY, Fortin JA (1992a): Growth, nutrition and response to water stress of *Pinus pinaster* inoculated with 10 dikaryotic strains of *Pisolithus* sp. Tree Physiol 10:153–167.

Lamhamedi MS, Bernier PY, Fortin JA (1992b): Hydraulic conductance and soil water potential at the soil root interface of *Pinus pinaster* seedlings inoculated with different dikaryons of *Pisolithus* sp. Tree Physiol 10:231–244.

Lehto T (1984): The effect of liming on the mycorrhizae of Scots pine. Fol For 609:1–20.

Leyton L (1952): The effect of pH and form of nitrogen on the growth of sitka spruce seedlings. Forestry 25:32–40.

Liss B, Blaschke H, Schütt P (1984): Vergleichende Feinwurzeluntersuchungen an gesunden und erkrankten Altfichten auf zwei Standorten in Bayern—ein Beitrag zur Waldsterbensforschung. Eur J For Pathol 14:90–102.

Livens FR (1991): Chemical reactions of metals with humic material. Environ Pollut 70:183–208.

Livingston WH, Blaschke H (1984): Deterioration of mycorrhizal short roots and occurrence of *Mycelium radicis atrovirens* on declining Norway spruce in Bavaria. Eur J For Pathol 14:340–348.

Lovett GM, Kinsman JD (1990): Atmospheric pollutant deposition to high-elevation ecosystems. Atmos Environ 24A:2767–2786.

Markkola AM, Ohtonen R (1988): The effect of acid deposition on fungi in forest humus. In Jansen AE, Dighton J, Bresser AHM (eds): Ectomycorrhiza and Acid Rain. Bilthoven, The Netherlands, pp 122–126.

Marks GC, Foster RC (1967): Succession of mycorrhizal associations on individual roots of radiata pine. Aust For 31:194–201.

Marschner H (1986): Mineral nutrition of higher plants. London: Academic Press.

Marschner H (1991): Mechanisms of adaptation of plants to acid soils. Plant Soil 134:1–20.

Marschner H, Häussling M, George E (1991): Ammonium and nitrate uptake rates and rhizosphere pH in non- mycorrhizal roots of Norway spruce <*Picea abies* (L.) Karst.>. Trees 5:14–21.

Marschner H, Römheld V, Horst WJ, Martin P (1986): Root-induced changes in the rhizosphere: Importance for the mineral nutrition of plants. Z Pflanzenernähr Bodenkd 149:441–456.

Marschner P (1994): Einfluss der Mykorrhiziierung auf die Aufnahme von Blei bei Fichtenkeimlingen. Ph.D. Thesis Universität Göttingen, Germany.

Martin F, Delaruelle C, Hilbert J-L (1990): An improved ergosterol assay to estimate fungal biomass in ectomycorrhizas. Mycol Res 94:1059–1064.

Mason PA, Last FT, Pelham J, Ingleby K (1982): Ecology of some fungi associated with an ageing stand of birches (*Betula pendula* and *B. pubescens*). For Ecol Manage 4:19–39.

Matzner E, Khanna PK, Meiwes KJ, Lindheim M, Prenzel J, Ulrich B (1982): Elementflüsse in Waldökosystemen im Solling—Datendokumentation. Göttinger Bodenkd Ber 71:1–267.

Matzner E, Cassens-Sasse E (1984): Chemische Veränderungen der Bodenlösung als Folge saisonaler Versauerungsschübe in verschiedenen Waldökosystemen. Ber Forschungszentrum Waldökosysteme Universität Göttingen 2:50–60.

Matzner E, Prenzel J (1992): Acid deposition in the German Solling area—Effects on soil solution chemistry and Al mobilization. Water Air Soil Pollut 61:221–234.

Mayer R, Heinrichs H, Seekamp G, Fassbender HW (1980): Die Bestimmung repräsentativer Mittelwerte von Schwermetallkonzentrationen in den Niederschlägen und iim Sickerwasser von Wald-Standorten des Solling. Z Pflanzenernähr Bodenkd 143:221–231.

McAfee BJ, Fortin JA (1987): The influence of pH on the competitive interactions of ectomycorrhizal mycobionts under field conditions. Can J For Res 17:859–864.

McCreight JD, Schroeder DB (1982): Inhibition of growth of nine ectomycorrhizal fungi by cadmium, lead, and nickel in vitro. Environ Exp Bot 22:1–7.

McQuattie CJ, Schier GA (1992): Effect of ozone and aluminum on pitch pine (*Pinus rigida*) seedlings —anatomy of mycorrhizae. Can J For Res 22:1901–1916.

Mehrer I, Mohr H (1989): Ammonium toxicity; description of the syndrome in *Sinapis alba* and the research for its causation. Physiol Plant 77:545–554.

Meier S (1991): Quality versus quantity: optimizing evaluation of ectomycorrizae for plants under stress. Environ Pollut 73:205–216.

Meier S, Grand LF, Schoeneberger MM, Reinert RA, Bruck RI (1990): Growth, ectomycorrhizae and nonstructural carbohydrates of loblolly pine seedlings exposed to ozone and soil water deficit. Environ Pollut 64:11–27.

Meier S, Robarge WP, Bruck RI, Grand LF (1989): Effects of simulated rain acidity on ectomycorrizae of red spruce seedlings potted in natural soil. Environ Pollut 59:315–324.

Melin E (1924): Über den Einfluss der Wasserstoffionenkonzentration auf die Virulenz der Wurzelpilze von Kiefer und Fichte. Bot Not 1924:38–48.

Menge JA, Grand LF (1978): Effect of fertilization on production of epigeous basidiocarps by mycorrhizal fungi in loblolly pine plantations. Can J Bot 56:2357–2362.

Menge JA, Grand LF, Haines LW (1977): The effect of fertilization on growth and mycorrhizae numbers in 11-year-old loblolly pine plantations. For Sci 23:37–44.

Meyer FH (1966): Mycorrhiza and other plant symbioses. In Henry SM (ed): Symbiosis. New York: Academic Press, pp 171–244.

Meyer FH (1967): Feinwurzelverteilung bei Waldbäumen in Abhängigkeit vom Substrat. Forstarchiv 38:286–290.

Meyer FH (1984): Mykologische Beobachtungen zum Baumsterben. Allg Forstzeitschrift 39:212–228.

Meyer FH (1985a): Einfluss des Stickstoff-Faktors auf den Mykorrhizabesatz von Fichtensämlingen in Humus einer Waldschadensfläche. Allg Forstz 40:208–219.

Meyer FH (1985b): Die Rolle des Wurzelsystems beim Waldsterben. Forst- Holzwirt 40:351–358.

Meyer FH (1987): Das Wurzelsystem geschädigter Waldbestände. Allg Forstz 42:754–757.

Meyer J, Schneider BU, Werk K, Oren R, Schulze E-D (1988): Performance of two *Picea abies* (L.) Karst. stands at different stages of decline. Oecologia 77:7–13.

Meyer M (1992): Untersuchungen zur Restabilisierung geschädigter Waldökosysteme im norddeutschen Küstenraum (Fallstudie Wingst II). Ph.D. Thesis, Universität Göttingen, Germany. Ber Forschungszentrum Waldökosysteme Universität Göttingen A94:1–276.

Mitchel RJ, Cox GS, Dixon RK, Garrett HE, Sander IL (1984): Inoculation of three *Quercus* species with eleven isolates of ectomycorrhizal fungi. II. Foliar nutrient content and isolate effectiveness. For Sci 30:563–572.

Modess O (1941): Zur Kenntnis der Mykorrhizabildner von Fichte und Kiefer. Symb Bot Ups 5:1–147.

Niederer M, Wieser U (1988): Phosphate uptake and carbohydrate metabolism as parameters for mycorrhizal activity. In Jansen AE, Dighton J, Bresser AHM (eds): Ectomycorrhiza and Acid Rain. Bilthoven, The Netherlands, pp 169–170.

Nietfeld H (1993): Ph.D. Thesis, Universität Göttingen, Germany. Ber Forschungszentrum Waldökosysteme Universität Göttingen (in press).

Nihlgård B (1985): The ammonium hypothesis—an additional explanation to the forest dieback in Europe. Ambio 14:2–8.

Nye PH (1981): Changes of pH across the rhizosphere induced by roots. Plant Soil 61:7–26.

Nylund JE (1987): The ectomycorrhizal infection zone and its relation to acid polysaccharides of cortical cell walls. New Phytol 106:505–516.

Nylund JE (1988): The regulation of mycorrhiza formation—Carboyhdrate and hormone theories reviewed. Scand J For Res 3:465–479.

Nylund JE, Wallander H (1989): Effects of ectomycorrhiza on host growth and carbon balance in a semi-hydroponic cultivation system. New Phytol 112:389–398.

Oelbe-Farivar M-T (1985): Physiologische Reaktionen von Mykorrhizapilzen auf simulierte saure Bodenbedingungen. Ph.D. Thesis, Universität Göttingen, Germany.

Ohenoja E (1978): Mushrooms and mushroom yields in fertilized forests. Ann Bot Fenn 15:38–46.

Ohenoja E (1988): Behaviour of mycorrhizal fungi in fertilized forests. Karstenia 28:27–30.

Olsthoorn AFM, Keltjens WG, Vanbaren B (1991): Influence of ammonium on fine root development and rhizosphere pH of Douglas-fir seedlings in sand. Plant Soil 133:75–81.

Orlov AY (1957): Observations on absorbing roots of spruce (*Picea excelsa* Link.) in natural conditions. Botanicheskii Zhurnal 42:1172–1180.

Pachlewski R, Chrusciak E (1986): Effect of lead and zinc on the growth of some mycorrhizal fungi in vitro. Acta Mycologia 22:73–80.

Pankow W, Niederer M, Wesel U, Schmid B (1988): Physiological monitoring of mycorrhizal activity in a spruce stand. In Jansen AE, Dighton J, Bresser AHM (eds): Ectomycorrhiza and Acid Rain. Bilthoven, The Netherlands, pp 60–67.

Pankow W, Niederer M, Wieser U, Schmid B, Boller T, Wiemken A (1989): Biochemical symptoms of stress in the mycorrhizal roots of Norway spruce (*Picea abies*). Trees 3:65–72.

Paulus W, Bresinsky A (1989): Soil fungi and other microorganisms. In Schulze E-D, Lange OL, Oren R (eds): Forest decline and air pollution. Ecological studies 77, Berlin: Springer-Verlag, pp 110–120.

Persson H (1988): Root growth and root damage in Swedish coniferous stands. In Jansen AE, Dighton J, Bresser AHM (eds): Ectomycorrhiza and Acid Raid. Bilthoven, The Netherlands, pp 53–59.

Puritch GS, Barker AV (1967): Structure and function of tomato leaf chloroplasts during ammonium toxicity. Plant Physiol 42:1229–1238.

Puxbaum H, Gabler K, Smidt S, Glattes F (1991): A one-year record of ozone profiles in an alpine valley (Zillertal/Tyrol, Austria, 600-2000 m a.s.l.). Atmos Environ 25A:1759–1765.

Rapp C (1991): Untersuchungen zum Einfluss von Kalkung und Ammoniumsulfat-Düngung auf Feinwurzeln und Ektomykorrhizen eines Buchenaltbestandes im Solling. Ph.D. Thesis, Universität Göttingen, Germany, Ber Forschungszentrum Waldökosysteme Universität Göttingen, A72:1–293.

Rastin N, Schlechte G, Hüttermann A (1990): Soil macrofungi and some soil biological, biochemical and chemical investigations on the upper and lower slope of a spruce forest. Soil Biol Biochem 22:1039–1047.

Reich BP, Schoettle AW, Stroo HF, Amundson RG (1986): Acid rain and ozone influence mycorrhizal infection in tree seedlings. J Air Pollu Contr Assoc 36:724–726.

Reich PB, Schoettle; A.W., Stroo HF, Troland J, Amundsen RG (1985): Effects of O_3, SO_2 and acidic rain on mycorrhizal infection in northern red oak seedlings. Can J Bot 63:2049–2055.

Ritter G, Tölle H (1978): Stickstoffdüngung in Kiefernbeständen und ihre Wirkung auf Mykorrhizabildung und Fruktifikation der Symbiosepilze. Beitr Forstwirtschaft 12:162–166.

Ritter T (1990): Fluoreszenzmikroskopische Untersuchungen zur Vitalität der Mykorrizen von Fichten [*Picea abies* (L.) Karst.] und Tannen (*Abies alba* Mill.) unterschiedlich geschädigter Bestände im Schwarzwald. Ph.D. Thesis, Universität Tübingen, Germany.

Roelofs JGM, Kempers AJ, Houdijk ALFM, Jansen J (1985): The effect of air-borne ammonium sulphate on *Pinus nigra* var. maritima in the Netherlands. Plant Soil 84:45–56.

Rühling A, Baath E, Nordgren A, Söderström B (1984): Fungi in metal contaminated soil near the Gusum brass mill, Sweden. Ambio 13:34–36.

Rühling A, Söderström B (1990): Changes in fruitbody production of mycorrhizal and litter decomposing macromycetes in heavy metal polluted coniferous forests in North Sweden. Water Air Soil Pollut 49:375–387.

Rühling A, Tyler G (1991): Effects of simulated nitrogen deposition to the forest floor on the macrofungal flora of a beech forest. Ambio 20:261–263.

Rygieviwz PT, Bledsoe CS, Zasoski RJ (1984): Effects of ectomycorrhizae and solution pH on [^{15}N]ammonium uptake by coniferous seedlings. Can J For Res 14:885–892.

Salmanowicz B, Nylund JE (1988): High performance liquid chromatography determination of ergosterol as a measure of ectomycorrhiza infection in Scots pine. Eur J For Pathol 18:291–298.

Scherfose V (1990): Feinwurzelverteilung und Mykorrhizatypen von *Pinus sylvestris* in verschiedenen Bodentypen. Ph.D. Thesis Universistät Hannover, Germany. Ber Forschungszentrum Waldökosysteme Universität Göttingen A62, 1–166.

Schier GA, McQuattie CJ, Jensen KF (1990): Effect of ozone and aluminum on pitch pine (*Pinus rigida*) seedlings: growth and nutrient relations. Can J For Res 20:1714–1719.

Schlechte G (1986) Zur Mykorrhizapilzflora in geschädigten Forstbeständen. Z Mykol 52:225–232.

Schulze E-D (1989): Air pollution and forest decline in a spruce (*Picea abies*) forest. Science 244:776–783.

Schütt P, Blaschke H, Hoque E, Koch W, Lang KJ, Schuck HJ (1983): Erste Ergebnisse einer botanischen Inventur des "Fichtensterbens". Forstwiss Centralbl 102:158–166.

Shafer SR, LF Grand RI Bruck, AS Heagle (1985): Formation of ectomycorrhizae on *Pinus taeda* seedlings exposed to simulated acidic rain. Can J For Res 15:66–71.

Shaw PJA, Dighton J, Poskitt J, Mcleod AR (1992): The effects of sulphur dioxide and ozone on the mycorrhizas of Scots pine and Norway spruce in a field fumigation system. Mycol Res 96:785–791.

Simmons GL, Kelly JM (1989): Influence of O_3, rainfall acidity, and soil Mg status on growth and ectomycorrhizal colonization of loblolly pine roots. Water, Air, Soil, Pollut 44:159–171.

Skärby L, Troeng E, Boström CA (1987): Ozone uptake and effects on transpiration, net photosynthesis and dark respiration in Scots Pine. For Sci 33:801–808.

Slankis V (1967): Renewed growth of ectotrophic mycorrhizae as an indication of an unstable symbiotic relationship. Proceedings of the 14th International Union of Forest Research Organ Congr, Vol. 5:84–99.

Slankis V (1971): Formation of ectomycorrhizae of forest trees in relation to light, carbohydrates, and auxins. In Hacskaylo E (ed.): "Mycorrhizae." U.S. Department of Agriculture Miscellaneous Publication 1189, pp 151–167.

Slankis V (1974): Soil factors influencing formation of mycorrhizae. Ann Rev Phytopathol 12:437–457.

Smidt S, Herman F (1992): Ecosystem studies at different elevations in an alpine valley. Phyton (Horn, Austria) 32:177–200.

Smith HS (1990): Air pollution and forests. Interactions between air contaminants and forest ecosystems. 2nd ed., New York: Springer.

Strnad V (1987): Lead (II), cadmium (II) and copper (II) complexation by water fulvic acid. Trace metal speciation in model and natural water. In Lindberg SE, Hutchinson TC (eds): Heavy metals in the environment. Edinburgh: CEP Consultants, Vol. 1, pp 314–316.

Stroo HF, Alexander M (1985): Effect of simulated acid rain on mycorrhizal infection of *Pinus strobus* L. Water, Air, Soil, Pollut. 25:107–114.

Stroo HF, Reich PB, Schoettle AW, Amundsen RG (1988): Effects of ozone and acid rain on white pine (*Pinus strobus*) seedlings grown in five soils. II. Mycorrhizal infection. Can J Bot 66:1510–1516.

Taylor GJ (1988): The physiology of aluminum phytotoxicity. In Sigel H, Sigel A (eds): Metal ions in biological systems. Vol 24. Aluminum and its role in biology. New York: Dekker, pp 123–163.

Temple PJ, Riechers GH, Miller PR, Lennox RW (1993): Growth responses of ponderosa pine to long-term exposure of ozone, wet and dry acidic deposition, and drought. Can J For Res 23:59–66.

Termorshuizen AJ, Ket PC (1991): Effects of ammonium and nitrate on mycorrhizal seedlings of *Pinus sylvestris*. Eur J For Path 21:404–413.

Termorshuizen AJ, Schaffers AP (1987): Occurrence of carpophores of ectomycorrhizal fungi in selected stands of *Pinus sylvestris* in the Netherlands in relation to stand vitality and air pollution. Plant Soil 104:209–217.

Termorshiuzen AJ, Schaffers AP (1991): The decline of carpophores of ectomycorrhizal fungi in stands of *Pinus sylvestris* L. in The Netherlands: possible causes. Nova Hedwigia 53:267–289.

Termorshiuzen AJ, Schaffers AP, Ket PC, ter Stege EA (1988): The significance of nitrogen pollution on the mycorrhizas of *Pinus sylvestris*. In Jansen AE, Dighton J, Bresser AHM (eds): Ectomycorrhiza and Acid Rain. Bilthoven, The Netherlands, pp 133–139.

Thompson GW, Medve RJ (1984): Effects of aluminum and manganese on the growth of ectomycorrhizal fungi. Appl Environ Microbiol 48:556–560.

Thomson J, Matthes-Sears U, Peterson RL (1990): Effect of seed provenance and fungal species on bead formation in roots of *Picea mariana* seedlings. Can J For Res 20:1746–1752.

Tietema A, Deboer W, Riemer L, Verstraten JM (1992): Nitrate production in nitrogen-saturated acid forest soils — Vertical distribution and characteristics. Soil Biol Biochem 24:235–240.

Tyler G (1985): Macrofungal flora of Swedish beech forests related to soil organic matter and acidity characters. For Ecol Manage 10:13–29.

Tyler G, Balsberg-Pahlsson AM, Bengtsson G, Baath E, Tranvik L (1989): Heavy-metal ecology of terrestrial plants, microorganisms and invertebrates. Water, Air, Soil, Pollut. 47:189–215.

Tyler G, Berggren D, Bergkvist B, Falkengren-Grerup U, Folkeson L, Rühling A (1987). Soil acidification and metal solubility in forests of South Sweden. In Hutchinson TC, Meema KM (eds): Effects of atmospheric pollutants on forests, wetlands and agricultural ecosystems. Nato ASI Series, Vol. G16. Berlin: Springer-Verlag, pp 347–359.

Ulrich B (1986): Die Rolle der Bodenversauerung beim Waldsterben: Langfristige Konsequenzen und forstliche Möglichkeiten. Forstwiss Centralbl 105:421–435.

Ulrich B (1989): Effects of acid deposition on forest ecosystems in Europe. Adv Environ Sci 2:189–272.

van Breemen N, Mulder J, van Grinsven JJM (1987): Impacts of acid atmospheric deposition on woodland soils in the Netherlands: II. Nitrogen transformations. Soil Sci Soc Am J 51:1634–1640.

van Breeman N, Verstraten JM (1991): Soil acidification / N cycling. In Heij GJ, Schneider T (eds): Acidification research in the Netherlands. Stud Environmental Science 46. Amsterdam: Elsevier, pp 289–347.

van den Driessche R (1978): Response of Douglas fir seedlings to nitrate and ammonium nitrogen sources at different levels of pH and iron supply. Plant Soil 49:607–623.

van Dijk HFG, Roelofs JGM (1988): Effects of excessive ammonium deposition on the nutritional status and condition of pine needles. Physiol Plant 73:494–501.

Väre H, Markkola AM (1990): Aluminium tolerance of the ectomycorrhizal fungus *Suillus variegatus*. Symbiosis 9:83–86.

Vincent J-M (1987): Vergleichende Untersuchungen an Feinwurzelm und Mykorrhiza in geschädigten Buchenbeständen (*Fagus sylvatica* L.). Ph.D. Thesis, Universität München, Germany.

Vincent J-M (1989): Feinwurzelmasse und Mykorrhiza von Altbuchen (*Fagus sylvatica* L.) in Waldschadensgebieten Bayerns. Eur J For Pathol 19:167–177.

Vogt KA, Grier CC, Vogt DJ (1986): Production, turnover and nutrient dynamics of above— and below-ground detritus of world forests. Adv Ecolog Res 15:303–377.

Vogt KA, Moore EE, Vogt DJ, Redlin MJ (1983): Conifer fine root and mycorrhizal root biomass within the forest floors of Douglas-fir stands of different ages and site production. Can J For Res 13:429–437.

Voiry H (1981): Classification morphologique des ectomycorrhizes du Chêne et du Hêtre dans le nord-est de la France. Eur J For Pathol 11:284–299.

von Alten H, Fischer B (1987): Mycorrhization of spruce in forest dieback stands in West Germany. In Sylvia DH, Hung LL, Graham JH (eds): Mycorrhizae in the next decade: practical applications and research priorities. Proceedings of the 7th NACOM, p 107.

von Alten H, Rossbach B (1988): Mycorrhiza development in four forest dieback stands in West Germany. Agr Ecosyst Environ 28:13–19.

Walker RF, McLaughlin SB (1991): Growth and root system development of white oak and loblolly pine as affected by simulated acidic precipitation and ectomycorrhizal inoculation. For Ecol Manage 46:123–133.

Wallander H, Nylund JE (1991): Effects of excess nitrogen on carbohydrate concentration and mycorrhizal development of *Pinus sylvestris* L seedlings. New Phytol 119:405–411.

Wallander H, Nylund JE (1992): Effects of excess nitrogen and phosphorus starvation on the extramatrical mycelium of ectomycorrhizas of *Pinus sylvestris* L. New Phytol 120:495–503.

Wiebe S, Blaschke H (1988): Zusammenhänge zwischen Kronenschäden und dem Stärkegehalt der Wurzeln bei *Fagus sylvatica* L. Eur J For Pathol 18:421–425.

Willenborg A, Schmitz D, Lelley J (1990): Effects of environmental stress factors on ectomycorrhizal fungi in vitro. Can J Bot 68:1741–1746.

Winterhoff W, Krieglsteiner GJ (1984): Gefährdete Pilze in Baden-Württemberg. Beih Veröff Naturschutz Landschaftspflege Baden-Württemberg 40:1–120.

Wöllmer H, Kottke I (1990): Fine root studies in situ and in the laboratory. Environ Pollut. 68:383–407.

Zel J, Blatnik A, Gogala N (1992): In vitro aluminum effects on ectomycorrizal fungi. Water, Air, Soil, Pollut 63:145–153.

Zel J, Gogala N (1989): Influence of aluminium on mycorrhizae. Agric Ecosyst Environ 28:569–573.

Zhao Z, Haselwandter K, Glatzel G (1990): Untersuchungen über Zusammenhänge zwischen Oberboden-pH-Werten und Buchenmykorrizen im Wienerwald. Centralbl Forstwesen 107:113–125.

Zimmermann R, Oren R, Schulze ED, Werk KS (1988): Performance of two *Picea abies* (L.) Karst. stands at different stages of decline II. Photosynthesis and leaf conductance. Oecologia 76:513–518.

Zöttl HW (1990): Remarks on the effects of nitrogen deposition to forest ecosystems. Plant Soil 128:83–89.

ALUMINUM AND HEAVY METAL STRESS: FROM THE RHIZOSPHERE TO THE WHOLE PLANT

DOUGLAS L. GODBOLD

Forstbotanisches Institut, Universität Göttingen, Büsgenweg 2, D-37077 Göttingen, Germany

1. INTRODUCTION

The rhizosphere is the soil area most strongly influenced by roots. Due to selective uptake or discrimination, the concentrations of mineral elements in the rhizosphere may differ considerable from those of bulk soil (Knoche, 1993). Similarly, due to the action of roots, the pH of the rhizosphere may also differ in both the bulk soil (Marschner et al., 1991) and along the root length (Gijsman, 1990b). Such changes in pH and the concentration of mineral elements may strongly effect the toxicity of metal ions to roots. The rhizosphere is also inhibited by a number of microorganisms both free living and symbiotic, which may influence the mineral nutrition and water uptake of plants, as well as the toxicity of metal ions toward them. Considering the importance of Al and heavy metal stress in forest ecosystems and in agriculture, as well as the large volume of literature that is devoted to this subject, it is remarkable how little is really known about the effects of Al and heavy metals in the rhizosphere. This lack of knowledge is due in part to the complexity of the subject, and also to the difficulties involved in applying data obtained in a simplified laboratory experiment to the more complicated real life field situation. The roots of most forest trees of temperate regions are almost entirely ectomycorrhizal. Although our knowledge of the function and physiology of the root–mycorrhizal symbiosis is rapidly increasing, our understanding of the role of ectomycorrhizae in

Effects of Acid Rain on Forest Processes, pages 231–264

nutrient uptake (Vogt et al., 1991) and resistance to metal stress is still in its infancy. This is a result of the complexity of the system, which makes experimental investigation difficult. The problem of finding a suitable culture system to investigate the role of ectomycorrhizae in nutrient uptake and resistance to metal stress has proved to be a major challenge (Jentschke et al., 1991a). As a consequence, many experiments were carried out in unsuitable culture systems. Thus our road to a better understanding may have pitfalls from a number of artifacts yet to be identified.

Much of our knowledge of the effects of Al and heavy metals on tree roots was determined in nonmycorrhizal systems. In fact, much of our understanding of the effects of metal stress are based on work with crop plants. The toxicity and tolerance of plants to metal stress has been subject to a number of reviews (Taylor 1988 a,b; Balsberg Pahlsson, 1989). This chapter describes the possible mechanisms of toxicity of Al and heavy metals determined on mycorrhizal and nonmycorrhizal trees and an attempt is made to bring them into the context of rhizosphere processes. Much of the work described was carried out with Norway spruce (*Picea abies*), a tree of major economic importance, and subject to decreasing decline in northern and central Europe.

2. ESTIMATING METAL TOXICITY

2.1. Root Elongation

One of the major questions that arises in discussions of the toxicity of Al and heavy metals is what concentrations are toxic to plants. One of the most common parameters used to estimate toxicity was inhibition of root growth by metals. Root growth was studied in both long- and short-term experiments, and the effects of metals were documented. Descriptions of the effects of metals on roots range from observations of changes in appearance (blackening or stunting of roots) to inhibition of root elongation rate over hours. Root growth was shown to be inhibited in a large number of tree species by Al and a range of heavy metals. In *Pinus strobus, Picea glauca,* and *P. abies* root growth was inhibited by Cd and Zn (Mitchell and Fretz, 1977), in *P. strobus* and *P. glauca* by Ni and Cu (Lozano and Morrison, 1982), and in *Picea sitchensis* by Cd, Cu, Ni, and Pb (Burton et al., 1984, 1986). Root growth was reduced in *Fagus sylvatica* grown in Pb contaminated soils (Kahle and Breckle, 1986). Schaedle et al. (1989) attempted to group trees species according to their Al sensitivity by using the effects of Al on root growth and nutrient uptake. As the investigators themselves stated, such an undertaking is fraught with difficulties as the response to both Al and heavy metals will be modified by many physical, chemical, and biological factors, as well as the techniques used and the parameters measured (Godbold and Kettner, 1991a). However, if species are grown under similar conditions, meaningful comparisons can be obtained (see Cronan, Chapter 2). Inhibition of root elongation is one of the most sensitive parameters for measuring the effect of metals on plants (Neitzke and Runge, 1987; Klotz and Horst,

1988a; Godbold, 1991b). A summary of some of the work carried out on the effect of various metals on root elongation of *P. abies* seedlings is shown in Table 7.1. Large differences were found among the metals in their effect on root elongation. For these metals the order of toxicity is $CH_3Hg > Hg > Tl^{3+} = Pb = Tl^{+} > Cu > Cd > V > Al$. For all of the metals inhibition of root elongation increased at higher external concentrations (Knoche, 1988; Godbold, 1991b). As recently discussed by Tyler et al. (1989), metals form stable complexes with a number of ligands. The average stability of metals with various ligands is in the order $Pb > Cu > Zn > Cd > Mn$. In general, the stability decreases with increasing electronegativity, but may be modified by a number of factors including the pH. Thus a similarity between electron negativity and toxicity in terms of inhibition of root elongation can be shown. However, this relationship is not always true and may be modified by a number of chemical and biological factors. In the case of CH_3Hg and Hg we can see that the toxicity of a metal can be modified. Methylation of Hg increases its toxicity by over 100-fold. Similarly, differences in toxicity can be found between different ionic species of the same metals. Knoche (1988) showed that thallium as Tl^{3+} inhibited root elongation more rapidly than Tl^{+}. As will be discussed more extensively, this was also shown to be important for the toxicity of Al. The speciation of a metal together with the modifying effect of other ions must be taken into account when considering the toxicity of metals in the rhizosphere.

2.2. Root Growth

Root elongation studies give a good indication of the initial effects of exposure to metals, but give little indication of changes in the morphology of root systems. Malone et al. (1978) showed that the growth of lateral roots of *Zea mays* and *Glycine max* was more sensitive than primary root growth to Cd and Pb. In sitka spruce (*P. sitchensis*) exposed to Cd, Cu, or Pb, in addition to a reduced root

TABLE 7.1. Percentage Inhibition of Root Elongation of *Picea abies* Seedlings After a 7-day Exposure to Metals in Nutrient Solutions

Element	Concentration (μM)	Percentage Inhibition
Be	5	0
Cd	1	23
Cu	10	100
V	5	12
Tl^{+}	0.1	19
Tl^{3+}	0.1	27
Pb	0.1	26
Hg	0.01	31
CH_3Hg	0.001	100
Al	800	100

TABLE 7.2. Primary Root Length and the Number of Side Roots of *P. abies* Seedlings Exposed to 800 μ*M* Al for 39 days[a]

	Primary Root (cm)		Number of Side Roots	
Days	Control	Al[b]	Control	Al
9	15.5(±0.3)	12.6(±1.1)	69(±4)	62(±7)
16	19.0(±0.3)	14.1(±0.7)*	68(±4)	60(±4)
26	20.8(±0.9)	14.5(±0.9)*	145(±15)	150(±21)
39	26.8(±1.2)	13.6(±1.3)*	210(±27)	179(±18)

[a] Standard error (±SE) is denoted by values in parentheses.
[b] An asterisk (*) denotes significant difference ($p < 0.05$).

growth, changes in root branching were also shown (Burton et al.,1984). In *P. abies* seedlings exposed to Al, changes in root morphology were concentration dependent. Exposure to 800 μ*M* Al strongly inhibited the growth of the primary root of the *P. abies* seedlings (Table 7.2). At this level of Al the length of the primary root only slightly increased over the duration of the treatment period. In contrast, the number of side roots was less affected by exposure to Al. Only after a 39-day exposure to Al were we able to observe a decrease in the number of side roots compared to the control.

Exposure to lower concentrations of Al (400 μ*M*) for 8 weeks produced a different response of the root system (Tables 7.3 and 7.4). At this level of Al, the length growth of the primary and secondary roots was not affected, and only the length of the tertiary roots was reduced (Table 7.3). The number of secondary roots formed was unaffected by Al, whereas the number of tertiary roots formed increased strongly after exposure to Al. Estimation of the distribution of secondary and tertiary root into length classes (Table 7.4) shows that exposure to 400 μ*M* Al results in the formation of an increased number of short tertiary roots, as demonstrated by

TABLE 7.3. Total, Primary, Secondary, and Tertiary Root Length and Number of Root Tips per Plant of *P. abies* Seedlings Exposed for 8 weeks to 400 μ*M* $AlCl_3$[a]

Treatment	Primary	Secondary	Tertiary
	Root Length		
Control	114(±4)	559(±60)	136(±27)
400 μ*M* of Al	109(±4)	567(±58)	83(±23)
	Root Tip Number		
Control	1	53(±4)	68(±11)
400 μ*M* of Al	1	54(±6)	113(±18)

[a] Standard error (±SE) is denoted by values in parentheses and $n = 32$.

TABLE 7.4. Distribution of Secondary and Tertiary Roots in Length Classes of *P. abies* Seedlings Exposed for 8 weeks to 400 μ*M* of $AlCl_3$

	Length Class (mm)						
Treatment	0–0.9	1–1.9	2–4.9	5–9.9	10–19.9	20–29.9	>30
				Secondary			
Control	12.9	15.3	27.3	23.3	7.8	4.4	9.0
400 μ*M* Al	33.5	17.4	15.5	13.0	5.5	2.5	12.7
				Tertiary			
Control	33.8	30.7	30.3	3.5	1.1		
400 μ*M* Al	87.1	8.7	3.0	0.9			

the increase in the number of tertiary roots and the dramatic increase in the percentage of roots in the 0–1-mm class. Comparison of the changes in root morphology induced by the two levels of Al used suggest that the response of the root system to Al is concentration dependent. Higher concentrations of Al inhibits root growth of the whole root system, whereas lower concentrations of Al affect primarily lower order root classes.

In forest stands near Witzenhausen in the state Hesse, Eichhorn (1987) observed interference points on the roots of mature spruce trees that were on acidified soils. The interference points were points where root growth at the apex had ceased and length growth continued through secondary roots. Interference points were more common in subvital than in vital roots, in damaged than in healthy trees, and also in the lower mineral soil compared to the upper mineral soil. Similar interference points can be observed in *P. abies* seedlings in response to pH and Al treatment (Table 7.5). At both pH 5 and 4 interference points on the roots were found at 400 μ*M* Al, but not at the lower Al concentration. At pH 3 no increase in the number of interference points in response to A1 was found, however, the low pH value alone induces a number of interference points. Exposure to heavy metals also results in changes in root morphology. With the exception of tertiary roots exposed to Pb, a 4-week exposure to 0.5 μ*M* Pb, 1 n*M* CH_3Hg, or 10 n*M* Hg significantly reduced

TABLE 7.5. Number of Interference Points on Roots of *P. abies* Seedlings After a 4-week Exposure to Aluminum at Various pH Levels

Al Concentration	Number of Interference Points (%)		
(μ*M*)	pH3	pH4	pH5
0	5	0	0
100	0	0	0
400	5	25	35

TABLE 7.6. Total, Primary, Secondary, and Tertiary Root Length and Number of Root Tips per Plant of *P. abies* Seedlings Exposed for 4 Weeks to 1 n*M* CH_3Hg, 10n*M* Hg, or 0.5 μ*M* $PbCl_2$[a]

Treatment	Total	Primary	Secondary	Tertiary
		Root Length		
Control	58.6(±2.8)[a]	15.5(±0.6)[a]	36.9(±2.0)[a]	6.2(±0.8)[a]
1 n*M* CH_3Hg	36.0(±2.0)[b]	9.0(±0.3)[c]	23.3(±1.5)[c]	3.7(±0.4)[bc]
10 n*M* Hg	41.5(±3.1)[b]	12.2(±0.4)[b]	26.3(±2.9)[bc]	3.1(±0.5)[c]
0.5 μ*M* Pb	45.0(±2.4)[b]	11.4(±0.4)[b]	28.6(±1.8)[b]	5.1(±0.7)[ab]
		Root Tip Number		
Control	127(±2)[a]	1	55(±3)[a]	73(±6)[a]
1 n*M* CH_3Hg	81(±5)[c]	1	43(±2)[b]	38(±4)[bc]
10 n*M* Hg	81(±5)[c]	1	45(±2)[b]	35(±4)[c]
0.5 μ*M* Pb	101(±7)[b]	1	47(±2)[b]	54(±7)[b]

[a]Standard error (±SE) is denoted by values in parentheses and $n = 32$. Treatments with no common indexes in columns are significantly different ($p = 0.05$).

the primary, secondary, and tertiary root length of *P. abies* seedlings (Table 7.6). The number of root tips formed for both secondary and tertiary roots per plant was significantly reduced by exposure to these metals. A comparison of the relative decrease in the number of secondary and tertiary roots formed indicates that the formation of tertiary roots is more inhibited than the formation of secondary roots by these metals. The distribution of the root lengths of both secondary and tertiary roots in different root length classes was not affected by the Pb, CH_3Hg, or Hg

TABLE 7.7. Distribution of Secondary and Tertiary Roots in Length Classes of *P. abies* Seedlings Exposed for 4 weeks to 1 n*M* CH_3Hg, 10 n*M* Hg, or 0.5 μ*M* $PbCl_2$

	Length Class (mm)						
Treatment	0–0.9	1–1.9	2–4.9	5–9.9	10–19.9	20–29.9	>30
				Secondary			
Control	34.1	23.7	21.1	9.6	4.3	0.9	6.3
1 n*M* CH_3Hg	34.4	20.6	19.9	9.7	7.6	3.6	4.1
10 n*M* Hg	38.1	21.5	20.1	10.0	4.5	1.5	4.3
0.5 μ*M* Pb	30.1	23.0	23.4	10.9	5.1	2.2	5.4
				Tertiary			
Control	72.6	21.3	5.9	0.2			
1 n*M* CH_3Hg	66.0	26.4	7.0	0.5	0.1		
10 n*M* Hg	74.4	20.0	5.4	0.2	0.1		
0.5 μ*M* Pb	66.9	22.9	7.1	0.1	0.1		

treatment (Table 7.7). This result indicates that the reduction in total lateral root length is mainly due to a decrease in the number of lateral roots formed. Furthermore, this also suggests that the formation of lateral roots is more sensitive than the elongation of already formed roots. This response of the root system to heavy metals contrast strongly to the response to Al. This finding may be due to the relatively high tolerance of *P. abies* to Al (Godbold, 1991b), as well as the more extracellular mechanism of toxicity of Al compared to heavy metals (see below). However, this may also be a reflection of the level of stress to which the plants were exposed. Low levels of stress inhibit length elongation of side roots promoting an increase in root initiation. With increasing stress the primary root becomes affected, until eventually growth of all roots is inhibited. Thus the response of the plants will be a function of the tolerance of the plant to the metal and the degree of metal stress imposed.

3. WHOLE PLANT MODELS OF METAL TOXICITY

Root growth often provides a good basis for estimating the toxicity of metals, however, for some metals, for example, Mn (Langheinrich et al., 1992), root growth is not the first affected or most sensitive parameter. In a number of studies, nutrient uptake was more sensitive to Al than root elongation (Göransson and Eldhuset, 1987; Raynal et al., 1990). Thus integration of a number of physiological factors to estimate toxicity in terms of the whole plant may provide a better and more reliable approach. In Sections 3.1–3.3 whole plant models of Al, Pb, and Hg are presented. These models attempt to explain the toxicity mechanism of the metals to *P. abies*.

3.1. Aluminum

3.1.1. Aluminum Uptake

After exposure to a range of aluminum (Al) concentrations, the Al content of the roots increased with increasing supply of Al (Fig. 7.1). However, an increase in Al supply from 100 to 800 μM resulted in only a 40% increase in the Al content of the roots. Thus roots become rapidly saturated with Al. The uptake of Al by roots is strongly dependent on the pH of the external solution (Fig. 7.2), the Al content of the roots is higher at the higher pH values. Root cortex cell walls are known to be a major site for the accumulation of Al (Godbold et al., 1988b; Stienen and Bauch, 1988; Schröder et al., 1988; Schlegel et al., 1992). In *Picea ruben* Al was shown to have a high affinity for root cortex cell walls, which was pH dependent (Cronan, 1991). The rapid saturation and the pH dependence of the Al content of roots is due to the binding of Al to cell walls. At low pH values carboxylate groups in the cell wall are protonated, which prevents binding of Al. Grauer and Horst (1992) modeled Al binding in cell walls and predicted maximum uptake at about pH 4.4. In the needles, aluminum accumulation is less dependent on the pH of the external medium. The Al levels in the roots exceed those of the needles over 10-fold.

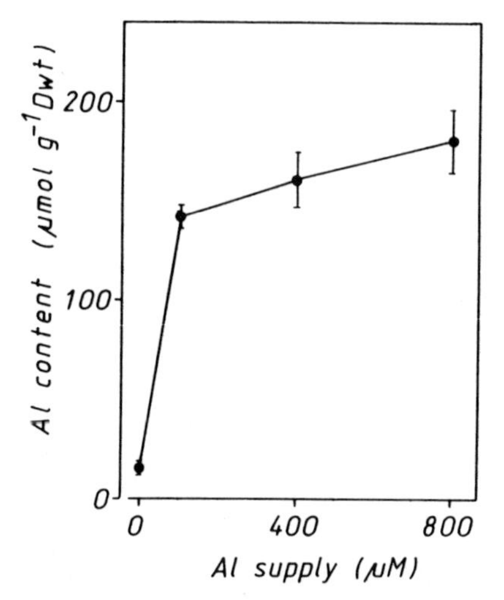

Fig. 7.1. Aluminum content of roots of *P. abies* seedlings exposed to a range of Al supply for 2 weeks (I = SE).

3.1.2. *Mineral Nutrition*

Magnesium deficiency is one of the most important symptoms of forest decline in European forests. A decrease in the contents of Mg and Ca in the roots and needles in response to Al exposure was shown in a number of tree species, including red spruce (*P. ruben*) (Thornton et al., 1987), sugar maple (*Acer saccharum*) (Thornton et al., 1986), and European beech (*Fagus sylvatica*) (Bengtsson et al., 1988). Figure 7.3 shows the decrease in the Mg and Ca content of roots and needles of *P. abies*

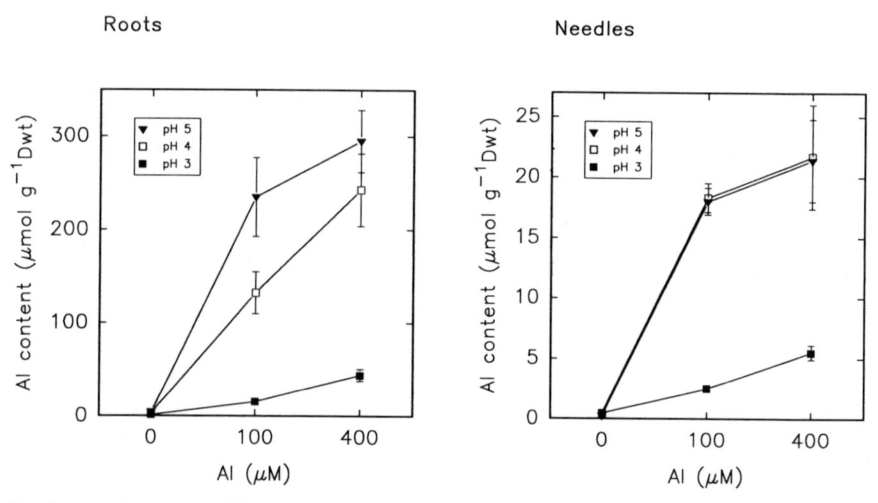

Fig. 7.2. Aluminum content of roots and needles of *P. abies* seedlings exposed to 100 and 400 µ*M* Al at a range of pH values (I = SE).

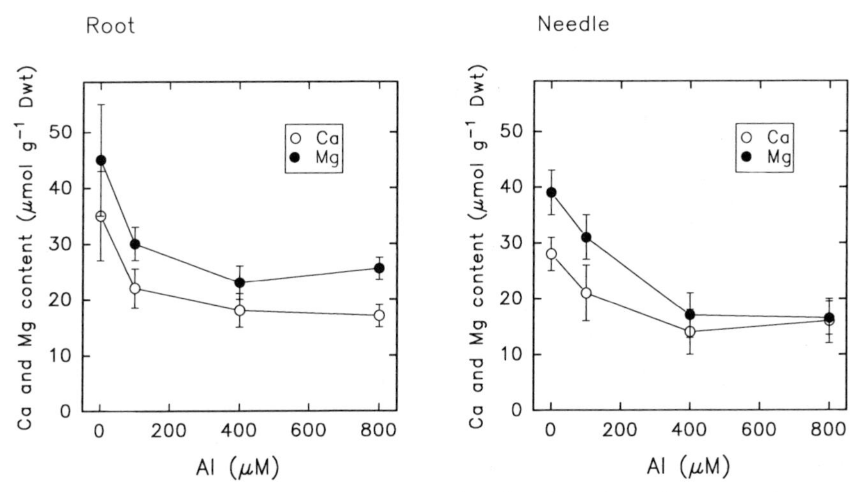

Fig. 7.3. Calcium and magnesium content of roots and needles of *P. abies* seedlings exposed to a range of Al supply for 2 weeks (I = SE).

seedlings with increasing Al supply. The Ca and Mg content of the roots decreased strongly at 100 μM Al supply. An increase in Al supply above 100 μM did not drastically decrease the Ca and Mg content of the roots. For both the roots and the needles the pattern of response to Al with increasing concentration is similar. The decrease in the content of Ca and Mg is better related to the accumulation of Al in the roots than the external concentration of Al. In a number of studies the decrease in Ca and Mg in the roots was shown to be due to a displacement of Al in the cortex apoplast (Godbold et al., 1988b; Schröder et al., 1988; Rengel and Robinson, 1989a; Jentschke et al., 1991c). The displacement of Ca and Mg in the root apoplast was postulated as one to the major lesions for Al toxicity. However, in investigations of the interaction between pH and Al, Godbold (1991a) also showed that Al reduces the content of Mg in needles of *P. abies* at pH values where no significant accumulation of Al in the root is found. This finding suggests that Al may inhibit Mg uptake by more than one mechanism.

However, not all investigators found inhibition of Ca and Mg uptake to be the major lesion in Al toxicity. In *Picea ruben* and *Pinus rigida,* Al reduced P translocation to the foliage (Cumming et al., 1986; Cumming and Weinstein 1990a). This was suggested to be due to Al binding P in the apoplast of the roots. A similar reaction was suggested to occur in *Fagus sylvatica* roots (Bengtsson et al., 1988; Jensen et al., 1989).

3.1.3. Gas Exchange

Exposure to Al decreases carbon dioxide (CO_2) uptake in both Norway spruce (*P. abies*) and red spruce (*P. ruben*) [Schlegel, 1989; Godbold, 1991b; Schlegel (unpublished)]. The decrease in CO_2 uptake was shown to be almost entirely due to a decreased level of chlorophyll in the needles and not due to the direct effects of Al

(Schlegel and Godbold, 1991). Transpiration was shown to be both increased or decreased by exposure to Al. A common feature seems to be that at lower levels of Al, which do not cause severe damage, transpiration is increased (Schlegel, 1989; Godbold, 1991b; Cumming and Weinstein, 1992a). In *P. abies* the increase in transpiration after exposure to Al does not appear to be due to the direct effects of Al (Schlegel and Godbold, 1991) or Mg deficiency (Godbold, 1991b). Schwartzenberg (1989) found higher cytokinin contents in needles of *P. abies* exposed to Al. A higher supply of cytokinins may increase transpiration by increasing stomatal openings (Radin et al., 1982). The increase in cytokinin production could be due to the increased number of root tips after exposure to Al as shown above.

3.1.4. A Model of Aluminum Toxicity in Norway Spruce

Aluminum inhibits root growth and the uptake of divalent cations. These two reactions are the primary events that cause all other changes in physiology in response to Al; lower levels of chlorophyll and changes in gas exchange (Fig. 7.4). Inhibition of root growth is due to inhibition of cell division and/or cell elongation. An effect of Al on cell division was shown by several authors (Clarkson, 1965; Morimura et al., 1978; Horst et al., 1982). Inhibition of cell division may be due to disruption of DNA synthesis by Al (Matsumoto and Morimura, 1980; Wallace and Anderson, 1984). The displacement of Ca in the apoplast by Al provides a possible mechanism of Al inhibition of cell elongation. Both Klimashevska and Dedov (1975) and Matsomoto et al. (1977a) described a decrease in cell wall elasticity due to Al binding to carboxylate groups in the cell wall. In a number of *Vigna unguiculata* genotypes, inhibition of root elongation could be explained by induction of Ca deficiency in the root tips (Horst, 1987). Bennet et al. (1991) postulated that Al inhibits root growth through a primary Al sensor, caused by the disruption of Ca homeostasis in the root tips. In the *P. abies* seedlings exposure to Al strongly decreases the Mg and Ca contents of both roots and needles. The inhibition of Mg and Ca uptake can be explained by both a lower loading of divalent cations in the root apoplast and a direct inhibition of cation uptake at the plasmalemma. In a number of studies, Al was shown to displace Mg in the root apoplast (Godbold et al., 1988b; Schröder et al., 1988; Rengel and Robinson, 1989a). Decreased loading of the apoplast with Mg may result in lower concentrations at the plasmalemma, and hence decreased Mg uptake. In studies with ^{45}Ca, evidence was found that the uptake of divalent cations into the symplast may be strongly influenced by the degree of loading of the apoplast (Godbold, 1991b). The accumulation of Al in the roots is strongly pH dependent and may be due to a decrease in the number of unassociated carboxylate groups in the cell wall at the lower pH values. At the lower pH values, despite the lower Al contents in the root, a significant decrease in the Mg content of the needles due to Al was found (Godbold, 1991a). If at a low pH we assume that the remaining Al is bound intracellularly or to the surface of the plasma membrane, this would suggest that Al not only inhibits the Mg uptake by displacement in the root apoplast but may also inhibit Mg uptake directly at the plasmalemma. Matsumoto and Yamaya (1986) and Suhayda and Haug (1986)

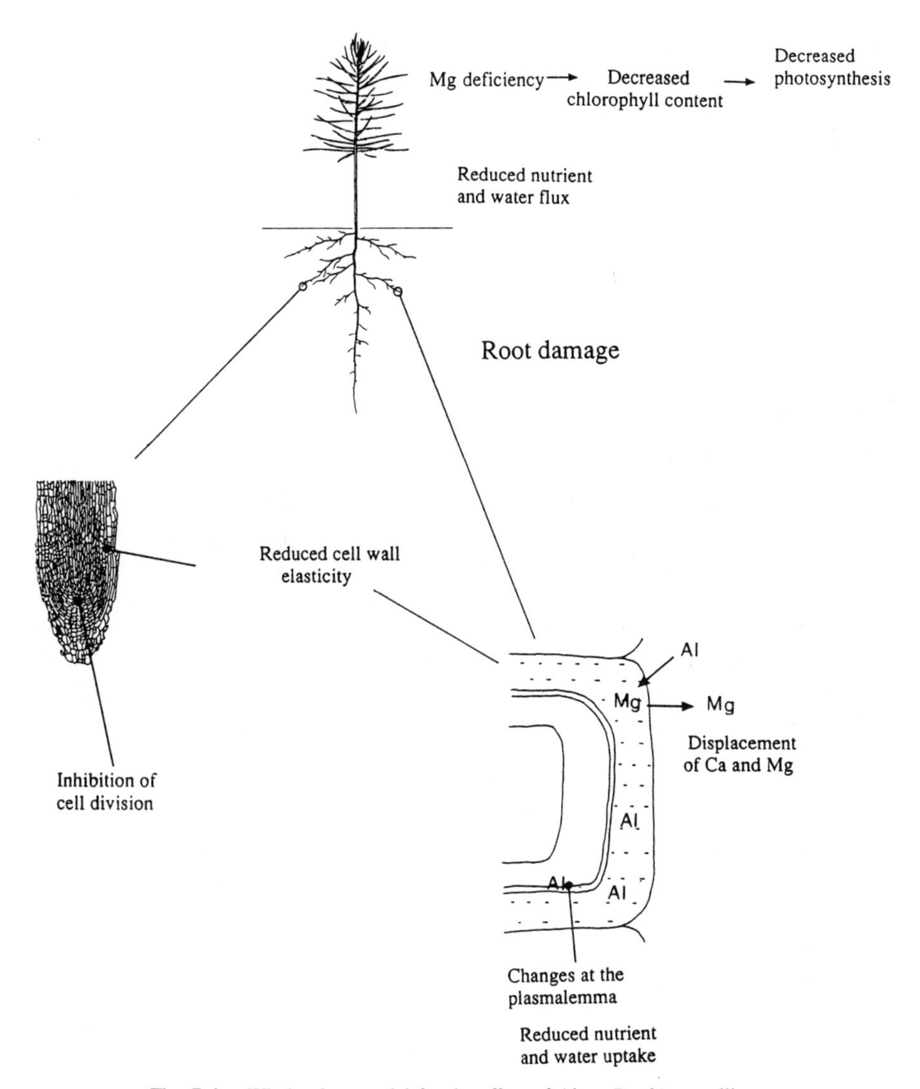

Fig. 7.4. Whole plant model for the effect of Al on *P. abies* seedlings.

showed that Al inhibits membrane bound ATPases, and also changes membrane fluidity and surface charge (Suhayda and Haug, 1986). These results suggest that Al inhibits Mg uptake by both displacing Mg in the root apoplast, thus lowering the Mg concentration at the plasmalemma, and by directly inhibiting Mg uptake at the membrane. If loading of the apoplast is important for uptake of Mg (Godbold, 1991b; Rengel, 1990; Marschner, 1991), then similarly, accumulation of Al in the apoplast will also lead to higher Al concentrations at the plasmalemma and impair membrane function. Thus a lower cation exchange capacity of the cell wall may be an advantage to restrict Al uptake, as previously stated for *Hordeum vulgare* cultivars (Foy et al., 1967) and genotypes of *Asplenium trichomanes* and *Polypodium vulgare* (Büscher and Koedam, 1983). However, it is less likely that the absolute cation exchange capacity is important, but rather the relative infinities of the cell wall for Al, Ca, and Mg (Grauer and Horst, 1992).

The changes in the roots impair the flux of nutrients to the needles, resulting in a induction of nutrient deficiency, particularly of Mg. Nutrient deficiency reduces chlorophyll synthesis that subsequently decreases photosynthesis. Magnesium deficiency not only reduces assimilation of CO_2, but also lowers the translocation of assimilates to the roots (Schlegel, 1989) thus, in the long term, further inhibiting root growth.

3.2. Lead

3.2.1. Lead Uptake

The levels of lead (Pb) in roots and needles after a 9-week exposure to various Pb concentrations are shown in Figure 7.5. The level of Pb in roots increased with an increasing Pb supply. The Pb levels in the roots were higher than those in the needles by a factor of 100–300. Lead levels in old needles exceed those of the young needles and, with the exception of between 0.5 and 2 μM Pb, increased with an increasing Pb supply. These results agree with other studies on Pb uptake (Hardiman et al., 1984) and studies on Norway spruce with other metals (Schlegel, 1985). If the biomass of the roots and needles is considered, approximately 95% of the Pb

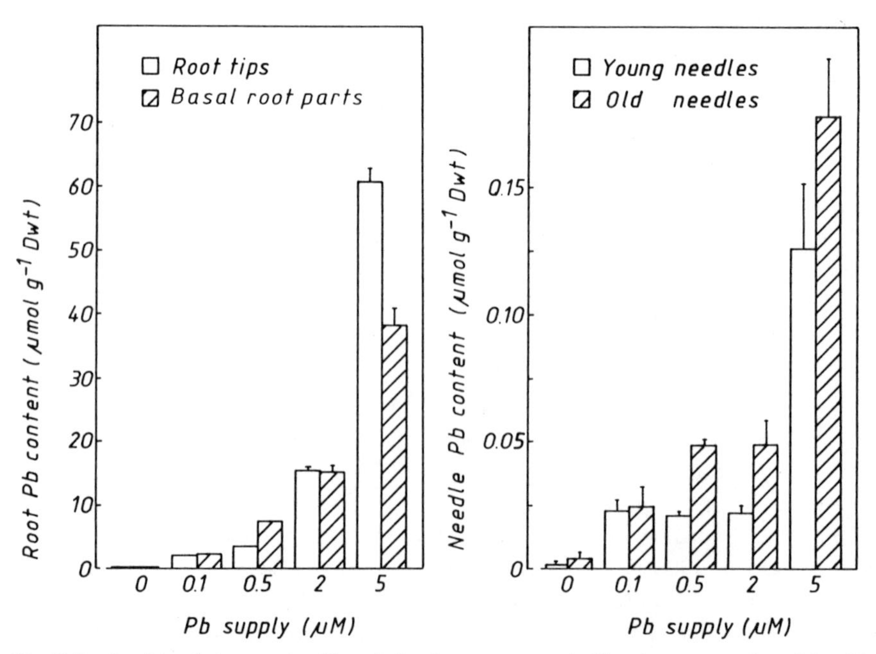

Fig. 7.5. Lead levels in root tips (5 mm), basal root parts, and old and young needles of *P. abies* seedlings after a 9-week exposure to different concentrations of Pb in nutrient solution (I = SE) (n = 12).

TABLE 7.8. Lead Contents in Roots of *P. abies* Seedlings Exposed to 0.5 μ*M* $PbCl_2$ in Nutrient Solutions at a Range of pH Values[a]

pH	μmol g^{-1} Dwt
2.5	0.23(±0.01)
3.2	1.51(±0.01)
4.0	6.36(±0.11)

[a]Standard error (±SE) is denoted by values in parentheses.
[b]From Godbold and Kettner, 1991b.

taken up accumulates in the roots. Several investigators suggested that the endodermis functions as a barrier to the radial transport of Pb in the root (Tanton and Crowdy, 1971; Hardiman et al., 1984; Jentschke et al., 1991b). In roots of *Anthoxanthum oderatum* (Quereshi et al., 1986) and *P. abies* (Jentschke et al., 1991b), the highest levels of Pb were determined in cell walls of the cortex or cortex and endodermis, respectively. The Pb levels in the roots are strongly dependent on the pH of the nutrient solution (Table 7.8). At lower pH, the Pb levels in the roots is decreased. As for Al, a greater uptake at higher pH can be in part explained by a higher number of nonprotonated carboxylate groups in the cell walls at a higher pH.

3.2.2. *Mineral Nutrition*

Exposure to Pb resulted in no clear effect on the mineral levels in the roots of *P. abies* seedlings (Godbold and Kettner, 1991b). However, in the root tips (Fig. 7.6)

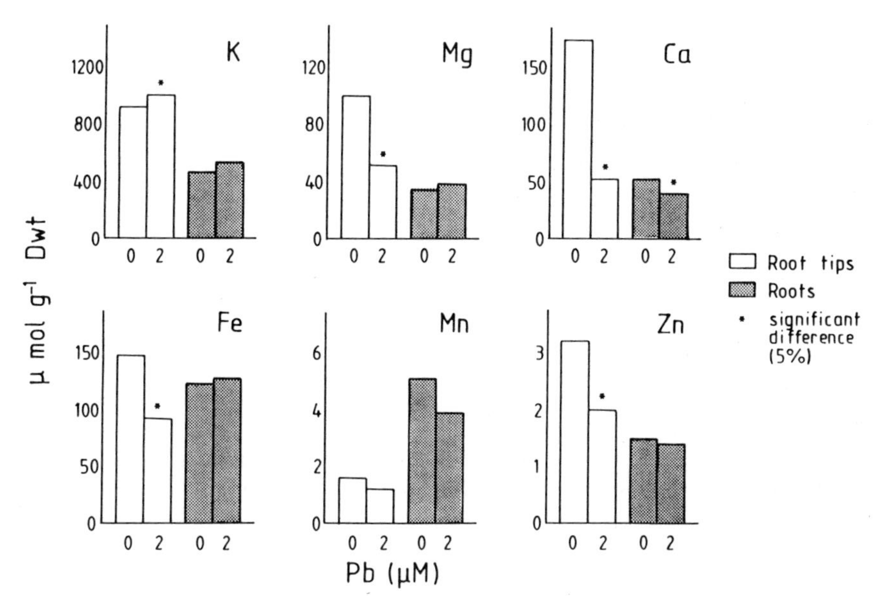

Fig. 7.6. Element levels in root tips (5 mm) of *P. abies* seedlings after a 9-week exposure to different concentrations of Pb in nutrient solution (I = SE) (n = 3).

TABLE 7.9. Calcium and Manganese Levels in Needles of *P. abies* Seedlings Exposed to Lead[a]

μmol g^{-1} Dwt	PbCl$_2$			
(μ*M*)	Ca	Mn	Ca	Mn
	Young Needles		Older Needles	
0.0	63(±2)^{a}	18.9(±1.0)^{a}	114(±5)^{a}	30.5(±2.2)^{a}
0.1	60(±3)^{a}	11.2(±0.6)^{b}	100(±3)^{b}	23.9(±0.9)^{b}
0.5	51(±2)^{b}	8.0(±0.3)^{c}	85(±3)^{c}	16.9(±0.8)^{c}
2.0	36(±2)^{c}	6.4(±0.4)^{d}	72(±6)^{c}	8.8(±0.6)^{d}
5.0	41(±6)^{bc}	9.5(±1.6)^{bcd}	78(±6)^{bc}	20.9(±3.2)^{bc}

[a] Standard error (±SE) is denoted by values in parentheses. Treatments with no common indexes are significantly different ($p = 0.05$).

the levels of Ca, Mg, Zn, and Fe were lower after exposure to Pb concentrations. In young and old needles of *P. abies* seedlings the Ca and Mn levels decreased after exposure to a range of Pb concentrations (Table 7.9). Higher levels of K and Zn were found in young and old needles at concentrations of 5 μ*M* Pb. Sieghardt (1988) found lower Mn levels in needles of *P. abies* growing on a Pb mine spoil heap compared to trees growing on noncontaminated soil nearby. A decrease in the level of Mn was also found in *P. sitchensis* exposed to Cd and Cu (Burton et al., 1986) and *P. abies* exposed to Hg (Godbold and Hüttermann, 1988).

3.2.3. Gas Exchange

Gas exchange effects are rarely found in intact plants exposed to realistic concentrations of Pb. In *P. abies* seedlings a decrease in CO_2 uptake was found only after exposure to 5 μ*M* Pb (Godbold, 1991b). In *Medicago sativa* and *Trifolium pratense* exposed to 1 m*M* Pb inhibition of photosystems I and II was shown (Becerril et al., 1988). Similarly, the inhibitory effects of Pb were shown in isolated chloroplast or enzymes (Miles et al., 1972; Hampp et al., 1973; Wong and Govindjee, 1976). However, as shown above, Pb concentrations in the needles of *P. abies* do not reach high levels, and as argued for Hg (Godbold and Hüttermann, 1988) direct effects in the needles cannot be expected.

3.2.4. A Model of Lead Toxicity in Norway Spruce

One of the major effects of Pb is the inhibition of root growth (Fig. 7.7). Lead inhibits both elongation of primary root length and formation of side roots. Lead may inhibit primary root growth by inhibiting cell division and/or cell elongation. In coleoptile segments of *Triticum vulgare,* Pb reduced the plastic and elastic extensibility of cell walls (Burzynski and Jakob, 1983). In root tips the levels of Ca decreased after exposure to Pb. In root tips of Norway spruce approximately 40% of the Ca taken up is used for growth of the root tips (Häussling et al., 1988). The inhibition of root growth after exposure to Pb may be due to a decrease in Ca in the

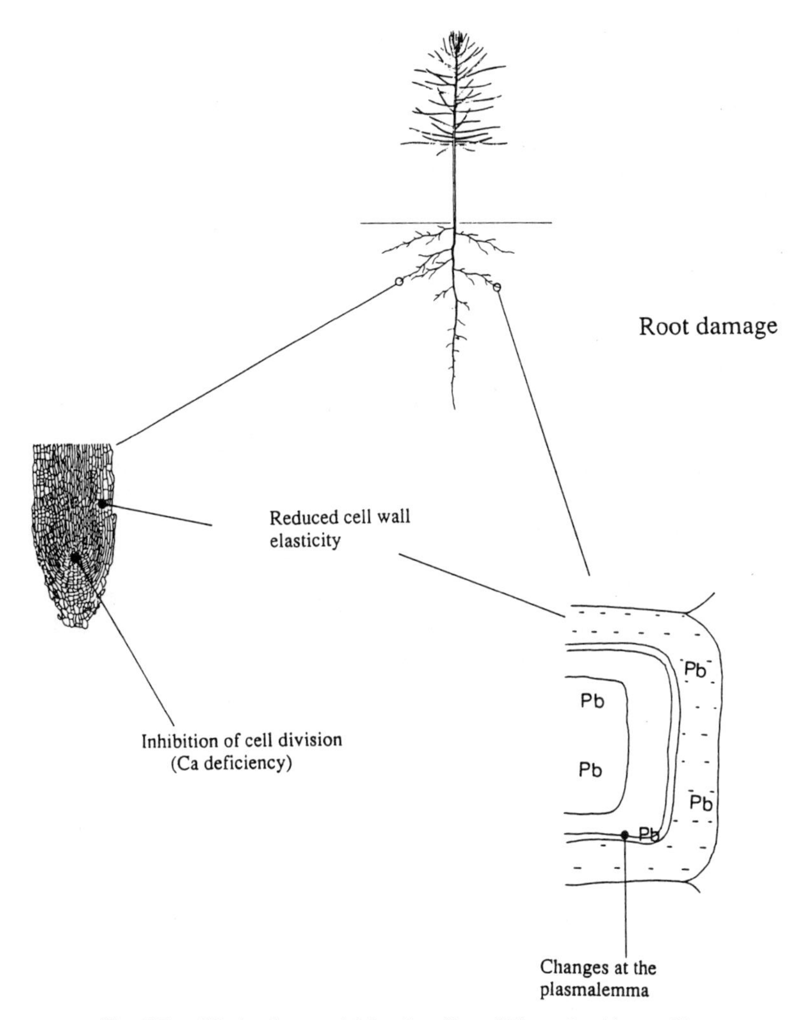

Fig. 7.7. Whole plant model for the effect of Pb on *P. abies* seedlings.

root tips leading to a decrease in cell division or cell elongation. Mechanisms in which the homeostasis of Ca in the root tips acts as a stress signal may also be involved (Cumming and Tomsett, 1992).

In the roots of *Populus* exposed to Pb, accumulation of Pb was found at the endodermis (Ksiazek and Wozny, 1990). High levels of Pb were shown at the endodermis of spruce roots grown in Pb containing nutrient solutions (Jentschke et al., 1991b) and collected from forest sites (Godbold et al., 1987). As lateral roots are initiated in the pericycle, high levels of Pb in the adjacent endodermis layer may influence their formation. Although the effects of Pb on other factors, such as phytohormone levels, cannot be excluded.

In *P. abies* the root tips are important sites for Ca uptake (Häussling et al., 1988).

These investigators also showed that in basal root zones, at sites where lateral roots penetrate the endodermis, the apoplastic pathway is important for the radial transport of Ca. This finding suggests that the lower levels of Ca and Mn in the needles could be due to a decrease in the number of root tips and sites for apoplastic solute flux through the endodermis. However, direct effects of Pb on the plasma membrane cannot be excluded. In roots of *Zea mays* ATPases and membrane potential were effected by low levels of Pb (Kennedy and Gonsalves, 1987, 1989).

3.3. Mercury

3.3.1. Mercury Uptake

After a 7-week exposure of roots to mercury (Hg) at an external concentration below 1000 n*M* Hg, higher levels Hg were found in the roots exposed to CH_3HgCl than to $HgCl_2$ (Fig. 7.8). At 1000 n*M* Hg the reverse was found. In needles equal Hg levels were found at 1 and 10 n*M* Hg irrespective of the level and form of Hg supplied. Only at higher external concentrations do needle Hg levels reflect those of the roots. If the accumulation of Hg by roots and needles over time is followed (Table 7.10), roots exposed to both $HgCl_2$ or CH_3HgCl become rapidly saturated. In contrast, a slow accumulation of Hg could be found in needles. Independent of the Hg form supplied and the external concentration, the Hg level in roots exceeded that of the needles by a factor of over 100.

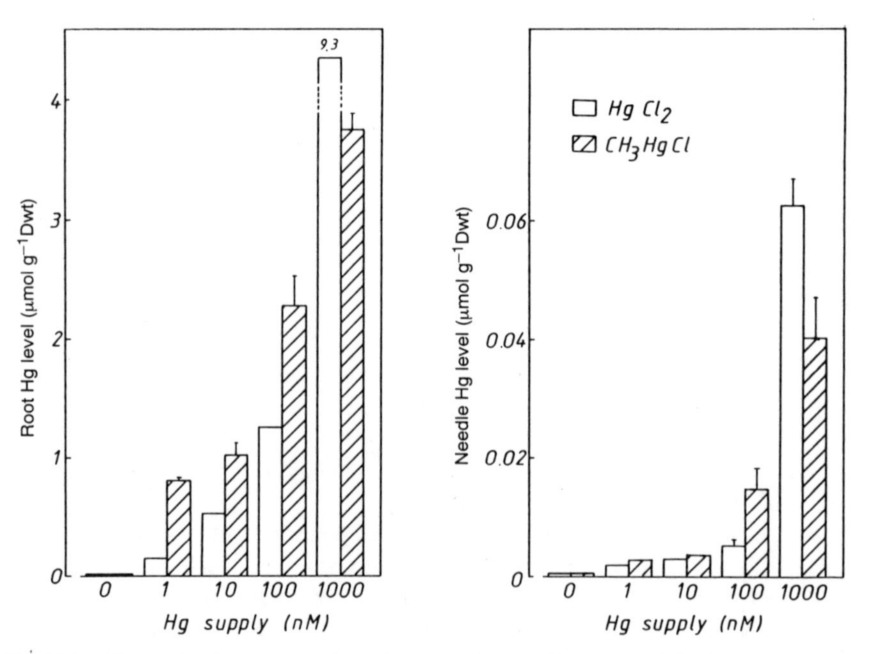

Fig. 7.8. Mercury levels in roots and needles of *P. abies* seedlings exposed for 7 weeks to nutrient solutions containing $HgCl_2$ or CH_3HgCl (±SE) (Godbold and Hüttermann, 1988).

TABLE 7.10. Mercury Levels in Roots and Needles of *P. abies* Seedlings After and up to a 39-day Treatment With 10 n*M* $HgCl_2$ or 1 n*M* CH_3HgCl in Nutrient Solutions[a,b]

	Hg Content (nmol g^{-1} Dwt)			
	Roots		Needles	
Days	1 n*M* CH_3Hg	10 n*M* Hg	1 n*M* CH_3Hg	10 n*M* Hg
9	760(±120)	580(±40)	1.0(±0.2)	0.12(±0.70)
16	820(±100)	640(±40)	2.8(±1.0)	0.33(±0.10)
26	1010(±90)	780(±60)	2.0(±0.6)	0.33(±0.15)
39	770(±80)	660(±40)	2.7(±0.6)	0.66(±0.10)

[a] Standard error (±SE) is denoted by values in parentheses.
[b] From Godbold, 1994.

3.3.2. *Mineral Nutrition*

After a 7-week exposure to Hg, the mineral content of root tips was found to be more strongly affected than that of the whole roots (Godbold, 1991c). The effects of exposure to $HgCl_2$ and CH_3HgCl on the mineral levels of 5-mm root tips are shown in Table 7.11. After exposure to 1000 n*M* $HgCl_2$ the levels of K, Mg, and Ca decreased. Exposure to 1000 n*M* $HgCl_2$ increased the levels of Fe in the root tips. In roots exposed to 1 n*M* CH_3HgCl, in 5-mm root tips the levels of K and Mg decreased, whereas those of Fe increased. At 10 n*M* CH_3HgCl an increase in Fe, Ca, and Zn was statistically significant. In needles (Table 7.11) exposure of the roots to 10 n*M* $HgCl_2$ significantly decreased the contents of Zn and Mn. In addition, at 1000 n*M* $HgCl_2$ the Mg and Ca level of the needles was also significantly lower. Exposure to 1 and 10 n*M* CH_3HgCl decreased the levels of all mineral elements determined.

The above data show that the effects of $HgCl_2$ and CH_3HgCl on the mineral nutrition of *P. abies* are similar. However, the levels of $HgCl_2$ that cause such changes in element levels exceed those of CH_3HgCl by about a factor of 100. The loss in roots of primarily cytosolic and vacuolar elements, such as K, Mg, and Mn, may be due to changes in membrane permeability, similar to those suggested by other authors (Shieh and Barber, 1973; De Filippis, 1979). In short-term experiments used to determine membrane integrity changes in response to Hg (Godbold, 1991c), it was shown that $HgCl_2$ directly affects the plasma membrane increasing the membrane permeability to ions. In comparison to $HgCl_2$, CH_3HgCl does not appear to act directly at the plasma membrane.

3.3.3. *Gas Exchange*

After a 7-week exposure to $HgCl_2$ and CH_3HgCl, the uptake of CO_2 was more strongly affected than transpiration (Godbold and Hüttermann, 1988). The degree of inhibition of CO_2 uptake for both forms of Hg was dependent on the external supply.

TABLE 7.11. Levels of Mineral Elements in 5-mm Root Tips and Needles of *P. abies* Seedlings After a 7-week Treatment With Various Concentrations of $HgCl_2$ or CH_3HgCl in Nutrient Solutions[a,b]

Treatment	K	Mg	Ca	Fe	Zn	Mn
			Root Tips			
Control	1016[a]	91[a]	56[a]	140[a]	2.4[a]	7.7[a]
10 n*M* $HgCl_2$	997[a]	110[b]	52[a]	133[a]	2.4[a]	5.0[a]
1000 n*M* $HgCl_2$	189[b]	34[c]	42[b]	228[a]	2.5[a]	2.4[a]
Control	1205[a]	128[a]	86[a]	96[a]	2.6[a]	4.1[a]
1 n*M* CH_3HgCl	334[b]	49[b]	71[a]	158[ab]	3.2[a]	1.2[a]
10 n*M* CH_3HgCl	198[c]	56[b]	148[b]	256[b]	4.9[b]	1.9[a]
			Needles			
Control	288[a]	54[a]	96[a]	1.5[a]	0.8[a]	15[a]
10 n*M* $HgCl_2$	277[a]	53[a]	71[a]	1.6[a]	0.4[b]	9[b]
1000 n*M* $HgCl_2$	292[a]	34[b]	54[b]	1.7[a]	0.4[b]	3[c]
Control	283[a]	53[a]	83[a]	1.0[a]	0.7[a]	15[a]
1 n*M* CH_3HgCl	240[b]	44[b]	59[b]	0.8[ab]	0.4[b]	6[ab]
10 n*M* CH_3HgCl	194[c]	26[c]	38[c]	0.7[b]	0.3[c]	4[b]

[a]Treatments with no common indexes are significantly different ($p > 0.05$).
[b]From Godbold (1994).

These changes were again induced at a CH_3HgCl concentration about 100 times lower than $HgCl_2$. At the lower levels of Hg, Godbold and Hüttermann (1988) estimated that inhibition of CO_2 uptake by Hg can be explained by a decrease in chlorophyll levels and stomatal closure. The changes in chlorophyll levels appeared not to be due to the direct effects of Hg in the needles, but rather to a decreased water and nutrient supply.

3.3.4. A Model of Mercury Toxicity in Norway Spruce

Both $HgCl_2$ and CH_3HgCl are extremely toxic compounds that inhibit root growth and primary processes at very low concentrations (Fig. 7.9). The higher toxicity and rate of accumulation of CH_3HgCl may be due to its higher lipid solubility (Carty and Malone, 1979), which allows CH_3HgCl to pass through biological membranes (De Filippis, 1978). Both $HgCl_2$ and CH_3HgCl appear to have a similar mechanism of action; both cause loss of membrane integrity. However, whereas $HgCl_2$ directly affects the plasma membrane, CH_3HgCl may primarily affect the metabolism in the cytoplasm, which subsequently affects membrane integrity. In the roots of *Zea mays,* Hg affected transmembrane potential and H-ion efflux, and at high concentrations, it affected ATPase activity (Kennedy and Gonsalves, 1987, 1989).

The root damage induced by Hg results in a decrease in water and nutrient uptake

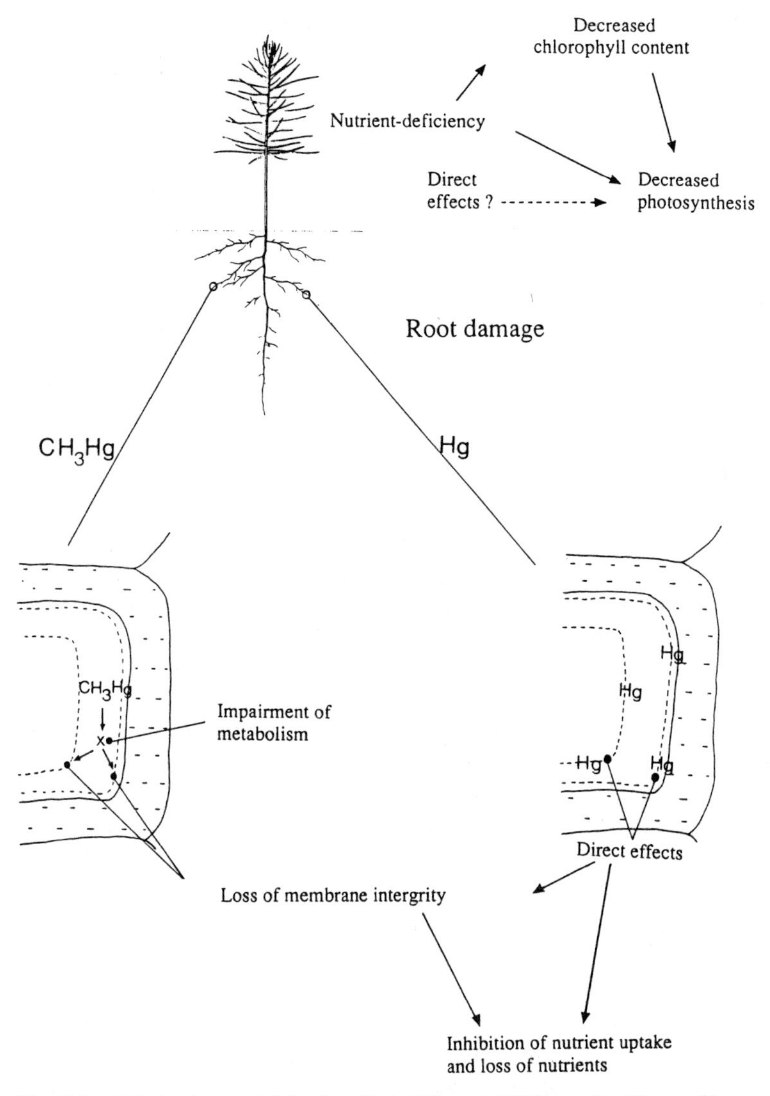

Fig. 7.9. Whole plant model for the effect of Hg and CH_3Hg on *P. abies* seedlings.

by the roots and a lower supply to the needles, which results in lower photosynthesis and transpiration. Davis et al. (1986) showed that active root tips are a prerequisite for keeping stomata open. At low Hg concentrations the decrease in photosynthesis is primarily due to lower chlorophyll levels due to the reduced supply of nutrients and water (Godbold and Hüttermann, 1988). Thus at a low Hg concentration root damage is the primary mechanism for all other physiological changes.

4. FACTORS THAT MAY MODIFY METAL TOXICITY

4.1. Metal Speciation

The influence of speciation on the toxicity of Al has received much attention and is extremely important when considering the toxicity of Al in the rhizosphere. The pH of the rhizoplane and rhizosphere can differ considerably from that of the bulk soil (Marschner, 1991), and increase in rhizosphere pH was suggested as a possible tolerance mechanism to Al. In roots of mature Norway spruce a pH increase of 0.5 pH units was found when the rhizoplane (pH 4.4) was compared to the bulk soil (pH 3.9). The speciation of Al changes dramatically with small changes in pH (Kinraide, 1991). Table 7.12 shows the Al species calculated to be present in a nutrient solution at a range of pH values. At lower pH values the dominant form of Al is Al^{3+}, however, as the pH increases, mononuclear aluminum hydroxy species begin to form and at pH 5 formation of polynuclear species (Al_{13}) is predicted. Which species of Al are toxic to plants is the center of much debate. While in most cases Al^{3+} is commonly considered to be the major toxic species of Al, Kinraide (1991) suggested that only for wheat has the toxicity of Al^{3+} been shown. Red clover, lettuce and turnip were not sensitive to Al^{3+}, but were instead sensitive to mononuclear aluminum hydroxy species (Kinraide and Parker, 1990). Similarly, in soybean, subterranean clover, alfalfa, and sunflower, inhibition of root growth was better correlated with the activity of the mononuclear species $Al(OH)^{2+}$ and $Al(OH)_2^+$ than with Al^{3+} (Alva et al., 1986b). A difference in sensitivity between grasses and dicotyledonous plants was suggested on the basis of these results (Kinraide, 1991). Several studies showed Al_{13} to be highly toxic to plants (Parker et al., 1989; Grauer and Horst, 1990). In soybean and wheat, Al_{13} was 10-fold more toxic than Al^{3+}. Aluminum sulphate complexes and Al fluoride complexes are not considered to be phytotoxic (Cameron et al., 1986). However, attempts to identify the rhizotoxic Al species is fraught with difficulties (Kinraide, 1991; Kinraide and Ryan, 1991). The toxicity of Al^{3+} may be ameliorated by protons (Grauer and Horst, 1990; Kinraide and Parker, 1990), thus giving an appearance of sensitivity to mononuclear aluminum hydroxy species. Also, due to differences in the pH value between the root apoplast and the root exterior, as well as the effects of the surface charge of the membrane, the Al speciation at the cell surface may differ to that found in the bulk solution (Kinraide and Ryan, 1991). The effects of Al on root

TABLE 7.12. Aluminum Species at Various pH Levels[a]

	Aluminum Species (μM)					
pH	Al	$AlSO_4$	AlOH	$Al(OH)_2$	$Al(OH)_3$	$Al_{13}(OH)_{32}$
3.2	378	17				
4.0	344	17	27	10		
5.0			9	37	7	25

[a] Aluminum species were calculated using the WAKO calculation programme (Prenzel, 1990).

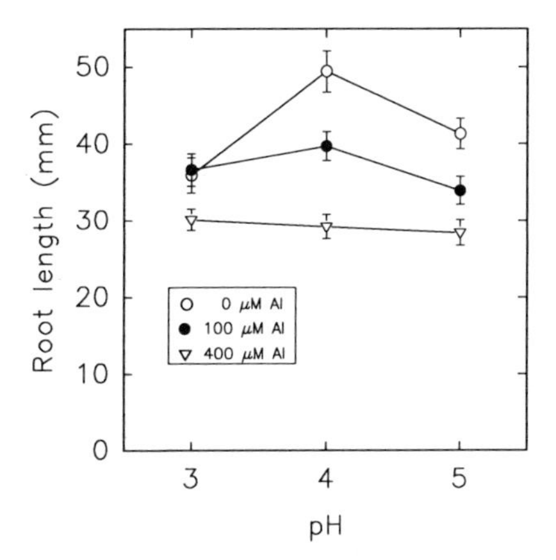

Fig. 7.10. Length of primary roots of *P. abies* seedlings exposed to 100 and 400 μ*M* Al at a range of pH values (*I* = SE).

elongation of *P. abies* seedlings at a range of pH values are shown in Fig. 7.10. At pH 4 and 5, 100 μ*M* Al reduced root growth, whereas at pH 3 root growth was not affected by Al. The effects on root growth can be compared to the speciation of Al as shown in Table 7.11. While interpretation of these data is, as discussed above, beset with difficulties, at first sight the data indicate that *P. abies* is sensitive to a range of Al species. Alternatively, the high toxicity of Al at pH 5 may reflect the high loading of Al in the root apoplast as discussed above. Thus the speciation of Al may be of secondary importance to the accumulation next to cell membranes and their subsequent impairment.

4.2. NH_4/NO_3 Ratios

The relative uptake of NH_4 or NO_3 by plant roots strongly affects the pH of the rhizosphere and apoplasm. For a number of tree species, and for both seedlings and mature trees, it was shown that when NH_4 uptake dominates, the pH of the rhizosphere decreases, and when NO_3 uptake dominates, the pH of the rhizosphere increases (Rollwagen and Zasoski, 1988; Gijsman, 1990a; Marschner et al., 1991; Olsthoorn et al., 1991). The rhizosphere and rhizoplane pH may differ from the pH of the bulk soil by up to 2 pH units (Marschner et al., 1991). Alkalization of the rhizosphere was prominent at the root tip in *P. abies* and *Pseudotsuga menziesii* (Gijsman, 1990b; Marschner et al., 1991), and was acidified in the older root parts of *Pseudotsuga menziesii* roots (Gijsman, 1990b). In *P. menziesii* the degree of acidification or alkalization was dependent on the water content of the soil (Gijsman, 1990b). In a dry soil with low NH_4 mobility the pH of the rhizosphere increased due to a dominance of NO_3 uptake. The increase in rhizosphere pH could alter the toxicity of Al by the formation of less toxic species or by precipitation of less soluble Al forms. An increase in rhizosphere pH was suggested as a possible mechanism of Al tolerance (Taylor and Foy, 1985a,b). However, investigations of the effect of a nitrogen source on the toxicity of Al are inconclusive. Several investigations showed no difference (Taylor, 1988c) in the toxicity of Al with NH_4 or NO_3 as the nitrogen source, or even a lower toxicity with NH_4 (Van Praag et al., 1985; Klotz and Horst, 1988b). *Picea abies* seedlings were slightly more sensitive to Al grown under a high NO_3 than low NO_3 concentration (Godbold et al., 1988a). In *Glycine max* the lower Al sensitivity was attributed to a low apoplastic pH, reducing the Al accumulation by the cell walls (Klotz and Horst, 1988a). This interpretation agrees fully with the model of Al toxicity described above. In mycorrhizal *Pinus rigida,* seedlings were found to have a lower uptake and toxicity of Al in the presence of NH_4 compared with NO_3 (Cumming and Weinstein, 1990c).

Almost no information is available concerning the effects of a nitrogen source on the toxicity of heavy metals. In mycorrhizal *P. abies* seedlings lower Pb contents were determined in cortex cell walls in seedlings grown in the presence of NH_4 compared with NO_3 (Marschner, personal communication), and thus some interactions can be expected.

4.3. Cations and Ionic Strength

There is a large body of evidence showing that the chemical milieu can strongly influence the toxicity of Al and heavy metals. This evidence was shown for both crop plants and trees. Some of the effects shown may be due to changes in ionic strength altering the activity of Al, however, there is also clear evidence showing specific ion effects. Alva et al. (1986a) found that an increase in the ionic strength of the nutrient solution at similar Al activities reduced the toxicity of Al to *Glycine max* and *Trifolium subterraneum*. This effect of ionic strength on Al toxicity was, however, less than that induced by increasing the Ca content of the nutrient solutions. In wheat (*Triticum aestivum*) cations reduced the toxicity of Al in the order Ca $> Mg > \geq Sr >> K = Na$ (Kinraide and Parker, 1987). In *Lotus pedunculatus* a decrease in Al toxicity was shown at high osmotic strength that could not be explained by a change in Ca/Al ratios (Blamey et al., 1991). An ameliorative effect of cations, in particular Ca and Mg, was also shown in a number of tree species. Inhibition of root growth by Al in *P. abies* (Rost-Siebert, 1985), *Fagus sylvatica* (Neitzke and Runge, 1987), *Quercus rubra* (Kruger and Sucoff, 1989), and *Gleditsia tricanthos* (Wolfe and Joslin, 1989) seedlings was found to be reduced by Ca. Jorns (1988) showed that Al induced vacuolization of cortex cells of *P. abies* could be reduced by the addition of Ca and Mg. Considering the importance of Mg deficiency in forest decline and the ability of Mg to ameliorate Al toxicity, it is a surprise that investigation of Al–Mg interactions are rare for tree species. In an Al sensitive variety of wheat, Mg was more effective than Ca in restricting Al uptake and maintaining membrane function (Keltjens and Dijkstra, 1991). Magnesium amelioration of Al toxicity was also shown in other varieties of wheat (Hecht-Buchholz and Schuster, 1987; Edmeades et al., 1991). Edmeades et al. (1991) suggested that Mg and Ca may ameliorate Al toxicity by preventing Al from binding to cell walls, and that lesions due to Al are formed only after Al reacts with the cell wall. This idea agrees for the most part with the model of Al toxicity described above. If loading of Ca and Mg in the cell walls maintains high concentrations of these elements in the vicinity of the plasmalemma, then it is reasonable to assume that accumulation of Al in the cell walls will also result in high Al concentrations at the plasmalemma. The high concentrations of Al may then affect changes in plasma membrane structure and changes in the cytoplasm. The ability to prevent displacement of Ca and Mg by Al was linked in ryegrass cultivars to higher Al tolerance (Rengel and Robinson, 1989b).

An influence of cations and ionic strength on the toxicity of heavy metals was also shown. In *Alyssum bertolonii* inhibition of root growth by Ni was reduced by Mg (Gabbrielli and Pandolfini, 1984). In *Festuca ovina* and *Hordeum vulgare* Ca reduced the inhibition of root growth by Pb and Pb uptake (Garland and Wilkins, 1981), where as in *Zea mays* and *Glycine max* Zn and Fe citrate reduced damage due to Cd (Malone et al., 1978). In *Betula* sp. seedlings Zn toxicity and Zn uptake was reduced by high levels of Ca (Brown and Wilkins, 1985). In *P. abies* inhibition of root growth by both Hg and Pb was less in more concentrated nutrient solutions than in dilute nutrient solutions (Godbold and Hüttermann, 1986; Godbold, unpublished). This could be due to both the effect of ionic strength and specific ion effects.

4.4. Organic Substances

In general, the toxicity of Al and heavy metals is reduced in the presence of organic substances. This finding is particularly important in forest soils that may have a high organic matter content, and in root rhizospheres, where large amounts of organic substances (mucilage, exudates, and sloughed-off cells) may be released by the roots. Hue et al. (1986) could show that a wide range of organic acids can complex and reduce Al inhibition of root elongation. Arp and Strucel (1989) reported that Al oxalate was less toxic to *Picea mariana* than inorganic Al. In *Fagus sylvatica* fulvic acid reduced the effect of Al on Ca and P uptake (Asp and Berggren 1990). In *Phaseolus vulgaris, Zea mays, Beta vulgaris,* and *Hordeum vulgare,* Cd uptake was controlled by the activity of Cd^{2+} in solution and was thus reduced by addition of humic acids (Tyler and McBride, 1982; Cabrera et al., 1988) or ethylenediaminetetraacetic acid (EDTA) (Greger and Lindberg, 1986). Similarly, in *P. abies* seedlings the uptake and the effect on root elongation of Pb, Hg, and methyl Hg was reduced by the addition of humic acid to nutrient solutions (Grote, 1988). However, for methyl Hg the addition of humic acid delayed the onset of methyl Hg toxicity but did not eliminate it. The ameliorating effect of humic acid was also pH dependent, a strong effect was found at pH 4, but at pH 3.3 little or no effect was shown.

4.5. Mycorrhizae

Under natural conditions roots of *P. abies* and most other temperate tree species are almost entirely mycorrhizal. It has now become well established that mycorrhization can have strong effects on the morphology and physiology of trees. Although our understanding of the role of ectomycorrhizae in mineral nutrient and water acquisition is poor (Vogt et al., 1991), there is evidence to suggest an involvement of ectomycorrhizae in the uptake of mineral nutrients particularly N (Finlay et al., 1989) and P (Jones et al., 1990). It was suggested, however, that this function of ectomycorrhizae may only be important on nutrient poor sites (Marschner, 1992). Similarly, it was suggested that ectomycorrhizae may regulate the uptake of metals and ameliorate metal toxicity (Kottke and Oberwinkler, 1986). In a number of cases growth and survival of several species of pine growing on strip mine soils was improved by inoculation with *Pisolithus tinctorius* (Marx, 1975; Marx and Altman, 1979; Berry, 1982). Although this improvement in growth could be for a large variety of reasons, such work provoked interest in amelioration of metal toxicity by ectomycorrhizae (Wilkins, 1991). The ectomycorrhizae were suggested to act as a physical or chemical barrier that restricts the uptake and transport of metals. Possible mechanisms involve blocking the movement of metal through the root apoplast (Ashford et al., 1989; Kottke, 1991a), or internal sequestering of metals in the symplast, bound in nontoxic forms. To bind and sequester a number of binding sites were suggested (Jones and Hutchinson, 1986) in the case of Al binding to polyphosphate bodies (Vare, 1991; Kottke, 1991a). In roots of *Fagus sylvatica* Al, Fe, and Ti accumulated in the fungal sheath, and indicated a barrier function of the mycorrhizae (Donner and Heyser, 1989). Recent work, however, shows an unbroken apoplastic transport pathway in *Pinus sylvestris* colonized by *Suillus bovinus* (Behrmann and Heyser, 1992). Kottke (1991b) could also show that in mycorrhizal roots Ce and La can pass freely in the apoplast into the root cortex. The idea that ectomycorrhizae may restrict metal uptake stems from the observation that mycorrhizal tree seedlings often have lower levels of metals in leaves and needles than nonmycorrhizal seedlings. In *Pinus banksiana* and *Picea glauca* grown in soils amended with metals, levels of Cd, Ni, Pb, and Cu were lower in the needles of plants inoculated with *Suillus luteus* than nonmycorrhizal plants at low metal levels in the soil (Dixon and Buschena, 1988). In these soils growth was not improved by mycorrhization, although mycorrhizal seedlings tended to have higher root and shoot weights. At higher metal levels, a strong decrease in mycorrhizal colonization was shown and there was no difference in metal contents between mycorrhizal and nonmycorrhizal tree seedlings. Cumming and Weinstein (1990a) found lower levels of Al in needles of *Pinus rigida* inoculated with *Pisolithus tinctorius* than nonmycorrhizal plants. This result was accompanied by higher needle biomass and improved P nutrition in the mycorrhizal plants. Cumming and Weinstein (1990a) suggested that interference with P uptake may be an important Al lesion. In needles of *Pinus sylvestris* inoculated with a number of mycorrhizal species and strains, in most cases lower levels of Cd (Colpaert and Van Assche, 1992a) and Zn (Colpaert and Van Assche, 1992b) were found compared to nonmycorrhizal seedlings. However, one species (*laccaria laccata*) did not affect Zn uptake and another (*Thelephora terrestris*) increased Zn accumulation in the needles. The amount of metals transported to the needles was dependent on the species or strain of mycorrhizae and on the mycorrhizae maintaining an intact extramatrical mycelium. In *Betula* seedlings, mycorrhization with *Amanita muscaria* and *Paxillus involutus* reduced the Zn contents of the leaves, and in terms of growth, improved tolerance to

Zn (Brown and Wilkins, 1985; Denny and Wilkins, 1987a). In subsequent work, Zn was found to accumulated in the extramatrical mycelium, and it was suggested that Zn binding to the extramatrical mycelium prevented Zn accumulation in the root (Denny and Wilkins, 1987b). The Zn was bound to extra-hyphal slime. However, not all studies showed that inoculation with mycorrhizae reduces metal uptake or improves tolerance to metals. Inoculation of *Picea abies* with *Paxillus involutus, Paxillus theiogalus,* or *Lactarius rufus* did not reduce uptake of Al or Pb by the roots (Jentschke et al., 1991a–c). Furthermore, Jentschke et al. (1991a and b) could show displacement of Mg in the root cell walls by Al in both mycorrhizal and nonmycorrhizal roots. In further work it could be shown that the uptake of Pb by roots of *P. abies* was even increased by some mycorrhizal fungi (Marschner et al., 1993). Colonization of roots with *Laccaria laccata* or *Amanita muscaria* greatly increased the Pb contents of root cortex cell walls. In a major piece of work Jones and Hutchinson (1986, 1988a,b) and Jones et al. (1988) investigated the effect of *Scleroderma flavidum, Lactarius hibbardae, Lactarius rufus,* and *Laccaria proxima* on the Cu and Ni tolerance of *Betula papyrifera.* Only inoculation with *S. flavidum* increased the Ni tolerance of the seedlings. Although the presence of *S. flavidum* did not reduce Ni accumulation in the leaves or influence the decreased uptake of P, Mg, Ca, and Fe induced by Ni, the growth of the seedlings was improved. Jones and Hutchinson (1988a) attributed the increase in Ni tolerance to the larger root systems of the *S. flavidum* infected plants that could retain Ni. The increased Ni tolerance in the *S. flavidum* infected plants correlated with high root levels of P (Jones and Hutchinson, 1988b). However, the Cu tolerance of *B. papyrifera* was decreased by these fungi. In a series of papers, inoculation with *Pisolithus tinctorius* was initially in terms of growth found to increase the tolerance of *Pinus rigida* to Al (Cumming and Weinstein, 1990a–c). Inoculated seedlings had lower Al contents and higher P contents in needles. But in subsequent work differences in Al accumulation in needles could not always be shown (Cumming and Weinstein, 1990b). This anomaly and some of those found in other work can in part be explained by the results of Hentschel et al. (1993) who found an amelioration of Al toxicity symptoms in *Picea abies* after inoculation with *Paxillus involutus* that was dependent on the duration of exposure to Al. Although after a 5-week exposure to Al, levels of Mg were depleted to deficiency levels, symptoms of Mg deficiency (chlorosis) were not expressed until 10 weeks. Although it was not measured, this was suggested to be due to a phytohormone effect. Higher levels of cytokinins were determined in mycorrhizal trees seedlings (Wullschleger and Reid, 1990; Kraigher et al., 1992), were shown to delay at the onset of senescence and chlorosis (Richmond and Lang, 1957), and Pan et al. (1989) showed that exogenous cytokinins ameliorated Al toxicity symptoms of the shoot. Such effects may explain the results of other experiments. Unlike the *Picea rigida–Pisolithus tinctorius* symbiosis (Cumming and Weinstein, 1990a) the increased growth of the shoots in the *Picea abies–Paxillus involutus* symbiosis was not due to better P nutrition. The increased growth of the mycorrhizal plants must be taken into account when considering the lower metal content in the shoots. In needles of *Picea abies* mycorrhizal with *Paxillus involutus* a lower Al content (μmol g^{-1} Dwt) was found compared to nonmycorrhizal seedlings. However, recalculation of their own data and the data of Cumming and Weinstein (1990a), Hentschel et al. (1993) showed similar total Al content in the needles of mycorrhizal and non-mycorrhizal seedlings. Hence, the lower Al content in the needles of mycorrhizal seedlings were due to dilution by growth.

As discussed above the nitrogen source can alter the toxicity of Al. In mycorrhizal *Pinus rigida* seedlings lower uptake and toxicity of Al was found in the presence of NH_4 compared to NO_3 (Cumming and Weinstein, 1990c). Under NO_3 nutrition the toxicity of Al was reduced by mycorrhization. The reduced toxicity was suggested to be due to improved P nutrition through preferential NH_4 uptake by the mycorrhizal fungus. These results can also be explained if mycorrhization aids the maintenance of a low apoplastic pH, thus preventing Al accumulation in the apoplast. Mycorrhization was shown to alter the stoichometry of ion uptake (Rygiewicz et al., 1984a,b). Thus, although there is evidence to support amelioration of metal toxicity by mycorrhizae, the effect of the mycorrhizae is very species and strain specific, and there seems to be link to host–mycorrhiza compatibility (Denny and Wilkins, 1987a). However, amelioration may only be transitory and is not necessarily due to exclusion of metals. In some cases mycorrhization may increase metal toxicity.

4.6. Soil Solutions and Rhizospheres

From the results presented and discussed in this chapter it is clear that the conditions surrounding the root are dominant for the effects of metal toxicity seen in all other plant parts. Although the exact mechanisms by which Al and heavy metals are toxic will continue to be discussed, it is generally agreed that factors such as pH and competing cations are of great importance for the expression of toxicity. In a number of forest sites, levels of Al (Ulrich, Chapter 1; Cronan, Chapter 2) and some heavy metals (Lamersdorf et al., 1991) in the soil solution were shown to be high enough to decrease the vitality of the trees. In forest soils at Camel's Hump state forest, VT, up to 30% of aqueous Al was found to be present as Al_{13} (Hunter and Ross, 1991), and as shown above, Al_{13} may be toxic to spruce. Estimates of the degree of metal stress for forest ecosystems were based on the concentration of elements in the bulk soil solution. However, the speciation and concentrations of elements in the soil solution in the bulk soil may vary considerably to that of the rhizosphere. Extracts of organic substances from the rhizospheres of *Zea mays* and *Triticum aestivum* complexed Co, Zn, and Mn added to soils (Merckx et al., 1986). In the rhizosphere of roots of *Hordeum vulgare* an increase in pH was accompanied by a decrease in the solubility of Cu and Zn, but a decrease in Mn and Fe (Youssef and Chino, 1989). For forest trees there are few measurements of the rhizosphere chemistry available. In pot experiments Al was shown to accumulate in the rhizosphere of *Quercus rubra* roots, but not in the rhizosphere of *Gleditsia tricanthos* (Ohno, 1989). By using a newly developed microtechnique, Knoche and co-workers determined the element concentration in the rhizospheres of adult Norway spruce roots. Knoche (1993) showed that in *P. abies* roots in Solling, Al accumulated in the rhizosphere of long roots and decreased in the rhizosphere of short roots, however, in both cases Al/Ca ratios were more unfavorable in the rhizosphere than in the bulk soil. Rhizoplane/root cortex Al concentrations of up to 1.8 m*M* were determined. The same concentration of Al was estimated to occur in the rhizosphere of long roots of Norway spruce using a rhizosphere model (Nietfeld, 1993). Nietfeld (1993) could show that the level of Al that accumulates is dependent on the water flux of the roots, and can fluctuate considerably. Thus even in relatively homogenous soils the levels of Al and heavy metals in the rhizosphere may vary considerably for different type of functional roots, dependent on their water uptake. Such a relationship could explain why in acidified soils the growth of water absorbing sinker roots are more affected than lateral roots (Murach, personal communication).

The evidence of more unfavorable conditions in the rhizosphere compared to the bulk soil is supported by determinations of Al and heavy metals in the root cortex cell walls of Norway and red spruce collected in the field (Godbold et al., 1987; Godbold et al., 1988b; Schlegel et al., 1992; Knoche, 1993). In these studies high levels of Al, and in some cases Pb, were determined in cell walls. Based on the models described above high levels of Al and Pb in the apoplast will inevitably lead to metal stress. Thus the conditions in the rhizosphere will be of major significance for the rooting strategy (stress avoidance) and vitality of trees. Beyond reasonable doubt the effects of unfavorable conditions in the rhizosphere are now being observed in the declining forests of Europe and North America.

ACKNOWLEDGMENTS

The author wishes to thank the German Federal Ministry of Research and Technology (grant OEF 2019) and the European Union (grant STEP-CT90-0059) for financial support.

REFERENCES

Alva AK, Asher CJ, Edwards DG (1986a): The role of calcium in alleviating aluminum toxicity. Aust J Agric Res 37:375–382.

Alva AK, Blamey FPC, Edwards DG, Asher CJ (1986b): An evaluation of aluminum indices to predict aluminum toxicity to plants grown in nutrient solutions. Commun Soil Sci Plant Anal 17:1271–1280.

Arp PA, Strucel I (1989): Water uptake by black spruce seedlings from rooting media (solution, sand, peat) treated with inorganic and oxalated aluminum. Water Air Soil Pollut 44:57–70.

Ashford AE, Allaway WG, Peterson CA, Cairney JWG (1989): Nutrient transfer and the fungus–root interface. Aust J Plant Physiol 16:85–97.

Asp H, Bergren D (1990): Phosphate and calcium uptake in beech (Fagus sylvatica) in the presence of aluminum and natural fulvic acids. Physiol Plantarum 80:307–314.

Balsberg Påhlsson AM (1989): Toxicity of heavy metals (Zn, Cu, Cd, Pb) to vascular plants. Water, Air, Soil, Pollut 47:287–319.

Becerril JM, Munoz-Rueda A, Aparicio-Tejo P, Gonzalez-Murua C (1988): The effects of cadmium and lead on photosynthetic electron transport in clover and lucerne. Plant Physiol Biochem 26:357–363.

Behrmann P, Heyser W (1992): Apoplastic transport through the fungal sheath of Pinus sylvestris/Suillus bovinus ectomycorrhizae. Bot Acta 105:427–434.

Bengtsson B, Asp H, Jensén P, Berggren D (1988): Influence of aluminum on phosphate and calcium uptake in beech (Fagus sylvatica) grown in nutrient solution and soil solution. Physiol Plant 74:299–305.

Bennet RJ, Breen CM (1991): The aluminum signal: New dimensions to mechanisms of aluminum tolerance. In Wright RJ, Baligar VC, Murrmann RP (eds): Plant–Soil Interactions at low pH. Dordrecht: Kluwer, pp 703–716.

Berry CR (1982): Survival and growth of pine hybrid seedlings with Pisolithus ectomycorrhizae on coal spoils in Alabama and Tennessee. J Environ Qual 11:709–715.

Blamey FPC, Edmeades DC, Asher CJ, Edwards DG, Wheeler DM (1991): Evaluation of solution culture techniques for studying aluminum toxicity in plants. In Wright RJ, Baligar VC, Murrmann RP (eds): Plant–Soil Interactions at low pH. Dordrecht: Kluwer, pp 905–912.

Brown MT, Wilkins DA (1985): Zinc tolerance of mycorrhizal *Betula*. New Phytol 99:101–106.

Burton KW, Morgan E, Roig A (1984): The influence of heavy metals upon the growth of sitka-spruce in South Wales forests. II. Greenhouse experiments. Plant Soil 78:271–282.

Burton KW, Morgan E, Roig A (1986): Interactive effects of cadmium, copper and nickel on the growth of sitka spruce and studies of metal uptake from nutrient solutions. New Phytol 103:549–557.

Burzynski M, Grabowski A (1983): Influence of lead on NO_3^- uptake and reduction in cucumber seedlings. Acta Soc Bot Pol 53:77–86.

Büscher P, Koedam N (1983): Soil preference of populations of genotypes of Asplenium trichomanes L. and Polypodium vulgare L. in Belgium as related to cation exchange capacity. Plant Soil 72:275–282.

Cabrera D, Young SD, Rowell DL (1988): The toxicity of cadmium to barley plants as affected by complex formation with humic acid. Plant Soil 105:195–204.

Cameron RS, Ritchie GSP, Robson AD (1986): Relative toxicities of inorganic aluminum complexes to barley. Soil Sci Soc Am J 50:1231–1236.

Carty AJ, Malone SF (1979): The chemistry of mercury in biological systems. In Nriagu JO (ed): The biogeochemistry of mercury in the environment. North Holland: Elsevier Biomedical Press, Chapter 1, pp 433–479.

Clarkson DT (1965): The effect of aluminum and some trivalent metal cations on cell division in the root apices of Allium cepa. Ann Bot 29:309–315.

Colpaert JV, Van Assche JA (1992a): Effects of cadmium on ectomycorrhizal pine (Pinus sylvestris) seedlings. In Read DJ, Lewis DH, Fitter AH, Alexander IJ (eds): Mycorrhizas in Ecosystems. Wallingford, CAB Intern. England, p 375.

Colpaert JV, Van Assche JA (1992b): Zinc toxicity in ectomycorrhizal Pinus sylvestris. Plant Soil 143:201–211.

Cronan CS (1991): Differential adsorption of Al, Ca and Mg by roots of red spruce (*Picea rubens* Sarg.). Tree Physiol 8:227–237.

Cumming JR, Eckert RT, Evans LS (1986): Effect of aluminum on ^{32}P uptake and translocation by red spruce seedlings. Can J For Res 16:864–867.

Cumming JR, Tomsett AB (1992): Metal tolerance in plants: Signal transduction and acclimation mechanisms. In Ariano DC (ed): Biochemistry of trace metals. Boca Raton: CRC Press, pp 329–364.

Cumming JR, Weinstein LH (1990a): Aluminum-mycorrhizal interactions in the physiology of pitch pine seedlings. Plant Soil 125:7–18.

Cumming JR, Weinstein LH (1990c): Nitrogen source effects on Al-toxicity in nonmycorrhizal and mycorrhizal pitch pine (*Pinus rigida*) seedlings. 1. Growth and nutrition. Can J Bot 68:2644–2652.

Cumming JR, Weinstein LH (1990b): Utilization of $AlPO_4$ as a phosphorus source by Ectomycorrhizal *Pinus rigida* Mill seedlings. New Phytolog 116:99–106.

Davies WJ, Metcalfe J, Lodge TA, da Costa AR (1986): Plant growth substances and the regulation of growth under drought. Aust J Plant Physiol 13:105–125.

De Filippis LF (1978): Localisation of organomercurials in plant cells. Z Pflanzenphysiol 88:133–146.

De Filippis LF (1979): The effect of heavy metal compounds on the permeability of Chlorella cells. Z Pflanzenphysiol 92:39–49.

Denny HJ, Wilkins DA (1987a): Zinc tolerance in Betula spp. III. Variation in response to zinc among ectomycorrhizal associates. New Phytol 106:535–544.

Denny HJ, Wilkins DA (1987b): Zinc tolerance in *Betula* spp. IV. The mechanism of ectomycorrhizal amelioration of zinc toxicity. New Phytol 106:545–553.

Dixon RK, Buschena CA (1988): Response of ectomycorrhizal Pinus banksiana and *Picea glauca* to heavy metals in soils. Plant Soil 105:265–271.

Donner B, Heyser W (1989): Buchenmykorrhizen: Möglichkeiten der Elementselektion unter besonderer Berücksichtigung einiger Schweremetalle. Forstwiss Centralbl 108:150–163.

Edmeades DC, Wheeler DM, Christie RA (1991): The effects of aluminum and pH on the growth of a range of temperate grass species and cultivars. In Wright RJ, Baligar VC, Murrmann RP (eds): Plant–Soil Interactions at low pH. Dordrecht: Kluwer, pp 913–924.

Eichhorn J (1987): Vergleichende Untersuchungen von Feinwurzelsystemen bei unterschiedlich geschädigten Altfichten (*Picea abies* Karst.). PhD Thesis, Universitat Göttingen, Göttingen, Germany.

Finlay RD, Ek H, Odham G, Söderström B (1989): Uptake, translocation and assimilation of nitrogen from ^{15}N-labelled ammonium and nitrate sources of Fagus sylvatica infected with Paxillus involutus. New Phytol 113:47–55.

Foy CD, Flemming AL, Burns GR, Arminger WH (1967): Characterization of differential aluminum tolerance among varieties of wheat and barley. Soil Sci Soc Am Proc 31:513–521.

Gabbrielli R, Pandolfini T (1984): Effect of Mg^{2+} and Ca^{2+} on the response to nickel toxicity in a serpentine endemic and nickel-accumulating species. Physiol Plantarum 62:540–544.

Garland CJ, Wilkins DA (1981): Effect of calcium on the uptake and toxicity of lead in Hordeum vulgare L. and Festuca ovina L. New Phytol 87:581–593.

Gijsman AJ (1990a): Rhizosphere pH along different root zones of Douglas-fir (Pseudotsuga menziesii), as affected by source of nitrogen. Plant Soil 124:161–167.

Gijsman AJ (1990b): Soil water content as a key factor determining the source of nitrogen (NH_4^+ or NO_3^-) absorbed by Douglas-fir (Pseudotsuga menziesii) and the pattern of rhizophere pH along its roots. Can J For Res 21:616–625.

Godbold DL (1991a): Aluminum decreases root growth and Ca and Mg uptake in *Picea abies* seedlings. In Wright RJ, Baligar VC, Murrmann RP (eds): Plant–Soil Interactions at low pH. Dordrecht: Kluwer, pp 747–753.

Godbold DL (1991b): Die Wirkung von Aluminium und Schwermetallen auf *Picea abies* Sämlingen. Schrift Forstl Fak Universitat Göttingen, 104, Frankfurt a.m.: Sauerländers Verlag.

Godbold DL (1991c): Mercury induced root damage in *Picea abies* seedlings. Water Air Soil Pollut 56:823–831.

Godbold DL (1994): Mercury in forest ecosystems: Risks and research needs. In Watras CJ, Huckabee JW (eds): Mercury pollution. Integration and synthesis. Chelsea: Lewis Publishing, in press.

Godbold DL, Dictus K, Hüttermann A (1988a): Influence of aluminum and nitrate on root growth and mineral nutrition of Norway spruce (*Picea abies*) seedlings. Can J For Res 18:1167–1171.

Godbold DL, Fritz E, Hüttermann A (1987): Lead contents and distribution in spruce roots. In Mathy P (ed): Air pollution and ecosystems. Dordrecht: Reidel, pp 864–869.

Godbold DL, Fritz E, Hüttermann A (1988b): Aluminum toxicity and forest decline. Proc Natl Acad Sci USA 85:3888–3892.

Goldbold DL, Hüttermann A (1986): The uptake and toxicity of mercury and lead to spruce (*Picea abies* Karst.) seedlings. Water Air Soil Pollut 31:509–516.

Godbold DL, Hüttermann A (1988a): Inhibition of photosynthesis and transpiration in relation to mercury-induced root damage in spruce seedlings. Physiol. Plantarum 74:270–275.

Godbold DL, Kettner C (1991a): Use of root elongation studies to determine aluminum and lead toxicity in *Picea abies* seedlings. J Plant Physiol 138:231–235.

Godbold DL, Kettner C (1991b): Lead influences root growth and mineral nutrition of *Picea abies* seedlings. J Plant Physiol 139:95–99.

Göransson A, Eldhuset TD (1987): Effects of aluminum on growth and nutrient uptake of Betula pendula seedlings. Physiol Plant 69:193–199.

Grauer UE, Horst WJ (1990): Effect of pH and nitrogen source on aluminum tolerance of rye (Secale cereale L.) and yellow lupin (Lipinus luteus L.). Plant Soil 127:13–21.

Grauer UE, Horst WJ (1992): Modelling cation amelioration of aluminum phytotoxicity. Soil Sci Soc Am J 56:166–172.

Greger M, Lindberg S (1986): Effects of Cd^{2+} and EDTA on young sugar beets (Beta vulgaris). I. Cd^{2+} uptake and sugar accumulation. Physiol Plantarum 66:69–74.

Grote R (1988): "Wechselwirkung zwischen verschiedenen Schwermetallen und Huminsäure in bezug auf das Wachstum von Fichtenkeimlingen (*Picea abies* Karst.)." Diplomarb, Universitat Göttingen, Göttingen, Germany.

Hampp R, Ziegler H, Ziegler I (1973): Die Wirkung von Bleiionen auf die $^{14}CO_2$-Fixierung und die ATP-Bildung von Spinatchloroplasten. Biochem Physiol Pflanzen 164:126–134.

Hardiman RT, Jacoby B, Banin A (1984): Factors affecting the distribution of cadmium, copper and lead and their effect upon yield and zinc content in bush beans. Plant Soil 81:17–27.

Häussling M, Jorns CA, Lehmbecker G, Hecht-Buchholz Ch, Marschner H (1988): Ion and water uptake in relation to root development in Norway spruce (*Picea abies* (L). Karst.). J Plant Physiol 133:486–491.

Hecht-Buchholz Ch, Schuster J (1987): Response of Al-tolerant and Al-sensitive kearney barley cultivars to calcium and magnesium during Al stress. Plant Soil 99:47–61.

Hentschel E, Godbold DL, Marschner P, Schlegel H, Jentschke G, (1993): The effect of Paxillus involutus on the aluminum sensitivity of Norway spruce seedlings. Tree Physiol 12:379–390.

Horst WJ (1987): Aluminum tolerance and calcium efficiency of cowpea genotypes. J Plant Nutr 10(9–16):1121–1129.

Horst WJ, Wagner A, Marschner H (1982): Mucilage protects root meristems from aluminum injury. Z Pflanzenphysiol 105:435–444.

Hue NV, Craddock GR, Adams F (1986): Effect of organic acids on aluminum toxicity in subsoils. Soil Soc Am J 50:28–34.

Hunter D, Ross DS (1991): Evidence for a phytotoxic hydroxy-aluminum polymer in organic soil horizons. Science 251:1056–1058.

Jensen P, Petterson S, Drakenberg T, Asp H (1989): Aluminum effects on vacuolar phosphorus in roots of beech (*Fagus sylvatica* L.). J Plant Physiol 134:37–42.

Jentschke G, Fritz E, Godbold DL (1991b): Distribution of lead in mycorrhizal and non-mycorrhizal Norway spruce seedlings. Physiol Plant 81:417–422.

Jentschke G, Godbold DL, Hüttermann A (1991a): Culture of mycorrhizal tree seedlings under controlled conditions: Effects of nitrogen and aluminum. Physiol Plant 81:408–416.

Jentschke G, Schlegel H, Godbold DL (1991c): The effect of aluminum on uptake and distribution of magnesium and calcium in roots of mycorrhizal Norway spruce seedlings. Physiol Plant 82:266–270.

Jones MD, Dainty J, Hutchinson TC (1988): The effect of infection by Lactarius rufus or Scleroderma flavidum on the uptake of ^{63}Ni by paper birch. Can J Bot 66:934–940.

Jones MD, Durall DM, Tinker PB (1990): Phosphorus relationships and production of extramatrical hyphae by two types of willow ectomycorrhizae at different soil phosphorus levels. New Phytol 115:259–267.

Jones MD, Hutchinson TC (1986): The effect of mycorrhizal infection on the response of *Betula papyrifera* to nickel and copper. New Phytol 102:429–442.

Jones MD, Hutchinson TC (1988a): Nickel toxicity in mycorrhizal birch seedlings infected with Lactarius rufus or Scleroderma flavidum. I. Growth and gas exchange characteristics. New Phytol 108:451–460.

Jones MD, Hutchinson TC (1988b): Nickel toxicity in mycorrhizal birch seedlings infected with Lactarius rufus or Scleroderma flavidum. II. Uptake of nickel, calcium, magnesium, phosphorus and iron. New Phytol 108:461–470.

Jorns CA (1988): Aluminium bei Sämlingen der Fichte (*Picea abies* (L.) Karst.) in Nährlösungskultur. In Ulrich B (ed): Ber Forschungszentr Waldökosyst, Göttingen, A, 42.

Kahle H, Breckle SW (1986): Wirkungen ökotoxischer Schwermetalle auf Buchenjungwuchs. Statuskolloquium "Luftverunreinigungen and Waldschäden," Düsseldorf: 84–90.

Keltjens WG, Dijkstra WJ (1991): The role of magnesium and calcium in alleviating aluminum toxicity in wheat plants. In Wright RJ, Baligar VC, Murrmann RP (eds): Plant–Soil Interactions at low pH. Dordrecht: Kluwer, pp 763–768.

Kennedy CD, Gonsalves FAN (1987): The action of divalent zinc, cadmium, mercury copper and lead on the trans-root potential and H^+ efflux of excised roots. J Exp Bot 38:800–817.

Kennedy CD, Gonsalves FAN (1989): The action of divalent Zn, Cd, Hg, Cu and Pb ions on the ATPase activity of a plasma membrane fraction isolated from roots of Zea mays. Plant Soil 117:167–175.

Kinraide TB (1991): Identity of the rhizotoxic aluminum species. In Wright RJ, Baligar VC, Murrmann RP (eds): Plant–Soil Interactions at low pH. Dordrecht: Kluwer, pp 717–728.

Kinraide TB, Parker DR (1987): Cation amelioration of aluminum phytotoxicity in wheat. Plant Physiol 83:546–551.

Kinraide TB, Parker DR (1990): Apparent phytotoxicity of mononuclear hydroxy-aluminum to four dicotyledonous species. Plant Physiol 79:283–288.

Kinraide TB, Ryan PR (1991): Cell surface charge may obscure the identity of the rhizotoxic aluminum species. Curr. Top Plant Bio Physiol. 10:94–106.

Klimashevskii EL, Dedov VM (1975): Localisation of the mechanisms of growth inhibiting action of Al in elongating cell walls. Fiziol Rast (Sofia) 22:1183–1190.

Klotz F, Horst WJ (1988a): Genotypic differences in aluminum tolerance of soybean (Glycine max L.) as affect by ammonium and nitrate nitrogen nutrition. J Plant Physiol 132:702–707.

Klotz F, Horst WJ (1988b): Effect of ammonium and nitrate nitrogen on aluminum tolerance of soybean (Glycine max L.). Plant Soil 111:59–65.

Knoche D (1988): Die Wirkung von Beryllium, Cadmium, Vanadium und Thallium auf das Keimwurzellängenwachstum von Fichtensämlingen (*Picea abies* (L.) Karts.). Diplomarb, Universitat Göttingen, Göttingen, Germany.

Knoche D (1993): Entwicklung einer röntgenmikroanalytischen Methode zur Quantifizierung des Rhizosphärenchemismus von Waldbäumen. Ph.D. Thesis, Universitat Göttingen, Göttingen, Germany.

Kottke I (1991a): Reactions and interactions of mycorrhizae and rhizosphere. In Ulrich B (ed): International congress on forest decline research: State of knowledge and perspectives. Karlsruhe: Kernforschungzentrum, pp 405–414.

Kottke I (1991b): Electron energy loss spectroscopy and imaging techniques for subcellular localization of elements in mycorrhiza. Methods Microbiol 23:369–382.

Kottke I, Oberwinkler F (1986): Mycorrhiza of forest trees. Trees 1:191–194.

Kraigher H, Strnad M, Hanke DE, Batic F (1992): Cytokinin content in needles of Norway spruce (*Picea abies* (L.) Karst.) inoculated with two strains of the ectomycorrhizal fungus Thelephora terrestris Fr. In Tesche M (ed): Air pollution and interactions between organisms in forest ecosystems, IUFRO Abstracts. Dresden: TU Dresden, Germany, p 122.

Kruger E, Sucoff E (1989): Growth and nutrient status of Quercus rubra L. in response to Al and Ca. J Exp Bot 40:653–658.

Ksiazek M, Wozny A (1990): Lead movement in poplar adventitious roots. Biol. Plantarum 32(1):54–57.

Lamersdorf NP, Godbold DL, Knoche D (1991): Risk assessment of some heavy metals for the growth of Norway spruce. Water, Air, Soil, Pollut 57:535–543.

Langheinrichs U, Tischner R, Godbold DL (1992): The effect of variation in nitrogen source on the response of spruce (*Picea abies*) seedlings to excessive manganese supply. Tree Physiol 10:259–271.

Lozano FC, Morrison IK (1982): Growth and nutrition of white pine and white spruce seedlings in solutions of various nickel and copper concentrations. J Environ Qual 11:437–441.

Malone CP, Koeppe DE, Miller RJ (1978): Root growth in corn and soybeans: effects of cadmium and lead on lateral root initiation. Can J Bot 56:277–281.

Marschner H (1991): Mechanisms of adaptation of plants to acid soils. Plant Soil 134:1–20.

Marschner H (1992): Nutrient dynamics at the soil–root interface (rhizosphere). In Read DJ, Lewis DH, Fitter AH, Alexander IJ (eds): Mycorrhizas in Ecosystems. Wallingford, CAB Intern. England, pp 3–12.

Marschner H, Häussling M, George E (1991): Ammonium and nitrate uptake rates and rhizosphere pH in non-mycorrhizal roots of Norway spruce [*Picea abies* (L.) Karst.]. Trees 5:14–21.

Marschner P, Hentschel E, Klam A, Jentschke G, Godbold DL (1993): Einfluss von Aluminium und Blei auf Mykorrhizapilze und mykorrhizierte Fichtenkeimlinge. In Tesche M, Feiler S (eds): Proceedings IUFRO. Air pollution and interactions between organisms in forest ecosystems. Dresden, Germany, pp 228–237.

Marx DH (1975): Mycorrhizae and the establishment of trees on strip-mined land. Ohio J Sci 75:288–297.

Marx DH, Altman JD (1979): Pisolithus tinctorius ectomycorrhiza improve survival and growth of pine seedlings on acid coal spoil in Kentucky and Virginia. Reclam Rev 2:23–37.

Matsumoto H, Morimura S (1980): Repressed template activity of chromatin of pea roots treated by aluminum. Plant Cell Physiol 21:951–959.

Matsumoto H, Morimura S, Takahashi E (1977a): Less involvement of pectin in the precipitation of aluminum in pea root. Plant Cell Physiol 18:325–335.

Matsumoto H, Morimura S, Takahashi E (1977b): Binding of aluminum to DNA in pea root nuclei. Plant Cell Physiol 18:987–993.

Matsumoto H, Yamaya T (1986): Inhibition of potassium uptake and regulation of membrane-associated Mg^{2+}-ATPase activity of pea roots by aluminum. Soil Sci Plant Nutr 32:179–188.

Merckx R, van Ginkel JH, Sinnaeve J, Cremers A (1986): Plant-induced changes in the rhizosphere of maize and wheat. Plant Soil 96:95–107.

Miles CD, Brandle JR, Daniel DJ, Chu-Der O, Schnore PD, Uhlik DJ (1972): Inhibition of photosystem II in isolated chloroplast by lead. Plant Physiol 49:820–825.

Mitchell CD, Fretz TA (1977): Cadmium and zinc toxicity in white pine, red maple and Norway spruce. J Am Soc Hort Sci 1021:81–84.

Morimura S, Takahashi E, Matsumoto H (1978): Association of aluminum with nuclei and inhibition of cell division in onion (Allium cepa) roots. Z Pflanzenphysiol 88:395–401.

Nietfeld H (1993): Modellierung der Ionendynamik in der Rhizosphäre. Ph.D. Thesis, Universitat Göttingen, Göttingen, Germany.

Neitzke M, Runge M (1987): Entwicklung und Mineralstoffgehalt junger Buchen in Abhängigkeit von den Aluminium- und Kalziumgehalten der Nährlösung. Teil 1: Entwicklung. Bot Jahrb Syst 108:403–415.

Ohno T (1989): Rhizosphere pH and aluminum chemistry of red oak and honeylocust seedlings. Soil Biol Biochem 21:657–660.

Olsthoorn AFM, Keltjens WG, Van Baren B, Hopman MCG (1991): Influence of ammonium on fine root development and the rhizosphere pH of Douglas fir seedlings in sand. Plant Soil 133:75–81.

Pan WL, Hopkins AG, Jackson WA (1989): Aluminum inhibition of shoot lateral branches of Gycine max and reversal by exogeneous cytokinin. Plant Soil 120:1–9.

Parker DR, Kinraide TB, Zelazny LW (1989): Aluminum speciation and phytotoxicity in dilute hydroxy-aluminum solutions. Soil Sci Soc Am J 46:993–997.

Prenzel J (1990): WAKO—A program for the calculation of complexation equilibria in aqueous solutions. Introduction and example. In Ulrich B (ed): Ber Forschungszentr Waldökosyst, Göttingen, B17, pp 1–15.

Quereshi JA, Hardwick K, Collin HA (1986): Intracellular localization of lead in a lead tolerant and sensitive clone of Anthoxanthum odoratum. J Plant Physiol 122:357–364.

Radin JW, Parker LL, Guinn G (1982): Water relations of cotton plants under nitrogen deficiency: V. Environmental control of abscisic acid accumulation and stomatal sensitivity to abscisic acid. Plant Physiol 70:1066–1070.

Raynal D, Thornton FC, Joslin JD, Schaedle M, Henderson G (1990): The sensitivity of tree seedlings to aluminum. III. Red spruce and loblolly pine. J Environ Qual 19:180–187.

Rengel Z (1990): Net Mg^{2+} uptake in relation to the amount of exchangeable Mg^{2+} in the donnan free space of ryegrass roots. Plant Soil 128:185–189.

Rengel Z, Robinson DL (1989a): Competitive Al^{3+} inhibition of net Mg^{2+} uptake by intact Lolium multiflorum roots. I. Kinetics. Plant Physiol 91:1407–1413.

Rengel Z, Robinson DL (1989b): Determination of cation exchange capacity of ryegrass roots by summing exchangeable cations. Plant Soil 116:217–222.

Richmond EA, Lang A (1957): Effect of kinetin on protein content and survival of detached *Xanthium* leaves. Science 125:650–651.

Rollwagen BA, Zasoski RJ (1988): Nitrogen source effects on rhizosphere pH and nutrient accumulation by Pacific Northwest conifers. Plant Soil 105:79–86.

Rost-Siebert K (1985): Untersuchung zur H- and Al-Ionentoxizität an Keimpflanzen von Fichte (*Picea abies* Karst) and Buche (Fagus silvatica L) in Lösungskultur In Ulrich B (ed): Ber Forschungszentr Waldökosyst, Göttingen 12.

Rygiewicz PT, Bledsoe CS, Zasoski RJ (1984a): Effects of ectomycorrhizae and solution pH on [^{15}N]ammonium uptake by coniferous seedlings. Can J For Res 14:885–892.

Rygiewicz PT, Bledsoe CS, Zasoski RJ (1984b): Effects of ectomycorrhizae and solution pH on [^{15}N]nitrate uptake by coniferous seedlings. Can J For Res 14:885–892.

Schaedle M, Thornton FC, Raynal DJ, Tepper HB (1989): Response of tree seedlings to aluminum. Tree Physiol 5:337–356.

Schlegel H (1985): Schwermetalltoxiziät bei Fichtenkeimlingen (*Picea abies* Karst) am Beispiel von Cadmium, Zink und Quecksilber. Diplomarb Universitat Göttingen, Göttingen, Germany.

Schlegel H (1989): Die Aufnahme von Aerosolen über Fichtennadeln und ihre physiologischen Wirkungen auf die Fichte (*Picea abies* Karst.). Ph.D. Thesis, Universitat Göttingen, Göttingen, Germany.

Schlegel H, Amundson RG, Hüttermann A (1992): Element distribution in red spruce (*Picea rubens*) fine roots: Evidence for aluminum toxicity at Whiteface Mountain. Can J For Res 22:1131–1138.

Schlegel H, Godbold DL (1991): The influence of Al on the metabolism of spruce needles. Water, Air, Soil, Pollut 57–58:131–138.

Schröder WH, Bauch J, Endeward R (1988): Microbeam analysis of Ca exchange and uptake in the fine roots of spruce: influence of pH and aluminum. Trees 2:96–103.

Schwartzenberg K von (1989): Der Cytokiningehalt in Nadeln unterschiedlich stark von "neuartigen Waldschäden" betroffenen Fichten (*Picea abies* (L.) Karst.), bestimmt mittels einer immunoenzymatischen Mehtode-ELISA. Ph.D. Thesis, Universitat Bonn, Bonn, Germany.

Shieh YJ, Barber J (1973): Uptake of mercury by Chlorella and its effect on potassium regulation Planta 109:49–60.

Sieghardt H (1988): Schwermetall- und Nährelementgehalte von Pflanzen und Bodenproben schwermetallhaltiger Halden im Raum Bleiberg in Kärnten (Österreich). II. Holzpflanzen. Z Pflanzenernähr Bodenk 151:21–26.

Stienen H, Bauch J (1988): Element contents in tissues of spruce seedlings from hydroponic cultures simulating acidification and deacidification. Plant Soil 106:231–238.

Suhayda CG, Haug A (1986): Organic acids reduce aluminum toxicity in maize root membranes. Physiol Plant 68:189–195.

Tanton TW, Crowdy SH (1971): The distribution of lead chelate in the transpiration stream in higher plants. Pest Sci 2:211–213.

Taylor GJ (1988a): The physiology of aluminum phytotoxicity. In Siegel H, Siegel A (eds): Metal ions in biological systems. Vol. 24. Aluminum and its role in biology. New York: Dekker pp 165–198.

Taylor GJ (1988b): The physiology of aluminum tolerance. In Siegel H, Siegel A (eds): Metal ions in biological systems. Vol. 24. Aluminum and its role in biology. New York: Dekker, pp 123–163.

Taylor GJ (1988c): Mechanisms of aluminum tolerance in Triticum aestivum L. (wheat). V. Nitrogen nutrition, plant-induced pH and tolerance to aluminum; correlation without causality? Can J Bot 66:695–699.

Taylor GJ, Foy CD (1985a): Mechanisms of aluminum tolerance in Triticum aestivum L. (wheat). I. Differential pH induced by winter cultivars in nutrient solutions. Am J Bot 72:695–701.

Taylor GJ, Foy CD (1985b): Mechanisms of aluminum tolerance in Triticum aestivum L. (wheat). I. Differential pH induced by spring cultivars in nutrient solutions. Am J Bot 72:702–706.

Thornton FC, Schaedle M, Raynal DJ (1986): Effects of aluminum on growth of sugar maple in solution culture. Can J For Res 16:892–896.

Thornton FC, Schaedle M, Raynal DJ (1987): Effects of aluminum on red spruce seedlings in solution culture. Environ Exp Bot 27:489–498.

Tyler G, Balsberg Pahlsson AM, Bengtsson G, Baath E, Tranvik L (1989): Heavy metal ecology of terrestrial plants, microorganisms and invertebrates. Water, Air, Soil, Pollut 47:189–215.

Tyler LD, McBride MB (1982): Influence of Ca, pH and humic acid on Cd uptake. Plant Soil 64:259–262.

Van Praag, HJ, Weissen F, Sougnez-Remy S, Carletti G (1985): Aluminum effects on spruce and beech seedlings. II. Statistical analysis of sand culture experiments. Plant Soil 83:339–356.

Vare H (1991): Aluminum polyphosphate in the ectomycorrhizal fungus Suillus variegatus (Fr) O Kunze as revealed by energy dispersive spectrometry. New Phytol 116:663–668.

Vogt KA, Publicover DA, Vogt DJ (1991): A critique of the role of ectomycorrhizae in forest ecology. Agric Ecosyst Environ 35:171–190.

Wallace SU, Anderson IC (1984): Aluminum toxicity and DNA synthesis in wheat roots. Agron J 76:5–8.

Wilkins DA (1991): The influence of sheathing (ecto-) mycorrhizas of trees on the uptake and toxicity of metals. Agric Ecosyst Environ 35:245–260.

Wolfe MH, Joslin JD (1989): Honeylocust (Gleditsia tricanthos L.) root response to aluminum and aluminum:calcium ratios. Plant Soil 119:181–185.

Wong D, Govindjee (1976): Effects of lead ions on photosystem I in isolated chloroplast: studies on the reaction center. Photosynthetica 10:241–254.

Wullschleger SD, Reid CPP (1990): Implication of ectomycorrhizal fungi in the cytokinin relations of loblolly pine (*Pinus taeda* L). New Phytol 116:681–688.

Youssef RA, Chino M (1989): Root-induced changes in the rhizosphere of plants. II. Distribution of heavy metals across the rhizosphere in soils. Soil Sci Plant Nutr 35:609–621.

MORPHOLOGY OF CONIFEROUS TREES: POSSIBLE EFFECTS OF SOIL ACIDIFICATION ON THE MORPHOLOGY OF NORWAY SPRUCE AND SILVER FIR

FRANZ GRUBER

Forstbotanisches Institut, Universität Göttingen, Büsgenweg 2, D-3400 Göttingen, Germany

1. INTRODUCTION

A widespread symptom of tree damage throughout middle and northern Europe in Norway spruce (*Picea abies* L. Karst.) and silver fir (*Abies alba* Mill.) is crown thinning. Thinning results from both an above-average loss of older sets of needles and an insufficient branching, especially preventitious branching inside the crown. The typical pattern of thinning is loss of needles from the inside to the outside and from the base to the top of the crown (Schütt, 1984; Liedeker et al., 1988).

Also prominent symptoms of crown thinning on spruce are the "Lametta (tinsel) syndrome" (Elstner, 1983; Schütt, 1984; Rehfuess and Rodenkirchen, 1984; Gruber, 1987; Magel and Ziegler, 1987) and "subtop thinning" (Schröter and Aldinger, 1985). "Tinsel twigs" are the older twigs of second and higher orders of the comb-like branched spruce. These twigs are characterized through an above-average loss of needles on the older parts of the naked and silvery twig axes; therefore, they look like silver tinsel on Christmas trees. Subtop thinning is a prominent needle lack (in

Effects of Acid Rain on Forest Processes, pages 265–324

the upper two and three fifths of the crown) below the top (upper one fifth) of the crown, which has no needle loss.

Essential principles of needle abscission and the branching system of Norway spruce and silver fir are necessary for a better understanding and interpretation of crown thinning symptoms in the decline of these trees as a possible consequence of soil acidification. It is important to know the regeneration behavior and response of a tree after stress because with this knowledge we can decide if the symptom was caused by acute or chronic stress. It is especially important to complete our knowledge of all morphological, morphogenetical, anatomical, and physiological features of tree species and their dependence on ecological factors (investigations on the *whole-tree* level).

The following concept of tree architecture is an ecologically dynamic one. This concept is based upon the genotypic determined blueprint, which is composed of single plant elements in successive ontogenetical stages and which will be realized as the "ideal type" under defined (e.g., optimal) environmental conditions. Changes of the conditions modify the ideal architecture into specific patterns. A goal of modern tree biology is to learn how to interpret these specific patterns and reactions and their dependence on the dynamic influences of the environment.

2. THE MECHANISM OF NEEDLE ABSCISSION IN SPRUCE

In Norway spruce we have to distinguish between needle fall via abscission zone and needle loss without the abscission apparatus (Fig. 8.1).

Needle fall via abscission apparatus occurs in a special primary abscission zone (see Napp-Zinn, 1966), which consists of three clearly differentiated layers (Neger, 1911; Campbell and Vines, 1938; Gruber, 1987, 1990), a shrinkage layer at the base of the green needle, a resistance layer on the needle cushion, and an intervening abscission layer (Fig. 8.2).

The green needle develops on the petiole of a footlike needle cushion on the twig (Fig. 8.3*a*). Figures 8.2*A* and 8.8 show a longitudinal section through the abscission zone of the needle. Between the thick-walled axially oriented cells of the hyaline layer (shrinkage tissue) and the lignified sclerenchyma layer of the petiole (resistance tissue), which is oriented at right angles to the needle axis, is the cutin layer (abscission tissue). The cutin layer is one cell thick and it is the zone where the abscission takes place. The cutin cells have thin primary and thick secondary walls, which consists of cutin or cutinlike material (Sudan III reaction; Gerlach, 1977). Needle and petiole are connected by the primary walls of the abscission layer cells, the leaf trace, and the cuticle. The development of the primary abscission zone (Figs. 8.4–8.8), which is called primary because of its early development in every young needle (Napp-Zinn, 1966), starts and runs synchronously with the differentiation of the developing needles early in May. About 8 weeks later at the end of July the abscission zone is fully differentiated even in the case where the needle may not abscise for several years (Fig. 8.8). The abscission of the needle is triggered by the

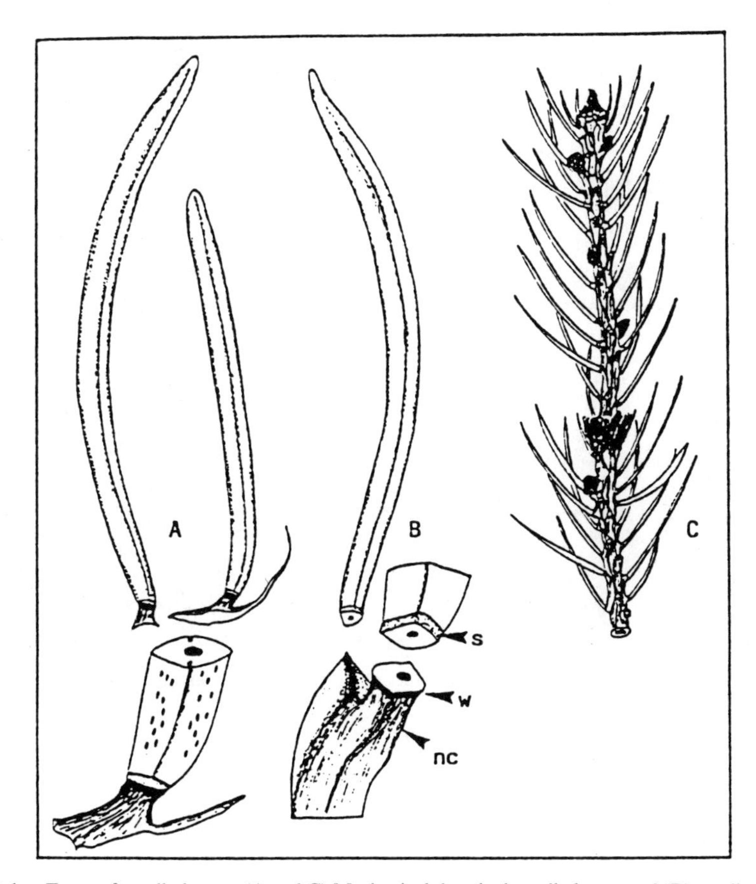

Fig. 8.1. Types of needle losses. (A and C) Mechanical detached needle losses and (B) needle loss via abscission zone (where s = shrinkage tissue; w = resistence tissue; and nc = shaft of the needle cushion).

shrinkage of the hyaline layer (Behrens, 1886; Neger, 1911; Neger and Fuchs, 1915; Campbell and Vines, 1938; Facey, 1956; Napp-Zinn, 1966), which can shrink to 80% of the cross-sectional area of the resistance layer (Fig. 8.3*b*). The shrinkage, which occurs only on desiccating needles, causes a tearing of the primary walls and the cuticle. In this stage the needle is only attached through the leaf trace; the needle drops off at the slightest touch when the leaf trace is broken (Fig. 2*A*–*D*). Thus, abscission of the spruce needle is a simple physical process, a "hygroscopic rhexolyse" (s. Pfeiffer, 1928). This process also operates in dead needles on dead twigs without physiological activity. Because of the physical nature of the abscission process the needle fall is a highly sensitive sensor, indicating and reacting to water stress on a twig, branch, or part of the crown.

Needle desiccation generally occurs, if

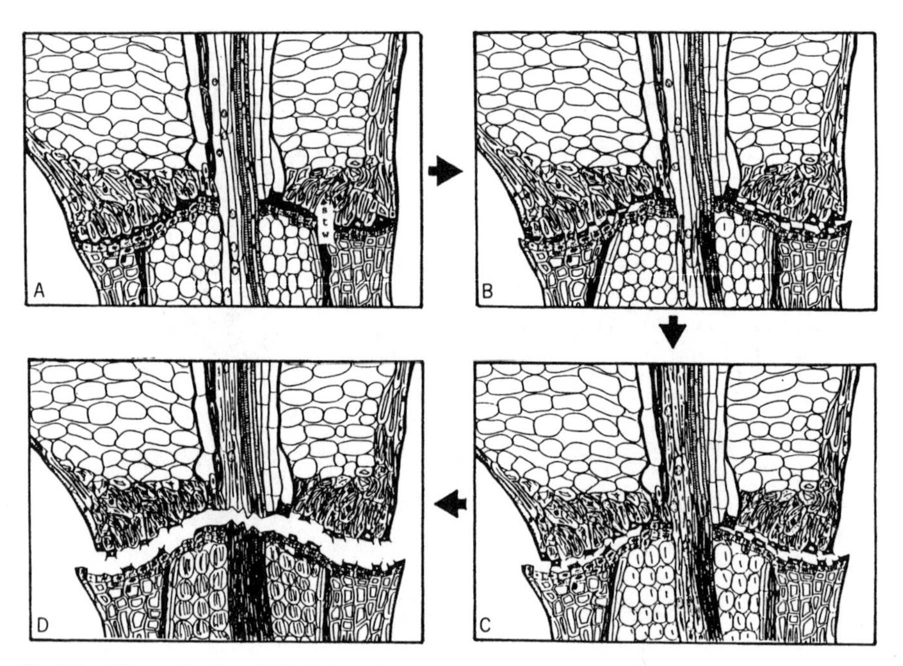

Fig. 8.2. The mechanism of abscission of needles of Norway spruce. (A) resting phase; (B) stretching phase; (C) tearing phase; and (D) breakage of the needle trace [where s = hyaline layer (shrinkage tissue); t = cutin layer (abscission tissue); and w = sclerenchyma layer (resistance tissue)].

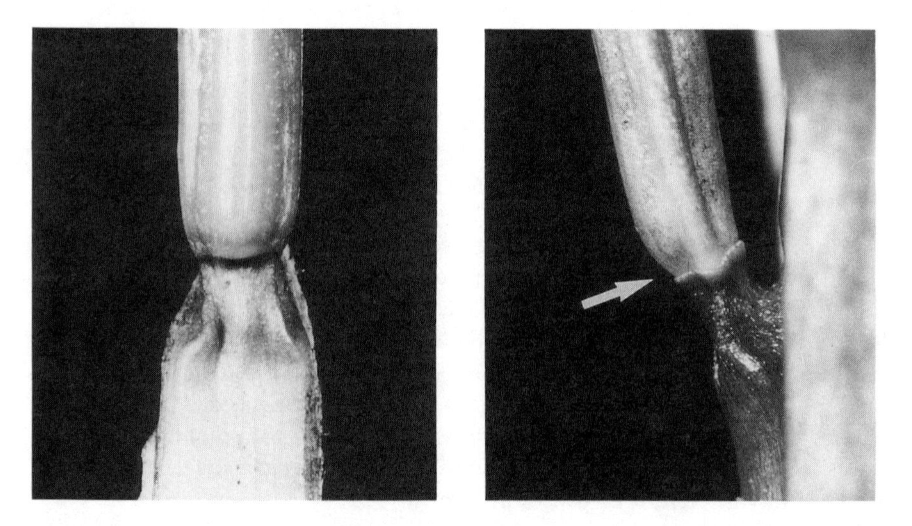

Fig. 8.3. Needle base. (*Left*) Living needle on the shaft of needle cushion. The black band marks the resistance layer. (*Right*) Shrunk needle base.

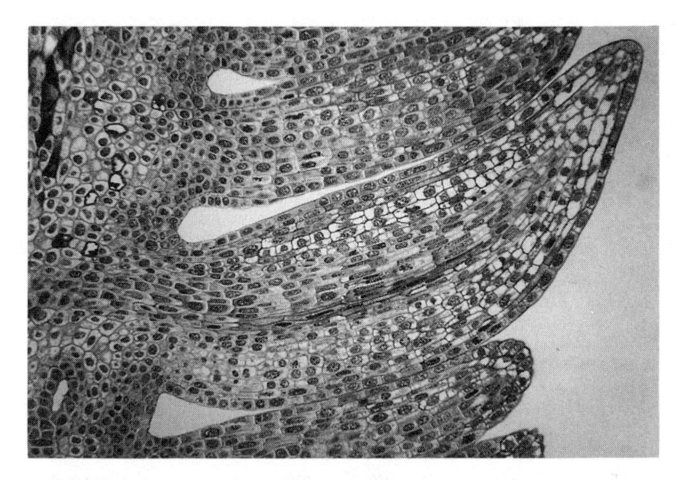

Fig. 8.4. Longitudinal section through a needle primordium at the beginning of needle growth (in early May).

- The whole or a part of the tree suffers from water deficiency (systemic water stress).
- The needle surface becomes irreparably damaged when cuticle waxes are eroding.
- The water transport system is blocked.
- An interruption of the conducting tissue between the needle trace and the shoot axis arises.

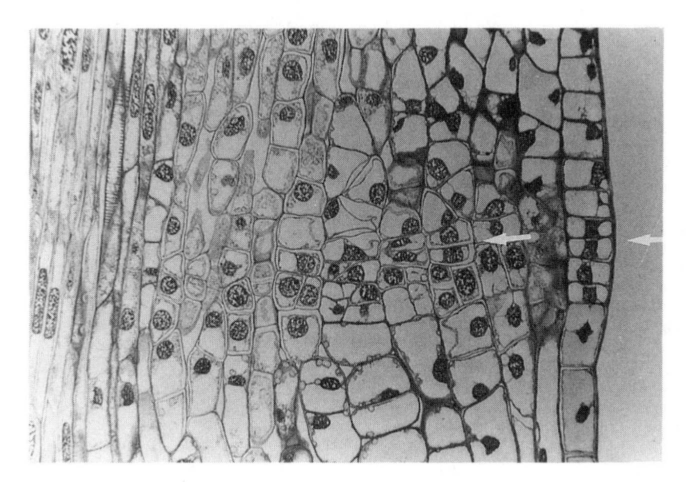

Fig. 8.5. Longitudinal section through the abscission zone (in late May). The region of the abscission zone is clearly indicated (arrows).

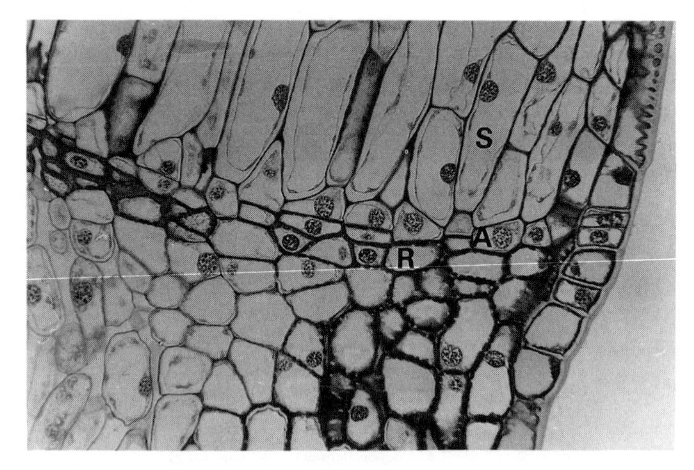

Fig. 8.6. Longitudinal section through the abscission zone during differentiation (middle of June). The three zones are already clearly distinguishable (where S = shrinkage tissue; A = abscission tissue; and R = resistance tissue).

A blockage within the transport system can be caused by the stoppage of the leaf trace in the region of the abscission zone (Campbell and Vines, 1938; Gruber, 1987). Thereby the spruce could have a regulating mechanism for adjusting the transpiring needle biomass to the actual water supply. Besides this the leaf trace stoppage also has a preventive function protecting the twig against desiccation and attacks of microorganisms. An interruption of the transport tracheids can be caused by a tearing of the leaf trace in the region of the axial cambium due to the secondary thickening growth (Markfeldt, 1885; Eames and McDaniels, 1947; Kestel, 1961).

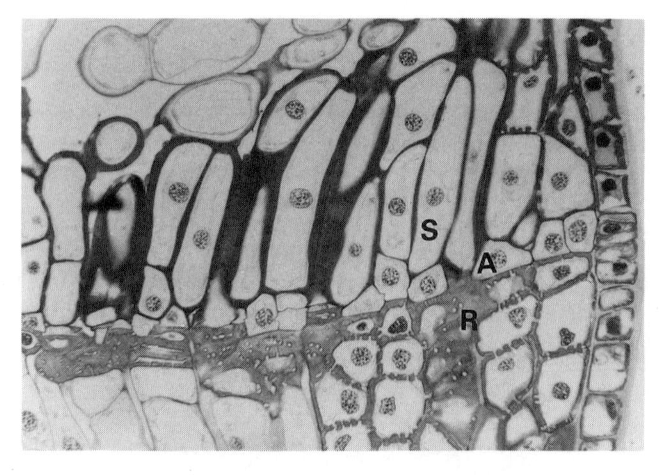

Fig. 8.7. Longitudinal section through a nearly fully developed abscission zone late in June (where S = shrinkage tissue; T = abscission tissue; and W = resistance tissue).

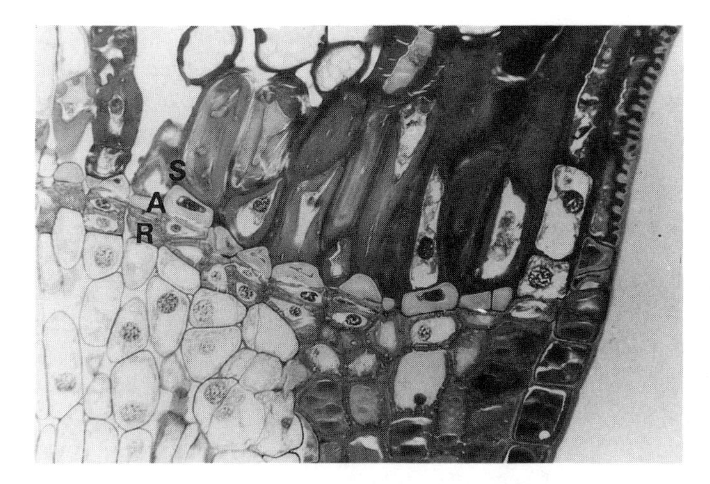

Fig. 8.8. Longitudinal section through a fully differentiated abscission zone in late July (where S = shrinkage tissue; T = abscission tissue; and W = resistance tissue).

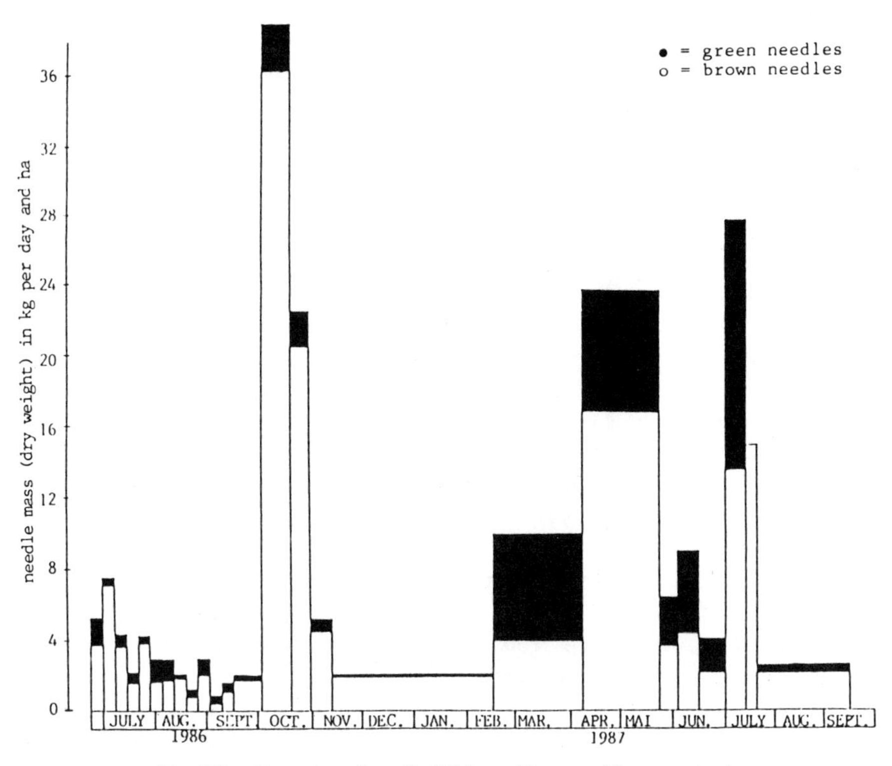

Fig. 8.9. Dynamics of needle fall in an 80 years old spruce stand.

This phenomenon does not play a significant role in needle abscission, because the thickening growth of the shoot axis of second and higher order, which carry most of the needle mass, is small. Therefore only a stretching of the old needle trace occurs, but no tearing. The needle loss without using the abscission apparatus is caused by mechanical forces, mainly by storms in autumn and spring, but also by damages due to frost, ice, and so on during winter.

Litter fall data show (Fig. 8.9; Gruber, 1990) that needle fall occurs during the whole year with two regular seasonally dependent maxima in autumn and in late winter–early spring. These periods occur after the dry periods in summer and winter. This kind of needle fall is normally caused by physiological aging of the older sets of needles. Due to the mechanical operating abscission apparatus needle fall can be delayed until spring. As Figure 8.9 shows, the trees can immediately react to a longer dry period in June with an abrupt increase of needle shedding in July. This kind of needle loss can be caused by acute deficiency of water supply. Therefore the rhythm of needle fall reflects the environment, mainly the climatic fluctuations, and needle fall of spruce via abscission apparatus seems to be a sensitive indicator of water stress.

Through the quantity and the type of shed needles we can suggest the causes of the needle fall. While shed needles via abscission zone point to natural aging or water stress, needles mechanically detached point to mechanical influences. The litter fall date showed that only 13% of the needle fall was caused mechanically and 87% were shed via abscission apparatus.

3. CROWN ARCHITECTURE AND BRANCHING TYPES OF NORWAY SPRUCE

3.1. The Modes of Shoot Formation

A tree runs through three modes and phases of shoot formation during its life cycle, primarily as a result of the increasing growth and changing the meristems' topography. These phases are proleptic (Rossmässler, 1863), regular (Marcet, 1975), and proventive (Büsgen and Münch, 1927; Braun, 1982; Hagemann, 1990), respectively, preventitious (Fink, 1983), proventitious (Liedeker et al., 1988; Gruber, 1990, 1992) shoot formations (Fig. 8.10). These phases differ in the duration of the resting time lying between leaf primordia induction or bud formation and their differentiation (= development) as leaves or shoots. This time interval will be defined as *phyllochron*. The phyllochron increases quite regularly with progressing ontogeny, which means increasing transport distances between the root and shoot pole. It can be beguiled artificially by reducing this transport distance, for instance, by grafting or crown pruning. Thus, the annual shoot morphology also changes during ontogenesis from very long mixed shoots (composed of pre- and neoformed parts or regular and proleptic shoots) in the "youth" to pure short shoots and preventitious buds in adult trees. Due to these different shoot formations and the dynamics of growth the whole crown architecture is changing continuously through

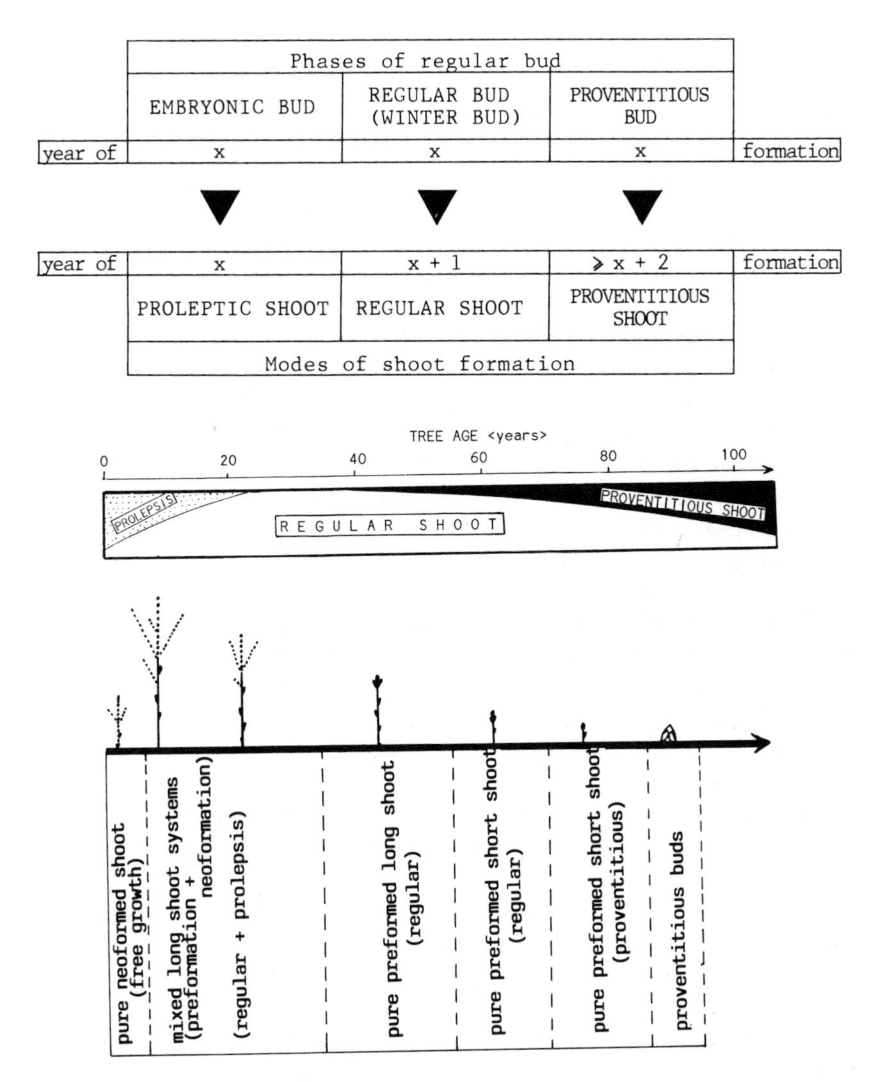

Fig. 8.10. Modes of shoot formation during the life cycle of a tree, for example, a Norway spruce.

the tree's ontogeny. This finding is the normal growth plan by which the spruce crown tries to increase or keep its foliage biomass constantly on a high level in spite of the fluctuating environmental factors. In this way the tree crown has a high morphogenetic and morphological flexibility and adjustment capacity. Furthermore, when evaluating the crown architecture of a tree we have to take into account three components determining the phenotype: the genetic factors, the environmental factors, and the ontogeny of the tree.

The concept of bud–shoot differentiation, which is based upon the bud–shoot differentiation process (the normal and regular growth modus on temperate woody plants with winter bud formation), is used in the following three formations (see Figs. 8.10 and 8.11; Gruber 1989):

Regular shoot formation (Figs. 8.10–8.12; see Marcet, 1975; Gruber, 1986, 1989) producing longer or shorter shoots in their dependence on topography (branching order, age, or longitudinal symmetry), takes place rhythmically every year from 1-year-old winter buds of temperate species. A regular winter bud is composed of the apex, the body of needle primordia, and the bud scales (Gruber, 1986, 1989). Because of this strong annual regularity of shoot differentiation it is called regular (see Marcet, 1975). Regular shoot formation can generally be defined as the shoot differentiation in year $x + 1$ from a winter bud, which is initiated and entirely developed in season x (Figs. 8.10 and 8.11). The phyllochron takes 1 year. Because the whole shoot is preformed as the body of needle primordia in the winter bud, this kind of differentiation is also called "preformation" or "fixed growth" (Pollard and Logan, 1974). It forms most of the trunk and the frame of the crown, and therefore is the core of the tree's architecture. It can be observed in the spruce crown in all phases of the trees' ontogeny.

The regular branching has the function to occupy and maximize the crown space and to defend the occupied crown space against crown expansion from neighboring trees. Therefore, regular shoot formation takes place at the periphery of the crown. Because of this promoted annual shoot formation at and near the shoot tips, this is called *distal directed shoot innovation* (Gruber, 1987). Note that the regular shoot differentiation can also be used for regeneration.

One modification of the regular shoot formation is the precocious flushing of developing winter buds well known as *prolepsis* in the European and international literature (Rossmässler, 1863; Späth, 1912; Troll, 1948; Kozlowski, 1964, 1971; Marcet, 1975; Guédès, 1979; Müller-Doblies and Weberling, 1984; Gruber, 1989, 1990, 1992; Hagemann, 1990).

Proleptic shoot formation is defined as the shoot differentiation in season x from a meristem or developing bud, which is initiated at the beginning of the same season x (Figs. 8.10 and 8.11). The phyllochron can last from few days to several months. This mode of differentiation or growth by spruce is an anticipated shoot differentiation and a mode of "neoformation," because all needles are neoformed in season x.

From the point of view of morphogenesis (knowledge of the morphogenetic cycle of the bud must be included; see Fig. 8.11) the proleptic shoot differentiation can be subdivided in early and late prolepsis (early and late neoformation). Early prolepsis is the early start of differentiation of needles and shoots early in the seasons respective morphogenetic bud cycle without first forming bud scales and resting needle primordia. Thus the first leaf primordia produced by the active apex do not differentiate as bud scales but as needles. Therefore, early proleptic shoots of *Picea abies,* which can be formed not only terminally (a synonym for terminal early prolepsis is "free growth": see Jablanczy, 1971; Pollard and Logan, 1974) but also lateral (syllepticly = "lateral free growth"), do not have any bud scales and compressed internodes at their bases (Fig. 8.13). Sylleptic shoots on *P. abies* are therefore special kinds of early proleptic shoots; they are lateral early proleptic ones.

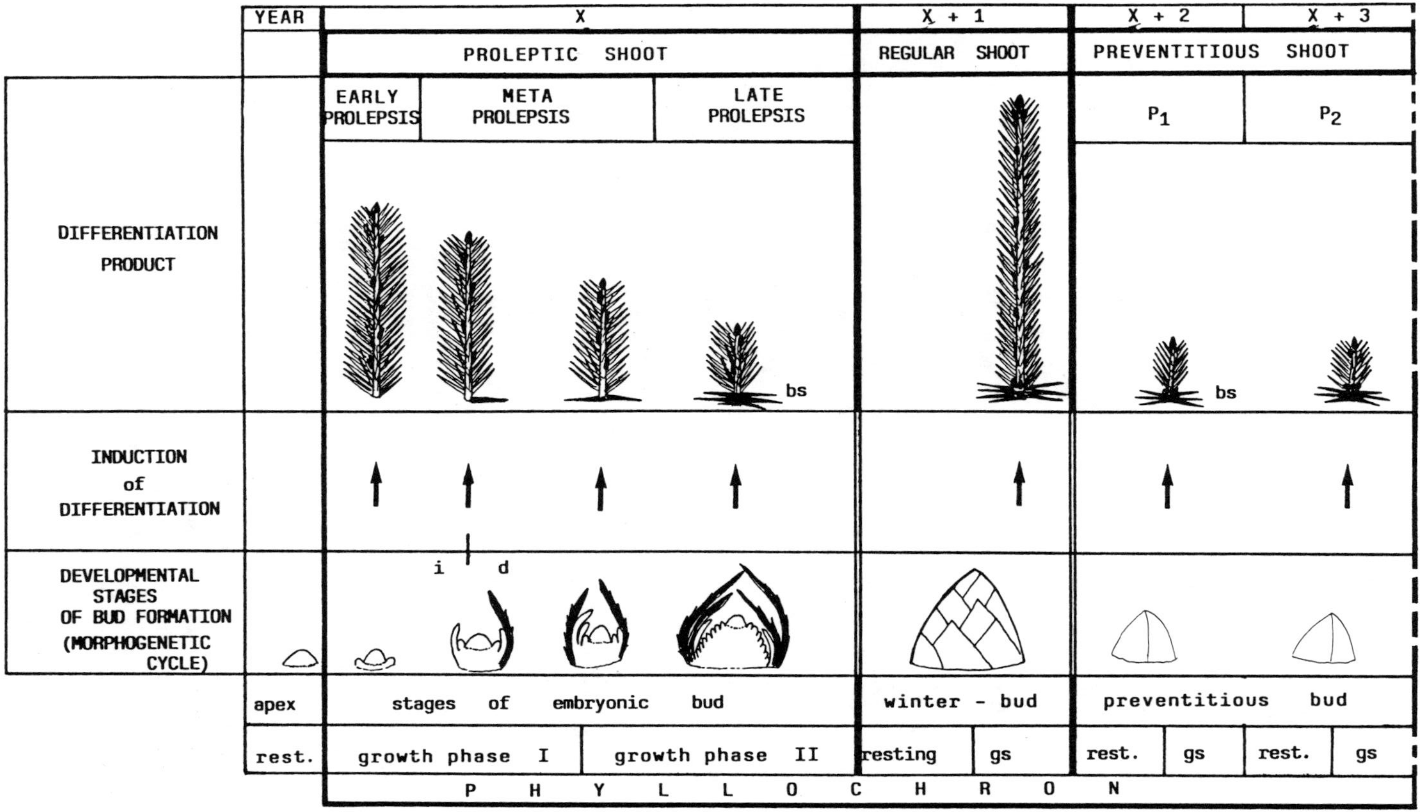

Fig. 8.11. Relations between the morphogenesis of buds (morphogenetic cycle), the induction of differentiation, shoot differentiation and morphology (where bs = bud scales; gs = growing season; d = definitive bud scales; and i = indefinitive bud scales).

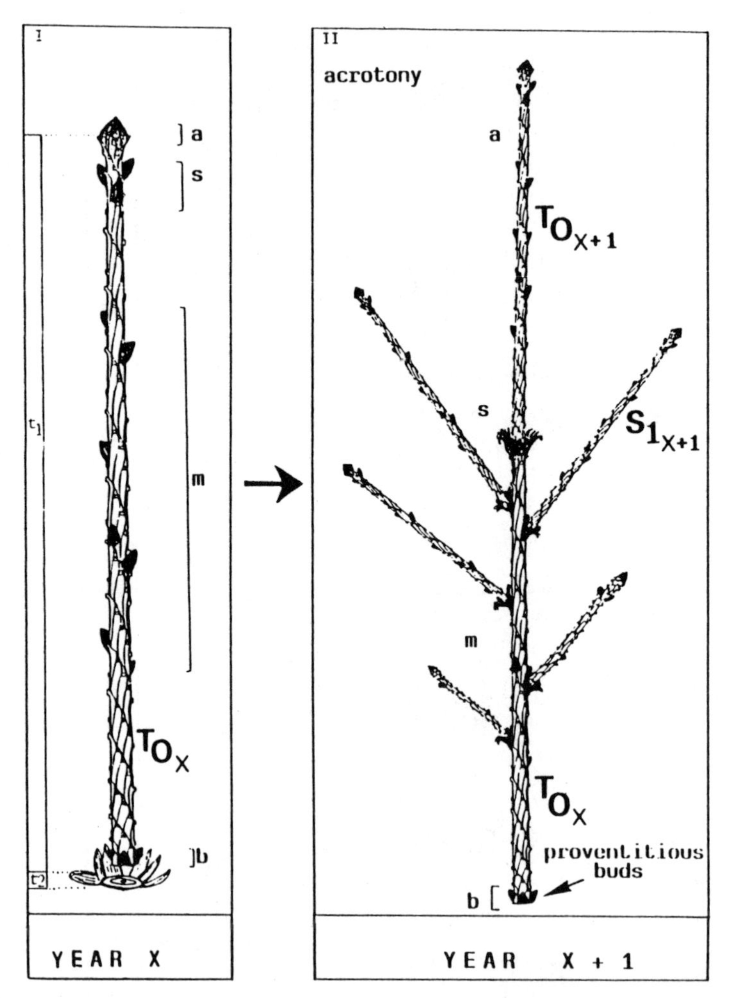

Fig. 8.12. Regular shoot formation and acrotonous branching (where a = apical, s = subapical, m = medial, b = basal, t_1 = elongated shoot portion of needles' insertion; and t_2 = compressed shoot basis of bud scales' insertion).

Late proleptic shoots on *Picea* first differentiate several or many bud scales followed by needle differentiation. The induction of needle differentiation is later in the season. Between early and the late proleptic shoot differentiation, or winter bud formation, all intermediate forms dependent on the developmental stage of the bud can occur (mata-proleptic shoots), namely, longer proleptic shoots with single or few bud scales at the base and short proleptic shoots with several bud scales (Fig. 8.11).

Proleptic shoot formation, which only appears in saplings and young trees of *Picea* and in stands with optimal growth conditions, develops from embryonic

Fig. 8.13. Early proleptic (terminal and lateral) branched leader of a young Norway spruce.

(undeveloped) buds (Gruber, 1986, 1989). This mode, which also depends on the genotype (Wühlisch, 1984; Gruber, 1986), allows a fast crown expansion and a quick regeneration of directly damaged or lost parts of the crown.

Due to shorter growing seasons at upper altitudes high-land provenances of spruce form no or only a negligible amount of proleptic shoots (see Holzer, 1977–1981). Therefore high-land provenances have a better frost resistance than low-land types, because the proleptic shoots, which also include "lammas shoots" (Kozlowski, 1964, 1971) are more prone to frost damage than regular shoots. By following the terminology used by Späth (1912) proleptic and lammas shoots differ only in their strong periodic differentiation; the first are controlled more exogenously, the second more endogenously.

In many cases of young vigorous trees the whole annual shoot system of leaders is a long mixed shoot consisting of preformed (regular shoot portion) and neoformed (terminal and lateral proleptic) portions. Such long mixed leader shoots often grow crooked (Fig. 8.14*a*) and become bushy (Fig. 8.14*b*) due to rapid growth and additional precocious (proleptic) branching.

With increasing tree age, preventitious shoot formation becomes more and more important, because the proportion of these shoots increases. *Preventitious shoot formation* is defined as the shoot differentiation that appears earliest in the season $x + 2$ (or later: $x + n$) of a suppressed so-called preventitious bud, which was initiated and developed in season x (Figs. 8.10, 8.11, and 8.17). Compared to a well-developed regular winter bud the preventitious bud is only composed of the apex enclosed in bud scales; the body of needle primordia, which would form the stretched shoot portion carrying the needles, is not developed. Not until before the preventitious starting shoot (first shoot emerging from the preventitious bud) grows out does the preventitious bud develop into the activated preventitious bud, which has the same structure as the regular winter bud except it is smaller (Figs. 8.17 and 8.18). The phyllochron is longer than 1 year. Preventitious shoots emerge from preventitious buds mainly at the base of annual shoots and on the upper side (epitonous) of horizontally oriented axes (Figs. 8.15, 8.18, and 8.19). Thus older ineffective twigs and branches can be replaced by younger effective ones. This

Fig. 8.14. Curvature (*a*) and bushing (*b*) of the leader due to an above average length growth and early proleptic branching.

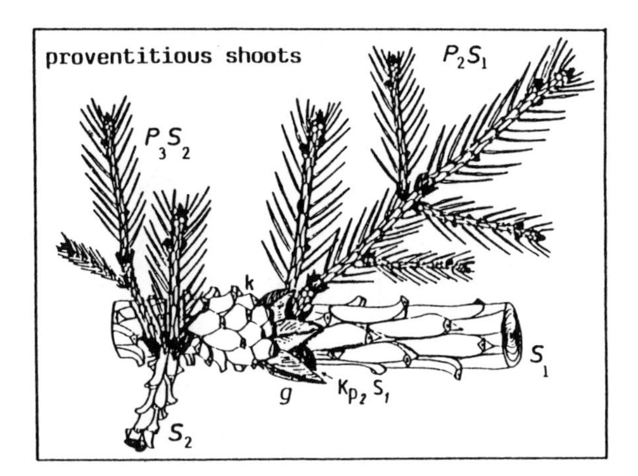

Fig. 8.15. Preventitious shoot formation: Preventitious shoots develop epitonous mainly from basal buds of shoots of the first and second order [where S_1 = regular shoot of first order; S_2 = regular shoot of second order; P_2S_1 = preventitious shoot of second order on S_1; P_3S_2 = preventitious shoot of third order on S_2; g = annual shoot boundary (bud scales); k = compressed needle cushions (Nadelkissen); and Kp_2S_1 = preventitious bud of second order].

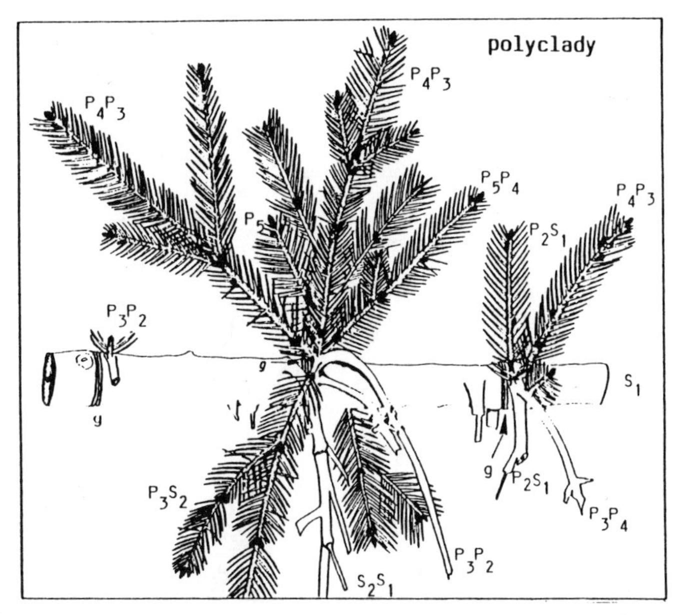

Fig. 8.16. Polyclady: Clusters of preventitious shoots of different age [where S_2S_1 = dead regular twigs of second order; P_2S_1 = preventitious twigs of second order; P_3P_2 = preventitious twigs of third order; P_4P_3 = preventitious twigs of fourth order; P_5P_4 = preventitious twigs of fifth order; and g = boundary between two annual length increments (shoot base scars)].

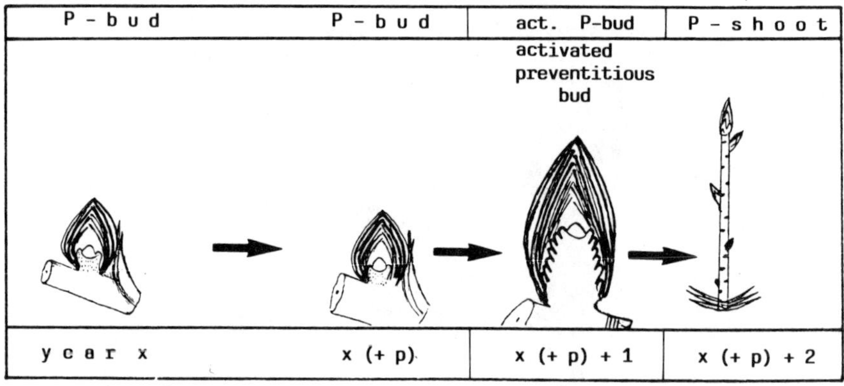

Fig. 8.17. Development from a preventitious starting shoot (where p = number of resting years as preventitious bud and P = preventitious).

continual substitution of ineffective branches by preventitious shoots and twigs, which occurs as *branching in successive series* (Figs. 8.18, 8.20; Veillon, 1976; Edelin, 1977; Gruber, 1987), is quite a normal process in vital crowns (see Rehfuess, 1983; Mohr, 1984; Gruber, 1987). The relocation of young shoots directly on the axes of lower order is ecologically very important and can be called *proximal starting shoot innovation* (Gruber, 1987). In this way it is possible to establish a highly efficient needle biomass in the inner part of the crown.

Crowns that suffer permanently from high biotic or abiotic (especially mechanical) stresses, such as wind, ice, snow, and frost, show a characteristic branching

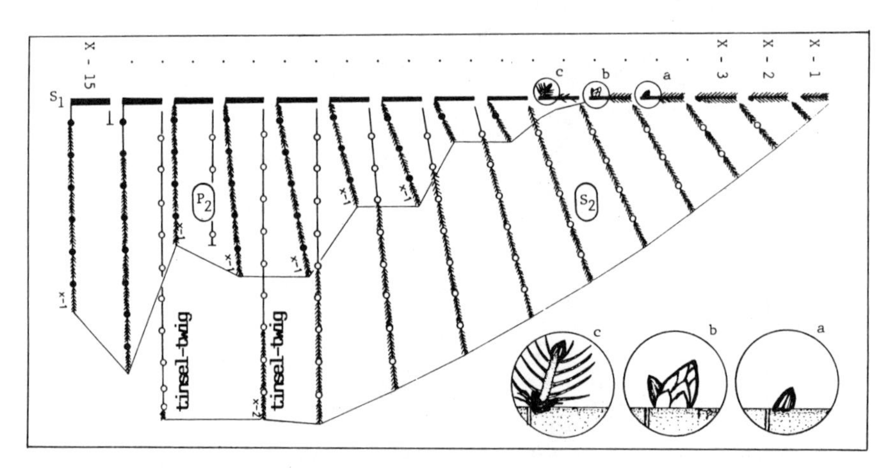

Fig. 8.18. Branching mode of a bough of the comb-spruce: Branch of the first order with two successive series S_2 and P_2 (where S_1 = regular branch axis of first order; S_2 = regular branch axis of second order; P_2 = preventitious branch axis of second order; a = preventitious bud; b = activated preventitious bud; and c = preventitious initial shoot).

Fig. 8.19. Orthotropically growing preventitious shoots emerging mainly from basal buds on the upper side of shoot axes of the first order and induced by cutting-off the branch tip (arrow).

pattern, the *polycladies* (Figs. 8.16 and 8.21). These are clustered preventitious shoots of different ages.

Preventitious (and also embryonic) buds are activated, if the correlation mechanism of suppressing shoot and leaf differentiation gets out of balance. A possible mechanism controlling this process can be explained by the auxin–cytokinin theory (Sachs and Thimann, 1967; Philips, 1975; Bryan and Lanner, 1981). An increasing preventitious (and proleptic) shoot formation would start according to this theory due to a promoted growth of roots (above average production of cytokinin) and a loss or injury of effective branches (below average production of auxin).

Beyond the ecological importance of preventitious bud formation, which is the result of strong acrotony, the spruce crown possesses a highly developed repair and

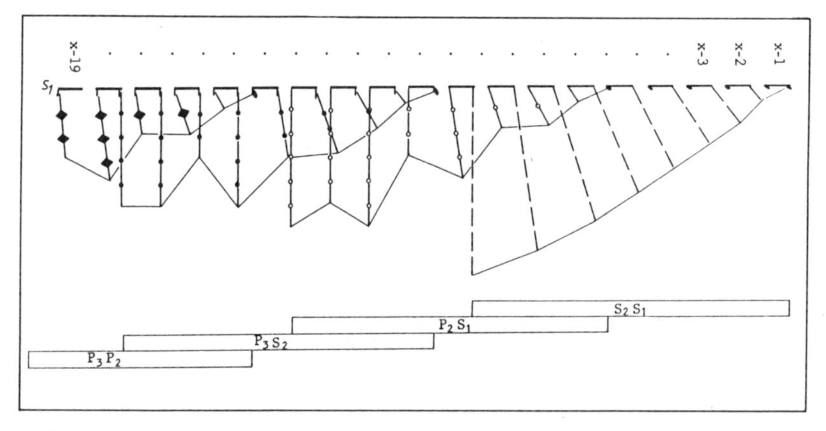

Fig. 8.20. "Branching in successive series" on Norway spruce (comb type). Branch with four successive series: one reglar (S_2S_1) and three preventitious series (P_2S_1 - P_3S_2).

Fig. 8.21. "Polycladies": Excessive cluster formations of preventitious shoots.

regeneration system for replacing biotically and abiotically damaged or lost efficient needles and branches. This system can also compensate senescent needles and those needles, twigs, and branches weakened by reproductive growth. Of further importance is the great flexibility of the crown structure and the enormous architectural adjustment to environmental factors, which preventitious shoot formation confers. In general, the phenotypic plasticity or flexibility and regeneration capacity of the spruce crown is very high.

Preventitious shoot formation is a widespread biological feature in most woody plants. Also, in all species of *Picea* investigated (e.g., *P. asperata, P. bicolor, P. brachytyla, P. breweriana, P. engelmannii, P. glauca, P. glehnii, P. mariana, P. likiangensis, P. neoveitchii, P. omorika, P. orientalis, P. pungens, P. rubra* (Liedeker et al., 1988), *P. schrenkiana, P. sitchensis, P. smithiana,* and *P. wilsonii*) this modus is more or less developed, as it was found by the author in younger trees (~ 20 years old) standing in arboreta in Göttingen. *Picea sitchensis* and *P. wilsonii* also can produce preventitious shoots on the trunk similar to the well-known "water sprouts" of *Abies* species.

The terminology and use of prolepsis and other forms of shoot differentiation given by Tomlinson and Gill (1973), Hallé et al. (1978), Wheat (1980), Tomlinson (1983), and Powell (1987), for example, is not adopted and must be rejected. After this definition, all shoots that derive from buds after a resting phase (e.g., from summer or embryonic buds, winter buds, and also from several year old preventitious buds) would be called proleptic. But its use in this sense is semantically uncorrect and cannot be used for the following reasons:

First, the original definition of "prolepsis" used by famous botanists and foresters (Rossmässler, 1863; Pax, 1890; Berthold, 1904; Späth, 1912; Büsgen and Münch, 1927; Goebel, 1933; Troll, 1937, 1948; Scharfetter, 1953; Kozlowski, 1964, 1971;

Marcet, 1975; Guédes, 1979) for a anticipated (e.g., second summer flush) is inadmissibly modified and violated (Müller-Doblies and Weberling, 1984; Gruber, 1986, 1989; Hagemann, 1990). The literal sense of the word is perverted from anticipation to delay in the sense of subsequent differentiation. The meaning of the term prolepsis, which comes of the Greek and means "anticipation or precocious shoot formation" only makes sense when we are referring to the behavior of the regular shoot formation (spring flush) of temperate woody plants as originally used (see Fig. 8.11). Regular and preventitious shoots emerging from resting winter buds cannot be called proleptic because they do not show anticipated differentiation. But, if we refer differentiation with resting periods to the behavior of a continuously growing tropical plant as Hallé et al. (1978) did, then we have to define the modified differentiations again with correct phylological terms, but in the sense of delayed (subsequent) and not of anticipated development.

Second, the absence of bud scales as a morphological sign for syllepsis is not given in any case. There are many species like *Tilia* or *Salix* that do not form sterile bud scales. Therefore, regular and preventitious shoots of these species also do not differentiate sterile bud scales at their bases; nevertheless, they are not sylleptic shoots. We see the same problem in *Araucaria, Taxus, Cryptomeria, Sequoiadendron, Thuja,* and other species that can develop delayed regeneration shoots from resting (latent) meristems of many years without forming bud scales (see Fink, 1984). Although remaining as resting meristems for several years, the shoot differentiation is neoformed because they do not develop resting leaf primordia.

Third, the use of the terms prolepsis and syllepsis implies that a big contrast exists between both terms. In reality, when we look at the phenomena from the morphogenesis of the bud and shoot differentiation and from a physiological point of view, if we include the knowledge of the morphogenetic cycles of buds and shoots our understanding of the different shoot forms and morphologies becomes very clear (see Romberger, 1963; Gruber, 1986, 1989, 1992), syllepsis and "lateral prolepsis" (in the classical sense) are only temporal graduations in leaf differentiation, that is, elongations of developing buds (embryonic buds) in different stages of bud development. These differences between an early and late induction are best defined and become easier to understand by using the terms early and late prolepsis. Then, early prolepsis, which includes syllepsis as synonym of lateral early prolepsis, is the induction of leaf differentiation early in season x, late prolepsis is the late induction in season x, regular shoot formation is the differentiation (development) in season $x + 1$, and preventitious shoot formation is the differentiation of leaves earliest in season $x + 2$ (see Fig. 8.11). Furthermore, no discrepancy exists between terminal and lateral differentiations, such as syllepsis and free growth or prolepsis and lammas shoots.

By using this system of terminology for shoot differentiation (see Figs. 8.10 and 8.11) we have two advantages: first, we have the morphogenetic, morphological physiological, and temporal continuum of the different shoot formations; second, we have the direct relation to the trees' ontogeny, adaptation strategy, and ecological reaction.

3.2. The Construction of Older Spruce Crowns

In Norway spruce we have to distinguish three basic branching types, which are based upon genetic differences. These types, which differ in the arrangement and position of the twigs of the second order, are the comb- (Fig. 8.22), brush-, and platelike spruce (Fig. 8.23) (Sylvén, 1909; Hassenberger, 1939; Heinkinheimo, 1920; Priehäuser, 1958; Rubner, 1936, 1943; Hofmann, 1968; Schmidt-Vogt, 1972, 1977, 1986).

The construction of the crown of a comblike spruce (cf. Bindseil, 1933) in a closed stand, for example, can be demonstrated as an ideal normal crown model. This model shows (Fig. 8.24) that the branching form changes from the top to the base of the crown simply by changing the proportions of regular and preventitious shoots and their arrangement. Thus, in principle, we have to differentiate between the juvenilelike branching form at the top of the crown, the comblike form, and the brushlike form in the middle and the platelike form at the base of the crown. Below the living crown is the zone of dead branches. The formation and extent of such a branching zone within a crown depends on environmental factors and genotype. Even within a single crown the slight modification and adjustment capacity of the

Fig. 8.22. Genetic differences within the branching phenotype: About 25-year-old grafts from about 250–300-year-old autochthonous Harz mountain spruces on the seed orchard Escherode: comb type.

Fig. 8.23. Genetic differences within the branching phenotype: About 25-year-old grafts from about 250–300-year-old autochthonous Harz mountain spruces on the seed orchard Escherode: plate type.

comblike spruce can be seen due to these different branching zones. In this case the changing of the branching architecture from the comb form to the plate form is an adaptation to the light conditions.

The study and investigation of many spruce crowns show that the portions of preventitious shoots increase with increasing tree and branch age, exposition, and mechanical (abiotic and biotic) stress.

A detailed crown analysis of many spruces indicated that older spruce crowns can consist of 40 to nearly 100% of shoots with preventitious origin (Fig. 8.25). Spruces on sheltered and in closed stands have smaller portions of preventitious shoots than spruces on exposed and extremely thinned stands.

Thinned spruce crowns have a higher proportion of preventitious shoots than crowns without needle losses (Fig. 8.26). We will get this result if we refer the proportion of the preventitious shoots to the assessed needle loss of the crown. This finding clearly shows that either regeneration of the crown is not yet finished or is insufficient. If we refer the proportion of preventitious shoots on a crown without needle loss, then we get the trend shown in Figure 8.27. We see a maximum of preventitious shoots at 41.7% needle loss. This result means that to this extent of

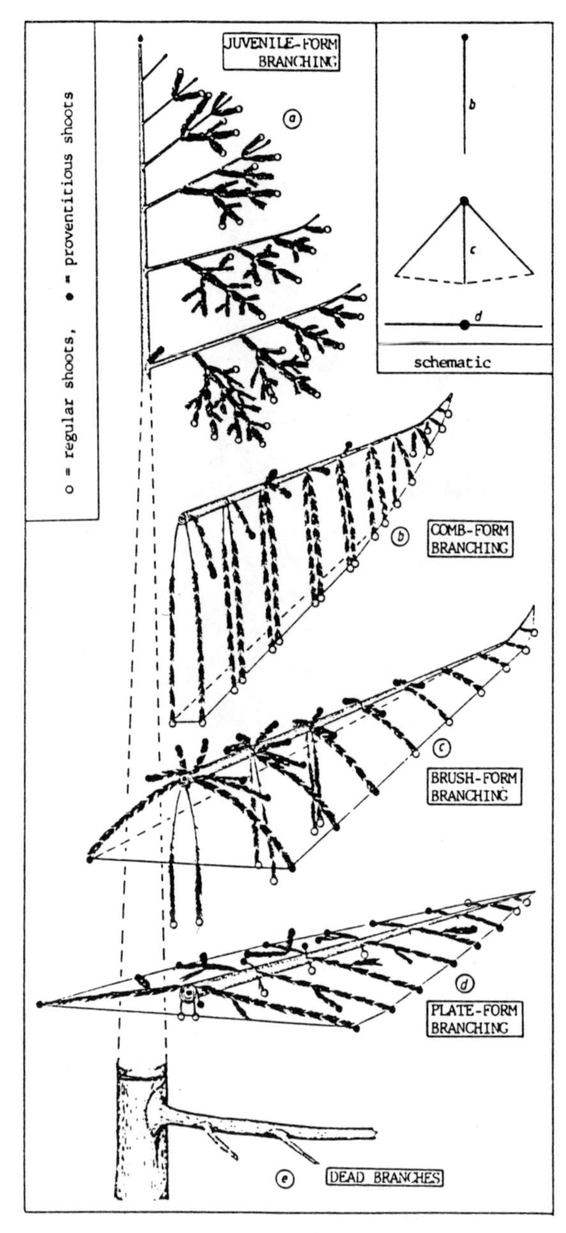

Fig. 8.24. Construction of the spruce crown (comb type within a closed stand) as an ideal normal model.

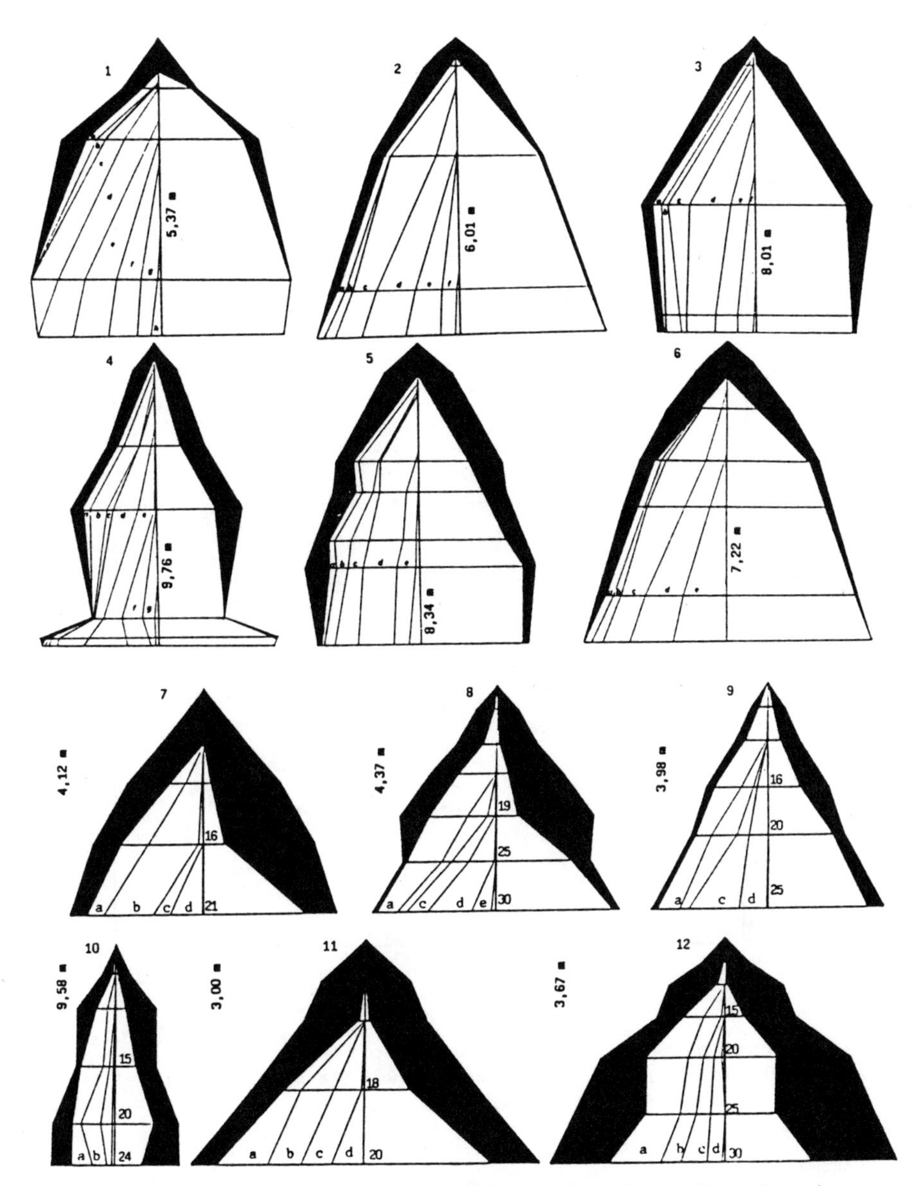

Fig. 8.25. Scale crown models: Distribution of living regular and preventitious shoots in spruce crowns. Left site: percentage of the number of living regular and preventitious shoots. Right site: percentage of the dry weight of living regular and preventitious shoots [where a = preventitious shoots of second order (P_2S_1); b = preventitious shoots of third order (P_3S_2); c = preventitious shoots of third order (P_3S_2); and g = preventitious shoots of seventh order (P_7P_6)].

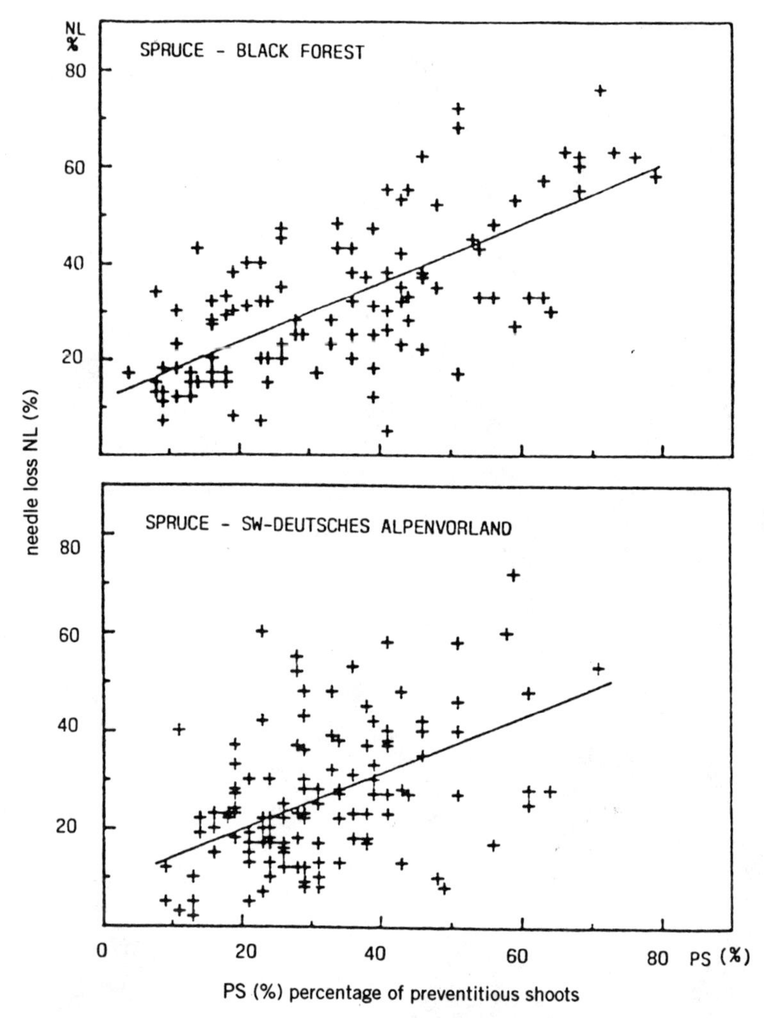

Fig. 8.26. Correlation between needle loss (NL) and assessed proportions of needles on preventitious shoots (PS) within the light crown of Norway spruce (Inventory 1988 in Baden Württemberg): Assessed proportion of PS referred to the needle statement of the bough.

needle loss a regeneration of the crown via preventitious shoots may be possible; higher needle loss causes the gradual degeneration of the tree (cf. Fig. 8.50: right and left path) because the regeneration is insufficient.

Detailed analyses of the age of preventitious shoots from boughs of the light crown showed that an increased preventitious shoot initiation had started in 1984, very likely as a consequence of the strong droughts in 1982 and 1983 (Fig. 8.28).

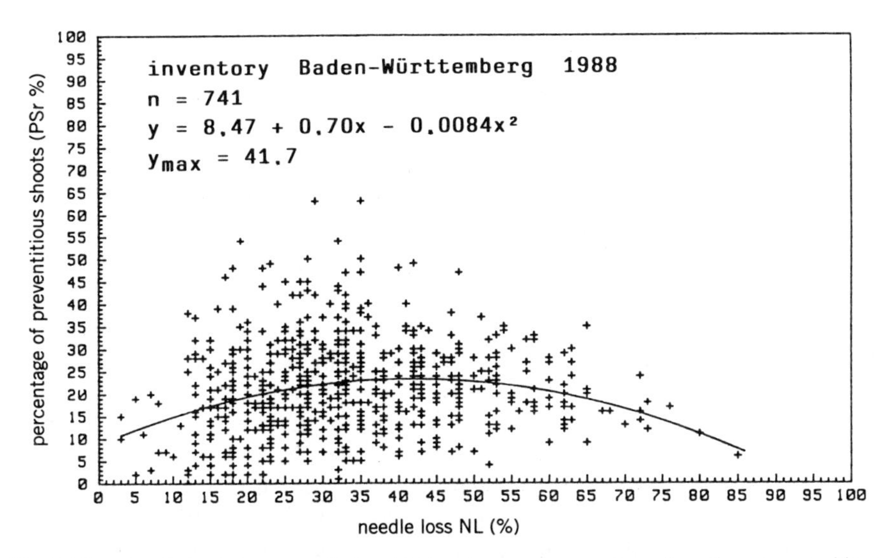

Fig. 8.27. Correlation between needle loss (NL) and assessed proportions of needles on preventitious shoots (PS_r) within the light crown of Norway spruce (Inventory 1988 in Baden Württemberg): Assessed proportion of PS_r referred to a bough with 100% needles: [PS_r = (100 - NL)/100 × PS].

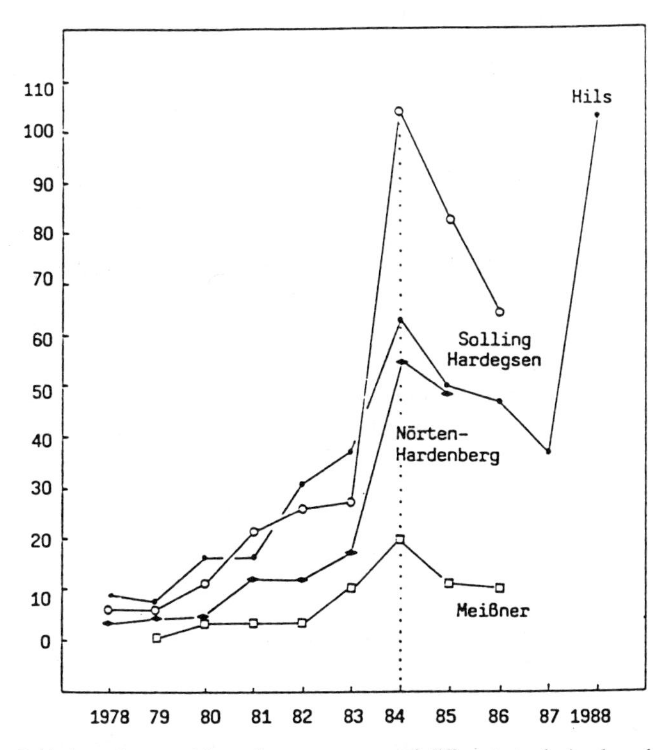

Fig. 8.28. Initiation of preventitious shoots on spruce of different stands (on branches of the light crown).

4. MODIFICATION AND PLASTICITY OF SPRUCE BRANCHING

Due to the restraints of the stand and because trees are stationary, a tree must have the capability of adjusting continuously to the permanently varying influences and changes of environmental factors. In particular, those tree species that live in extreme habitats with high, permanent exogenous, and above all mechanical stress, must possess a good adjustment strategy and regeneration or repair capacity. One important architectonic element is the preventitious shoot. Through these means the branching form can be modified secondarily, as well.

Thus, for example, the comblike spruce is able to change its original (primary) form into the unregular comblike, brushlike, bandlike, pipelike, or platelike branching form (Figs. 8.29 and 8.30*a–d*). This occurs when the environmental factors of the tree change suddenly or permanently in one direction. When the modifying stress abates, the tree crown is able to reform to the habit of its original branching form, which is the defined genetic blueprint (Fig. 8.29). Therefore, the plasticity of spruce crowns, especially that of the comblike spruce, is very high. The comblike spruce (comb type) must be called an unstable or flexible branching type with a wide architectural reaction capacity. It can be modified from the irregular comb form to the band form. In contrast to that the platelike spruce is a stable branching type with a narrow branching reaction capacity (Fig. 8.31).

The variability of the branching and crown phenotype of Norway spruce does not only depend on genetic differences but can also be determined by the influence of environmental factors (Fig. 8.31). Therefore, we cannot determine the genotype from the phenotype in all cases. This knowledge is very important for the stability of forests and for a sensible ecological and economic spruce silviculture.

5. THE HYDRAULIC ARCHITECTURE OF NORWAY SPRUCE

The hydraulic design of trees influences the movement of water from roots to needles (Tyree and Ewers, 1991). Thus different designs in different tree compartments (roots, stem, crown axis, branches, or twigs) could have different specific xylem conductivities, which can lead to different phenological features and symptoms because of different water and mineral supply. In Section 5.1 some aspects of the relationship between needle mass and conducting tissue, which may have an influence on the crown phenotype, may cause specific thinning symptoms, and are of interest for assessment of yield, are shown.

5.1. Relationships Between the Sapwood Area and Tree Compartments

Many investigations have shown (Shinozaki et al., 1964 a and b; Whitehead et al., 1984; Kaufmann and Troendle, 1981; Waring, 1980; Waring et al., 1982; Kaibiyainen et al., 1986; Oren et al., 1986; Eckmüllner, 1990) that there are strong correlations between the sapwood area and needle mass or needle area.

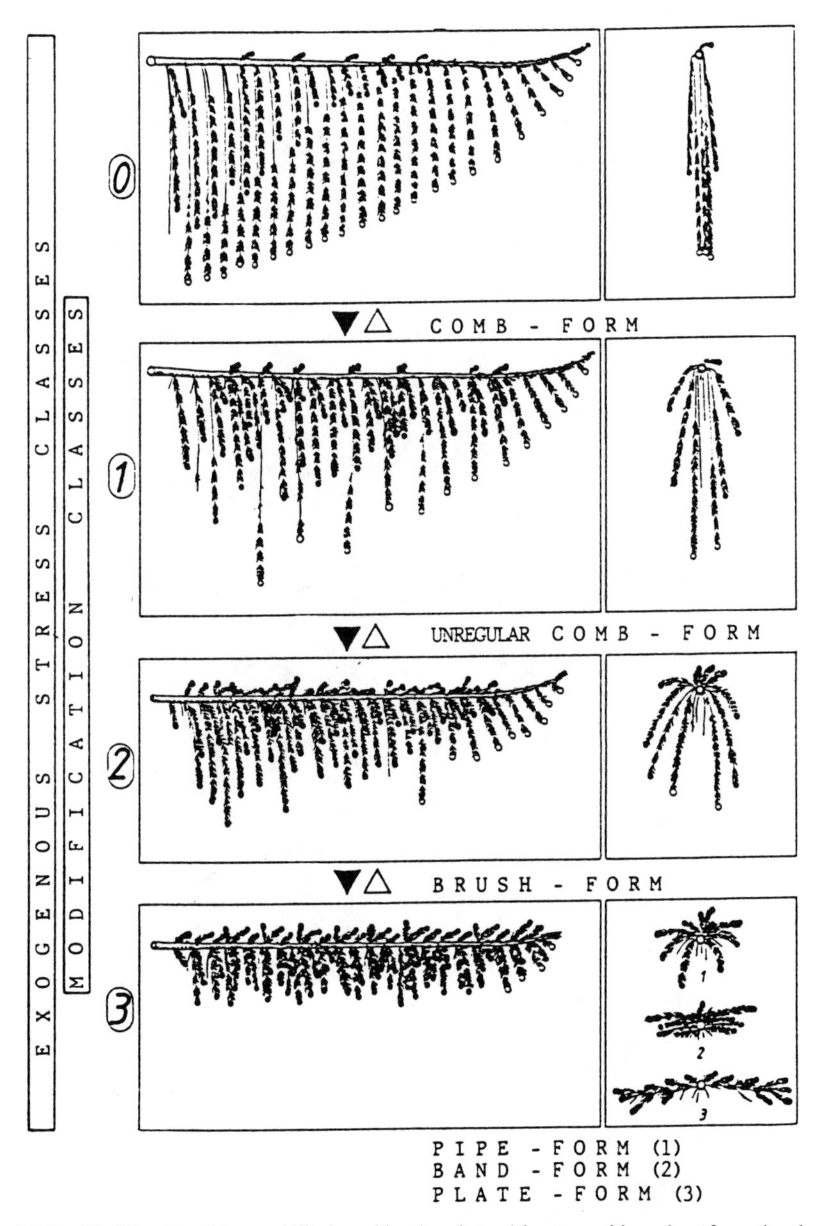

Fig. 8.29. Modification of the comblike branching (comb-type) by preventitious shoot formation due to exogenous stress factors. The branching form can change from form 0 to 3 under increasing stress, but can also recover by preventitious shoots from 3 to 0, when stress abates.

Fig. 8.30. Modification of the comblike branching type of Norway spruce. (*a*) "Branching in successive series" of the comb-type (the regular branching triangle is largely complete). (*b*) Substitution of the lost regular branches by a strong formation of preventitious shoots (unregular comb-form). (*c*) Entirely substitution and recovery by preventitious shoots (plate form). (*d*) Entirely substitution and recovery by preventitious shoots (pipe form).

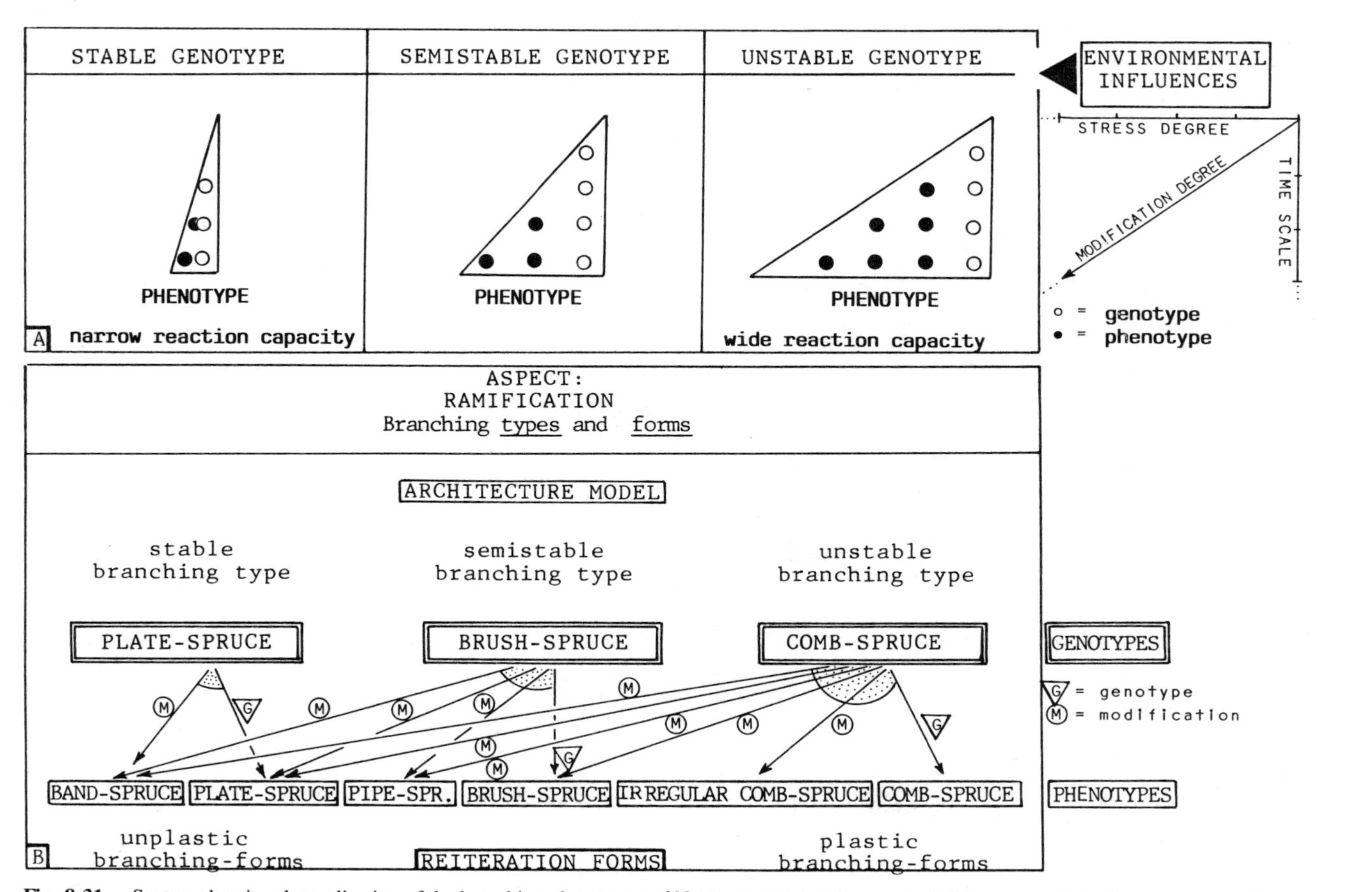

Fig. 8.31. System showing the realization of the branching phenotypes of Norway spruce. (A) = general diagram and (B) = branching phenotypes of Norway spruce.

Investigations of the relationship between sapwood area (in breast height and at crown base) and needle dry weight in 17–45-year-old spruce stands with and without crown thinning symptoms showed that there are more or less strong correlations between these parameters. However, the regression line changes with tree age (Fig. 8.32), and therefore the needle sapwood (g cm^{-1}) relationship increases, and the Huber value (see Section 5.2 for the sapwood needle relationship) decreases with tree age (Table 8.1). In addition, other relationships between sapwood and, for example, above-ground dry-biomass (Fig. 8.33), coarse root dry biomass (Fig. 8.34), and between brushwood and needle dry mass (Fig. 8.35) with high correlation coefficient could be found in the two 17-year-old stands.

Analogous to the construction of a river system is the construction of the water conducting system of the leafy crown and of the root crown. Figure 8.36 shows that there is a strong correlation between the sapwood of the stem and the sum of the sapwoods of branches above the stem disk. However, the correlation between the sapwoods of branches and the corresponding needle mass is not as strong as the sapwoods of the stem.

5.2. The Huber Values Along the Crown: Patterns in the Trunk and Branches

The xylem conductivity is dependent on the area of the cross-sectional conducting xylem (sapwood), the size and number of functioning tracheids, and the size and number of bordered pits. Huber (1928) devised a ratio (Huber value), which is the xylem cross-sectional area divided by the weight of the supported leaves [sapwood/needles, s/n in cm^2 g^{-1}]. These Huber values, according to the pipe model theory

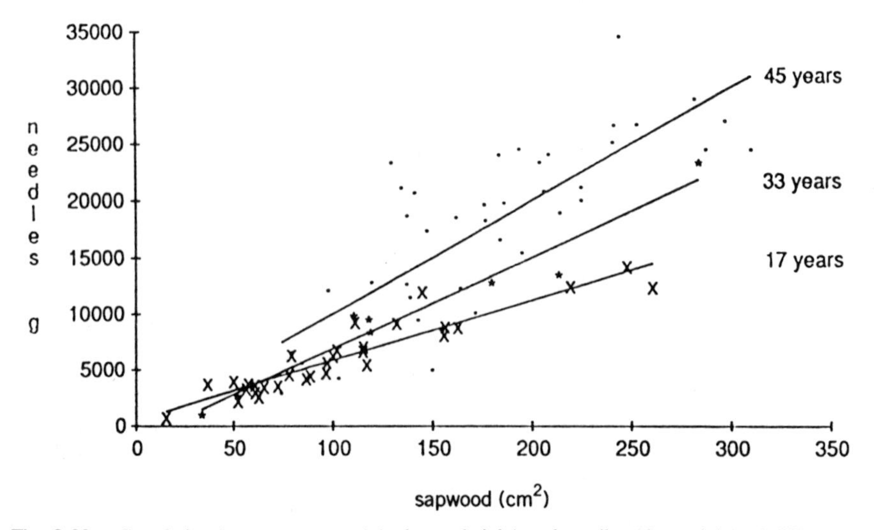

Fig. 8.32. Correlation between sapwood (at breast height) and needles (dry weight) of different old spruce in different stands.

TABLE 8.1. Regressions of the Relationship Between Sapwood (in BH) and Needle Dry Mass (Compare Fig. 8.32)[a]

	Age	Regression	Corr Coefficient	n/s ($g\ cm^{-2}$)	s/n ($cm^2\ g^{-1}$)
(1)	17	$y = 53.9\,x + 497.9$	$r = 0.93^{***}$	56–64	0.016
(2)	33	$y = 81.5\,x - 1241.8$	$r = 0.98^{***}$	56–77	0.015
(3)	45	$y = 100.5\,x - 26.4$	$r = 0.80^{***}$	100	0.010

[a] 1 = F65 and F103, Hils; 2 = Ebergötzen, 3 = R (Reinhausen); F1, F3, and F9 (Lange Bramke).

of tree form (Shinozaki et al., 1964a,b), should be constant throughout the tree, because this theory states that a unit of foliage requires a unit cross-sectional area of active pipes for its physiological and mechanical support. Huber (1928) and others (Ewers and Zimmermann, 1984) already reported that this value increased towards the top of the tree's trunk and that the trunk had much greater values than the lateral branches in *Picea* and *Abies*.

Similar features could also be found in ca. 45-year-old spruces of four different stands. The needle dry mass and the n/s and s/n relationship line (Huber value) along the trunk of the crown is seen in Figure 8.37. The minimum of the Huber value line is just below the maximum of needle mass line. This means that there is a structural bottleneck at this point of the crown axis (a unit of sapwood has to support a maximum of needles or a unit of needles is supported by a minimum of sapwood). If water stress occurs in the tree it is easily imaginable that the needles are shed first at and above this point of the crown. This stress may result in the well-known

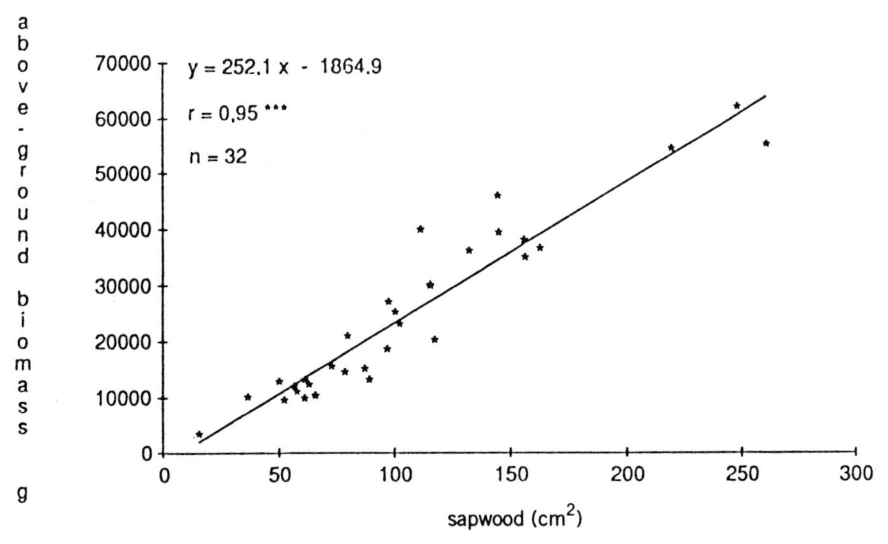

Fig. 8.33. Correlation between sapwood and above-ground biomass (dry weight) of 17-year-old spruces (F 103, F65, Hils).

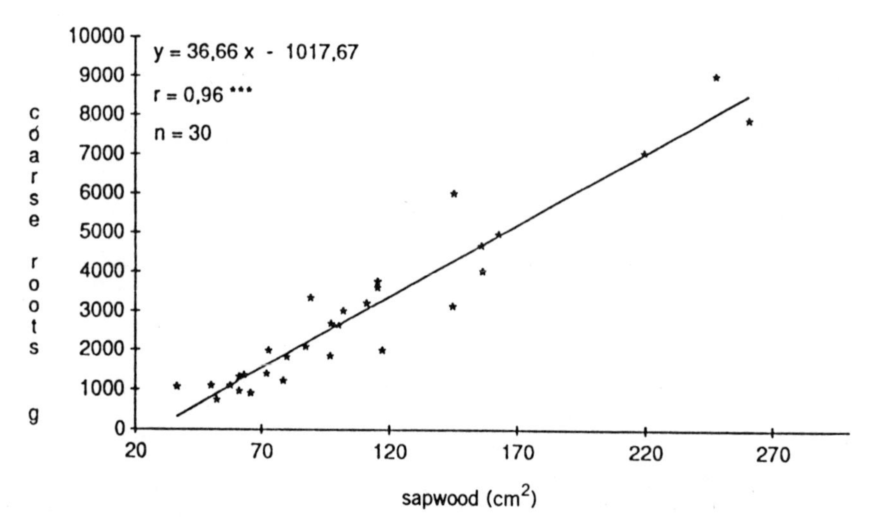

Fig. 8.34. Correlation between brushwood and needles (dry weight) of 17-year-old spruces.

symptoms of crown thinning from the inside to the periphery of the crown, stag-headedness (top thinning, top dying) or subtop thinning (subtop dying or window effect). Through this we have a direct causal relationship between a fundamental hydraulic feature and the reaction of the crown after water deficiency. Furthermore, we have a direct causal relationship between water stress and the operating mechanism of needle abscission.

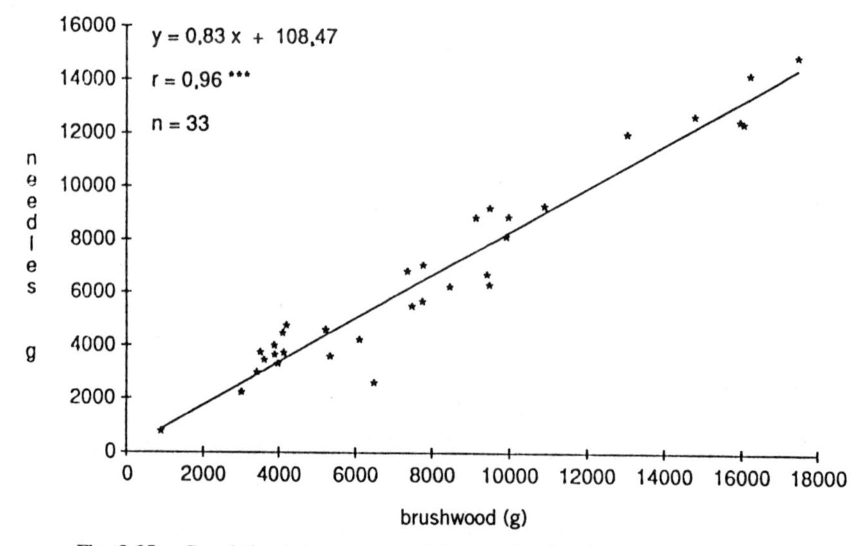

Fig. 8.35. Correlation between sapwood (at stem base) and coarse roots (dry weight).

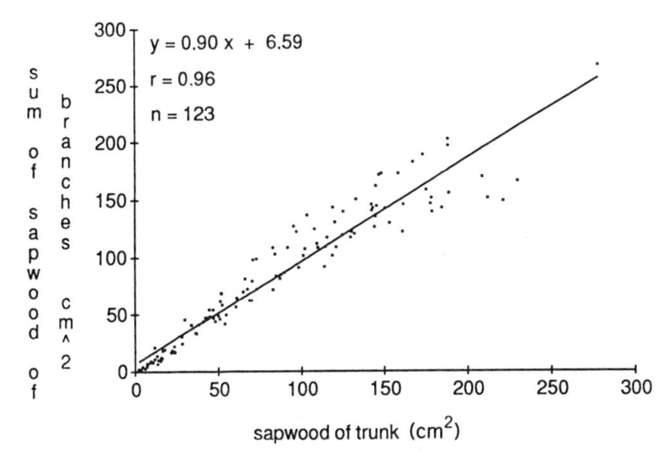

Fig. 8.36. Correlation between the sapwood of the trunk (at crown base) and the sum of sapwood of the branches (10-cm distance from the stem).

Figure 8.38*d* shows (exposed stand F9 on the ridge in Lange Bramke in the Harz mountains) how the tree is able to adopt by shifting its needle mass lower in the crown. This finding is ecologically very important and will be done by the proximal innovation mechanism formatting preventitious shoots. The n/s line changes synchronously.

Huber (1928) and Ewers and Zimmermann (1984) reported that in any case the Huber values of the trunk are much greater than of the branches. However, in Figure 8.39 several patterns of the Huber values could be found along tree crowns. The author of this chapter is convinced of the narrow relationship between these different hydraulic patterns and special thinning symptoms due to water stress.

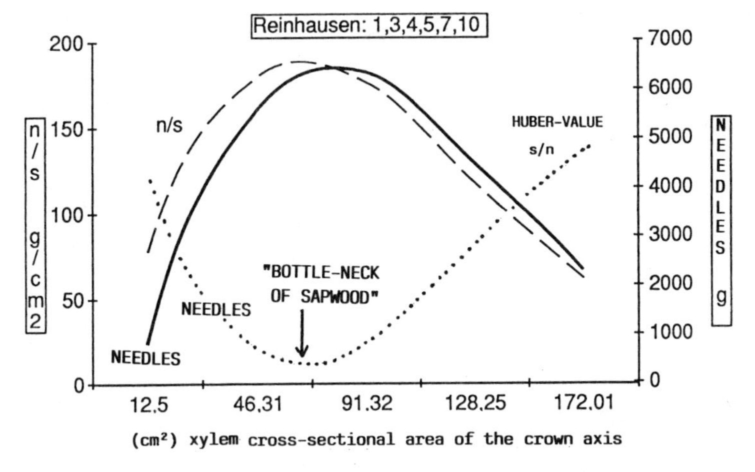

Fig. 8.37. Distribution of needles (*a*) and the n/s (needles/sapwood) relationship, respectively. Huber values (= s/n relationship) along the crown.

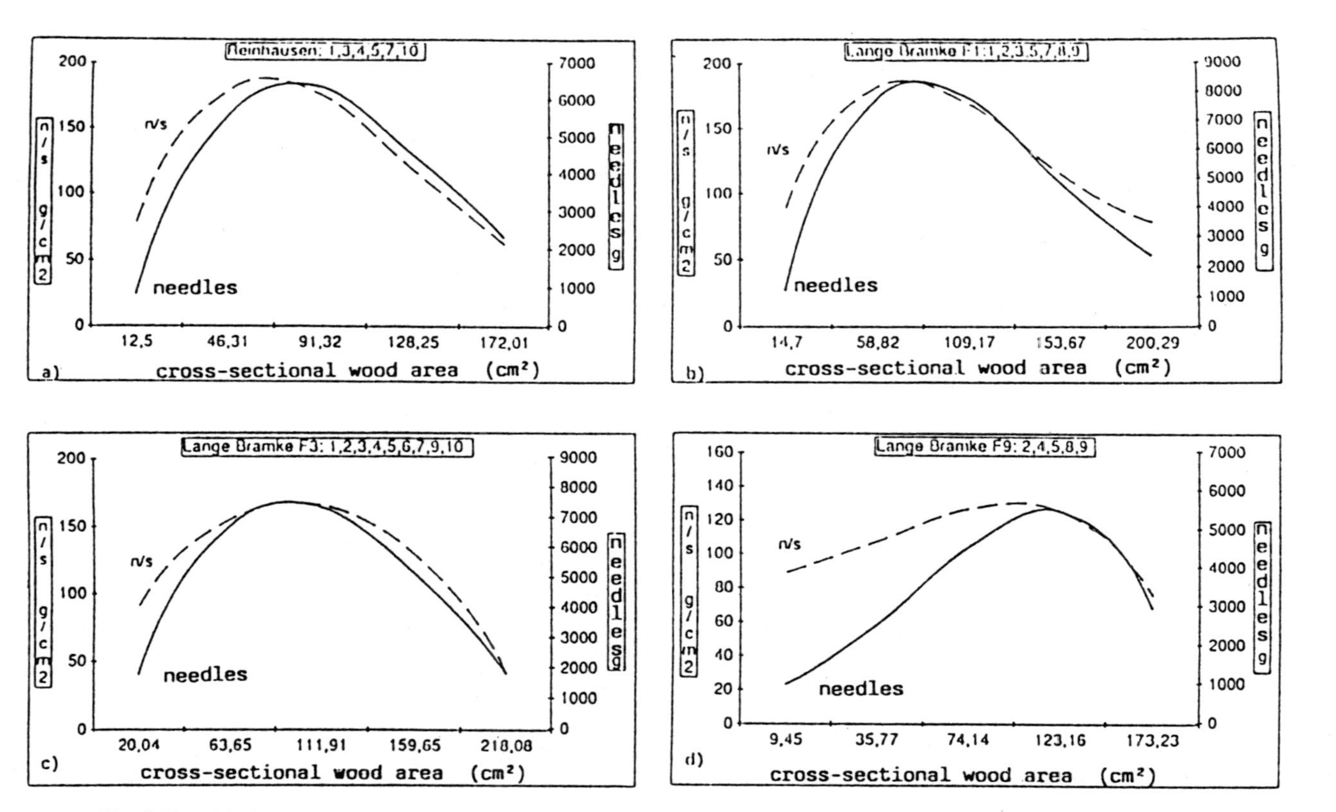

Fig. 8.38. Distribution of needles (right) and the n/s (needles/sapwood) relationship along the crown on four different stands.

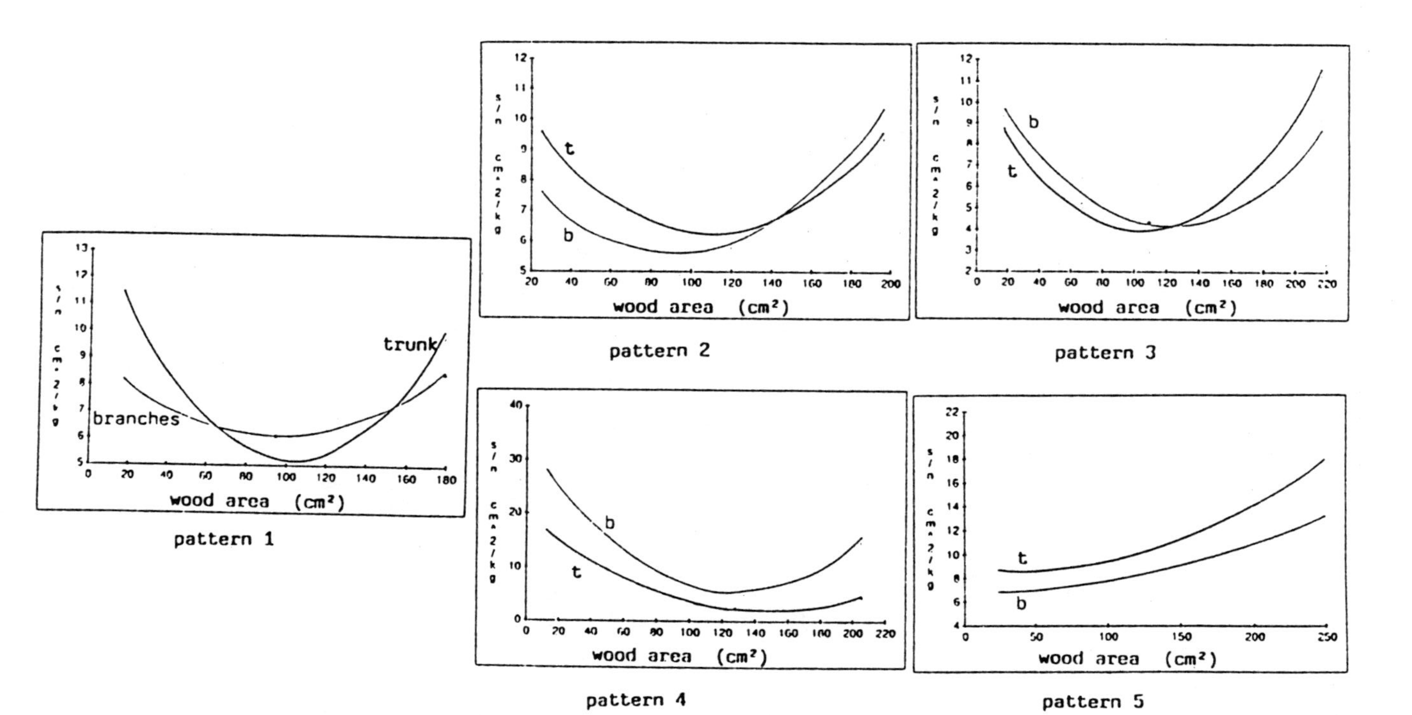

Fig. 8.39. Patterns of the distribution of the Huber values along the trunk (t) and in the branches (b) above.

6. THE ROOT ARCHITECTURE OF NORWAY SPRUCE

6.1. The Mechanisms of Root Differentiation

Contrary to the shoot the root does not form buds, where a part or the whole root increment developed per season is preformed. Therefore, it is impossible to determine the periodic length growth by outer morphological signs. Figure 8.40 shows the branching modus of an undisturbed growing root, and in comparison to this, the branching modus of a leader of *Betula* spec. is figured. The tip of the primary root grows geotropicly and monopodially by neoformed differentiation. Contemporaneous to this terminal differentiation, lateral distichous branching already occurs some millimeters or centimeters behind the root tip (neoformed syllepsis). Thus, the root is able to branch up to the fourth or fifth order within one growing season. The lateral root meristems develop endogenously just in front of the primary xylem rows that are mostly diarch constructed (Fig. 8.41). The primary xylem of the lateral roots is connected with the primary xylem of the mother roots (Fig. 8.42).

Root formation arises either from regularly (primary root meristems) or irregularly (secondary roots) induced meristems. The latter develops from the callus (Fig. 8.43) or cambial tissues (Fig. 8.44), or on shoots and stems (shoot born), and belongs to the category of *adventitious* roots.

The *regularly induced meristems* can be subdivided into three modes; these are roots, which develop from nonresting meristems (see Fig. 8.40), from short periodic resting meristems (short-term suppressed meristems), and from latent meristems (long-term suppressed meristems; Fig. 8.45*a,b*).

The first mode is the growth according to the genetic blueprint, that is, continuously differentiation and syllepticly branching as described above (distal root innovation; Fig. 8.40). This mode allows a rapid exploration of the rooting area. The second mode is exogenously controlled and occurs when the growth conditions are unfavorable (e.g., frost or drought). The third mode arises from latent meristems, which can be suppressed for several years. These meristems develop after a resting phase in the older proximal regions of woody roots (proximal root innovation; Fig. 8.45*a,b*). Anatomically, they can be identified by the "root trace" (Fig. 8.45*c,d*) being in direct connection with the primary xylem of the mother root. These meristems are analogous to the preventitious or latent shoots and are very important for recovery, competition, and exploitation of the occupied root space.

Root "inversions" are a further important possibility of root formation. These formations are defined as transformations from short non-woody (absorption roots) to long woody roots (conversions) or from long to short or shorter roots (reversions). By these means (adventitious, regular continuously, regular periodicly, latent root formation, conversions, and reversions) the root growth is very flexible to adjust to exogenous influences.

6.2. The Architectural Model of the Root System

Analogous to the crown models sensu Hallé et al. (1978), single root models were also developed (Jenik, 1978; Gruber, 1992). Three models, into which most temperate trees can be classified, are seen in Figure 8.46. These models are

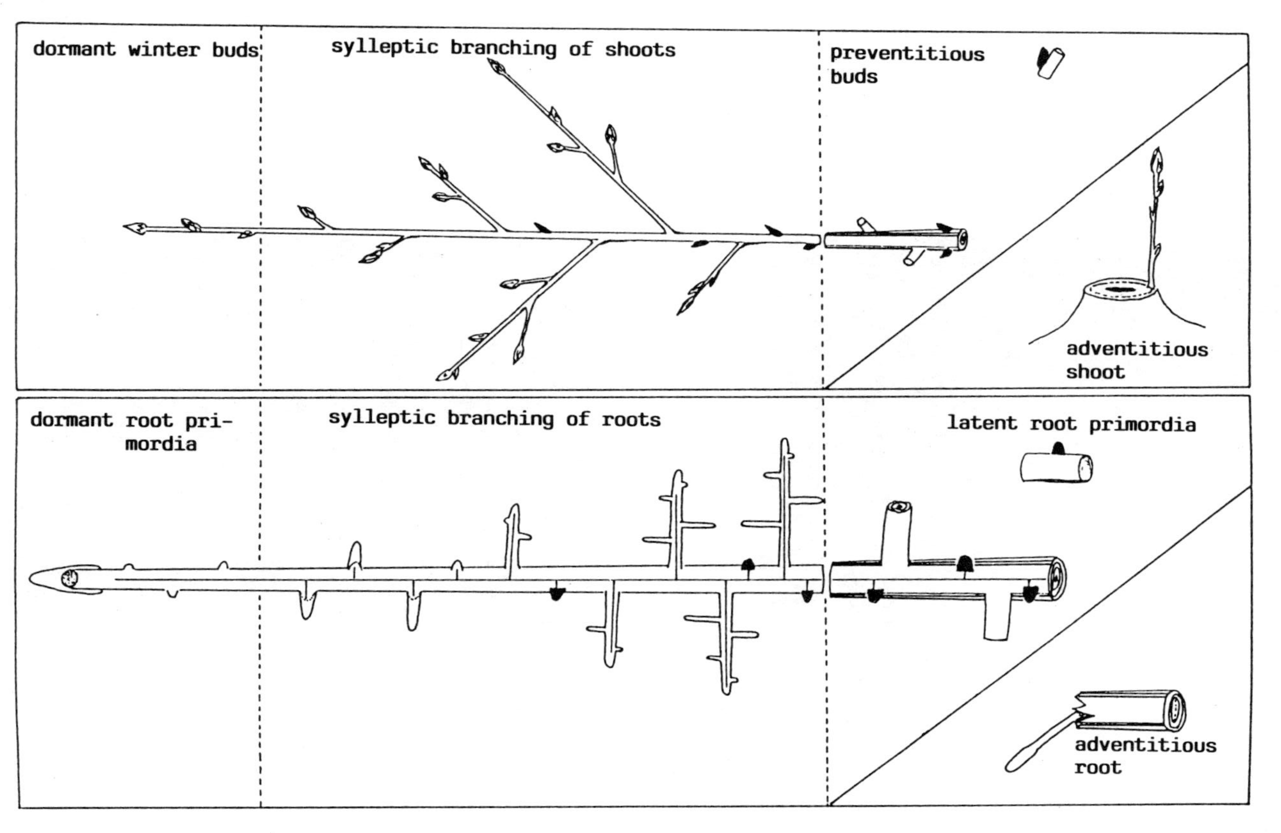

Fig. 8.40. Branching modus of root and shoot (in comparison).

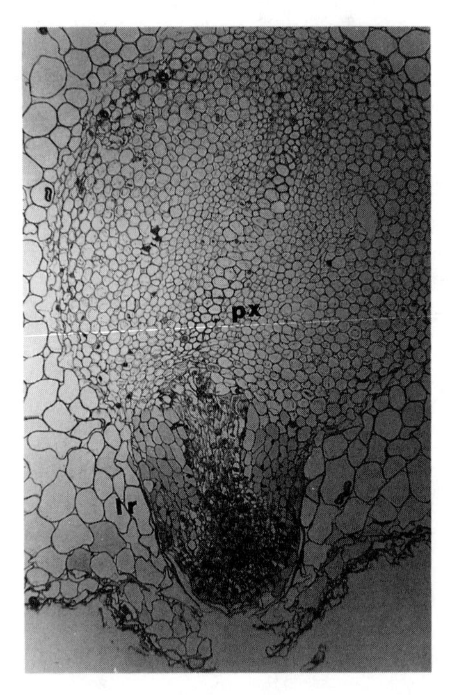

Fig. 8.41. Regular lataral root (lr) formation (sylleptic branching) in front of the diarch primary xylem (px).

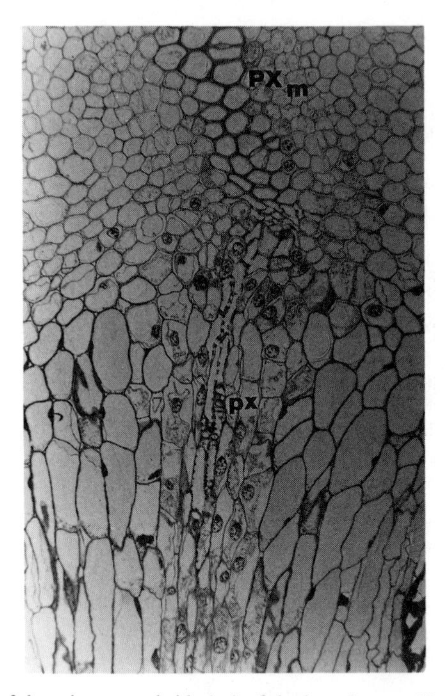

Fig. 8.42. Connection of the primary tracheids (px) of the lateral root with the primary xylem of the mother root (PX_m).

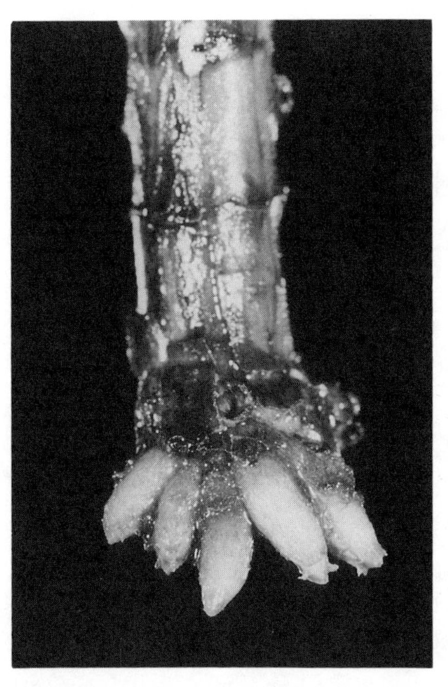

Fig. 8.43. Terminal adventious root regeneration from callus at the injury point at the root tip.

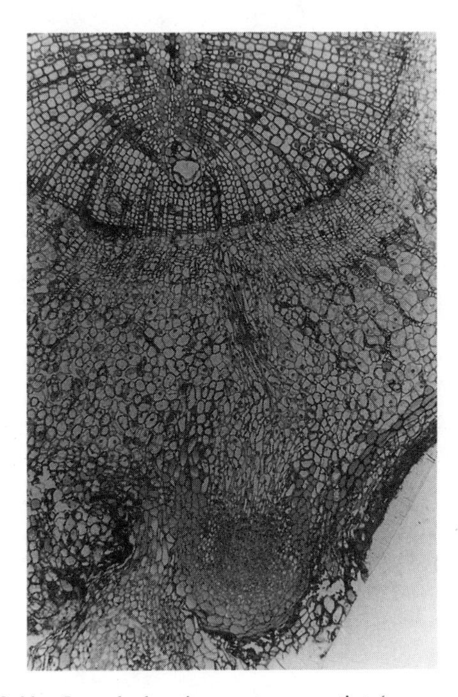

Fig. 8.44. Lateral adventious root regeneration (cross section).

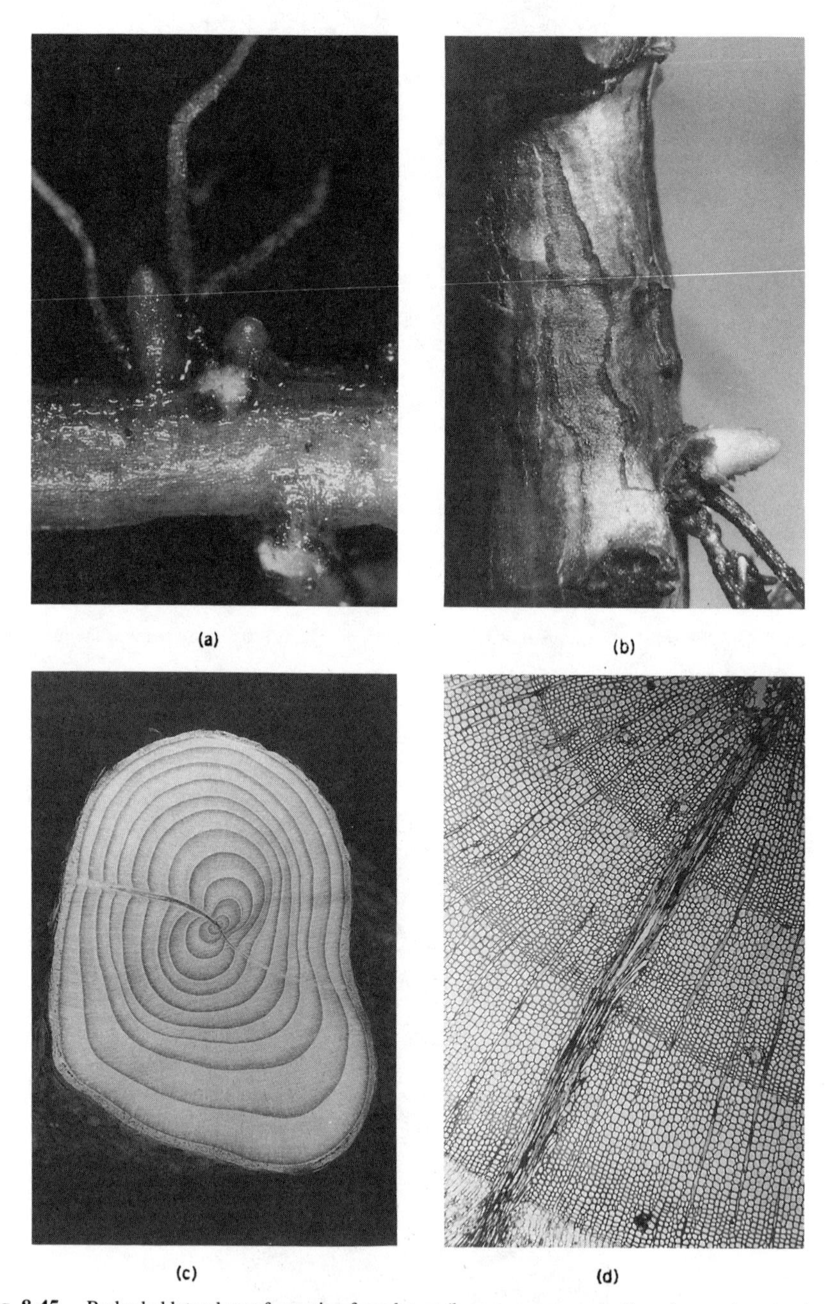

Fig. 8.45. Redarded lateral root formation from latent (long-term suppressed) meristems (proximal root innovation). (*a* and *b*) Lateral root tips emerging from several years old woody mother roots near old lateral unwoody roots. (*c*) Opposite primary "root traces". (*d*) Primary "root trace" (anatomical cross section).

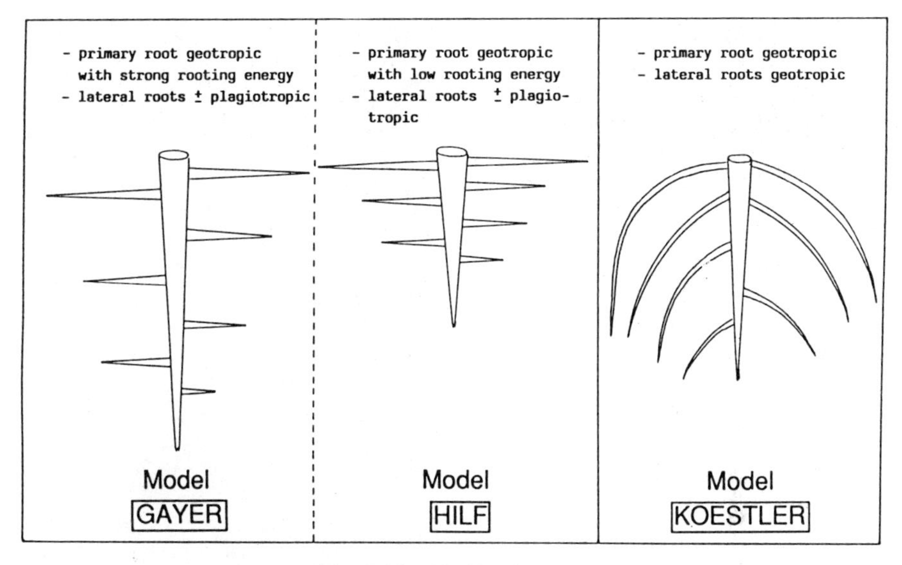

Fig. 8.46. Models of roots.

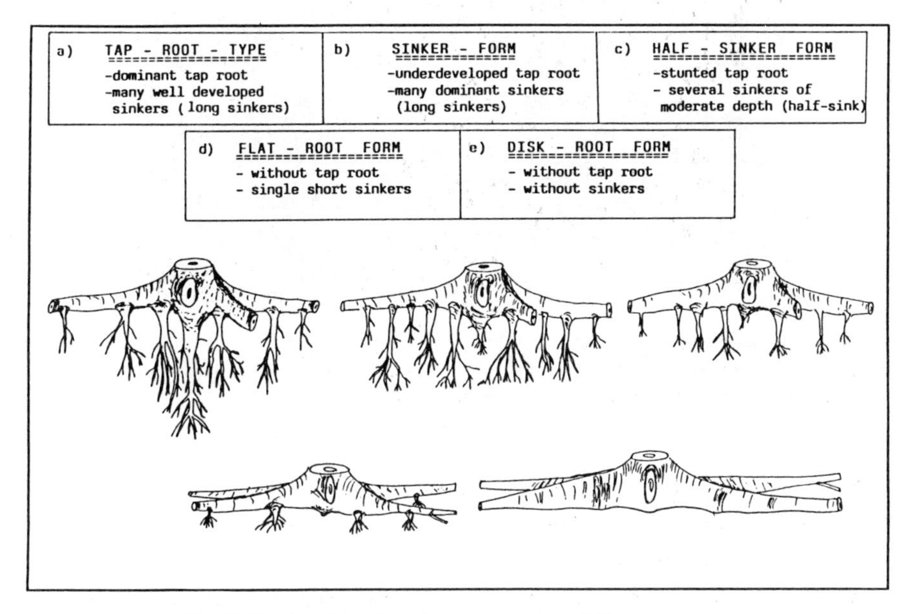

Fig. 8.47. Modification of the root system of Norway spruce.

Fig. 8.48. Root forms of Norway spruce. (*a*) Tap root of a young plant (about 10 years old) on sandy loam (natural regenerated). (*b*) Sinker-root form of an about 80-year-old spruce. (*c*) Disk-root form of an about 200-year-old spruce on natural stand in the Harz mountains.

Model 1: Exogenous controlled periodic growth.

Primary root positive geotropic.

Lateral roots plagiotrop.

Model 1a (model GAYER): Primary root dominant with heavy root energy, strong apical control. For example, *Pinus sylvestris, Abies alba*.

PATTERN	NAME	OCCURREN.
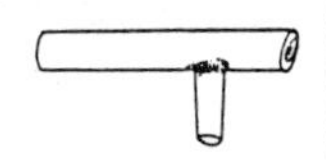	RIGHT ANGLE BRANCHING	all roots
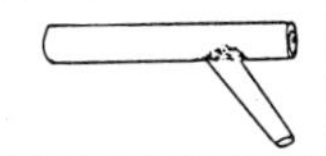	OBLIQUE BRANCHING	all roots
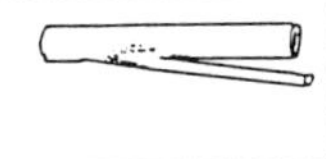	STEEP BRANCHING	all roots
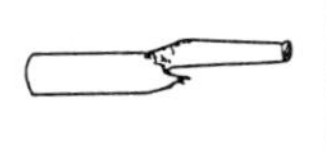	BAYONET (monochasium)	all roots
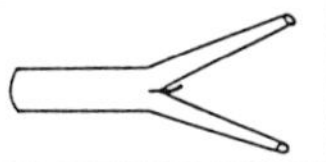	FORK (dichasium)	flat roots
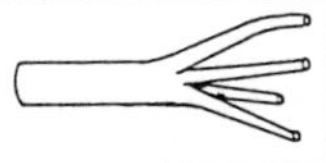	POLYCHASIUM	flat roots
	"BRANCHING OFF RIGHT OF WAY"	flat roots
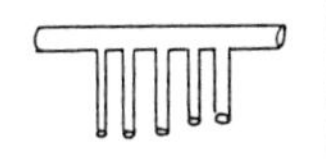	COMB	flat roots
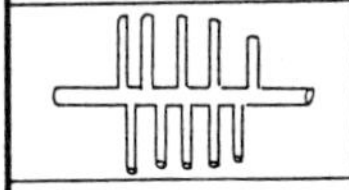	DOUBLE COMB	flat roots
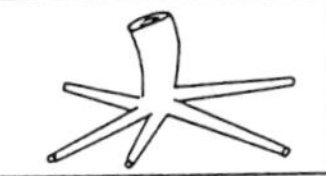	STAR	sinkers
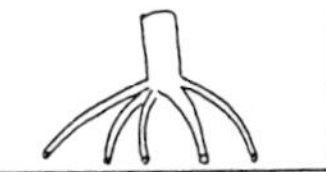	CLAW	sinkers
	BROOM	sinkers

PATTERN	NAME	OCCURRE.
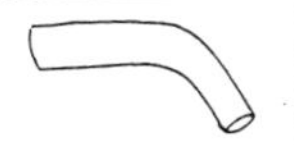	CURVE	all roots
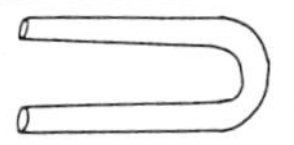	TURNING CURVE	all roots
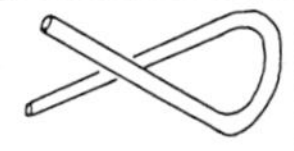	LOOP	all roots
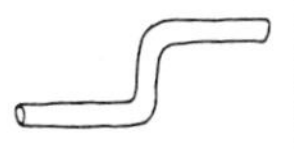	S - CURVE	sinkers
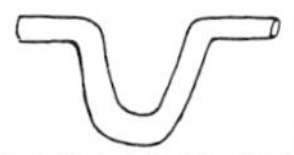	U - CURVE	all roots
	WRIGGLE FORM	sinkers
	SPIRAL	sinkers
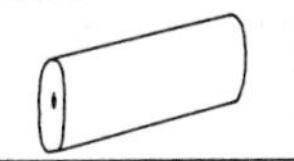	FLATTENING	sinkers
	COMPRESSING	sinkers
	THICKENING	flat roots
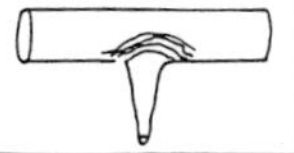	SINKER	sinkers
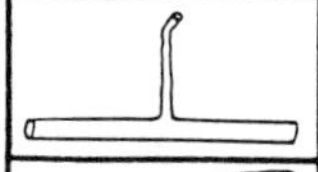	ORTHOTROPIC ROOT	deep roots
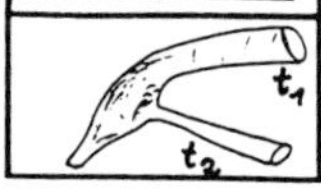 t_1 t_2	ROOT CONTACT	flat roots

Fig. 8.49. Branching and root form patterns of Norway spruce.

Model 1b (model HILF): Primary root with weak root energy, weak apical control therefore only under optimal growth conditions dominant, easy modifiable. For example, *Picea abies, Fraxinus exelsior.*

Model 2 (model KOESTLER): Exogenously controlled (periodic growth).

Primary root positive geotropic with weak apical control.

Lateral roots positive geotropic.

Norway spruce shows root growth belonging to model HILF (Figs. 8.46 and 8.47*a*). Therefore, Norway spruce (also older trees) is able to develop a root system with a dominant tap root (tap root type) under favorable soil conditions (see Aaltonen, 1920; Köstler, 1956; Scheffold, 1971; Fig. 8.47*a*). But this root system can be easily modified if soil conditions change.

6.3. The Modification of the Root System

The most prominent modifications in root systems are on the one hand the level of rooting depth and on the other hand the different patterns of branching and growth. Four modification forms are suggested for describing the variability of rooting depth (Fig. 8.48*b–e*). The first modification form is the sinker root form (Figs. 8.47*b* and 8.48), which is not considered here as a discrete rooting as described in the literature (Köstler et al., 1968). In comparison to the tap root type the primary root is underdeveloped because of unfavorable soil conditions just below the primary root; the sinkers, which are reiterations of the architectural model (reiterated tap roots) develop on those points in the soil where geotropic growth is not hindered or limited. The other modification forms are the half-sinker root (Fig. 8.48*c*), the flat root (Fig. 8.52*d*), and the disk root (Figs. 8.47*c* and 8.48*e*). The latter form is characterized only by horizontal roots growing in the upper 10 cm of the soil. Such root systems are viewed as very unstable forms, especially when storms and droughts occur.

Modifications of root branching and growth are illustrated in Figure 8.49. The causes of these modifications are manifold. The quantification of these different patterns allows a detailed characterization of the stand and soil (see Gruber, 1992).

7. THINNING (SYMPTOMS) AND REGENERATION OF NORWAY SPRUCE CROWNS

Crown thinning shows very different patterns. Many of these patterns are described by Schröter and Aldinger (1985), Gruber (1987; 1992), Lesinsky and Westman (1988), Lesinsky and Landmann (1988), Neumann und Pollanschütz (1988). Thinning may originate suddenly or gradually, may run acutely or chronically, and may result from a high loss of needles and/or insufficient branching. While high needle loss is a sign of a directly inflicted damage or stress, poor branching generally indicates a chronic stress or disease. Therefore, the regeneration behavior of the crown is very important.

In Figure 8.50 the development of branching and needle density of crowns under acute and chronic stress with and without the capacity of regeneration is illustrated, for example, on a comb-type spruce.

In describing the effects of acute stress on the crown it must be distinguished between crowns with and without a regeneration capacity. The response of a crown to acute stress is dependent on age and tree vigor, as well as the extent of the initial

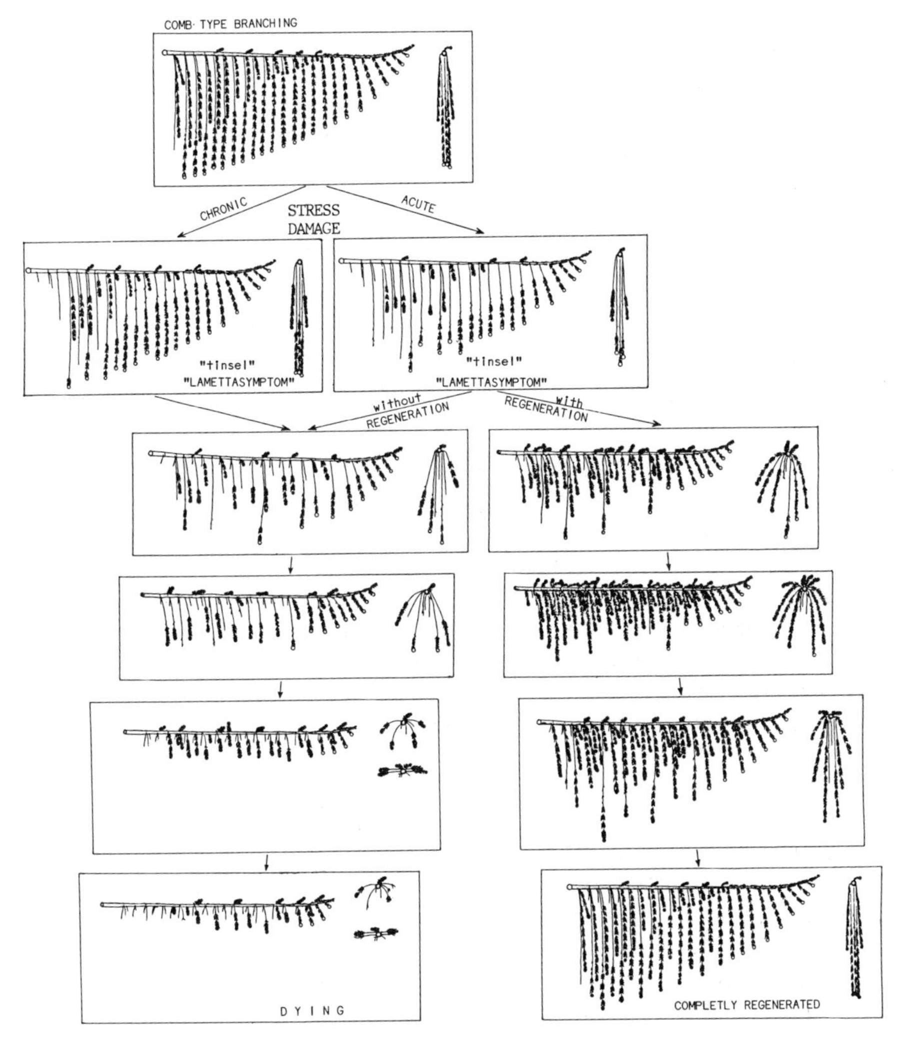

Fig. 8.50. Diagram showing the dynamics and alteration of the branching and needle statement of spruce crown due to chronic and acute stress with and without regeneration.

damage and the intensity of stress. This response is again regulated by differences in the environment and genotype.

A measure for the regeneration capacity is the activity of preventitious shoot formation of a species. The shorter the phyllochron (resting period) the higher the activity. Figure 8.51 shows the average phyllochron of Douglas fir, Norway spruce, and silver fir. It is interesting that Douglas fir possess the shortest (between 3 and 4 years) and silver fir the longest phyllochron (between 6 and 7 years). Norway spruce

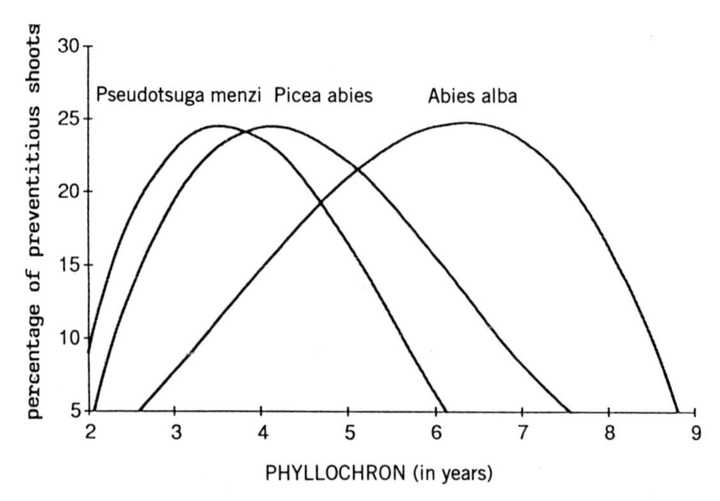

Fig. 8.51. Resting time of preventitious buds in three species.

shows about a 4-year resting interval. This sluggish regeneration capacity of silver fir may be one reason why this species is the most damaged tree species in Europe.

Due to its high regeneration capacity Norway spruce is normally able to withstand and overcome extreme climatic and biotic damage under natural conditions. Under acute (short-term) stress crown thinning is therefore normally only a temporary symptom, and the dynamics and variation of the foliation density in the crowns of trees with many sets of needles can be very high in their dependence on environmental factors (Münch, 1928). For example, Pechmann (1958) reported that needle loss above 60% are lethal in Norway spruce. The investigations on the relationship between needle loss and regeneration (preventitious) of shoots show that needle loss above about 40–55% cause a decrease of the regeneration capacity. One example is documented in Figure 8.27.

Longer drought periods, for example those in 1973 and 1976, may have significantly influenced the needle density for a long time. Typical symptoms of drought stress are top dying, subtop dying (window effect), and crown thinning. Such events can be recognized by studying the chronology of tree ring growth in connection with annual shoot (length) growth of stems and branches.

Investigations on the "window effect" of spruce crowns revealed that the gap of foliation and branching can be interpreted as a less heavy stagheadedness, but with following regeneration at the top. Conspicuous gaps coincided with the drought years 1973 and 1976. This situation was exactly reconstructed by the investigation of the annual height and diameter increment. Furthermore, it could be stated that the height increment of such thinned crowns fell abruptly to a lower level without the ability to spring back to the previous higher level (Fig. 8.52).

A strong growth of preventitious shoots is an indication of a high turnover in the crown. This requires an intact uptake and transport system for water and nutrients. It is therefore unlikely that spruces with such a branching pattern suffer from

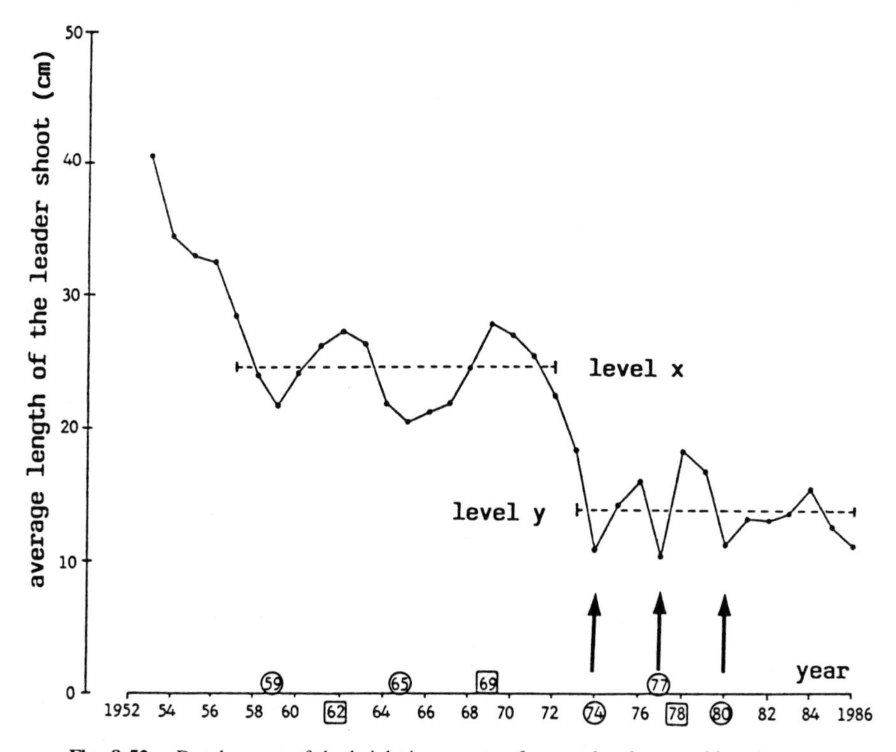

Fig. 8.52. Development of the height increments of exposed and crown-thinned spruces.

stronger or chronic root damage or stress. On the contrary, this indicates an acute (short-term induced) damage caused by biotic or abiotic stress factors. Therefore an increased formation of preventitious shoots cannot be a symptom of a destabilized ecosystem per se. However, the number of preventitious shoots is a measure for exogenous stress (Gruber, 1987; Liedeker et al., 1988), and when the shoot length and the percentage of needle statement are also taken into account, it may then be used to estimate the regeneration capacity of the tree. The regeneration capacity is a direct measure of a tree's vitality.

Under chronic stress crown thinning does not only develop with a gradual decrease in the number of annual sets of needles but also with a decrease in branching, especially in the inner part of a crown. This depends on an insufficient or lack of preventitious shoot formation. Consequently, the mass of needles and the number and length of new shoots decreases mainly on the older parts of the branches. As long as this is not the result of mechanical competition or shading due to the branching pattern or the social position of the tree in the stand, it must be assumed that this is due to a reduced capacity of regeneration.

With regard to the formation of the "tinsel syndrome" (Lamettasyndrom), which is the result of an above average needle loss on comblike spruce, we have to distinguish three forms. The first is a form of natural aging (see Fig. 8.19). Anatomical investigations at the base of 12–18-year-old tinsel branches showed (Table 8.2) that

TABLE 8.2. Differences Between the Number of Annual Length Increments and Annual Xylem Rings at the Base of "Tinsel Twigs"

	Number of Annual Length Increments	Number of Annual Diameter Increments	Failing Annual Xylem Rings
	14	7	7
	15	8	7
	17	7	10
	14	11	3
	12	6	6
	14	7	7
	16	9	7
	12	7	5
	18	8	10
	15	8	7
Mean	14.7	7.8	6.9

the ring growth had already stopped while the length growth continued to grow for some years. This early cessation of ring growth (between 7 and 12 rings could be recognized) is seen as a sign that the older branches of the second order are becoming gradually ineffective. These branches are replaced by preventitious shoots forming successive branching series. The second develops under acute stress, and is therefore a temporary but very spectacular symptom (see Fig. 8.50, right path). The third, which can arise under chronic stress, is a serious disease. This last symptom is only visible in the crown at the beginning of the disease, because the branching changes with time and duration of chronic stress influences (Fig. 8.50, left path). But in this stage the differentiation between an acute and chronic symptom is very difficult to determine.

The symptom of permanently increasing crown thinning from the inside to the outside of the crown (AFZ Farbbildhefte, 1984; Schütt, 1984), which is typical of the present spruce damage, may be attributed to a reduced preventitious shoot formation. Apart from possible genetic differences this poor branching and insufficient regeneration in the crown can be seen as a result of an insufficient supply of water and/or nutrients.

The causes of such a deficiency of water and/or nutrients may be attributed to a restricted function of the root system due to root damage (Hüttermann, 1983; Matzner et al., 1984; Matzner, 1988; Ulrich, 1988, 1989; Godbold, 1991). Therefore, we have to take into account also soil acidification due to the input of atmospheric pollutants as one possibility of a serious and important process, which is able to deformate the root architecture from deep to flat rooting, and which therefore makes a tree more sensitive to all natural stress influences, especially drought. Through this, the stability of trees can be reduced significantly. Before we can more fully consider the causes of crown-thinning symptoms, more research in root structure and function based on the tree as a whole must be carried out.

8. CROWN ARCHITECTURE AND REGENERATION OF SILVER FIR (*ABIES ALBA* MILL.)

The branching of fir is similar to the branching of spruce. The most important results gained from investigations of the crown architecture of silver fir from the Black Forest (Müllheim, Wehr, Freiburg) are the following:

As a rule silver fir shows only regular (predetermined) and preventitious shoot growth. Proleptic shoot formations are seldom or induced traumatically.

The regular axes of fir are produced by monopodial growth pattern (distal shoot innovation). Therefore the regular branching prevails at the periphery of the crown by producing long shoots, which is a rather inflexible shoot growth.

Fir also forms preventitious shoots on the trunk and on axes of the first and second order (Figs. 8.53 and 8.54). Due to the strong acrotony the preventitious shoots originate mostly from medial or basal preventitious buds.

Just as with Norway spruce the preventitious shoot formation on silver fir becomes increasingly important with tree age (especially height growth) and exposition. These preventitious shoots, which can be seen as "reiterations of branching," also play a great role in the regeneration, architectonic adaptation, and

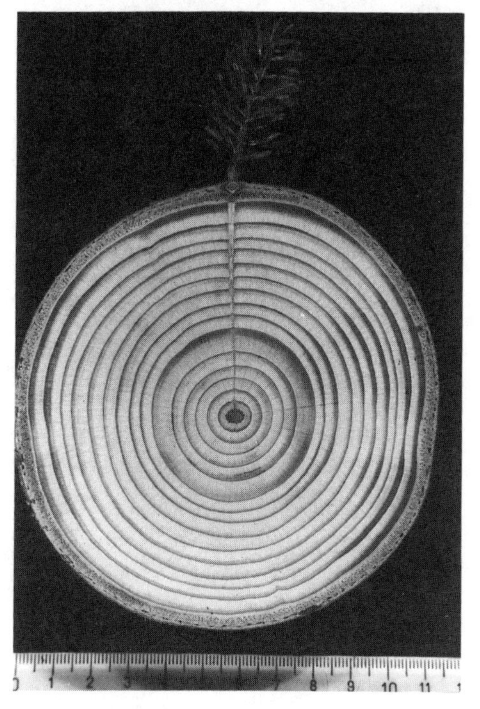

Fig. 8.53. Preventitious shoot: The 1-year-old lateral shoot is connected with the pith of the 16-year-old mother shoot by the primary "bud trace."

Fig. 8.54. Preventitious shoots epitonous along the bough axis.

Fig. 8.55. Suppressed tree with "a shade-secondary crown" (where pc = primary crown and sc = secondary crown).

crown space exploitation of silver fir. Adventitious shoot formations could not be observed.

Fir mostly form preventitious shoots only up to the second and third (fourth) order. Polycladies (preventitious shoots in clusters), as they occur on Norway spruce, are absent in silver fir. Nevertheless, the branching on fir can be described as "branching in successive series" similar to the branching pattern of the "plate spruce." In comparison to Norway spruce the number of such series is smaller.

In contrast to Norway spruce, silver fir is able to produce preventitious shoots not only on the axes of the primary crown, but also along the trunk as so-called "water sprouts". By these means fir can establish an entirely new "secondary crown" (Figs. 8.55 and 8.56), which develops gradually from the base to the top, or suddenly along the whole stem. This relocation of the young sprouts back to the axes of the primary crown on the one hand and along the trunk on the other hand (proximal shoot innovation) is a very important ecological adaptation capacity. Therefore, the regeneration behavior of a tree is important for the evaluation of the tree's vitality. In this respect the crown of silver fir is more flexible in regeneration and adjustment than Norway spruce.

Fig. 8.56. Light exposed tree with a "light-secondary crown" (where pc = primary crown and sc = secondary crown).

In regard to the development of the secondary crown two types are distinguishable:

1. The "shade-secondary crown" (Fig. 8.55): This crown forms on suppressed trees, whose primary crown is outcompeted for light by neighboring trees. The secondary crown is adjusted to the unfavorable light conditions in the understory. The diameter increment at dbh is poor.

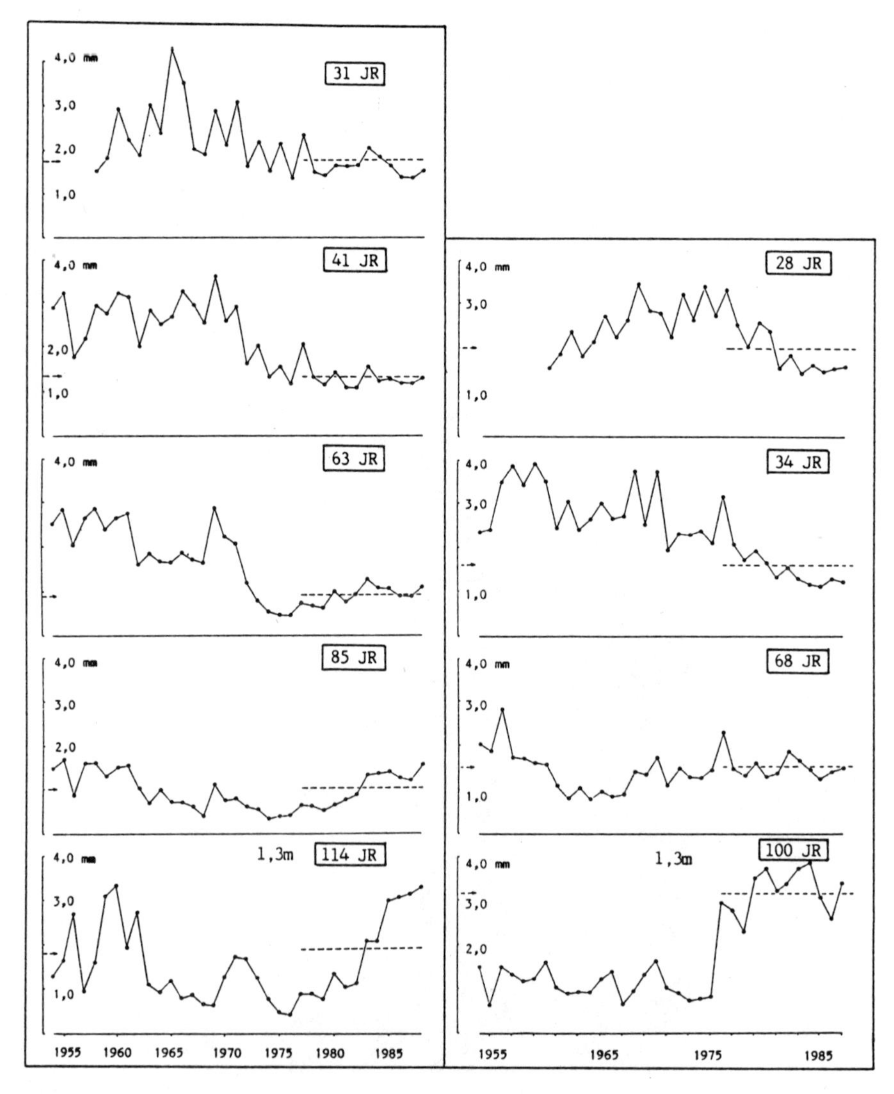

Fig. 8.57. Distribution of the annual thickening increment along the trunk of two firs with well-developed secondary crown (JR = annual rings).

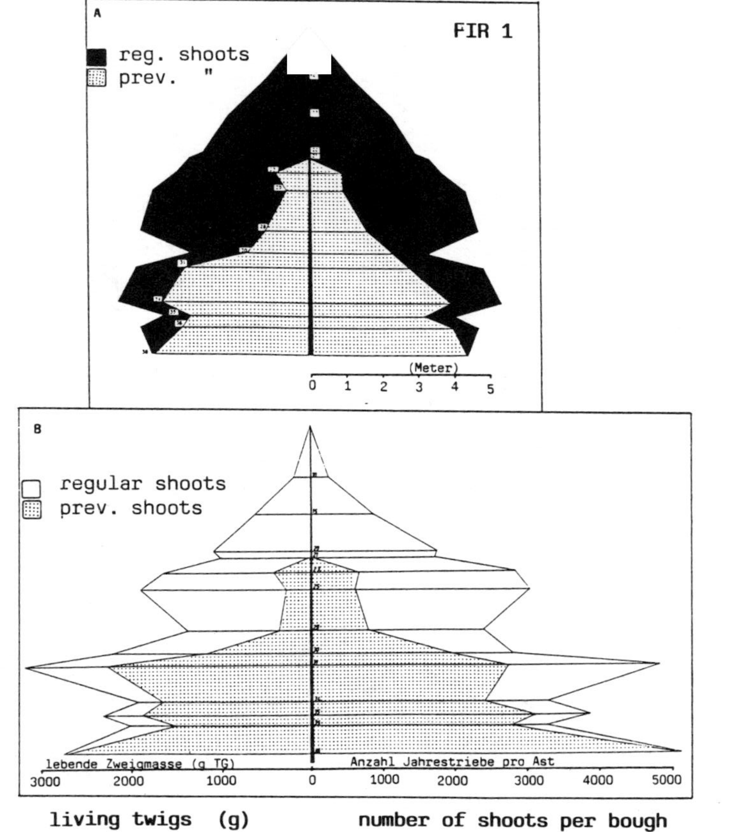

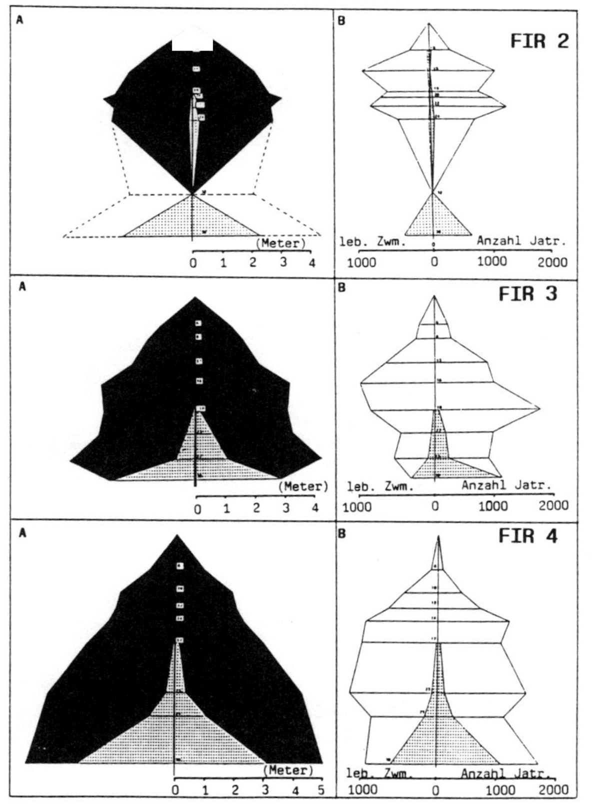

Fig. 8.58. Scale crown models: Distribution of living regular and preventitious shoots in fir crowns. Left site: Percentage of the number of living regular and preventitious shoots. Right site: Percentage of the dry weight of living regular and preventitious shoots. Fir 1: Without damage symptoms and Fir 2–4: With crown thinning symptoms.

TABLE 8.3. Proportions of Regular and Preventitious Shoots (in Percent) of Fir Crowns With and Without a Secondary Crown[a]

	Primary Crown		Inclusive Secondary Crown				
Tree	R	P_{pk}	R	P	P_{pk}	P_{sk}	$R + P_{pk}$
(A) Dry Weight of the Branches							
1	51.5	48.5					
2	59.6	40.4					
3	91.1	8.9					
4	91.9	8.1					
5	86.5	13.5					
6	63.6	36.4	26.0	74.0	15.0	59.0	41.0
7	13.7	86.3	7.8	92.2	49.2	43.0	57.0
8	5.7	94.3	2.2	97.8	35.7	62.1	37.9
(B) Number of Annual Shoots							
1	52.8	47.2					
2	54.1	45.9					
3	85.6	14.3					
4	82.9	17.1					
5	80.7	19.3					
6	45.7	54.3	14.2	85.8	16.8	69.0	31.0
7	9.6	90.4	5.6	94.4	52.9	41.5	58.5
8	9.8	90.2	2.7	97.3	24.9	72.4	27.6

[a] R = regular; pk = primary crown; P = preventitious; and sk = secondary crown.

2. The "light-secondary crown" (Fig. 8.56): This crown develops on trees when the canopy is opened or in exposed stands. The primary crown diminishes in the same measure as the secondary crown increases in growth. The growth increment of the trunk between the primary and secondary crown is poor; but it can be very considerable at breast height (Fig. 8.57) if the secondary crown is already well developed, although the primary crown can be heavily damaged.

Crown analyses revealed that the crown of silver fir may consist nearly 100% of preventitious shoots (Fig. 8.58, Table 8.3). Note that strongly damaged (thinned) crowns of silver fir show a deficiency in crown regeneration producing insufficiently preventitious shoots (Fig. 8.58, FIR 2–4). In comparison, well-needled crowns have a greater relative and absolute proportion of preventitious shoots in the living crown (Fig. 8.58).

A most important feature for assessing the regeneration capacity is the resting time of preventitious buds, because this time reflects the regeneration velocity. An average fir has resting times of preventitious buds between 5 and 7 years (mean 5.8, Fig. 8.51). In comparison to fir, spruce has resting times between 4 and 6 years (mean 4.8) and Douglas fir between 3 and 4 (mean 3.6) years (Gruber, 1992).

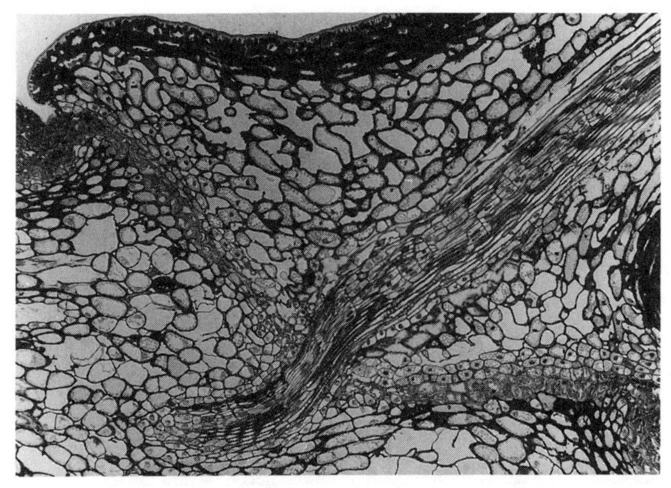

Fig. 8.59. Longitudinal section through a mature abscission zone of a few months old needle (end of June).

In comparison to Norway spruce, the abscission zone of silver fir is quite different (Fig. 8.59). The abscission zone shows three layers, with the abscission layer consisting of small living cells at the base of the green needle, as well as below the cork layer and the collenchym cells (Fig. 8.60). The abscission occurs in the abscission layer (Fig. 8.61).

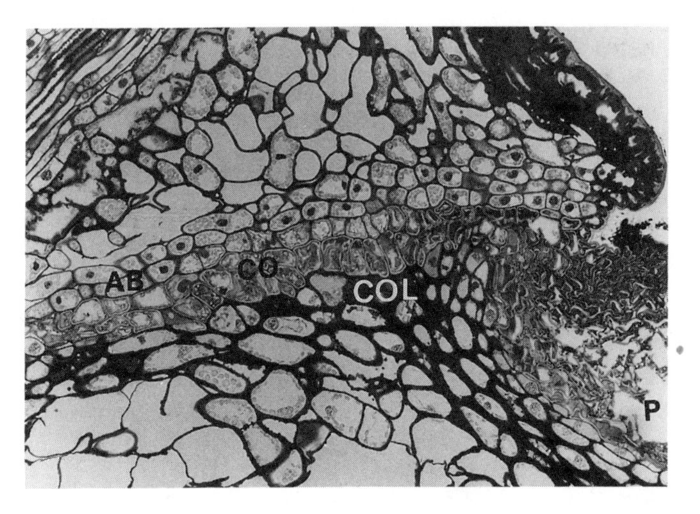

Fig. 8.60. Longitudinal section through a mature abscission zone of a few months old needle (end of June) (where AB = abscission layer, CO = cork layer, COL = collenchym cells, and p = periderm).

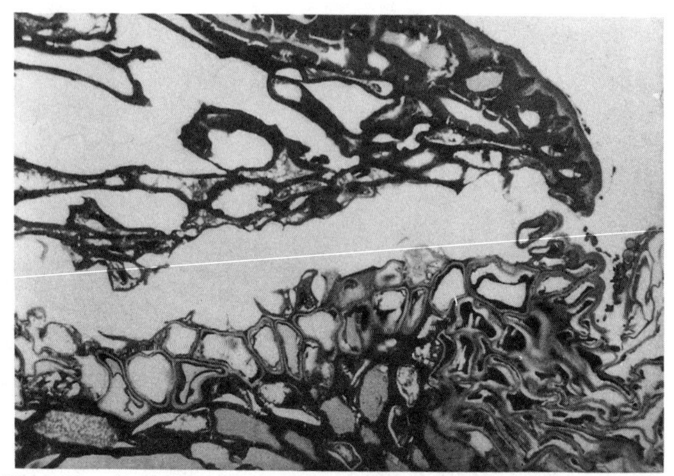

Fig. 8.61. Abscission of a several years old needle within the abscission layer.

ACKNOWLEDGMENTS

This work was promoted by the financial support of the German Federal Ministry of Research and Technology (BMFT) under 373074 and OEF 2019. The author thanks the Minister Dr. H. Riesenhuber.

REFERENCES

Aaltonen VT (1920): Über die Ausbreitung und den Reichtum der Baumwurzeln in den Heidewäldern Lapplands. Helsinki:

Anonymus (1984): Allgemeine Forstzeitschrift (AFZ)-Farbbildhefte Zum Erkennen von Immissionsschäden an Waldbäumen 1984, Zur Diagnose und Klassifizierung der neuartigen Waldschäden 1984.

Behrens J (1886): Über die anatomischen Beziehungen zwischen Blatt und Rinde der Coniferen. Ph.D. Thesis, Universität Kiel, Kiel.

Berthold C (1904): Physiologic der pflumzlihen Organisation BJII Leipzig, Engelmum.

Bindseil W (1933): Das bauökonomische Problem der Fichtenkrone. Mitt. Forstw Forstwiss 4:560–577.

Braun HJ (1982): Lehrbuch der Forstbotanik. Stuttgart, New York: Fischer.

Bryan JA, Lanner RM (1981): Epicormic branching in Rocky Mountain Douglas-fir. Can. J For Res 11–2:190–199.

Büsgen M, Münch E (1927): Bau und leben unseer Waldbäumen 3 Aufl Jenn G Fischer.

Campbell AH, Vines AE (1938): The effect of Lophodermellina macrospora (Hartig) Tehon on leaf-abscission in Picea excelsa Link. New Phytal 37:358–368.

Eames AI, McDaniels LH (1947): Introduction to plant anatomy. 2. Auflage, McGraw-Hill Book Co.

Eckmüllner O (1990): Benadelung und Splintflächen von Fichten aus Wuchsgebieten Österreichs Holz-Zentrbl 18:266–267.

Edelin C (1977): Images del' architecture des conifères. These, Montpellier.

Elstner E (1983): Baumkrankheiten und Baumsterben. Naturwiss Rsch 9/36:381–388.

Ewers FW, Zimmermann MH (1984): The hydraulic architecture of balsam fir (Abies balsamea). Physiol Plant 60:453–458.

Facey V (1956): Abscission of leaves in Picea glauca (Moench) Voss. and Abies balsamea L. North Dakota Acad Sci Proc 10:38–44.

Fink S (1983): The occurrence of adventious and preventitious buds within the bark of some temperate and tropical trees. Am J Bot 70:532–542.

Fink S (1984): Some cases of delayed or induced development of axillary buds from persisting detached meristems in conifers. Am J Bot 71:44–51.

Gerlach D (1977): Botanische Mikrotechnik. Stuttgart: Thieme.

Godbold DL (1991): Die Wirkung von Aluminium und Schwermetallen auf Picea abies Sämlinge. Schr. Forstl. Fak. Universität Göttingen u. Nieders. Forstl. Versuchsanst. 104. J.D. Sauerländer's Verlag, Frankfurt a.M.

Goebel K (1933): Organographie der Pflanzen. 3. Band. Jena: Fischer.

Gruber F (1986): Beiträge zum morphogenetischen Zyklus der Knospe, zur Phyllotaxis und zum Triebwachstum der Fichte (Picea abies (L.) Karst.) auf unterschiedlichen Standorten. Dissertation Göttingen und Ber. Forschungsz. Waldökos./Waldst. Reihe A, Bd. 25, 215 S. (1987).

Gruber F (1987): Das Verzweigungssystem und der Nadelfall der Fichte (Picea abies (L.) Karst.) als Grundlage zur Beurteilung von Waldschäden. Ber. Forschungsz. Waldökos./Waldst. Reihe A, Bd. 26

Gruber F (1989): The proleptic shoot formation of Picea abies (L.) Karst. Beitr Biol Pflanz 64:75–113.

Gruber F (1990): Branching system, needle fall and needle density of Norway spruce (*Picea abies*). Contributiones Biologiae Arborum 3 (136 Seiten). Birkhäusen: Berlin, Basel, Boston.

Gruber F (1992): Dynamik und Regeneration der Gehölze. Ber. Forschungsz. Waldökosysteme. Reihe A, Bd 86/Teil I: Ergebnisse (420 Seiten); Teil II: Fotografische Dokumentation und Tabellen (176 Seiten).

Guédes M (1979): Morphology of Seed Plants. Vaduz: Cramer.

Hagemann W (1990): Comparative morphology of acrogenous branch systems and phylogenetic considerations. II. Angiosperms. Acta Biotheor 38:207–242.

Hallé F, Oldeman RRA, Tomlinson PB (1978): Tropical trees and forests. Berlin, Heidelberg, New York.

Hassenberger K (1939): Fichtenformen und Fichtenrassen im Glatzer Schneegebirge Zschr. f. Forstwiss-Jagdwiss 71/3:113–140.

Heinkinheimo O (1920): Über die Fichtenformen und ihren volkswirtschaftlichen Wert. Commun Inst Quaest For Finl 2: 1–102.

Hofmann J (1968): Über die bisherigen Ergebnisse der Fichtentpyenforschung. Arch Forstwiss 17/2: 207–216.

Huber B (1928): Weitere quantitative Untersuchungen über das Wasserleitungssystem der Pflanze. Jahrb. f. wissenschaftl. Bot 67/5:877–959.

Hüttermann A (1983): Auswirkungen "saurer Deposition" auf die Physiologie des Wurzelraumes von Waldökosystemen, AFZ, 663–664.

Jablanczy A (1971): Changes due to age in apical development in spruce and fir. Can For Serv Biom Res Notes 27:10.

Jahn R, Scheffold K, Hauck U (1971): Wurzeluntersuchungen an Waldbäumen in Baden-Württemberg. Schriftenr. Landesforstv: Baden-Württ: Band 33.

Jenik J (1978): Roots and root systems in tropical trees: morphologic and ecologic aspects. In Tomlinson PB, and Zimmermann MH: Tropical trees as living systems. Cambridge: Cambridge University Press, pp. 323–349.

Kaibiyainen LK, Khair P, Sazonova TA, Myakelya A (1986): Balance of the water transport system in the Scotch Pine. III. Area of conducting xylem and needle mass. Lesovedenie 1:31–37.

Kaufmann MR, Troendle CA (1981): The relationship of leaf area and foliage biomass to sapwood conducting area in four subalpine forest tree species. For Sci 27(3):477–482.

Kestel P (1961): Der Anschluss der Blattspuren an das Holz bei den Coniferen. Ph.D. Thesis, Universität München, München, Germany.

Köstler JN (1956): Waldbaulische Beobachtungen an Wurzelstöcken sturmgeworfener Nadelbäume. Forstwiss Centralbl 88:65–91.

Köstler JN, Brückner E, Bibelriether H (1968): Die Wurzeln der Waldbäume, Hamburg, Berlin: Verlag Paul Parey.

Kozlowski TT (1964): Shoot growth in woody plants. Bot Rev 30:335–392.

Kozlowski TT (1971): Growth and development of trees. Vol. I: New York: Academic Press.

Lesinsky J, Westman L (1987): Crown injury types and their applicability for forest inventory. In Perry R et al. (ed): Acid rain. Scientific and technical advances. Westville Grange: Selper Ltd. pp. 657–662.

Lesinsky J, Landmann G (1988): Crown and branch malformation in conifers related to forest decline. Scientific basis of forest decline symptomatology, Cape JN, Mathy P (Ed.): Air pollution report series, No. 15, pp. 92–105.

Liedeker H, Schütt P, Klein RM (1988): Symptoms of forest decline (Waldsterben) on Norway and red spruce. Eur J For Pathol 18:13–25.

Magel E, Ziegler H (1987): Die "Lametta"-Tracht-ein Schadsymptom? Allg Forstzschr 27/28/29:731–733.

Marcel E (1975): Bemerkungen und Beobachtungen über den Augusttrieb. Schweiz Zschr Forstwiss 126: 214–237.

Markfeldt O (1885): Über das Verhalten der Blattspurstränge immergrüner Pflanzen beim Dickenwachstum des Stammes oder Zweiges. Flora 68:33–39, 81–90.

Matzner E (1988): Der Stoffumsatz zweier Waldökosysteme im Solling. Ber Fz Waldökos A 40:217 S.

Matzner E, Ulrich B, Murach D, Rost-Siebert K (1984): Zur Beteiligung des Bodens am Waldsterben. Ber Fz Waldökos./Waldst. 2.

Mohr H (1984): Baumsterben als pflanzenphysiologisches Problem. Biol Z 14/4:105–110.

Müller-Doblies D, Weberlineg F (1984): Über Prolepsis und verwandte Begriffe Beitr Biol Pflanz 59: 121–144.

Münch E (1928): Winterschäden an Fichte und anderen Gehölzen. Forst Jb 79:276 ff.

Napp-Zinn K (1966): Anatomie des Blattes. I. Blattanatomie der Gymnospermen. Berln: Gebr. Borntraeger.

Neger FW (1911): Zur Mechanik des Nadelfalls der Fichte. Naturwiss Zschr Forst- Landwirtsch 9:214–223.

Neger FW, Fuchs J (1915): Untersuchungen über den Nadelfall der Koniferen. Jb Wissensch Bot 55: 608–660.

Neumann M, Pollanschütz J (1988): Taxationshilfe für Kronenzustandserhebungen. Öster Forstz 6:27–37.

Oren R, Werk KS, Schulze ED (1986): Relationship between foliage and conducting xylem in Picea abies (L.) Karst. Trees 1:61–89.

Pax F (1899): Allgemeine Morphologie der Gewächse. Stuttgart: Enke.

Pechmann H von (1958): Über die Heilungsaussichten bei hagelbeschädigten Waldbeständen. Forstwiss Centralbl 357–373.

Philips IDJ (1975): Apical dominance. Ann Rev Plant Physiol 26:341–367.

Pfeiffer H (1928): Die pflanzlichen Trennungsgewebe. In: Handbuch der Pflanzenanatomie. Berlin: K. Linsbauer Bd. V.

Pollard DFW, Logan KT (1974): The role of free growth in the differentiation of provenances of black spruce (Picea mariana). Can J For Res 4:308–311.

Powell GR (1987): Syllepsis in Larix laricina. Analysis of tree leaders with and without sylleptic long shoots. Can J For Res 17:490–498.

Priehäusser G (1958): Die Fichtenvariationen und -kombinationen des Bayerischen Waldes nach phänotypischen Merkmalen mit Bestimmungsschlüssel. Forstwiss Centralbl 77:151–171.

Rehfuess KE (1983): Ersatztriebe an Fichten. Allg Forstz 41:1111.

Rehfuess KE, Rodenkirchen H (1984): Über die Nadelröte-Erkrankung der Fichte (Picea abies (L.) Karst.) in Süddeutschland. Forstwiss Centralbl 103:248–262.

Romberger JA (1963): Meristems, growth and development in woody plants. US Dep Ag For Ser Tech Bull 1293.

Rossmässler EA (1863): Der Wald, Leipzig, Heidelberg, Germany.

Rubner K (1936): Beitrag zur Kenntnis der Fichtenformen und Fichtenrassen. Thar Forstl Jb 87:101–176.

Rubner K (1943): Die praktische Bedeutung der Fichtentypen. Forstwiss Centralbl Thar Forst Jb 6:233–246.

Sachs T, Thimann KV (1967): The role of auxins and cytokinins in the release of buds from dominance. Am J Bot 54:136–144.

Scharfetter R (1953): Biographien von Pflanzensippen. Wien, Germany: Springer-Verlag.

Scheffold K (1971): Wurzelprofile im Altmoränengebiet des südwestdeutschen Alpenvorlandes, Wurzeluntersuchungen an Waldbäumen in Bad.-Württ., Schriftenreihe der LFV B.-W., Band 33, 47–86.

Schmidt-Vogt H (1972): Studien zur morphologischen Variabilität der Fichte (Picea abies (L.) Karst.). Allg Forst Jztg 143/7, 9, 11 Sonderdruck: 133–144, 177–186, 221–240.

Schmidt-Vogt H (1977): Die Fichte. Bd. 1, Berlin, Hamburg: Parey.

Schmidt-Vogt H (1986): Die Fichte. Bd. 2, Berlin, Hamburg: Parey.

Schröter H, Aldinger E (1985): Beurteilung des Gesundheitszustandes von Fichte und Tanne nach der Benadelungsdichte. AFZ 18:438–442.

Schütt P (1984): Der Wald stirbt an Stress. München, Germany: Bertelsmann.

Shinozaki K, Yoda K, Hozumi K, Kira T (1964a): A quantitative analysis of plant form. The pipe model theory. I. Basic analysis. Jpn J Ecol 14(3):97–105.

Shinozaki K, Yoda K, Hozumi K, Kira T (1964b): A quantitative analysis of plant form. The pipe model theory. II. Further evidence of the theory and its application in forest ecology. Jpn J Ecol 14(4):133–139.

Späth HL (1912): Der Johannistrieb. Berlin: Parey.

Sylvén N (1909): Studien über den Formenreichtum der Fichte, besonders die Verzweigungstypen derselben und ihren forstlichen Wert. Mitt Forst Versuchsanst Schwedens, H. 6:37–117.

Tomlinson PB (1983): Tree architecture. Am Sci 71:141–149.

Tomlinson PB, Gill AM (1973): Growth habits of tropical trees. Some guiding principles. Can J Bot 53: 129–143.

Troll W (1937): Vergleichende Morphologie höherer Pflanzen. Berlin, Germany:

Troll W (1948): Allgemeine Botanik. Stuttgart, Germany: Enke.

Tyree MT, Ewers FW (1991): The hydraulic architecture of trees and other woody plants. New Phytol 119:345–360.

Ulrich B (1988): Bodenkundliche Forschung in Zusammenhang mit den neuartigen Waldschäden. Allg Forstz 43:1171–1173.

Ulrich B (1989): Effects of acid deposition on forest ecosystems in Europe. Adv Environ Sci 4 (Springer)

Veillon J-M (1976): Architecture of New Caledonian species of Araucaria. In Tomlinson PB, Zimmermann MH (eds): Tropical Trees as living systems. Cambridge, London: pp. 233–245.

Waring RH (1980): Stem growth per unit leaf area: a measure of tree vigor. For Sci 26:112–117.

Waring RH, Schroeder PE, Oren R (1982): Application of the pipe model theory to predict canopy leaf area. Can J For Res 12:556–560.

Wheat D (1980): Sylleptic branching in Myrsine floridana (Myrsinaceae). Am J Bot 67/4:490–499.

Whitehead D, Edwards WRN, Jarvis PG (1984): Conducting sapwood area, foliage area and permeability in mature trees of Picea sitchensis and Pinus contorta. Com J For Res 14:940–947.

Wühlisch Gv (1984): Untersuchungen über das prädeterminierte und freie Trieblängenwachstum junger Fichten (Picea abies (L.) Karst.) als Voraussetzung für einen Frühtest. Dissertation, Universität Hamburg, Hamburg, Germany.

Zimmermann MH (1983): Xylem structure and the ascent of sap. Berlin, Heidelberg: Springer-Verlag.

EFFECTS OF POLLUTANTS AND ENVIRONMENTAL STRESS ON THE MORPHOLOGY OF DECIDUOUS TREES AND ON THE GROUND VEGETATION

ANDREAS ROLOFF

Institut für Forstbotanik, Technische Universität Dresden, D-01737 Tharandt, Germany

1. SYMPTOMS OF FOREST DECLINE IN HARDWOODS

1.1. Introduction

It is still very difficult to determine vitality and thereby the effects of acid disposition (e.g., in deciduous trees). The reason for this difficulty is due to the fact that until now most inventories considered only parameters such as "percentage leaf loss" and leaf coloring. Therefore, this chapter will consider the branching and the crown structure of trees in the assessment of tree vitality. This consideration is also important for the interpretation of aerial photographs.

Among deciduous trees the European beech (*Fagus sylvatica*) is of particular interest, as its distribution ranges throughout central Europe, and in many countries it represents the most important broad-leaved tree species. Furthermore, it grows naturally on extremely different sites.

1.2. Decline Symptoms of Tree Crowns: "Leaf Loss" Versus Crown Structure

The consideration of crown structures in the assessment of tree vitality has become increasingly important. That scientists are now aware of the problem of "leaf loss"

Effects of Acid Rain on Forest Processes, pages 325–351

is shown in many studies (Roloff 1986, 1989 a,b; Flückiger et al., 1986, 1989; Perpet, 1988; Gies et al., 1989; Möhring, 1989; Richter, 1989; Westman, 1989): The number of leaves, and above all the leaf size, are subject to considerable annual fluctuations, for example, as a result of drought and insect damage or flowering and fructification, respectively. Thus crown structures are to some degree inappropriate to the vitality assessment of deciduous trees. The correlation between "leaf loss" that is crown transparency and fructification has been well demonstrated (Flückiger et al., 1989). On the other hand, the foliage must also be considered. However, the leaf size varies greatly even within the same crown of a deciduous tree. Thus it is difficult to show a statistical significance in values between different trees (Roloff, 1986; Gies et al., 1989).

For this reason it is not surprising to find considerable disagreement between so-called "leaf loss" and crown structure (to be discussed later) occurring if vitality assessments of the same beech trees are compared (Table 9.1). Assessments may differ by up to two damage or vitality classes and agreement is only achieved in about 50% of the assessed trees (Hessische Forstliche Versuchsanstalt Anst., 1988; Athari and Kramer, 1989a).

Therefore it would be advantageous if the term "leaf loss" could be replaced by a different term (e.g., crown transparency), which does not lead to the misconception of shedded leaves. A deciduous tree showing a "leaf loss" of 30% does not mean that 30% of the leaves were shed. These leaves simply never existed at the beginning of the vegetation period (Roloff, 1986).

In this context, there is one more aspect that should be mentioned, which has become particularly apparent in the most recent investigations. There are a great variety of tree species in which the crown, with increasing shoot lengths, becomes more transparent with better growth (Fig. 9.1). In this case a vitality assessment on

TABLE 9.1. Disagreements Between a Vitality Assessment of Beech Trees Based Upon Leaf Loss Damage Classes and Crown Structures, Vitality Classes

Vitality Class (crown structure)	Damage Class (leaf loss)			Total
0	17	17	5	39
1	8	21	14	43
2	1	6	11	18
Total	26	44	30	100
0	21	35	7	63
1	0	17	13	30
2	0	1	6	7
Total	21	53	26	100

[a] Athari and Kramer, 1989 (n = 333).
[b] Hessische Forstliche Versuchanstalt, 1988 (n = 1052) (distribution in %).

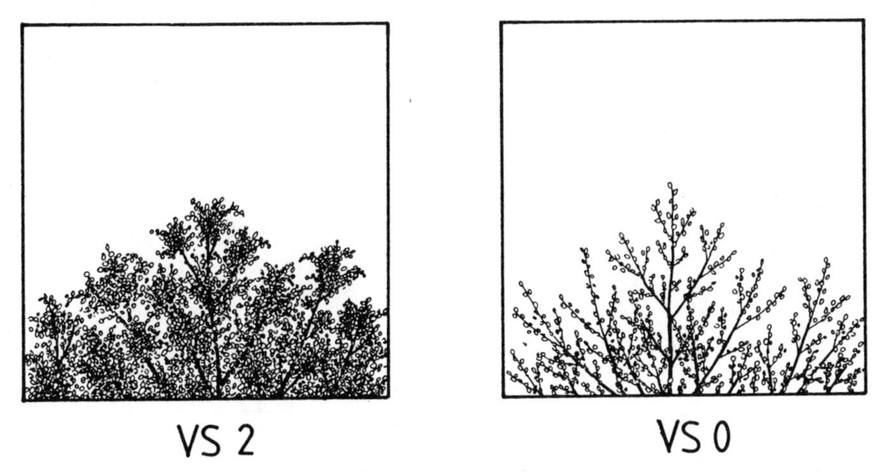

Fig. 9.1. Disagreement between a vitality assessment based upon crown transparency and crown structure (VS = vitality stage according to Roloff, 1989a): In many tree species (here: wild cherry) the crown becomes more transparent with better growth.

the basis of crown transparency versus crown structure is bound to produce exactly opposite results (Roloff, 1989a,b).

A problem inherent to present inventories for the determination of decline progress is discussed by Möhring (1989). By using a series of photographs taken over a 5-year period of the same crown parts of beech trees, Möhring could show that due to die-off and breakage of branches, decline trend is inadequately represented by "leaf loss." Thus, no change of damage class is noticed when the dying-off and shedding of branches happens simultaneously (Fig. 9.2). If the number of branches that break off is greater than the number of branches that die-off, a decrease of decline is determined. However, if the number of branches that die-off is greater than that of the branches shed, an increase of decline is recorded. In these cases, the true decline process cannot be assessed by "leaf loss."

1.3. Changes in the Crown Structure With Decreasing Vitality

In this section tree vitality is discussed in terms of growth potential, which in trees is expressed in shoot growth. Although various branching structures within one tree crown have long been known (see e.g., Büsgen, 1927; Thiebaut, 1981, 1988), their significance as a vitality indicator has only recently been discovered (Roloff, 1985).

1.3.1. Shoot Morphology

By careful observation of the branching pattern of a hardwood tree, closely packed grooves upon the shoot surface are conspicuously recognized (Fig. 9.3). The significance of these shoot-base scars was unfortunately ignored for a long time. They are the scars of the bud scales, which when closely packed originally encased the young shoot primordia, and hence mark the boundary between 2 years of growth exactly to

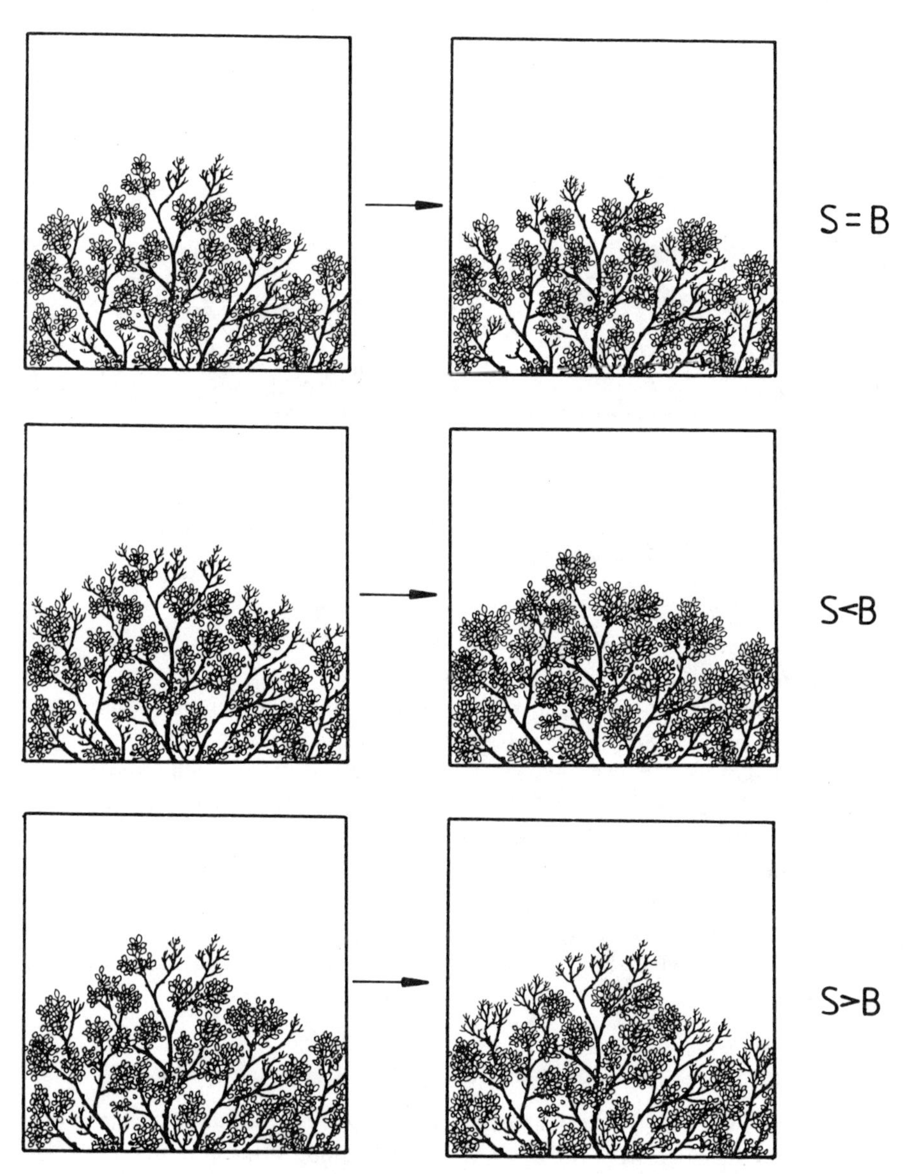

Fig. 9.2. Decline trend is often represented inadequately by "leaf loss" because of a different time course of die-off (S) and breakage (B) of branches.

a millimeter. In beech the rind does not turn to bark as it does in oak, but remains more or less smooth. Thus it becomes possible to retrace the development of any hardwood branching pattern for many years (in beech, e.g., over decades) and in this way to reconstruct its growth.

Further investigation of the branching pattern in most tree species distinguished

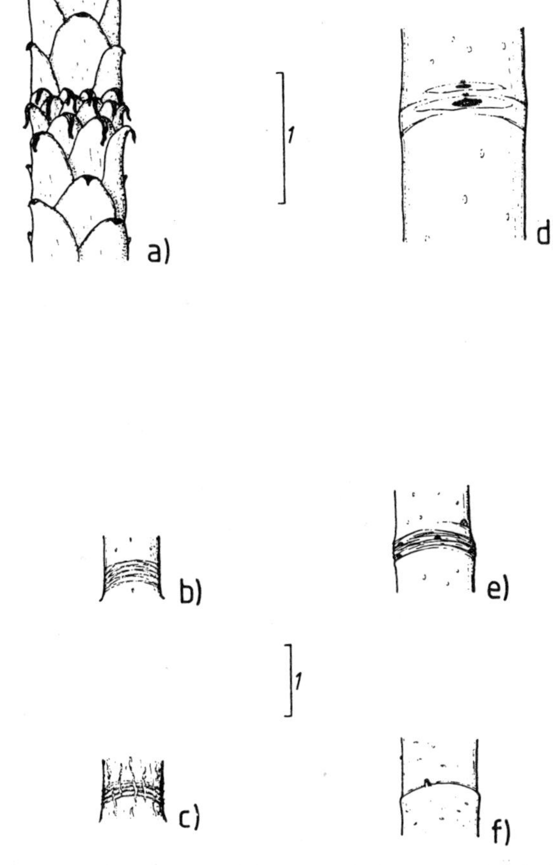

Fig. 9.3. Shoot base scars of Scotch pine (*a*), Sycamore (*b*), Norway maple (*c*), European ash (*d*), Oak (*e*), and Common willow (*f*) (scales in cm).

two kinds of shoots: short shoots and long shoots (Fig. 9.4). Short shoots are only a few millimeters or centimeters long, have only 3–5 leaves, and do not ramify in the following years, because they bear only small dormant lateral buds. The terminal bud of short shoots, however, either produces a short shoot again the following season and short-shoot chains are formed, or else it returns to forming a long shoot. Long shoots are clearly longer, show more leaves, and ramify in the following year.

In any tree species, because of maturity, the length of growth of the treetop shoots, and therefore the height increment of the tree, decreases after passing a culmination point. This finding reflects decreasing vitality in the tree. The lengths of treetop shoots are interpreted as a sign of vitality, because the strategy of a forest forming tree species striving to conquer new airspace must occur at the very top of the tree. On the other hand, the shoot lengths in the inner, lower, and lateral crown areas mainly depend on competition and light conditions, and therefore are unsuitable for the assessment of the vitality of a whole tree.

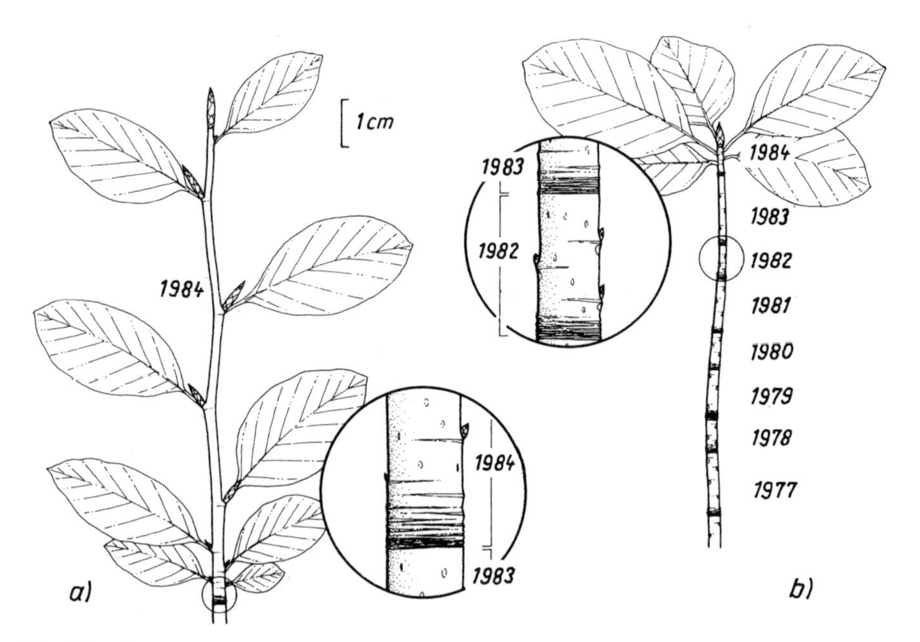

Fig. 9.4. Shoot morphology of beech. (*a*) long-shoot with shoot-base scar (circle), alternate distichous leaves and lateral buds, terminal bud; detail: shoot-base scar, the boundary between the annual shoots of 1983 and 1984. (*b*) A 9-year-old short-shoot chain without ramification and with a terminal cluster of leaves; detail: annual shoot boundaries, clearly marked by the shoot-base scars; dormant lateral buds (reserve buds).

1.3.2. *Model of Growth Phases*

Figure 9.5 shows how a typical ramification originates from a leader shoot of a vigorous beech. This ramification is similar in most other hardwoods with only few modifications. This "exploration phase" produces the branching structure that is best known and most frequently found: The terminal and upper lateral buds develop yearly into long shoots, the lower lateral buds develop short shoots, and finally, the lowest lateral buds do not shoot at all, but remain for years as very small dormant buds preserved for unusual events.

On every annual leader shoot the lengths of the lateral younger shoots decrease from top to bottom and the developing branching pattern is turned upwards or forwards. In this way an obviously storied branch system is developed and the annual shoot boundaries (marked by the interruptions of the black lines in Fig. 9.5) can be distinguished even from a distance by the steps of the branching pattern and by the abrupt change from long lateral shoots to short shoots. This exploration phase is the widespread appearance of the leader shoots in healthy vigorous trees until an old age, because this is the only way the treetop can fulfill its main purpose for the tree; that is, to steadily conquer new airspace, to fill it up with lateral shoots, and to be successful against rival trees.

In the "degeneration phase" (Fig. 9.6), however, the terminal bud develops into

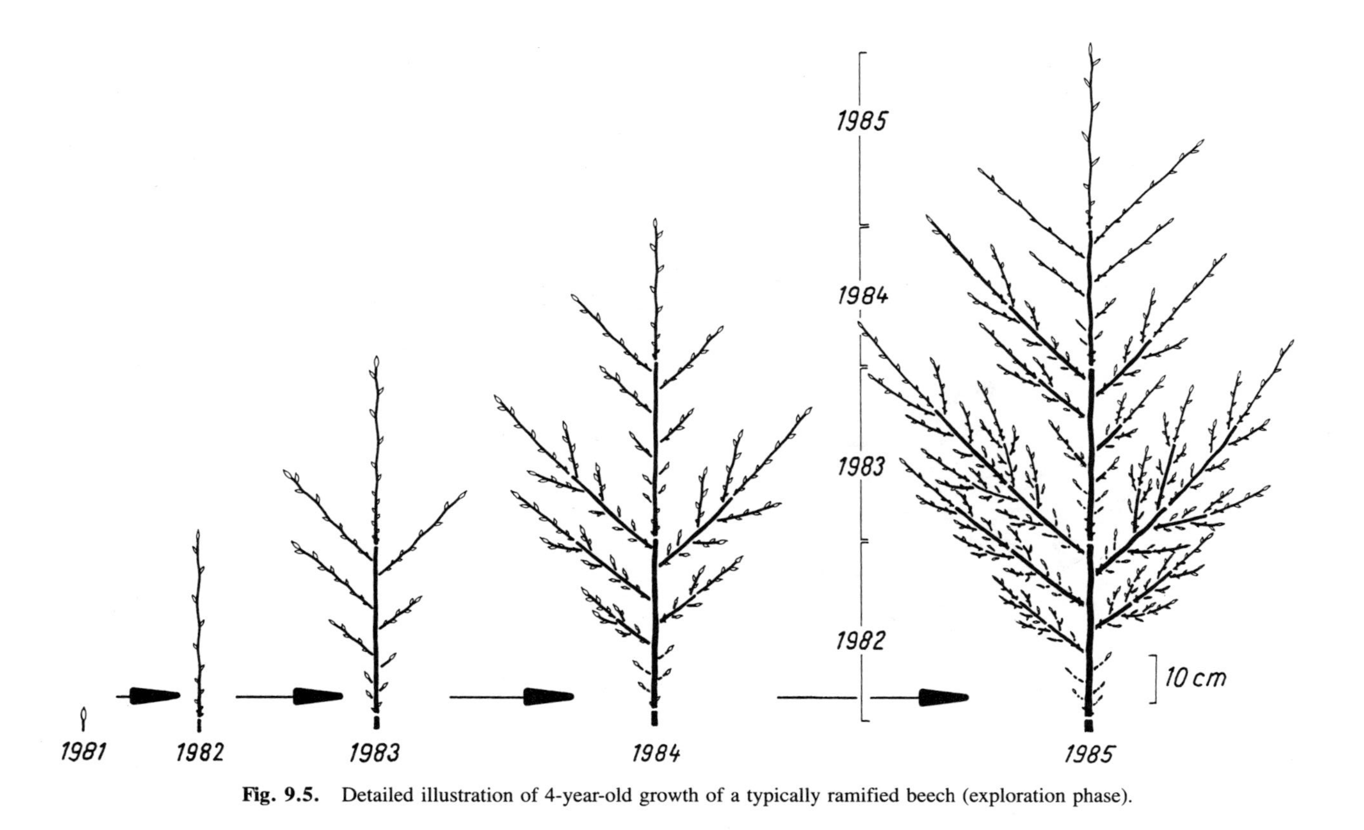

Fig. 9.5. Detailed illustration of 4-year-old growth of a typically ramified beech (exploration phase).

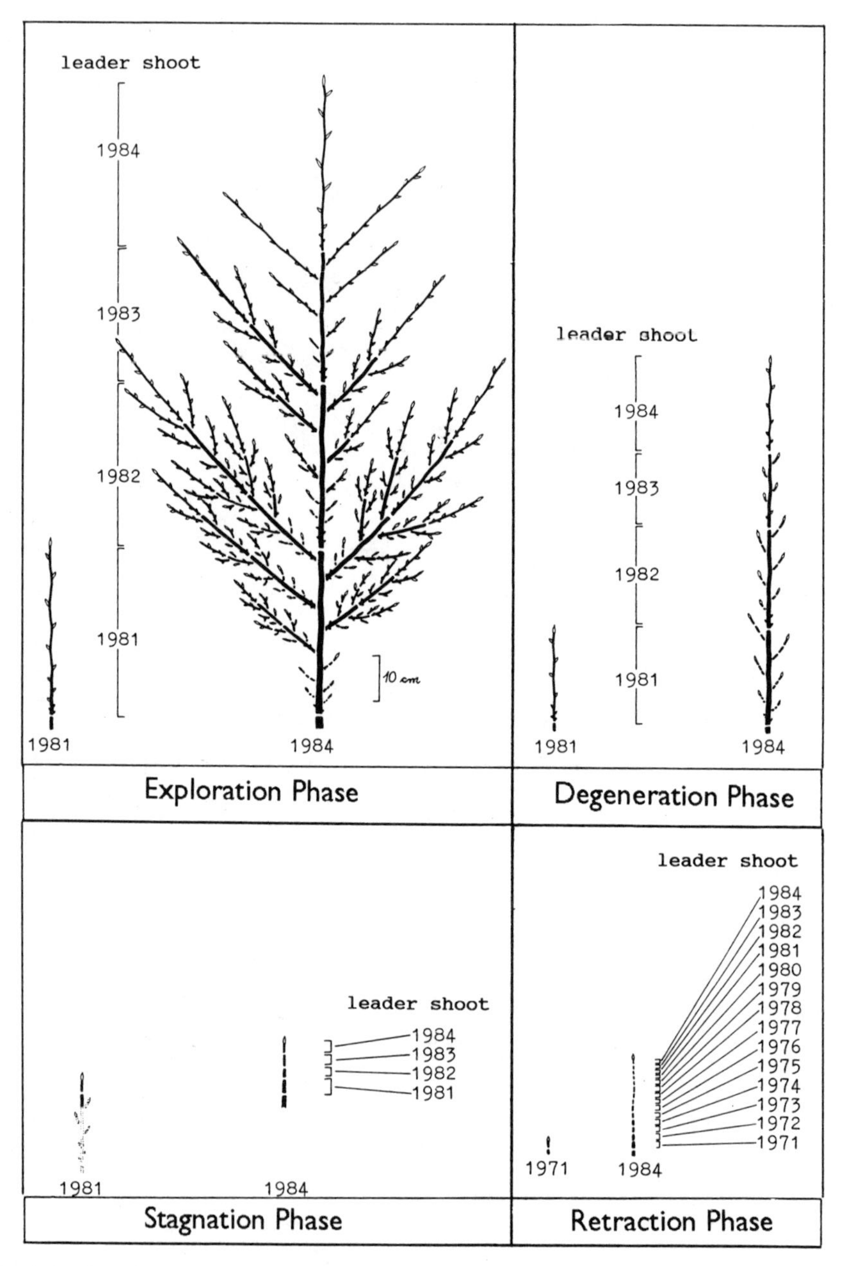

Fig. 9.6. Growth phases of beech (model): on the left the state of the leader in 1981, on the right its development and ramification after 3 years (1984) (annual boundaries marked by interruptions of the black lines).

shorter long shoots, but from nearly any lateral bud (from the uppermost, also) short shoots arise almost without exception. Thereby an obvious impoverishment of the branching pattern takes place and spears are formed in the periphery of the crowns, which may also be seen from a great distance.

In the course of further decreasing vitality, even the terminal bud changes into developing short shoots. In this "stagnation phase" ramification ceases because short shoots do not ramify. Because of the short annual length of these shoots, the length increment of the branch and the height increment, respectively, of the tree stagnates.

If this stagnation phase persists longer than a few years (if it is not only temporary), the branch or (if concerning the treetop shoots) the treetop dies back. This stage is called the "retraction phase."

As a result of their disadvantageous mechanical–static features (a dense cluster of leaves at the end of very delicate shoots), the short shoot chains cannot grow to any length or any age in the upper-crown area exposed to the wind. At this time, secondary factors determine the exact time of dieback. Typically, at this stage claws are formed in the crown periphery because the short shoot chains, which are getting longer, stretch toward the light.

Now it becomes clear why the different phases of growing are due to decreasing shoot lengths and, therefore, reflect a decreasing vitality.

Similar growth phases can be identified in every other investigated broadleaved tree species, with only little species-dependant modifications, for example, in oak, maple, ash, birch, willow, and others (Roloff, 1989a). The following vitality class system is developed based on these growth phases.

1.3.3. Vitality Classes

Healthy vigorous trees of "vitality class 0" (Fig. 9.7) show treetop shoots in the exploration phase: Both the main axes and part of the lateral twigs consist of long shoots. For this reason a regular netlike branching pattern is developed, which reaches deep into the interior of the crown. The crowns are equally closed and domed, and do not show any greater gap unless a stronger intervention has occurred in the stand, because such a gap is closed quickly by the intensive ramification. In this manner the newly conquered airspace is quickly filled by the harmonical branching pattern. In summer a dense foliage arises without any greater gap.

Weakened trees of the "vitality class 1" show treetop shoots in the degeneration phase. Thus, spears are formed rising above the canopy. The leaves on these spears are dense and go all around the spears (at the top of the lateral short shoots or short-shoot chains). The crowns make a frazzled impression on the outside and show a fastigiate appearance, because the airspace between the spears is not completely filled by leaves and twigs, and the crown has a spikey outline. Inside the crown the branching pattern, and therewith the foliage, is quite dense. In this vitality class straight percurrent main axes of the treetop branches are still dominant, but the crowns no longer look as intact as in class 0 because of the spears shooting out of the canopy.

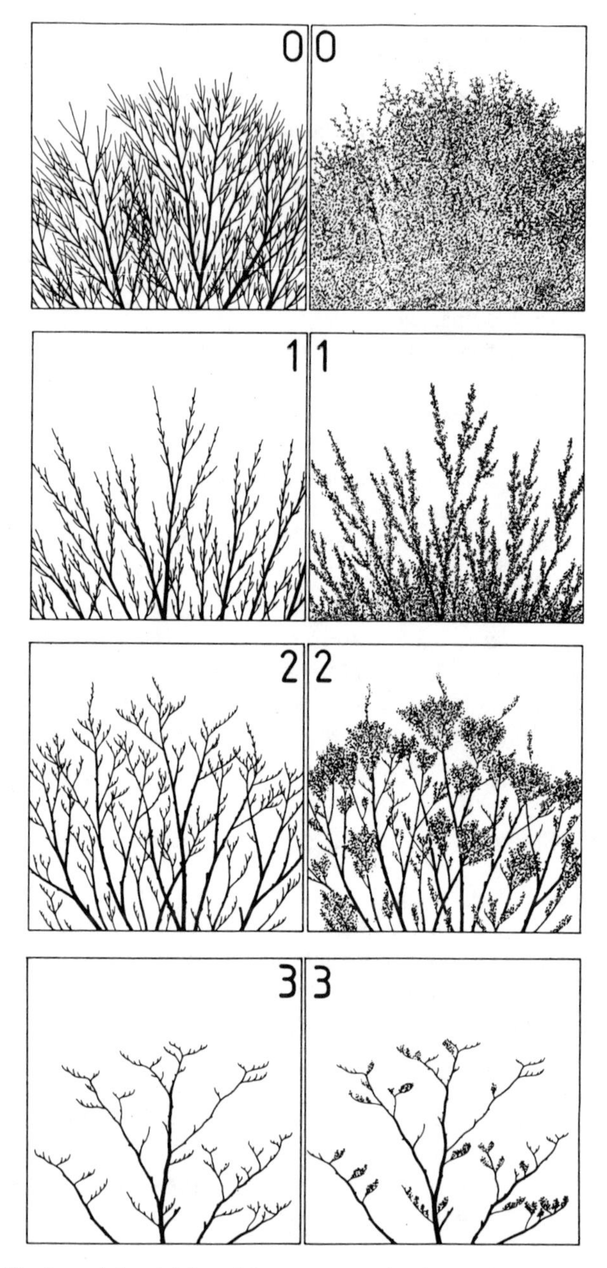

Fig. 9.7. Vitality classes in beech (view of the upper crown in winter and summer; based on the growth phases of Fig. 9.6).

In the obviously declining trees of "vitality class 2" the treetop shoots begin to build short shoots in the stagnation phase. The leafless state could be designated as the claw stage, because the short-shoot chains in the outside of the crowns grow longer, are predominant, and stretch clawlike to the light. These short-shoot chains, which are growing too long, break off in the summer in thunderstorms and heavy rains and strew the forest floor in declining beech stands. Under normal circumstances beech as well as other tree species rid themselves of part of their unimportant twigs in the inner and lower crown parts in this way. But, if the treetop shoots themselves are declining, the self-pruning of twigs progresses into the outskirts of the crown, and the crowns become thin from inside outwards. The cause for this occurrence is not premature leaf fall but broken short-shoot chains, lack of shoots, and dead buds and twigs. The branching pattern shows a bushy and lumpy accumulation in the periphery of the crowns. This accumulation causes summer and winter bushy crown structures and greater gaps. The crown periphery still hardly has any straight percurrent branches.

In the considerably damaged or dying trees of "vitality class 3" the crowns finally fall apart by the breaking off of larger branches and dieback of whole crown parts. The tree only seems to consist of more or less surplus subcrowns dispersed randomly in the airspace and forming whiplike structures. The treetop often is dying back or is already dead, because the treetop shoots grew in the retraction phase.

1.3.4. Drought Damage and Relationship of the Presented Vitality Criteria to Environmental Factors

After the previous explanations, it should be self-evident that such a vitality class system based on criteria of the branching structures solely goes back to a long-term, chronic diminution of vitality.

Drought damage, for example, can influence branching only temporarily and does not lead to a fundamental variation of the branching structure. Drought damage can be identified even after many years with the help of shoot base scars, because very short shoots are formed abruptly in the year after a dry summer (in species with determinated growth) and recover their original values again quickly the following season (Fig. 9.8).

On the other hand, drought damage can cause a fundamental passing thinness of the foliage because of the premature leaf-fall during the current summer and smaller and fewer leaves in the following season. But from a distance it is impossible to recognize the temporary short shoots caused by a past drought (e.g., from ground level in a forest) and they cannot effect a fundamental modification of the branching structure (Fig. 9.9), whereas a chronic decrease of the vitality (connected with a long-term decrease of the shoot lengths in the treetop) can cause a conspicuously different branching structure in hardwoods (Fig. 9.10). Of course, the best time for an exact assessment of vitality with the help of this method is after leaf-fall, in autumn and winter.

As only long-term influences can affect a fundamental modification of the branching pattern, this method is suitable to identify "forest decline" in hardwoods,

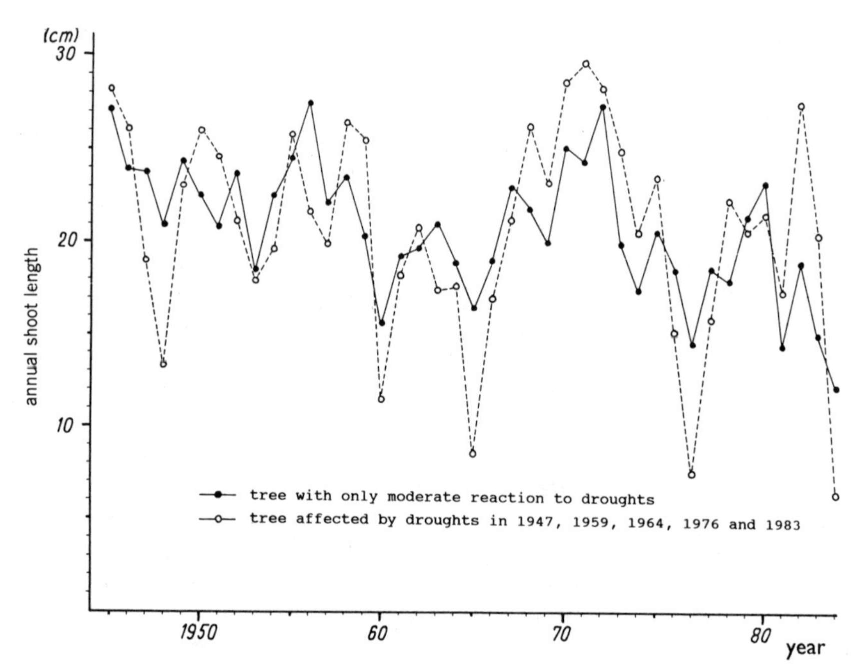

Fig. 9.8. Annual length increment of the leader shoots of two 120-year-old beech trees, one of which clearly shows drought damages (very short shoots formed abruptly in the year following an extremely dry summer and afterwards quick recovery), whereas the other remained nearly unaffected.

Fig. 9.9. A "hidden" drought damage in the branching pattern on the right (see arrows) does not lead to a fundamental modification of the branching structure, but to a "delay" of 1 year compared with the unaffected one on the left.

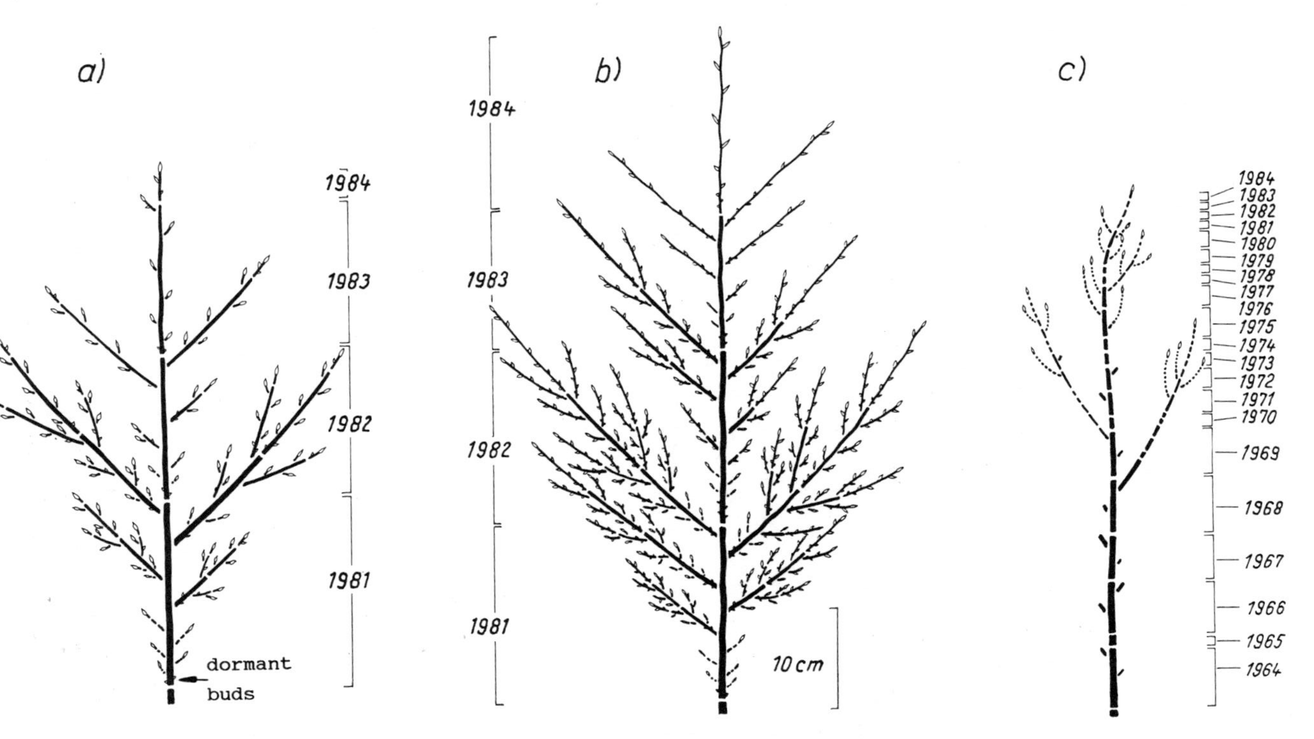

Fig. 9.10. Comparison of typical treetop branching patterns of a vigorous beech (*b*), a drought damaged beech (*a*), and a chronically declining beech (*c*).

because the recent rise in the decline of European and North American forests is supposed to be caused by existing and long lasting influences. Therefore, modifications of the branching structure are not exclusively specific to air pollution, but, for example, are due to such long-term negative factors affecting tree growth such as air pollution. It is similar in almost all other symptoms of forest decline that are known to date: They can be caused by air pollution, but may also be a result of other stress factors (Hanisch and Kilz, 1990; Hartmann et al., 1988).

Surprisingly, further investigations of beech trees have clarified how long the decline in this important European hardwood species has been going on (Fig. 9.11). Only the vigorous, healthy trees of vitality class 0 (vs 0) follow the age trend of the yield tables, whereas trees of the vs 1–3 decline many years since. Transforming the absolute values of the leader increment into values in percent of vs 0 (Fig. 9.12) emphasizes that trees of vs 1 + 2 have in most cases been in decline for about 10–15 years and trees of vs 3 decline for about 20–25 years. Another way to show the long-term differences between the vigor of trees of different vitality classes is the average annual length increment of the leader shoots during the last 10 years (Fig. 9.13). This diagram emphasizes again the expressiveness of the presented vitality class system.

This system, which is based on crown structures, may also be used successfully in aerial photographs, thus making it possible to cover a larger area in a shorter time (Fig. 9.14).

Recent research in 18 other hardwoods has shown the possibility of assessing the

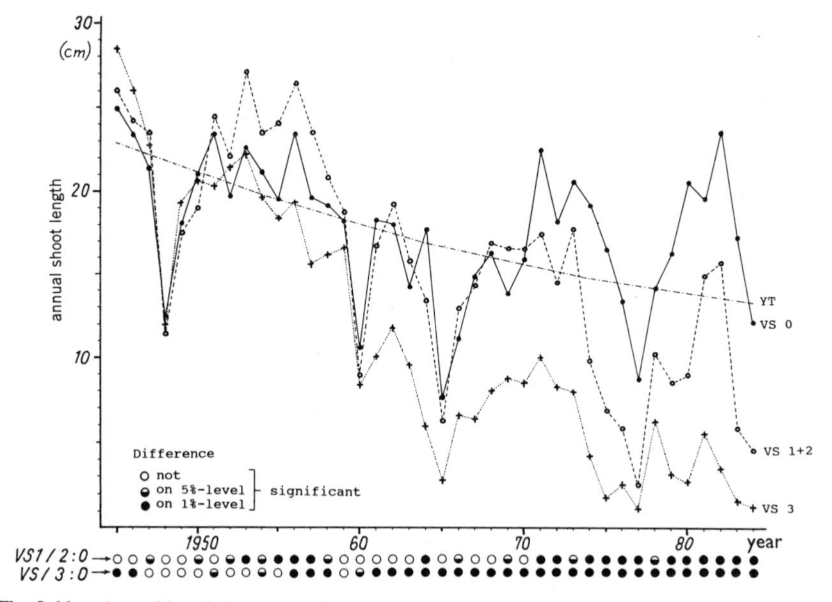

Fig. 9.11. Annual length increment of the leader shoots of 100–160-year-old beech trees during the last 40 years (classified by vitality classes according to Fig. 9.7: VS = vitality stage; YT = expected trend according to the yield tables; VS 0: n = 140, average age 131 years; VS 1 + 2: n = 279, average age 132 years; VS 3: n = 141, average age 134 years).

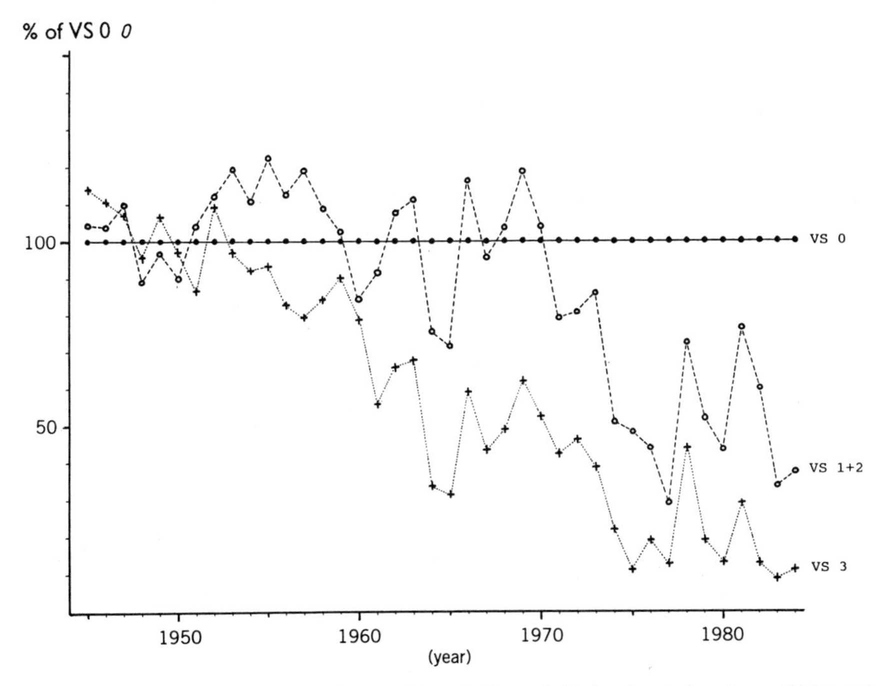

Fig. 9.12. Annual length increment of the leaders as in Figure 9.11, but the vitality classes (VS) 1 + 2 and 3 are in percent of the annual value of VS 0 (VS 0 in each year as 100%).

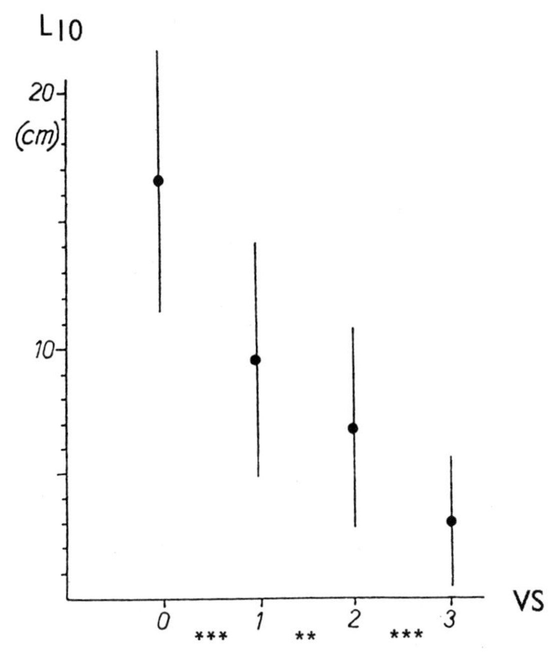

Fig. 9.13. Average annual length increment of the leaders of different vitality classes (VS) in the last 10 years (L_{10}) (mean value, standard deviation, and statistical significance of the difference, * significant on 5%-level, ** on 1%-level, *** on 0.1%-level).

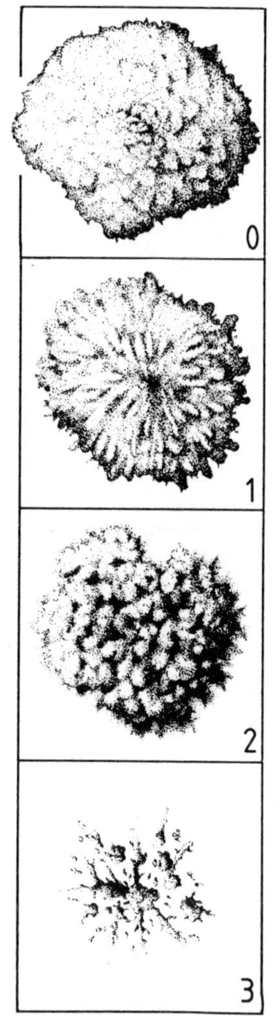

Fig. 9.14. Vitality classes (according to Fig. 9.7) in aerial photographs (beech).

vitality of other tree species of the northern hemisphere in the same way as reported here for beech trees. In these hardwoods only very few species-dependent modifications must be taken into account. This research was conducted for *Betula pendula, B. pubescens* (Roloff, 1989; Westman, 1989), *Acer platanoides, A. pseusoplatanus, A. saccharum, Aesculus hippocastanum, Alnus glutinosa, Carpinus betulus, Fraxinus excelsior, Fagus americana, Prunus avium, Quercus petraea, Q. robur, Robinia pseudacacia, Salix caprea, Tilia cordata, T. platyphyllos* (Roloff, 1989a).

The vitality class key for beech trees was approved and applied in a variety of investigations (Lonsdale, 1986; Longsdale and Hukman, 1988; Dobler et al., 1988; Hessische Forstliche Versuchanstalt, 1988; Gies et al., 1989; Stribley, 1993). Because vitality class 1 is the most important class for early diagnosis Perpet (1988) suggests an even more differentiated key, with a low, a medium, and a strong spearlike development of branches. The determination of a vitality quotient is postulated by Gies et al. (1989): This nondimensional number is calculated following the formula

$$V = \frac{\Sigma\frac{S}{E} + \Sigma\frac{K}{10}}{2n}$$

in which E represents the length of the apical shoot; K is the number of buds along the apical shoot; S is the length of the uppermost lateral shoot, and n is the number of measurements carried out per tree.

The quotient can be determined from photographs taken during the winter period. It shows strong correlation with the leaf biomass, the leaf surface index, the shoot length, and the number of buds (Gies et al., 1989).

Interestingly enough, on fertilized sites (set up in the early 1980s) in the Solling mountains, where there are plots that have been either acidified, limed, or left untreated since 1980, a differentiation of the crown structures of beech trees is just beginning to emerge (Fig. 9.15).

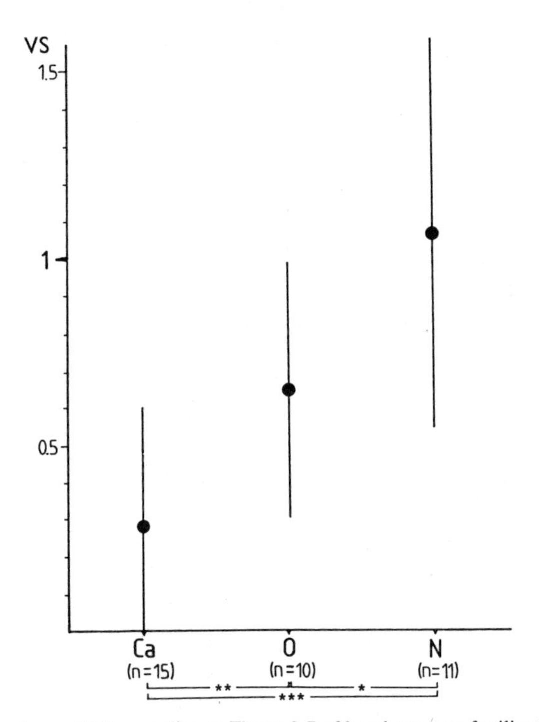

Fig. 9.15. Vitality classes (VS) according to Figure 9.7 of beech trees on fertilized sites in the Solling mountains in Germany (set up in the early 1980s) with acidified (N), limed (Ca), and untreated (O) plots (for explanation of the statistical significance see Fig. 9.13).

1.4. Effects on Radial Growth, Correlations With Root Development, and Genetic, as Well as Silvicultural, Consequences

Although a close correlation between vitality class (on the basis of crown structure) and radial growth of branches of the crown does exist (Fig. 9.16; Roloff, 1989a), it is difficult to correlate with radial growth at breast height (Gärntner and Nassauer, 1985; Wahlmann et al., 1986; Perpet, 1988; Mahler, 1988; Athari and Kramer, 1989a,b; Fischer and Rommel, 1989), particularly if the space of the trees investigated is not considered. Possible reasons for this generally unexpected discrepancy will not be discussed here (refer to Roloff, 1986, 1989a). However, it must be pointed out that the growth at breast height as a vitality indicator is not only problematic in the case of beech trees. On the other hand, close relationships were found if the space of the trees were taken into account and if investigated trees were of the same vitality class based on both "leaf loss" and crown structure.

The correlation between changes of the crown and root growth, which unfortunately require a considerable amount of research, are of particular interest. A legitimate assumption is that the presented modifications of the branching structure are related to similar modifications of the root system, because recent research has shown the high correlations between both of these components of the system tree. By proceeding on the hypothesis that the crown and root development of a tree are related, a tree can be considered as an integrated system of interdependant relationships. On three very differing sites, 24 beech trees of 2-m height were completely excavated and quantitatively analyzed (Roloff, 1989a; Roloff and Römer, 1989). For these trees (apart from soil data) the following parameters were determined: total leaf number, total leaf area, average leaf size, number of short and long shoots developed in different years, total shoot lengths, total root lengths exceeding a root diameter of 2 mm, total shoot dry weight, total root dry weight exceeding a root diameter of 2 mm, age of plant, and rooting depth.

In almost all cases unexpectedly high significant correlations between all crown and root parameters were found (Fig. 9.17). This correlation emphasizes the fact that neither phenomenon, the crown development nor the root–crown interaction, are purely accidental, but that the entire tree represents an integrated regulated system in which every change of one parameter immediately induces changes in other parameters of the crown and/or the root system. Neither the age of the tree (9–31 years) nor the site have a considerable influence on the relationships within the crown and between the crown and root of 2-m high beech trees.

Genetic consequences are presented by Müller-Starck and Hattemer (1989). In beech trees they found (on the basis of crown structure) a decreasing genetic variety (heterozygotic grade and genetic diversity, respectively) with a decreasing vitality of the tree, assessed by the method presented. Accordingly, only beech trees of high genetic variety survive stress without suffering obvious damage.

Finally, changes of the crown structure of stressed beech trees may also have silvicultural consequences (Dobler et al., 1988; Hessische Forstliche Versuchsanstalt, 1988). As the canopy becomes more transparent after thinning, it is difficult to selectively regulate natural regeneration of beech stands. A modification of thinning approaches and methods were suggested.

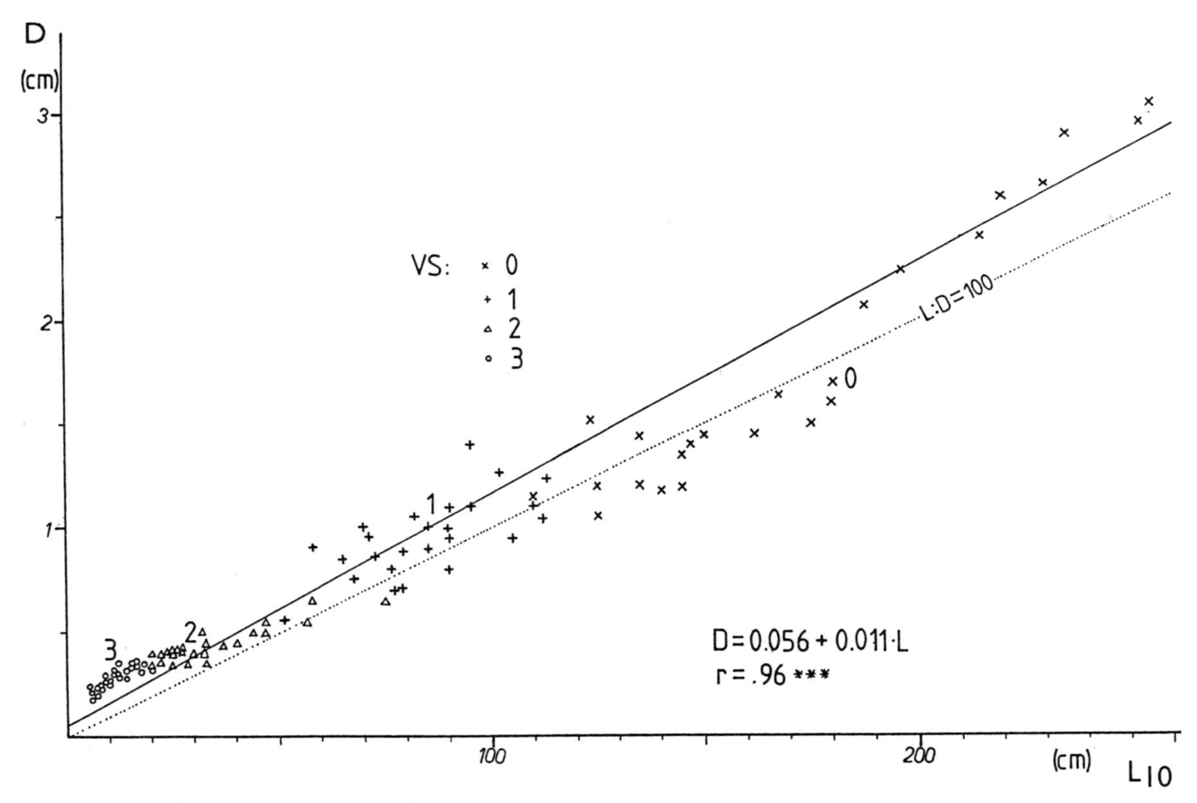

Fig. 9.16. Correlation between 10 years of growth of shoots (L 10) and their basal diameter (*D*) at the 10-year-old bud scale scar (VS = vitality class according to Fig. 9.7, *r* = correlation coefficient) (for explanation of the statistical significance see Fig. 9.13).

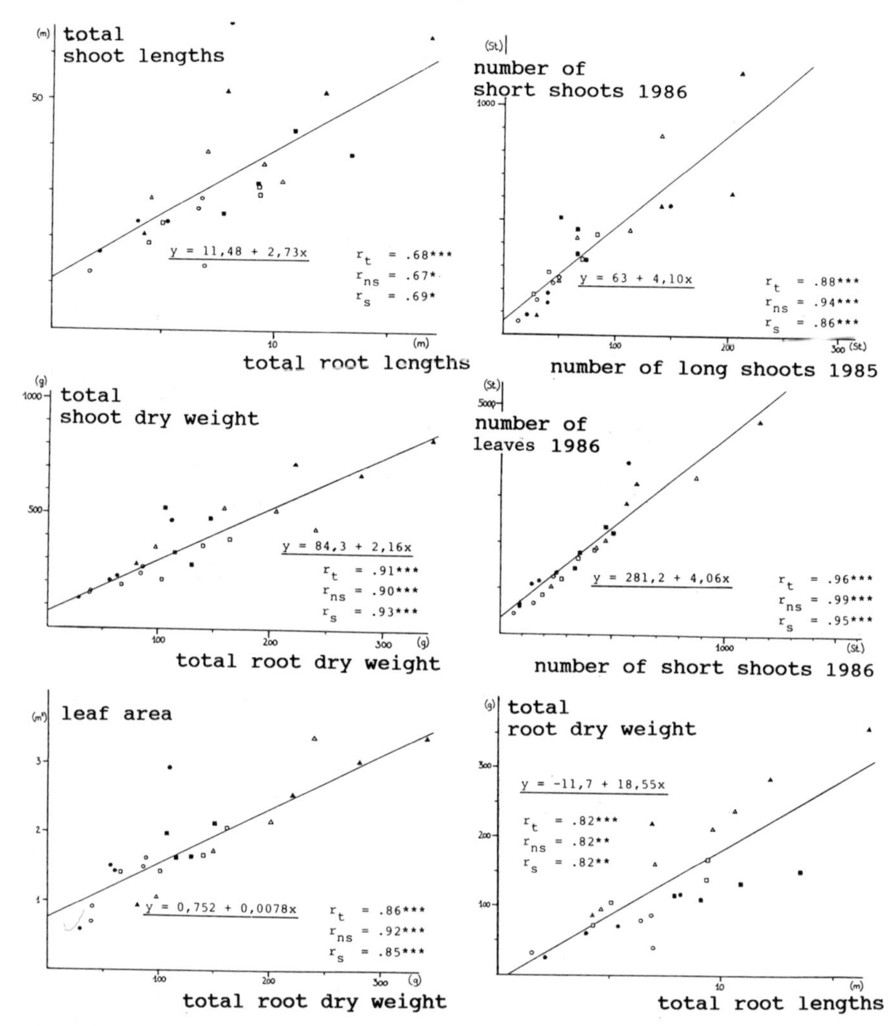

Fig. 9.17. Results and interrelationships of crown and root parameters in excavated 2-m high beech trees (y: linear regression, r_t: correlation coefficient total and for sylleptic (r_s) and non-sylleptic trees (r_{ns}), circle: site 1, triangle: site 2, square: site 3; full symbols: sylleptic trees, blank symbols: non-sylleptic trees) (for explanation of the statistical significance see Fig. 9.13).

2. EFFECTS OF ACID AND NITROGEN DEPOSITION ON GROUND VEGETATION

The changes in forest soil vegetation due to the increasing effect of input from the elements in recent years are very important factors to be considered by all plant sociologists and vegetation ecologists, even if there is a disagreement on the causes and effects. Meanwhile much research was carried out on this subject, which was based on two different methods: some groups reinvestigated sites on which a vegetational survey was carried out years or even decades ago. These investigators wanted to find out if the vegetation and flora had changed (e.g., Kaiser and Roloff, 1989). Others examined the changes in vegetation up to the root collars of the trees in the same beech tree stand.

The hypotheses of how the increasing input of elements affects the ground vegetation are quite varying. While decreases in the density of the canopy layer due to forest damages probably result in a reduction of the competition strength of the trees and in a better development of the herb layer (especially of nitrogen indicators), Wittig and Werner (1986), for example, suspect such a development is attributable to the earlier leaf fall of the trees and the resulting increase in light for the ground vegetation, particularly in late summer and fall. Kuhn et al. (1987) assume that forest stands get darker due to a better tree growth caused by a high nitrogen input that can hinder the development of the ground vegetation.

An increase in nitrogen indicators in the forest soil vegetation is quite obvious nowadays, even if a vegetation comparison over decades seems to be problematic. Research studies show an immigration of nitrogen indicators, especially *Sambucus nigra, Galeopsis tetrahit, Solanum dulcamara, Moehringia trinervia,* and *Urtica dioica,* and these studies even found an outbreak of these species in stands in which they could not be found earlier (e.g., Trautmann et al., 1970; Kowarik and Sukopp, 1984; Kuhn et al., 1987). Such a change in the nitrogen supply of forest stands can also be derived from an increase in the mean indicator value (according to Ellenberg et al., 1991) of the forest soil vegetation for the nitrogen supply. This is shown in Figure 9.18 and in a number of other publications (e.g., Wittig, 1986; Rost-Siebert

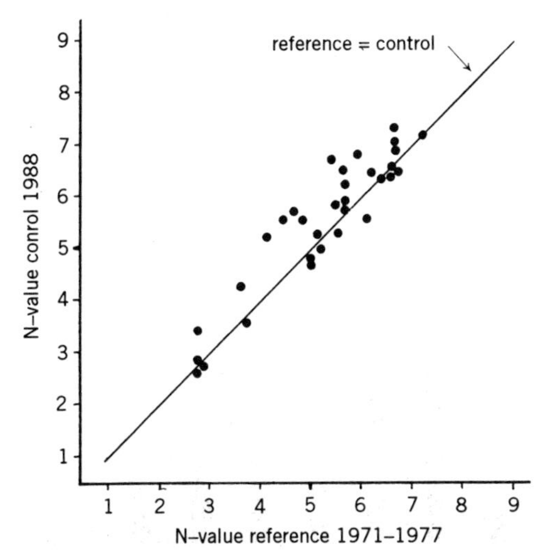

Fig. 9.18. Mean nitrogen indicator value for a comparison of 32 sites in southern lower Saxony; reference survey 1971–1977, control 1988 (*Note:* Most of the control surveys in 1988 show a higher nitrogen indicator value; value 1: nitrogen-poorest sites, value 9: sites with highest nitrogen supply).

and Jahn, 1988; Wehausen, 1989; Bürger, 1991; Philippi, 1991; Storm, 1991). This finding is mainly a result of the high nitrogen input from the atmosphere. Ellenberg (1985) obtains the same result by comparing the indictor values of both endangered ("Red List") and nonendangered species. He concludes that species on nitrogen-poor sites are especially endangered nowadays (Fig. 9.19).

While the effects of the high nitrogen input can be seen all over the forest soil vegetation, those of the acidification in beech tree stands are restricted mostly to the stem flow area. It has to be taken into consideration that this area, due to the concentration of precipitation there and many other factors (e.g., higher variation in temperature, reduced supply with litter, and less snow cover), has always been a special site. But research in beech tree stands in relative immission-free regions (e.g., Croatia and Corsica) shows that the phenomena noticed here today are mainly the results of air pollution (Glavac et al., 1985), for example, the decrease in the pH value, the increase in the concentration of heavy metals, and in the acidotolerant plants near the root collar (Glavac et al., 1970; Wittig and Neite, 1983; Wittig, 1986; Papitz, 1987).

At the same time a clear shift of the species combination compared to the remaining stand often occurs. As seen in Figure 9.20, there is an increase of the acidotolerant moss *Dicranella heteromalla* and a decrease of the acidophobic grass *Brachypodium sylvaticum* up to the root collar. For the tolerant *Deschampsia flexuosa* and the intolerant *Vicia sepium,* Neite and Wittig (1985) found a significant correlation between the cover at the root collar and the proportion of iron ions to the effective exchange capacity of the topsoil (Fig. 9.21).

More results are found as follows: compared to control vegetation sites the total cover of higher plants decreases in some cases to 10% near the root collar. This finding results in a great advantage in competition of especially acidotolerant

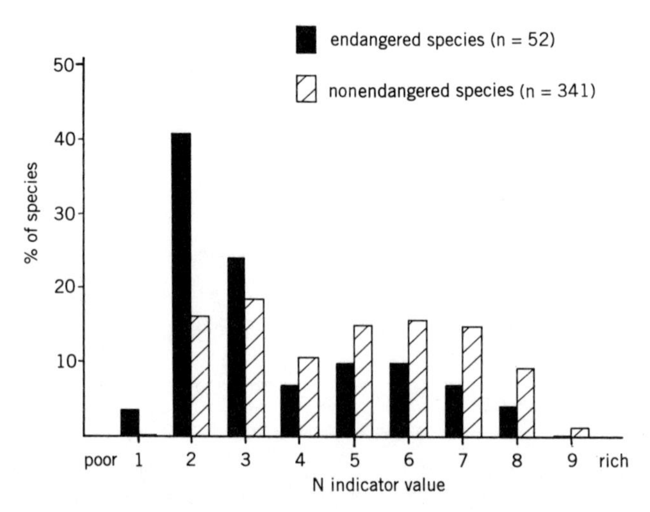

Fig. 9.19. Distribution of endangered (Red List) and nonendangered species of the forest ground vegetation with regard to their nitrogen indicator value (Ellenberg et al., 1991), after Ellenberg, 1985.

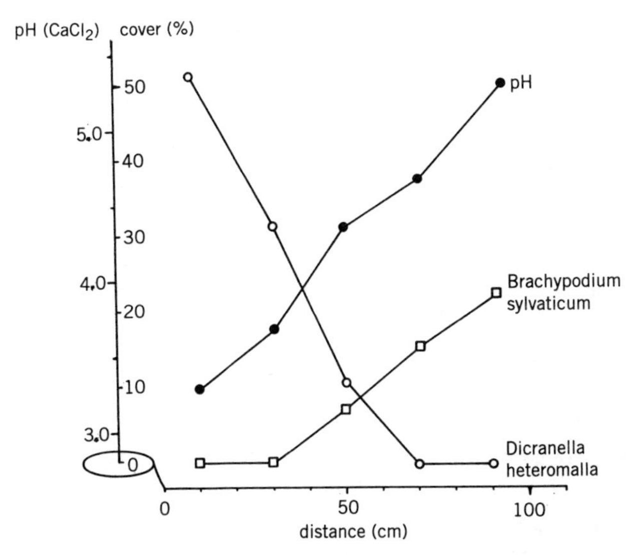

Fig. 9.20. Change of pH value and cover of *Brachypodium sylvaticum* and *Dicranella heteromalla* with an increasing distance from the stem of 160–180-year-old beech trees (Leine mountains, southern lower Saxony, 270 m above sealevel, $n = 30$).

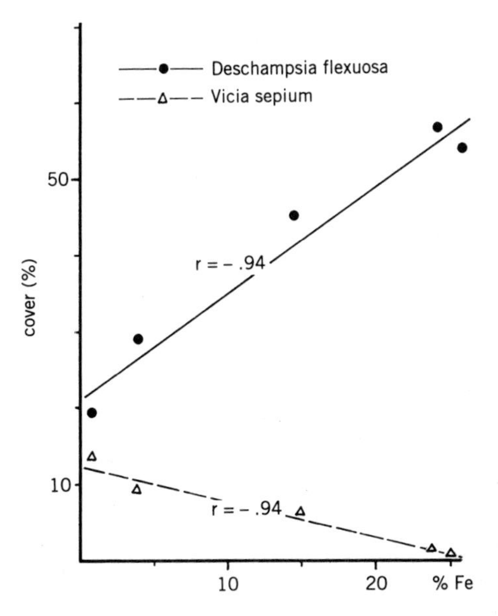

Fig. 9.21. Correlation between the share of the effective exchange capacity of the upper soil layer (%Fe) and the cover of *Deschampsia flexuosa* and *Vicia sepium* in the stem flow area of old beech trees (r = correlation coefficient; after Neite and Wittig, 1985).

mosses such as *Dicranella heteromalla* and *Hypnum cupressiforme,* which can cover up to 80% of the ground here.

Three plant groups in the root collar area are distinguishable and thus are usable as bioindicators (according to Jochheim, 1985):

1. The first group consists of acidophytes, which immigrate into the lime and melic grass-beech woods in the root collar area due to the decrease in other species. But these species disappear even in these areas in woodrush–beech woods; the following species belong to this group: *Agrostis tenuis, Luzula luzuloides, Deschampsia flexuosa,* and *Vaccinium myrtillus.*
2. The second group withdraws from the root collar area in lime and melic grass–beech woods (e.g., *Melica uniflora, Galium odoratum,* and *Carex sylvatica*).
3. The third group can be found more often in the root collar area of lime–beech woods than on control sites (e.g., *Poa nemoralis, Hieracium sylvaticum,* and *Phyteuma nigrum*).

Previous research shows that the former advantage of the beech trees in competition (i.e., a high stem flow with the concentration of precipitation in the central root area) has now changed into a disadvantage caused by the accumulation of noxious substances and the decrease of the pH value.

So the root collar area will probably function as a bioindicator for us to learn at an early stage about the stress of beech tree stands and the tendency for them to develop, which can already be seen at the root collar. At present, beech tree stands develop an increasing vertical and horizontal site heterogenity, which together with a decrease in the density of the canopy layer, temporarily leads to an increment of species, but in the long term leads to a decrease in species richness.

C. SUMMARY

In this chapter I discussed first the general methodical problems of vitality assessments of deciduous trees and existing disparities or contradictions were pointed out if assessments based on "leaf loss" and based on crown structures were compared. The necessity of considering the branching pattern is substantiated and the methods developed to date were presented.

With the help of the "shoot base scars," it becomes possible to reconstruct the crown development over the last 10 years, and in some species over decades. In every investigated (broad leaved) tree species, there are four growth phases to discriminate: exploration, degeneration, stagnation, and retraction. These phases are due to (statistically significant) decreasing shoot lengths, which result in fundamental modifications of the branching structure. Especially in the leafless state, these different branching structures in the treetop are perceived from a distance (and in aerial photographs, too) and are the basis of vitality assessment in four vitality classes. These classes were developed for 19 hardwood species. By using this approach, which is based upon branching structures, a long-term, chronic decrease of vitality can be recognized. Therefore, it is a practical method to use in detecting forest decline and to discriminate it from short-term influences (e.g., drought damages). Correlations between structural changes of the crown and radial growth as well as root development do exist as do genetic and silvicultural consequences.

Presently, plants are becoming more important as bioindicators of environmental stress, especially for acid and nitrogen deposition. The passive monitoring of plant communities in forest ecosystems is presented. Up-to-date examples demonstrate the use of flora and vegetation as indicators for high nitrogen input and acidification close to the root collar in beech tree stands. Their long-term development already appears in the stem flow area, "and will continue to grow" if the high input of noxious substances continues.

REFERENCES

Athari S, Kramer H (1989a): Problematik der Zuwachsuntersuchungen in Buchenbeständen mit neuartigen Schadsymptomen. Allg Forst Jagdztg 160:1–8.

Athari S, Kramer H (1989b): Beziehungen zwischen Grundflächenzuwachs und verschiedenen Baumparametern in geschädigten Buchenbeständen. Allg Forst Jagdztg 160:77–83.

Bürger R (1991): Luftschadstoffbedingte Veränderungen in der Artenzusammensetzung verschiedener Waldgesellschaften. Beih Veröff Naturschutz Landschaftspflege Baden-Württemberg 64: 100–116.

Büsgen M (1927): Bau und Leben unserer Waldbäume. 3rd ed. Jena:

Dobler D, Hohloch K, Lisbach B, Saliari M (1988): Trieblängen-Messungen an Buchen. Allg Forstz 43: 811–812.

Ellenberg H (1985): Veränderungen der Flora Mitteleuropas unter dem Einfluss von Düngung und Immissionen. Schweiz Z Forstwesn 136:19–36.

Ellenberg H, Weber HE, Düll K, Wirth V, Werner V, Paulissen D (1991): Zeigerwerte von Pflanzen in Mitteleuropa. Göttingen, Germany.

Fischer H, Rommel W-D (1988): Jahrringbreiten und Höhentrieblängen von Buchen mit unterschiedlicher Belaubungsdichte in Baden-Württemberg. Allg Forstz 44:264–265.

Flückiger W, Braun S, Flückiger-Keller H, Leonardi S, N Asche, Bühler U, Lier M (1986): Untersuchungen über Waldschäden in festen Buchenbeobachtungsflächen der Kantone Basel-Landschaft, Basel-Stadt, Aargau, Solothurn, Bern, Zürich und Zug. Schweiz Z Forstwesen 137:917–1010.

Flückiger W, Braun S, Leonardi S, Förderer L, Bühler U (1989): Untersuchungen an Buchen in festen Waldbeobachtungsflächen des Kantons Zürich. Schweiz Z Forstwesen 140:536–550.

Gärtner EJ, Nassauer KG (1985): Aktuelles zur Waldschadenssituation in Hessen. Allg Forstz 40:1265–1271.

Gies T, Braun H, Küchler AV (1989): Untersuchungen in mesotrophen Buchenwäldern des Frankfurter Oberwaldes. Poster Intern. Proceedings Kongress Waldschadensforschung Friedrichshafen/Bodensee 2.-6.10.89.

Glavac V, Krause A, Wolff-Straub R, 1970: Über die Verteilung der Hainsimse (Luzula luzuloides) im Stammfussbereich der Buche im Siebengebirge bei Bonn. Schriftenr Vegetationsk 5:187–192.

Glavac V, Jochheim H, Koenies H, Reinstädter R, Schäfer H (1985): Einfluss des Stammabflusswassers auf den Boden im Stammfussbereich von Altbuchen in unterschiedlich immissionsbelasteten Gebieten. Allg Forstz 40:1397–1398.

Hanisch B, Kilz E (1990): Monitoring of forest damage. Stuttgart, Germany:

Hartmann G, Nienhaus F, Butin H (1988): Farbatlas Waldschäden. Stuttgart, Germany:

Hessische Forstliche Versuchanstalt (1988): Waldschadenserhebung 1988. Hann-Münden.

Jochheim H (1985): Der Einfluss des Stammabflusswassers auf den chemischen Bodenzustand und die Vegetationsdecke in Altbuchenbeständen verschiedener Waldbestände. Ber Forschungszentr Waldökosysteme Göttingen 13:1–225.

Kaiser T, Roloff A (1989): Wandel von Flora und Vegetation unter dem Einfluss des Menschen—Beobachtungen im Schweinebruch bei Celle. Forstarch 60:115–122.

Kowarik I, Sukopp H (1984): Auswirkungen von Luftverunreinigungen auf die Bodenvegetation von Wäldern, Heiden und Mooren. Allg Forstz 39:292–293.

Kuhn N, Amiet R, Hufschmid N (1987): Veränderungen in der Waldvegetation der Schweiz infolge Nährstoffanreicherungen aus der Atmosphäre. Allg Forst Jagdztg 158:77–84.

Lonsdale D (1986): Beech Health Study. For Comm Res Dev Pap 149:1–15.

Lonsdale D, Hickman IT (1988): Beech health study. For Res 23:42–44.

Mahler G, Klebes J, Höwecke B (1988): Holzkundliche Untersuchungen an Buchen mit neuartigen Waldschäden. Allg Forst Jagdztg 159:121–125.

Möhring K (1989): Wuchsstörungen und Absterben in den Kronen einiger Buchen im Solling. Allg Forstz 44:113–116.

Müller-Starck G, Hattemer HH (1989): Genetische Auswirkungen von Umweltstress auf Altbestände und Jungwuchs der Buche. Forstarch 60:17–22.

Neite H, Wittig R (1985): Korrelation chemischer Bodenfaktoren mit der floristischen Zusammensetzung der Krautschicht im Stammfussbereich von Buchen. Oecol Plant 6:375–385.

Papitz A (1987): Veränderungen der Bodeneigenschaften im Stammfussbereich von Waldbäumen. Schweiz Z Forstwes 138:945–962.

Perpet M (1988): Zur Differentialdiagnose bei der Waldschadenserhebung auf Buchenbeobachtungsflächen. Allg Forst Jagdztg 159:108–113.

Philippi G (1991): Veränderungen der Kraut-und Moosschicht in Wäldern als Folge von Immissionen. Beih Veröff Naturschutz Landschaftspflege Baden-Württemberg 64:198–202.

Richter J (1989): Stand der Buchenschäden nach den Waldschadenserhebungen in Nordrhein-Westfalen. Allg Forstz 44:76–763.

Roloff A (1985): Schadstufen bei der Buche. Forst Holz 40:131–134.

Roloff A (1986): Morphologie der Kronenentwicklung von Fagus sylvatica L. (Rotbuche) unter besonderer Berücksichtigung möglicherweise neuartiger Veränderungen. Ber Forschungszentr Waldökosysteme Göttingen 18:1–177.

Roloff A (1989a): Kronenentwicklung und Vitalitätsbeurteilung ausgewählter Baumarten der gemässigten Breiten. Frankfurt, Germany:

Roloff A (1989b): Entwicklung und Flexibilität der Baumkrone und ihre Bedeutung als Vitalitätsweiser. Schweiz Z Forstwes 140:775–789, 943–963.

Roloff A, Römer H-P (1989): Beziehungen zwischen Krone und Wurzel bei der Buche (Fagus sylvatica L.). Allg Forst Jagdztg 160:200–205.

Rost-Siebert K, Jahn G 1988: Veränderungen der Waldbodenvegetation während der letzten Jahrzehnte—Eignung zur Bioindikation von Immissionswirkungen? Forst Holz 43:75–81.

Runkel M, Roloff A (1985): Schadstufen bei der Buche im Infrarot-Farbluftbild. Allg Forstz 40:789–792.

Storm C (1991): Immissionsbedingte Veränderungen in Wäldern des Kaiserstuhls. Beih Veröff Naturschutz Landschaftspflege Baden-Württemberg 64:117–133.

Stribley GH (1993): Studies on the health of beech trees in Surrey, England: Relationship between winter canopy assessment by Roloff's method and twig analysis. Forestry 66:1–26.

Thiebaut B (1981): Formation des rameaux. In Tessier du Cros E. (ed): Le Hetre. Paris: pp 169–174.

Thiebaut B (1988): Tree growth, morphology and architecture, the case of beech: Fagus sylvatica L. In Comm Eur Communities (ed): Scientific basis of forest decline symptomatology. Brüssel: pp 49–72.

Trautmann W, Krause A, Wolff-Straub R (1970): Veränderungen der Bodenvegetation in Kiefernforsten als Folge industrieller Luftverunreinigungen im Raum Mannheim–Ludwigshafen. Schriftenr. Vegetationsk 5:193–207.

Vincent J-M (1989): Feinwurzelmasse und Mykorrhiza von Altbuchen (Fagus sylvatica L.) in Waldschadensgebieten Bayerns. Eur J For Pathol 19:167–177.

Wahlmann B, Braun E, Lewark S (1986): Radial increment in different tree heights in beech stands affected by air pollution. IAWA Bull 7:285–288.

Wehausen V (1989): Vergleich alter und neuer Vegetationsaufnahmen in der Göttinger Umgebung. Diplarb Forstwiss Fachber Universitat Göttingen, Germany.

Westman L (1989): A system for regional inventory of damage to birch. Poster Intern. Kongress Waldschadensforschung Friedrichshafen/Bodensee 2.-6.10.89.

Wittig R (1986): Acidification phenomena in beech (*Fagus sylvatica*) forests of Europe. Water, Air, Soil Pollut 31:317–323.

Wittig R, Neite H (1983): Sind Säurezeiger im Stammabflussbereich der Buche Indikatoren für immissionsbelastete Kalkbuchenwälder? Allg Forstz 38:1232–1233.

Wittig R, Werner W (1986): Beiträge zur Belastungssituation des Flattergras-Buchenwaldes der Westfälischen Bucht—eine Zwischenbilanz. Düsseld Geobot Koll 3:33–70.

PROCESS HIERARCHY IN FOREST ECOSYSTEMS: AN INTEGRATIVE ECOSYSTEM THEORY

BERNHARD ULRICH

Institut für Bodenkunde und Waldernährung, Universität Göttingen, Büsgenweg 2, D-37077 Göttingen, Germany

1. THEORETICAL BACKGROUND

1.1. Definition of the Forest Ecosystem

An attempt to fix the limits of individual ecosystems as part of the ecosphere inevitably constrains our ability to understand gradual transitions in the ecosphere. The idea of naturally occurring distinctive interfaces is certainly fictional. However, the scientific research of ecosystems is not possible without such distinctions. Therefore the differentiation of ecosystems merely has a practical function and is carried out in accordance with the problem or the objective of the investigation. In this chapter, the problem is one of understanding the temporal dynamics of forest ecosystems under the influence of anthropogenic environmental changes (emissions or climatic changes) and management. In this context the spatial reference is the area being managed (compartment) and the temporal reference is centuries to millennia.

An ecosphere is commonly considered to be any space on the earth inhabited by life forms (Schaefer, 1993). In this chapter, ecosystems are considered as three-dimensional sections from the ecosphere, which are characterized by a certain structure and a specific energy, material cycle, and budget. A constituent part of these ecosystems are a priori all organisms as a whole.

Effects of Acid Rain on Forest Processes, pages 353–397

From the abiotic environment of the organisms, those variables that show interrelationships with organisms (organism-dependent variables with feedback) have to be included in the system. On the basis of these criteria the soil solution including its material content, representing the reaction vessel of the ecosystem for the turnover of mineral substances (ions), has to be considered as a component of the system: The element concentrations in the soil solution are highly dependent on the sink function of the roots and the source function of the mineralizers (microorganisms in the soil). The suction of the soil water and the ion activities in the soil water (soil solution) also define the chemical environment of the roots and the microorganisms. Thus, changes in the element concentrations of the soil solution considerably influence plants and microorganisms. The soil solution also undergoes chemical reactions with the mobilizeable element pool at the surfaces of the soil solid phase; the element concentrations in this pool are thus also organism dependent, consequently, this pool is also a system component. The weathering of minerals by contrast is only marginally influenced by the manifestations of living organisms, so that the ion release due to mineral weathering can be considered as a system transgressing input. The dividing line of the ecosystem with the lithosphere runs along the mineral surface and has a fractal dimension: The smaller the selected scale, even to atomic dimensions, the greater the surface area.

The element concentrations in the hydrosphere and atmosphere at a selected point in time are considered to be independent of the condition of the forest ecosystems in the time scale defined, and are presumed not to be anthropogenically influenced. Therefore the hydrosphere and the atmosphere, as well as the lithosphere, are constituents of the environment of the ecosystem.

Boundary layers exist, either as gases or solutions, to the atmosphere (troposphere) and to the soil solution at the surface of organisms. Within these boundary layers, which result from sink and source functions resulting from metabolism of the organisms, gradients of the chemical potential and temperature are formed. The soil air as a whole can be considered as a boundary layer. The steepness of these gradients decides whether the dependency of the material and energy fluxes from the gradients is linear (in case of flat gradients) or nonlinear (in case of steep gradients). In the case of linear fluxes the subsystems concerned may come close to the thermodynamic equilibrium, whereas for nonlinear fluxes a greater distance from equilibrium is to be expected. In Section 2.4.4 this criterion is used to define the attractor function of a flux equilibrium.

The surface of organisms exposed to the atmosphere and the soil solution also have a fractal dimension; its size is not defined. Dependent on the spatial–temporal dissolution of the observed flux (e.g., root uptake of water or nutrients), calculations are possible with fictive surface areas.

In terms of mathematical system theory, a specific forest ecosystem is defined by a specific set of state variables. These state variables are (Ulrich, 1987):

- The species of primary producers that use sun radiation as an energy source and carry out photosynthesis. In forest ecosystems, trees and ground vegetation are the most important components. Besides species composition the vitality of species is an important qualitative aspect of the state. An example is the use of leaf loss as an indicator of forest decline (see also Section 2.3).

- The species of secondary producers that use organic matter as an energy source. In forest ecosystems, decomposers (soil animals, microorganisms, etc.) and herbivores, as well as plant pathogens, are of major importance.
- Dead organic matter including soil humus as an energy source for decomposers.
- The soil solution with the dissolved ion pool as the soil compartment that is directly influenced by the sink and source functions of the organisms.
- The mobilizable material (ion) pool bound on surfaces of the solid soil phase and in soil organic matter that tends to equilibrate with the soil solution and is thus indirectly dependent of the activity of the organisms.

Since ecosystems are not homogeneous, state variables vary in space and time. Ecosystems are thus defined by mean values of state variables (species composition or soil state) and their variance in space and time. In practice, forest ecosystems are defined in the following three ways by the vegetation type (plant association), by the humus form as an indicator of the decomposer association, and by the soil type. The chemical state of the soil does not always show up in the soil type, however. Thus the soil type may be a unreliable indicator of the state of the soil solution and of the mobilizable ion pool.

The environment of a forest ecosystem thus consists of:

- The atmosphere that is represented by the climate (heat climate, humidity climate, chemical climate, and mechanical climate).
- The element pool that is bound in the interior of soil minerals and being mobilized by weathering. This soil environment is often characterized by the parent rock of soil formation. From the mineral composition of the soil or of the parent rock, weathering rates can be estimated by calculation (Sverdrup and Warfvinge, 1988). With this definition, the release of ions by weathering is considered as an input into the ecosystem.
- The seepage water that is leaving the rooted soil and is the main carrier of the output of dissolved ions. The seepage water can be collected with suction lysimeters below the rooted soil.
- Neighboring ecosystems. Another ecosystem begins if there is a change in state variables exceeding the spatial and temporal variance that is a characteristic feature for a given ecosystem.

The term "environment" is also applied to organisms; it then includes the soil solution and the mobilizable ion pool in the soil.

The climate shows a daily, seasonal, and interannual variability. A constant climate means that its variance is constant. The variance of the climatic factors driving forest ecosystems increases from the humid tropics to the arid and boreal zones. The climate affects the ecosystem in two ways. Climatic variance may either change physical and chemical parameters at the surface of organisms or in the soil. During phases of high climatic variance, organisms can be directly damaged [e.g., by late frost or by wind (wind-throw)]. Such effects correspond to disturbancies of the ecosystem.

The soil minerals, as the soil environment of the ecosystem, degrade by weathering at a slow rate. This process is irreversible. If the variance of the climatic factors is not too large, forest ecosystems may succeed in retaining the nutrients released by weathering in the mobilizable pool within the rooting zone. A characteristic feature of such ecosystems are soils with Bv or cambic B horizons. Such soil horizons reflect the formation of a 2:1 layer of clay minerals with a high cation echange capacity caused by weathering. Also, nutrient input from the atmosphere, especially nitrogen, may be retained in the ecosystem. This retention can result in a gradual increase of productivity and species richness up to a certain level, which is determined by the climate (climax stage; cf. Section 2.7, aggradation phase). In the long term, however, the weathering rate, and thus the input of nutrients into the system, slows down according to the decrease in mineral content of the soil and will finally cease. This change in the below-ground environment of the forest ecosystem can result in its degradation (see Section 2.7).

1.2. Ecosystem Processes

At an ecosystem level all processes that are influenced by the activity of organisms have to be considered. This includes biochemical and physiological processes as far as they have an influence on the vitality of an organism. This characterization of ecosystem processes allows exclusion of the basic physical and chemical processes whose nature is not influenced by the manifestation of life. The most interesting interactions are between organisms and their environment. These interactions occur through energy and material fluxes at interfaces. The fluxes change intensive state parameters at the surface of the organisms like temperature or chemical potentials (approximated by concentrations) of materials, which carry specific information. Energy and material fluxes are therefore connected with fluxes of information.

Fluxes are the result of ecosystem processes. The measurement or calculation of fluxes requires the definition of a measuring plane through which the flux passes. These measuring planes represent interfaces. Interfaces exist between organisms and their abiotic environment (atmosphere or soil solution) and between different organisms. The interfaces between organisms are described by the type of interaction between populations (commensalism, symbiosis, competition, consumption, parasitism, and predation). Additional interfaces have to be defined between the soil solution and the mobilizable element pool at the surface of the soil solid phase, as well as between the seepage water and hydrosphere. For terrestrial ecosystems the interface to the hydrosphere is placed at a level in the soil below the rooting zone, where by use of mathematical simulation models (e.g., Hauhs, 1985; Manderscheid, 1991), the seepage water output can be calculated and seepage water can be collected with lysimeter techniques for chemical analysis (Meiwes et al., 1984).

1.3. Relevance of Hierarchy Theory

Hierarchical structures were identified in ecosystems by various research groups (e.g., Pickett et al., 1989; Lenz and Schall, 1991). O'Neill et al. (1986) discuss the fact that all complex systems, including ecosystems, appear to be hierarchically structured as a natural consequence of evolutionary processes on thermodynamically open, dissipative systems. In applying hierarchy theory to ecosystems, the authors have generally focused on the compartments and not on process (e.g., the hierarchy of cell, tissue, genotype, population, biocenoses, ecosystem, and landscape; Pickett et al., 1989). O'Neill et al. (1986) attribute a dual hierarchical structure to ecosystems based on the two pillars of population and function. Pickett et al. (1989) seek the respective minimal structural unit (minimal structure) where disturbances become effective. Disturbances on level -1 of Table 10.1 are visually noticeable. Focusing on certain issues and on subsystems lead to the conclusion that hierarchies differ considerably, depending on the problem to be investigated.

The hierarchy theory gives general information concerning the relationships between the different levels of a hierarchy. This general information is exemplified by a figure from Müller (1992; Fig. 10.1). A section of a system is presented showing the hierarchical levels $+1$, 0, and -1 with an increasing spatial–temporal dissolution corresponding to doubling the number of holons.

Holons represent the fundamental units of hierarchically structured systems. Holons are self-organizing, autonomous entities, and are also part of superior organizational units. These entities include all subordinate subsystems. The spatial extent of the holons decreases with declining hierarchical levels, the rates of the processes increase (cf. Fig. 10.1). The interactions between the components of a holon are strong, whereas the interactions between holons of the same level are weak. Assymetrical relationships exist between holons of different levels ($+1/0$, $0/-1$). The superior level constrains the subordinate level; it determines the boundary conditions for the activities of the holons of the subordinate level. Each level in the hierarchy can be overridden by the next higher level; it is under the constraint or control of the next higher level. The effects of the subordinate on the superior level are weak as the superior level is able to tone down or buffer the effects. Therefore, the effects of the subordinate level on the superior level are described as signals, and the ability to tone down the signals is described as filtering. The superior level appears as a barrier against the effects of the subordinate level. Each level of the hierarchy corrects for errors or inefficiencies in the levels below.

This type of interaction between hierarchical levels is characteristic for stable systems close to a steady state or flux equilibrium. In nonstable systems the signal filtration diminishes and the hierarchy can collapse: The effects of the subordinate level on the superior then assume the characters of constraining parameters that cause change at the superior level. This can result in fluctuations and changes in the system.

The ability of the forest ecosystem to filter signals from the subordinate level results in its "resistance against deviations" (Webster et al., 1975), it gives the

TABLE 10.1. Temporal and Spatial Scale of Processes in Forest Ecosystems

Level	Process (holon)	Response Time	Compartment	Pattern
+4	Macroevolution	>Millenia	Continents	Species formation and extinction
+3	Succession or management	Centuries	Landscape	Transition phases of ecosystems
+2	System renewal	Centuries	Ecosystem	Stability phases
+1	Stand development (storage change of biomass and humus)	Decades	Stand (ecosystem section)	Age class, material budget of the soil
0	Material cycle	Year	Tree + neighbors	Material budget of the ecosystem
−1	Growth processes (leaves, fine roots, fruits, wood)	Weeks to Months	Tree + ground vegetation + soil horizons	Branching above-/below-ground, foliage
−1	Population dynamics (decomposers, phytophages)			Humus form, soil fabric
−2	Physiological processes (assimilation, uptake)	Hours	Leaf root	Carbon and ion allocation
−2	Mineralization	Days–Weeks	Aggregate	Soil water chemistry
−3	Biochemical processes	Seconds to Minutes	Cell	Biochemical pattern
−3	Soil chemical processes		Mineral surface	Buffer range heterogeneity

ecosystem elasticity against perturbations. In this chapter the term elasticity is used to express the ability of signal filtering and buffering at the various hierarchical levels.

The forest ecosystem is hierarchically structured according to the time scale of process categories (see Table 10.1). The spatial units are defined in which the processes occur. A holon is thus defined as a specific process that takes place in a specific spatial unit of the ecosystem. A hierarchical level is defined as the sum of holons in a specific process category. The sum of holons at each hierarchical level represents the whole ecosystem.

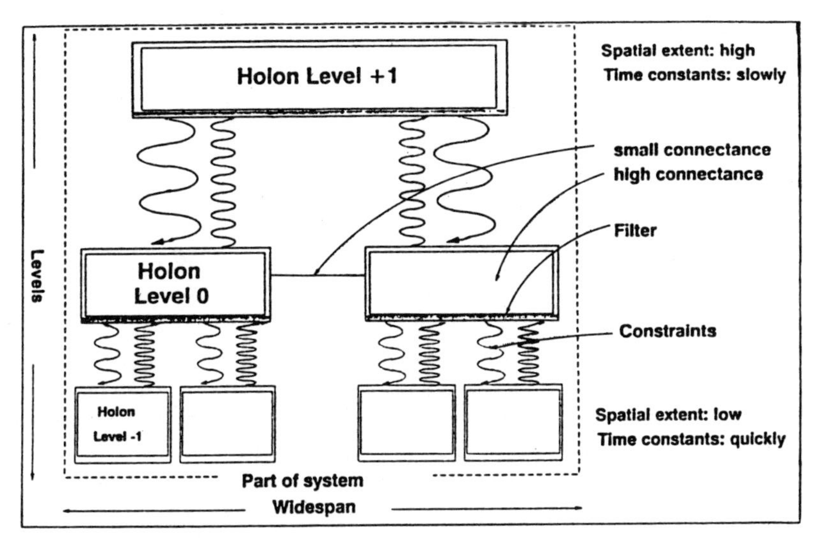

Fig. 10.1. Some characteristics of hierarchical structures (adapted from Müller, 1992).

The processes are initiated by a change in physiological or ecosystematic condition. These processes run their course until a new process-characteristic condition is reached and then subsides. Reinitiation is brought about by a further change of the condition. These changes can be caused by environment dynamics (e.g., the day/night rhythm or the seasons), or by ecosystem dynamics (e.g., due to aging and regeneration of system elements). The duration of the process from its initiation to its subsidence after the new condition is reached (response time) is selected as the time scale.

A specific process can run its course in a variety of performances and can lead to a variety of results. Nutrient uptake, for example, can take place with different nutrient ratios, which may result in different kinds of nutrient deficiencies. Nutrient deficiencies or the nutrient content in leaves are the patterns in which the performance of nutrient uptake shows up and can be judged. The kind of patterns that allow us to decide on the performance of processes are also indicated. These conclusions represent a very important tool for ecosystem analysis and modeling. Extensive information on these patterns is already available, especially from the scientific branches of physiology, morphology, ecology, soil science, forest nutrition and geobotany.

1.4. Relevance of Processes With Positive Feedback

To organisms, ranges of physicochemical parameter constellations can be associated with ranges where they can adapt and where they can be vital or at least can survive. These parameter constellations include, for example, temperature, light, soil parameters describing water, oxygen, nutrient, and acid–base status. The parameters involved are subjected to continuous changes. These changes are due to the variance of the climate as well as to ecosystem internal processes. It is hypothesized that stress results if the rate of change of the parameter constellation exceeds the ability of the organism to adjust to the changes. Even if the change remains within the range of vitality, the organisms may be subjected to stress. The adaptability is genetically controlled. Genotypes reflect the experiences of a species with respect to the rate of variation in their environment. If a parameter changes at a constant rate (i.e., the system is linear with respect to this parameter), and if the change is a component of the experience of the genotype, than the preconditions for a sufficient rapid adjustment are good. The situation is worse if a parameter of ecological significance changes exponentially due to positive feedback. The probability increases that the rate of adjustment is smaller than the rate of parameter change, therefore resulting in stress. Stress can result in diminished vitality. With respect to such a process, the system is in the range of nonlinearity.

This concept includes the possibility that organisms may recover from stress (a) if the process with positive feedback levels out and (b) if the parameter constellation

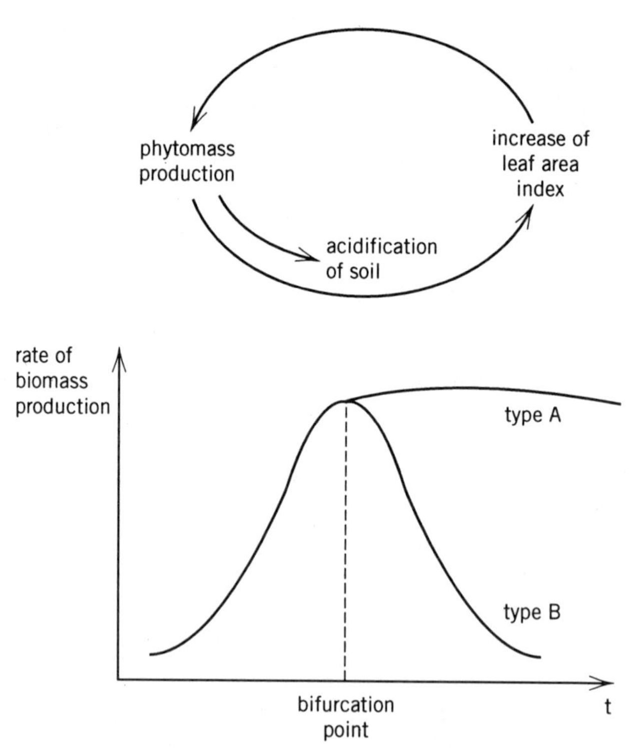

Fig. 10.2. Autocatalytic character of the growing up of plant stands.

remained within the range of vitality. Organisms with diminished vitality are less resistant to pathogens, less effective in competition, and less able to withstand enemies.

Growth processes (Fig. 10.2) are familiar examples of nonlinearity. These processes are described by S-shaped curves. The exponential part of the curve is due to positive feedback. The growth of populations is connected with changes in the physicochemical conditions in their environment. Also, these changes increase exponentially. Periods of exponential growth can therefore lead to the situation mentioned above: The vitality of the population, or of a part of it, may be diminished. The ability of the organisms of low vitality to filter the signals from the subordinate level may be exhausted. In this case the hierarchy is broken and the subordinate level constrains the superior level. Such a situation corresponds to a bifurcation point: A small environmental (e.g., climatical) effect, which is within the variance of the environmental factors, can have enormous consequences for specific components of the system. The population may begin to fluctuate. If it is the tree population, the system may be subjected to change.

2. SPACE AND TIME SCALES OF THE PROCESSES IN FOREST ECOSYSTEMS

In Table 10.1 the process categories that can be differentiated in forest ecosystems are ordered and compiled according to their reference to time and space. In Table 10.2 the constraints and signals acting between the hierarchical levels are summarized.

2.1. Hierarchical Level −3: Biochemical and Soil Chemical Reactions

Biochemical reactions (space reference: cells and cell organelles) and soil chemical reactions, such as the binding of water or cation exchange (space reference: surface of clay minerals and humus substances), are the processes of the shortest duration (seconds to minutes). These processes, biochemical as well as soil chemical reactions, represent the interfaces between the ecosystem and its environment. The signals that reach the ecosystem from its environment are represented by changes in physicochemical parameters like temperature, concentrations of materials in air, and concentrations of materials (including water) in deposition. The patterns, from which the type of process can be identified, are of a chemical nature (e.g., biochemical parameters).

The ability to filter environmental signals (temperature and concentrations of materials in environmental media) at the level of *biochemical reactions* is of primary importance with respect to the elasticity of the ecosystem against the variance of the above-ground and below-ground environment of the organisms. This aspect of elasticity will not be discussed, however.

In the *soil* the processes are initiated by a change in water content (water flux) or

TABLE 10.2. Constraints and Signals of Plant Related Processes in Stability Phases

Levels	Constraint	Process	Signal
+4	Adaptation, fitness	Evolution	
+3	Species composition	Succession or management and planting	Vitality decrease
+2	Regeneration	System renewal	Soil change
+1	Storage change of organic matter	Growth, utilization, humus accumulation/decomposition	Senescence of dominating trees
0	Steady nutrition	Material cycle	Decoupling
−1	Leaf area index, number of root tips, xylem area	Formation of organs (leaves, fine roots, fruits, wood)	Litter production, litter quality, increment
−2	Photosynthetic capacity, nutrient/water uptake	Assimilation water/nutrient uptake	Availability of assimilates
−3		Biochemical reactions	Stress hormones
−3		Soil water equilibria	Soil water potential
−3		Soil chemical reactions	Chemical potentials in soil water

Solid arrows represent constraints, broken arrows represent signals.

the ion concentration in the soil solution (diffusion, cation exchange, adsorption–desorption, dissolution, and precipitation). Such changes can be caused by inputs and by the sink and source functions of the organisms.

Patterns: The buffer range (Ulrich, 1981a) provides comprehensive evidence about the chemical reactions that are occurring. Different soil horizons can belong to different buffer ranges. Even within morphologically uniform horizons differences in the chemical condition of the soil may occur within small areas. The pattern is therefore manifested in the spatial distribution of the chemical condition of the soil. Different scale ranges can be distinguished (macroscale: within a soil form; mesoscale: within the rooting zone of a tree; microscale: within the rhizosphere or the soil aggregates). Changes in the chemical soil condition start in the microscale range and continue in case of a continuous driving force in to the superior scale ranges. Such a development is connected with the buildup of chemical gradients in the soil. The existence of chemical gradients at the microscale can be used to indicate the change tendency to which the soil as a whole is subjected. In forest soils a drift towards acidification as a result of acid input is apparent in the selective acidification and the nutrient depletion of the macropore walls (Horn, 1987; Hildebrand 1990, 1992). This example shows that surveys of the soils chemical heterogeneity can serve as early ecological indicators of the chemical change in the plant roots surroundings, decomposers, and mineralizers.

Elasticity: The soil stores water and compensates for the very irregularly occurring precipitation. It further serves as a reaction vessel in which the activities of the primary producers and secondary producers are coupled. Similar to water, the soil is also able to compensate spatial and temporal decoupling in the material cycle of nutrients and bases. Important parameters in this respect are the storages of mobilizable nutrients and the base saturation of the exchange complex. The quality of a soil is estimated by its balancing function in the water and ion cycle.

Constraints: The hydrological and chemical state of the soil is constrained by the environment (input of water, ions, and acidity) as well as by the sink and source functions of organisms (plant uptake, and mineralization).

2.2. Hierarchical Level −2: Sink and Source Functions of Organisms

The sink and source functions of the organisms are manifested in the assimilation and element uptake of green plants, as well as in the decomposition and microbial mineralization of organic substances in the soil.

Assimilation and water–ion uptake by plants: According to Table 10.1 and Figure 10.1, for ecosystem internal regulation the hierarchical level −2 constrains the biochemical processes and the soil hydrological and chemical processes, both of level −3 (the level −3 is also constrained by inputs from the environment). Response time: *Photosynthesis* and the water and nutrient uptake of plants subside after the closing of the stomata, thus exhibiting a diurnal rhythm. The space reference is given by the respective organs, namely, leaf and root, and integrates about the spatial units (cells) at the lower hierarchy level. The patterns show up in the allocation of the assimilates, and ultimately in the growth of the different plant organs. While at the stand level the production of leaves varies very little from one year to another, radial growth often reflects the climatic conditions for photosynthesis. The unknown factor is still the consumption of assimilates in the fine root system, which is not yet quantifiable. With respect to photosynthesis, biochemical signals indicate the necessity for adaptation to changing environmental conditions, which is provided by physiological mechanisms like stomata regulation (elasticity).

The patterns of *water uptake* show up in the water status of the plant, as measured by the leaf water potential, and in the water status of the soil (soil water potential). The soil water potential (level −3) is also influenced by environmental inputs such as rain and represents a signal from level −3 to level −2. Elasticity is provided both by physiological mechanisms (regulation of transpiration) and by the soil's capacity to store water.

The *ion uptake* by roots leads to a formation of gradients in the nutrient and acid–base status in the rhizosphere (soil surrounding the roots) at level −3, which act as signals. Methods for a determination of the spatial heterogeneity in the rhizosphere are being developed. Signals of acid load are represented by the concentrations of protons and cation acids like Al, Mn, Fe, and heavy metals and their ratios to nutrient cations (Ca, Mg, or K) in the soil solution. A temporally integrated pattern of ion uptake is provided by the element contents and element ratios in plant organs, particularly in leaves (indication of nutrient status; Bergmann, 1988). Extreme nutrient deficiency becomes apparent in a characteristic coloring of leaves. Patterns of this kind are used for diagnosis, and also as an additional characteristic (yellowing as a result of magnesium deficiency) for the evaluation of forest damage. With respect to nutrition, elasticity is provided by the ability of the plant to take up nutrients selectively or in surplus (luxury consumption), to use older needle classes as nutrient storage, to eliminate surplus ions by leaf fall, and by the storage capacity of nutrients in soil. Plant species differ greatly in their physiological tolerance with respect to acid load. In Norway spruce the principal mechanism of Al toxicity is the displacement of divalent cations in the root apoplast (Godbold, 1991). Since acid stress acts on the fine root, the patterns are also located there (element contents and ratios as indication of acid–base status; Murach and Wiedemann, 1988). Soils with base saturation greater than 15% can usually buffer an acid input to the soil solution by cation exchange, the soil solution may return to weak alkalinity. Acidifying soils may, however, exhibit great spatial heterogeneity with respect to base saturation (cf. Section 2.1).

Microbial mineralization: Response time: The duration of mineralization pushes triggered by a sudden occurrence of suitable substrate and more favorable conditions in the environment (e.g., rewetting of a dried soil) depends on the quality of the substrate. The easier this can be utilized, the faster the autocatalytic growth of the bacteria population and the depletion of the substrate takes place. A mineralization push is more likely to last for several weeks than for only a few days. The reference to space is determined by the local occurrence of the substrate. Most of the bacteria depletion takes place after plant parts are broken down due to grazing by animals (earth worms, enchytraeids, or arthropods; constraints of level −1) in the then formed soil aggregates or in animal droppings. In older aggregates the microbial biomass is accumulated at the aggregate surface (Augustin, 1992). The excretions of roots and dead fine roots are a further substrate. The space reference is thus given by soil aggregates and the rhizosphere, which again integrate about the space units of the lower hierarchy level.

Patterns: Under laboratory conditions mineralization can be observed in the formation of carbon dioxide (CO_2) and the heat development. In forests, mineralization is reflected in the chemical change of the soil solution (increase in nitrate concentration, decrease in alkalinity; Ulrich, 1981b). The space and time reference suggests that under favorable temperature and moisture conditions the mineralization varies according to animal activity. Investigations of the soil solution support a high element dynamic in a small spatial area triggered by mineralization pushes, which can be vertically perpetuated (Raben, 1988) by the shift of seepage water (mechanic dispersion). Microbial populations are themselves a food resource for decomposers (level −1) and products of their metabolism may act as signals.

Elasticity: Evolution has led microbial populations to have an enormous adaptability to different nutrition resources and different environmental conditions. In forest ecosystems with aerated soils, a flux equilibrium between litter production and mineralization of organic matter can be reached under all circumstances. As a consequence, soil organic matter accumulation always tends to approach a finite value that depends on environmental and ecosystem conditions.

2.3. Hierarchical Level −1: Seasonal Growth Processes

On a time scale of weeks or months the growth processes of plants, which are responsible for the formation of leaves, fine roots, blossoms, fruits and wood, as well as many population dynamic processes in animals, take place. Among the fauna the reducers (decomposers) together with the subsequent food chains, and sporadically the consumers (herbivores), are of particular importance. The spatial reference is given by the tree and the area marked by its crown, which is divided into strata (tree layer, shrub layer, herb layer, moss layer, and soil horizons). The spatial integration covers the different populations in a three-dimensional section of the ecosystem. Much information exists on climatic effects (external constraints) and ecosystem internal constraints for the population dynamics of plant pathogens, which will not be discussed here.

Patterns and elasticity of growth processes of plants: The signal from level −2 is the availability of assimilates. On this signal, trees can react through a high flexibility in the allocation of assimilates (e.g., limitation of wood formation). When trees form their organs, they follow a certain pattern from which it is possible to detect disturbances in the organ formation. In the absence of disturbance, the above-ground and below-ground branching follows a species specific construction scheme (broad-leafed trees and pine: Roloff, 1989, coniferous trees: Gruber, 1992). Deviations from the construction scheme point to disturbances. With respect to the branching pattern Roloff differentiates between exploration, degeneration, stagnation, and resignation phases as vitality stages that may comprise time spans of 10 or more years. Gruber differentiates chronic and acute stress effects as well as regeneration characteristics. The interpretation of the branching pattern in the root zone is still under investigation (Puhe, 1994). The ability to adapt to adverse soil conditions and to their changes is even greater in the root system than for the above-ground branching. Changes in the growth of the tree height and in the width of the annual rings provide a basis for more far reaching estimates of stress load. For nonannual flowering tree species the initiation of bloom is also influenced by the environment; although, the causal relationships were not clarified conclusively. Overall, these patterns allow differentiating and integrating statements to be made about the vitality of a tree, even those that occurred many years ago. Nevertheless, the forest damage inventory still relies on a much less differentiated pattern for assessment, namely, the loss of leaves (crown thinning). Leaf loss does not allow us to draw any conclusions about its causes or about the vitality of the tree to overcome the acute stress situation that is indicated by the leaf loss.

Leaf area (photosynthesis), number of root tips (water and nutrient uptake), and xylem area (water flow to the canopy) are examples of how constraints are exerted on the level −2.

Population dynamics of soil animals: By degrading plant litter, creating special environments for microbial populations in their droppings, and utilizing microbial populations as a food resource, decomposers constrain the microbial processes of mineralization.

Patterns and elasticity: In the course of ecological discussions, the population dynamics of soil inhabiting animals were greatly ignored. The population dynamics of decomposers is connected with the availability of the nutrition resources, thus they should reflect the dynamics of the processes involved in decomposition. In terms of the seasons, correlations between the course of decomposition and environmental factors were clearly demonstrated, whereas these correlations could not be found for the short-term dynamics of decomposition processes. The nature of the equations by which the development of populations can be described (e.g., the logistic equation) suggests that the population dynamics of decomposers, and consequently the decomposition process itself, shows chaos by way of determinative feedback: It is not possible to determine the initial conditions so precisely that the course of the process is predictable on the basis of initial and boundary conditions. A species can respond to fluctuations in the condition of the environment not only by changes in the population structure but also by adaptation of its patterns and physiological performance. Commonly, other species take over available niches and thus compensate the effects at the process level. Different patterns of behavior or similar patterns of behavior, respectively, can induce different effects (Wolters, 1989). It was suggested that spatial and temporal heterogeneity in the soil contributes to the compensation of local and seasonal differences in the performances of soil inhabiting animals (Wolters, 1991). This pattern also shows the possibilities, at the population level, of the ecosystem to compensate for a chaotic condition. We attempt to recognize patterns in order to determine trophically similar species (guilds) and food chains (Schaefer, 1991). Food chains are generally reflected in the humus form, which can therefore serve as a morphological pattern (mull: dominant role of soil burrowers; moder: dominance of arthropods; raw humus: low biological activity, by microorganisms, a dominance of fungi). The mixing activity of soil burrowers in mull increases the elastic properties of the soil. Bioturbation is of great significance for minimizing spatial heterogeneity in the soil.

2.4. Hierarchical Level 0: Material Cycle

In the preceding hierarchical levels, the observed processes take place within or between subsystems. Only at the level of the material cycle does the behavior of the entire ecosystem come into perspective. At this level theoretical approaches used to describe the state of the ecosystem, and not only its special components, must be discussed.

2.4.1. *Material Budget Equation*

At the level of the material cycle all organisms can be aggregated into two functional groups, the effects of which are mutually counterbalanced (Ulrich, 1981b, 1987): primary producers that can form organic substance from inorganic substances by utilizing the energy of the sun, and secondary producers that utilize organic substance as an energy resource. The role of organisms in the material budget of ecosystems can be described by the material budget equation, which combines the material turnover of the primary and the secondary production. In its simplest form, the turnover can be described by the following balance equation (10.1):

Forward reaction: → photosynthesis and ion uptake: primary production, formation of phytomass
Backward reaction: ← respiration and mineralization: secondary production, mineralization of organic substances

$$CO_2 + H_2O + xM^+ + yA^- + (y - x)H^+ + h\nu \rightleftharpoons CH_2OM_xA_y + O_2 \quad (10.1)$$

where x and y represent stoichiometric coefficients, M^+ is a cation, and A^- is an anion of unit charge. The compound $CH_2OM_xA_y$ represents the primary and secondary producers, as well as litter and soil organic matter, which constitute the entire organic matter in the ecosystem. The cations M^+, the anions A^-, and the protons H^+ are present in the soil within the dissolved and mobilizable ion pool. The material budget equation is a broadening of the assimilation or respiration equation by the addition of cations M^+ and anions A^- taken up into or released from the organic substance. As the equation shows, the turnover of cations and anions is linked to a proton turnover in order to maintain electric neutrality. This turnover indicates that the uptake of nutrients into the plant not only results in a decrease of nutrients in the soil, but can also lead to a change of the acid–base condition of the soil. In general, the plant is able to take up more cations than anions, inducing a release of protons into the soil: Due to the production of plant material the soil becomes acidified. In the course of the recycling of organic substances by secondary producers apart from the nutrients being released, the protons released due to ion uptake are recycled to the effect that the acid–base status of the soil is maintained.

The ion budget, according to the material budget equation, is the basis for calculating acidification or alkalinization of the environment of organisms due to their sink and source function for ions.

On the level of the material cycle, the amount and quality of plant litter and the activity of decomposers represent signals that may lead to decouplings of the material cycle. A decoupling means that either the formation (forward reaction) or mineralization of organic matter (backward reaction) has a greater rate. A decoupling may also be caused by the variance of the microclimatic conditions that constrain the microbial activity. A decoupling of the material cycle results in a change in nutrient concentrations and the acid–base status of the soil solution. These signals can pass through levels -2 and -1 up to the level of the material cycle again. Part of the elasticity of the ecosystem at these levels is the adjustment of the rate of processes like mineralization, nutrient uptake, litter production, and decomposition, which make up the material cycle. On the level of the material cycle the elasticity is thus guaranteed by negative feedback loops involving the levels -1, -2, and -3. This elasticity may be demonstrated by two examples. First, a greater rate of mineralization indicates that at the level of the ion cycle there is an acidification push and a surplus of nitrate and other nutrients in the soil solution. This constrains (increases) plant growth (level -1), nutrient uptake (level -2), and consumes protons (level -3). Thus, the system is shifted towards a balanced ion cycle and a neutralization of the acidity created during the acidification push. Second, a higher rate of plant growth indicates that at the level of the ion cycle there is a deficiency of nutrients in the soil solution. This constrains (limits) nutrient uptake and plant growth. Again, the system is shifted towards a balanced ion cycle.

The patterns are represented by the material budget of the ecosystem, that is, the input–output relationships. Examples of material budgets from case studies are given by Ulrich (Chapter 1).

2.4.2. Application of the Mathematical System Theory

Ecosystems are open systems that exchange energy and materials with their environment. On the level of the material cycle ecosystems can be described as dynamic systems in the sense of the mathematical system theory. Dynamic compartmental systems E are defined by the quintupel

$$E = (\mathbf{u}, \mathbf{y}, \mathbf{x}, f, g) \qquad \text{Ludyck, 1977} \qquad (10.2)$$

where $\mathbf{u}$ represents the input vector, $\mathbf{y}$ is the output vector, $\mathbf{x}$ is the state vector, f is the transfer function, and g is the output function. In addition, the "state of rest" of the system, which is comparable to an equilibrium state, has to be defined.

The input vector consists of all independent variables that have an influence on components of the state vector. The output vector consists of all dependent variables that connect the system with its environment, and that in turn are influenced by the input. The state vector represents all state variables of the system that are influenced by the input. The transfer functions allow calculation of the change in state variables due to an input. The output function allows the calculation of the output as a function of changes in state variables.

The state variables are defined in Section 1.1. The input and output vectors consists of energy and materials. It is impossible to assess the input and output vectors totally. The problem is solved by specifying the aim of ecosystem analysis: A target has to be defined, and those variables (inputs and outputs) must be selected that cannot be neglected with respect to the target. All variables excluded are in practice set to zero. With respect to many aims of forest ecosystem analysis of practical importance, inputs and outputs can be measured or assessed with sufficient precision to draw conclusions.

The "state of rest" can be defined by equal rates of the forward and backward reaction of the balance equation (10.1).

The case studies discussed by Ulrich (Chapter 1) show, in agreement with theoretical expectations, that the output of cations with the seepage water is controlled quantitatively by the production or the input of mobile anions, and qualitatively by the chemical soil state (buffer range). This finding means that the output function can be assessed in a first approximation, if the atmogenic deposition is specified (scenario) and the soil state is known.

2.4.3. *Flux Equilibria*

The "state of rest," as defined by equal turnover rates of primary and secondary producers, represents a very special condition. If ecosystems succeed in obtaining a complete linkage of primary and secondary production, then they would have reached steady state (flux equilibrium): They would perpetually be able to generate plant material that is recycled within the ecosystem while neither the nutrient status nor the acid–base status of the soil would change. The "state of rest" of a forest ecosystem is characterized by high turnover rates including the growth and decline of organisms, but, if a sector of sufficient size is considered, the variance of the state vector with respect to time and space is constant. This characterization represents the most extreme form of flux equilibrium and homeostasis that ecosystems are theoretically capable of reaching; it is the premise for a temporally unlimited continuity of the system. Naturally, the temporal continuity of forest ecosystems is finite. The causes are decouplings of the primary and secondary production in space and time, which are due to the variance of the climate (temporal decoupling) and to the variance of the structure of the ecosystem (spatial decoupling).

The "state of rest" is thus an ideal state that may never be reached in reality. As a consequence the stability of forest ecosystems is time limited. With small rates of change forest ecosystems can be persistent for many regeneration cycles, however. At the time scale considered (see Section 1.1) such ecosystems can be considered as stable. The problem of how the state of the ecosystem (an ecosystem in the stability phase or in the transition phase) can be assessed still remains.

In the "state of rest" the output vector of materials must necessarily be equal to the input vector. The measurement of the inputs and outputs can therefore serve as a criterion to determine whether the ecosystem is stable, or in which respect it deviates from flux equilibrium. For example, in flux equilibrium the output of ions from the ecosystem with the seepage water and of gases (e.g., N compounds) will be in balance with the input of the same element. For this reason the input–output budget of an ecosystem is an appropriate means to judge whether the ecosystem is in a stable phase or in a transition phase. From the comparison of element inputs and outputs it is possible to determine accumulation or loss of materials in the ecosystem and in the soil (Ulrich, Chapter 1). Knowledge of changes in soil nutrient and base storage allows us to draw conclusions on the changes in physicochemical parameters (hierarchical level −3). If the filtering ability of the superior levels for the signals from the chemical soil state become exhausted, the hierarchy may be broken and a change of the system initiated.

The material flux budgets are therefore a key to prognosis (Ulrich, 1989a, 1991). The methodology to determine accurately the material budget of forest and agro ecosystems, as well as that of larger regions (Steiger and Baccini, 1990) has advanced enough so that conclusions can be drawn about control factors for the ecosystems.

2.4.4. *Attractor Function of Flux Equilibria*

According to the thermodynamics of irreversible processes, steady states, or more generally states of a flux equilibrium, act as attractors. Flux equilibria tend to maintain the state of the system and are thus a necessary condition for stability. The temporal variance of climatic factors (intra- and interannual variability) continuously induces new signals on the hierarchical level −3. The signals may be strong enough to constrain the superior levels up to the material cycle (level 0). At this level the effect becomes measurable as a decoupling (deviation of the ion budget from zero). Changes in the patterns of population dynamics of decomposers and phytophages, as well as in the above- and below-ground branching of trees, including leaf and fine root losses, may be involved.

This theory states that subsequent to a deviation, ecosystems strive to remain on trajectories within the attractor sphere of the steady state or to return to it. The attractor function of a steady state is higher, the closer the steady state approaches thermodynamic equilibrium. In order to assess stability it is therefore necessary to measure the distance of the system from equilibrium. The first question to be answered is whether a compartment can be defined in terrestrial ecosystems, in which chemical thermodynamics can be applied.

We consider the turnover of materials in terrestrial ecosystems. In the following paragraph we restrict our view to water and ions, since the requirements of water, nutrients, and the acid–base status of the soil are the main soil factors influencing organisms living in aerated soils. These organisms represent sinks and sources of water and ions, and the soil represents the reaction vessel. The soil, and here the soil solution, is thus the compartment to which the thermodynamic criteria should be applied. In case of primary producers (plants) the water and ion fluxes pass through the root surface. Around 90% of the mass of secondary producers is represented by microorganisms in the soil, where the water and ion fluxes pass through the cell surfaces. The flux of ions and water through root and cell surfaces create gradients of chemical potentials in the surrounding soil (e.g., in the rhizosphere). These gradients can be used to describe the distance of the soil state from thermodynamic equilibrium (cf. Section 1.1). Weak gradients that allow linear relationships between the gradients and the fluxes across the surfaces indicate a small distance from thermodynamic equilibrium. These gradients result in a simple temporal behavior of the system. Plants can adjust to such a behavior, that is, the elasticity of organisms is sufficient to filter the signals from subordinate hierarchy levels. Steep gradients indicate nonlinear relationships between gradients and fluxes, great distance from equilibrium, and a complex temporal behavior of the system. The variance of the climate may induce an exponential change in parameters at the root surface to which the organism cannot adjust. This finding indicates that signal filtering has failed at this level of hierarchy.

Depending on the elasticity of the soil (rate and capacity of buffer mechanisms), the steepness of the gradients developing in the soil as a consequence of water and nutrient uptake (as an example) can differ. For agricultural plants in soils of good nutrient and base status it was demonstrated that the rates of nutrient uptake are linear functions of the thermodynamic forces (the gradients). Such functions are used in mathematical models to simulate nutrient uptake (Nye and Tinker, 1977). In general, it can be stated that soils with a good buffering capacity with respect to changes in their nutrient, acid–base, water, and oxygen status, (i.e., of high elasticity) allow a closer approximation to linear fluxes in nutrient uptake then soils of low elasticity. Ecosystems with such soil properties can withstand a greater variance of the climate, the feedback loop between the hierarchical levels 0 and -3 can operate more effectively, and can more easily avoid the hierarchy break. The steady state as defined by the "state of rest" acts as a strong attractor, which is summarized in Table 10.3.

Open systems that operate close to equilibrium can organize themselves according to the principle of minimizing entropy production (Prigogine, 1979). Whether or not this has any meaning for ecosystems is not known (Ulrich, 1991).

Nonlinearity of processes in the water and nitrogen cycle: The largest temporal and spatial variability exists, even under favorable conditions, for the concentrations of nitrogen species (NH_4^+ and NO_3^-) in soil solution and for the binding strength of water in soil (suction). The relationship between content, suction, and conductivity of water in soil are strongly nonlinear, resulting in a strong decrease of water availability in soil with increasing water consumption by transpiration. The approximation to linear water fluxes in soils is therefore restricted to good soil conditions (high water storage close to field capacity) and regular precipitation during the vegetation period. Since the transport of solutes across larger distances occurs as mass flow, nonlinear dependency of water flow from soil suction refers also to the flow of nutrients. Plants are more or less adopted to this nonlinearity and its consequences for water uptake by regulating transpiration through stomata closure and decreasing the growth rate (elastic strain) without necessarily being damaged (plastic strain).

The nitrogen species in soil solution are only controlled by biological activity and not by chemical equilibria in the soil. Nitrate concentrations in soil solution with low temporal variation that are sufficient for unlimited plant growth therefore require a continuous production of easily decomposable organic matter with a high nitrogen content (which may be provided by the ground vegetation), as well as a continuous activity of decomposers and mineralizers. Due to the limited buffer abilities in soil, resulting in nonlinear dynamics, water and nitrogen are the growth factors that often limit the biomass production by primary producers. Even in forests of the highest yield class, optimal nitrogen nutrition can double the rate of

TABLE 10.3. Soil State and Forest Ecosystem Stability

	Soil State Close to Equilibrium	Soil State Far From Equilibrium
SOIL STATE		
Biological	Soil burrowers (mull)	Missing of soil burrowers (organic top layer)
Base saturation	High	Low
Nutrient storage	High	Low
Spatial heterogeneity	Low	High
Water capacity	High	Low
Aeration	Good	Bad
EXAMPLES		
	Beech forest on lime stone Mixed mountain forest on soils rich in silicates	Tropical rain forest on poor soil
MATERIAL CYCLE		
	Compensation function of soil as store and buffer	Short circuit, since compensation function of soil is missing
STABILITY		
Elasticity in respect to weather conditions, to biomass utilization and to regeneration	Dry periods, acidification pusches, and nutrient losses can be buffered by the soil	Ecosystem persistent only at minimal climatic variability (rain forest) and with minimal biomass export, since buffering in soil is missing

biomass production by primary producers. In agricultural ecosystems, this method was used to maximize production.

2.5. Hierarchical Level +1: Stand Development

Aging or dying off of organisms and rejuvenation causes a characteristic cyclic development of forest ecosystems. These ecosystems pass through phases that are characterized by the development of the dominating trees: juvenile phase, thicket, pole stage, timber tree, senescence (opening of the canopy), and regeneration. The stand development reflects a trajectory on which the system moves. The time scale of these development phases are years to decades. The spatial scale can differ greatly. In pristine forests of high stability these phases may be restricted to small patches (mosaic-cycle concept; Remmert, 1991), the minimum size given by the crown diameter of a dominating tree. In forests managed according to age classes, these phases are represented in different forest compartments of 10–30-ha size.

Constraints and signals: The dominant tree layer plays a key role due to their regulating function for the stand climate (microclimate). In modern system theory terminology (Haken, 1981), it "enslaves" the other populations in the system: the plant populations by regulating light, the decomposer populations by regulating temperature, humidity, and the food resource. The dying off of individuals of the prevailing tree layer allows sunlight to penetrate to the deeper strata; this affects a (micro)climate change, which sets off impulses (signals) for new developments with respect to primary and secondary producers. This may result in a shift in species composition (regeneration flora or clear felling flora), which is significant at the level of system renewal (+2). Connected with the stand development are significant changes in the storage of above-ground biomass. The development in the secondary producers can lead to a change in the storage of organic matter in the soil.

Patterns: Morphologically, the process of stand development shows up in the development phases mentioned. These phases reflect patterns of the material budget of the ecosystem that are of great significance for its stability. In the early phases, till the timber tree is reached, the ecosystem can be far from a "state of rest" (closed material cycle). At a greater distance from a state of equilibrium is a characteristic feature and even a necessary condition to build up new structures (i.e., the timber tree). In the initial stages of the stand development the change in microclimate may induce organic matter, nitrate, and nutrient cation losses from the soil. In the thicket and pole stage phases organic matter accumulation takes place in the biomass and eventually even in litter. These changes in organic matter storage in the ecosystem are connected with changes in the nutrient and acid–base status of the soil, that is, with constraints reaching the hierarchical level −3. The changes in nutrient and acid storage in soil can be calculated if the material budget equation is applied, assuming that the kind and amount of change in organic matter are known. The material budget of the soil allows us to judge the deviation of the ecosystem due to stand development from the "state of rest" and thus the risk of destabilization of the ecosystem while moving on this trajectory.

An important aspect with respect to destabilization of the ecosystem is the rate of soil change. During the pole stage, tree growth is in its exponential phase (cf. Fig. 10.2). This means, however, that the cation uptake, and the proton production in the soil connected with it, increase exponentially. Acid soils of low nutrient cation availability and acid buffer capacity (low elasticity) can lead to a rate of change in physicochemical conditions at the root surface to which the roots cannot adjust. The consequences can be a hierarchy break that shows up first in root branching and a large fraction of subvitale fine roots. In a pole stage, if the total tree population is involved, this differentiation into tree classes (dominant and suppressed) can fail. In this case the whole tree stand may decline. In the decline secondary stressors (wind, wet snow, and droughts) may be involved. The leveling off of the exponential growth phase (i.e., the transition of a pole stage in a timber tree) can be considered as a bifurcation point in the stand development (cf. Fig. 10.2).

2.6. Hierarchical Level +2: System Renewal

The renewal of a forest ecosystem is triggered (signal) by the senescence of dominating trees or, in the case of managed forests, by tree cutting. Whether the signal leads to regeneration or not depends on the structure of the forest: If there are codominant or suppressed vital trees, they may close the gap (elasticity). The time scale is in the range of one to several centuries. The spatial scale should reflect all stages of stand development characteristic for the ecosystem. In forests managed according to the age class principle it is usually impossible to reach this condition within an area of comparable site conditions. To construct yield tables (which reflect one aspect of stand development), yield science has used a broad ecosystem definition with respect to site conditions and has taken its examples of stand development from large geographical regions.

Patterns: Persistence reflects the ability of the ecosystem to regenerate itself. During the aging and regeneration phase the impulses for the deviation from the "state of rest" (closed material cycle, see Section 2.4.3) due to the variance in the ecosystem environment (climate factors) are superimposed by ecosystem internal changes of the microclimate. Temporary changes in the species composition of the forest floor vegetation are common. In forest ecosystems with high elasticity the species that form the dominant tree layer immediately regrow again and the species composition of the tree layer remains unaltered. Such ecosystems stay in the attractor field of the flux equilibria, which forms during the timber tree phase and can be maintained from decades to centuries. By contrast, forest ecosystems with less elasticity can undergo a cyclic succession; that is, the initial composition of the tree species does not return until after a pioneer phase of other tree species. The boreal coniferous forests at the polar tree line, which are exposed to a strong climatic stress are an example. Based on Holling (1973) and Webster et al. (1975) the "oscillation ability" of forest ecosystems after the regeneration is induced can be described as resilience. Highly resilient ecosystems stay in the same attractor field during regeneration, less resilient ecosystems, however, oscillate between different attractors. This phenomenon is demonstrated in Figure 10.3. Forest ecosystems can exist in different stability phases (I and II), which are characterized by different state vectors and by steady states (states of rest) that function as attractors. Autocatalytic processes (positive feedback, see Section 2.7) can force the system to leave the attractor field I: The system is then in a transition phase (i.e., in the development towards a new attractor II). Transition phases are characterized by changes in state variables like species composition, storages in organic matter (biomass and humus), nutrient, and acid–base status of the soil. Before changes in the species composition become visible, the input–output budget indicates the kind and direction of change in storages and soil state. The knowledge of these changes is important in order to assess the new attractor field. In a cyclic succession the system can return to its previous attractor field. This return is indicated in Figure 10.3 as input controlled retrogradation. The input can be due to species immigration (from areas of different phases of stand development), to silicate weathering (increase of mobilizeable nutrient cation storage and base saturation in soil), and to atmospheric deposition (increase in mobilizeable nitrogen and trace element storage).

In Figure 10.4 a scheme is given representing the stability and transition phases of forest ecosystems (Ulrich, 1987, 1991, 1992). Only ecosystems in stability phases are capable of system renewal. A deciding property of the system for it to be

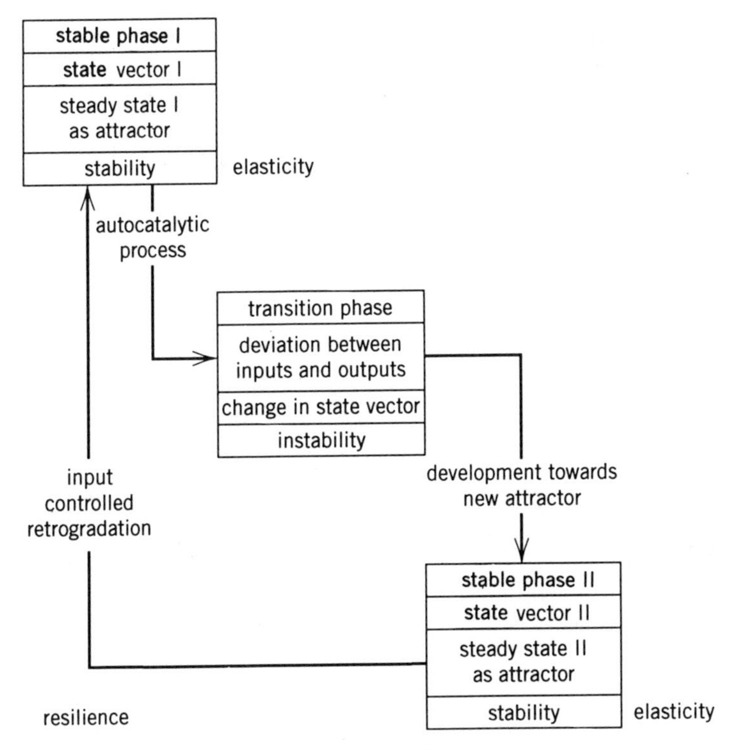

Fig. 10.3. Destabilization and retrogradation of forest ecosystems.

persistent is its elasticity, that is, its ability to filter signals from the subordinate hierarchical levels. As pointed out, this elasticity is due to physiological properties of the organisms, to the soil state, and to the structure of the ecosystem. These various components of elasticity are interrelated. For example, on acid soils plant species dominate that are acid tolerant, but often their litter is poorly decomposable. These properties of the litter result in the accumulation of an organic top layer in which leaf litter decomposition takes place. Thus on acid soils biodiversity is reduced in respect to primary as well as secondary producers. A further consequence is a lower turnover rate of organic matter, as well as nutrients and a simpler structure of the ecosystem (tendency for single-layered tree stands). Due to these interrelationships the chemical soil state (buffer range) is used in Figure 10.4 as an expression of ecosystem elasticity.

Two steady phases are defined, one with high elasticity (phase 2), one with low elasticity (phase 4). These phases represent two maxima of the distribution of elasticity. Vegetation science differentiates between forest types on the basis of plant associations that vary according to climate and site (e.g., Ellenberg, 1978; Hartmann and Jahn, 1967; Mayer, 1984). The humus form (mull, moder, and raw humus) can be used to characterize the kind of decomposer association. These forest types correspond to the stability phases 2 and 4 in Figure 10.4. Phases 2 and 4 can

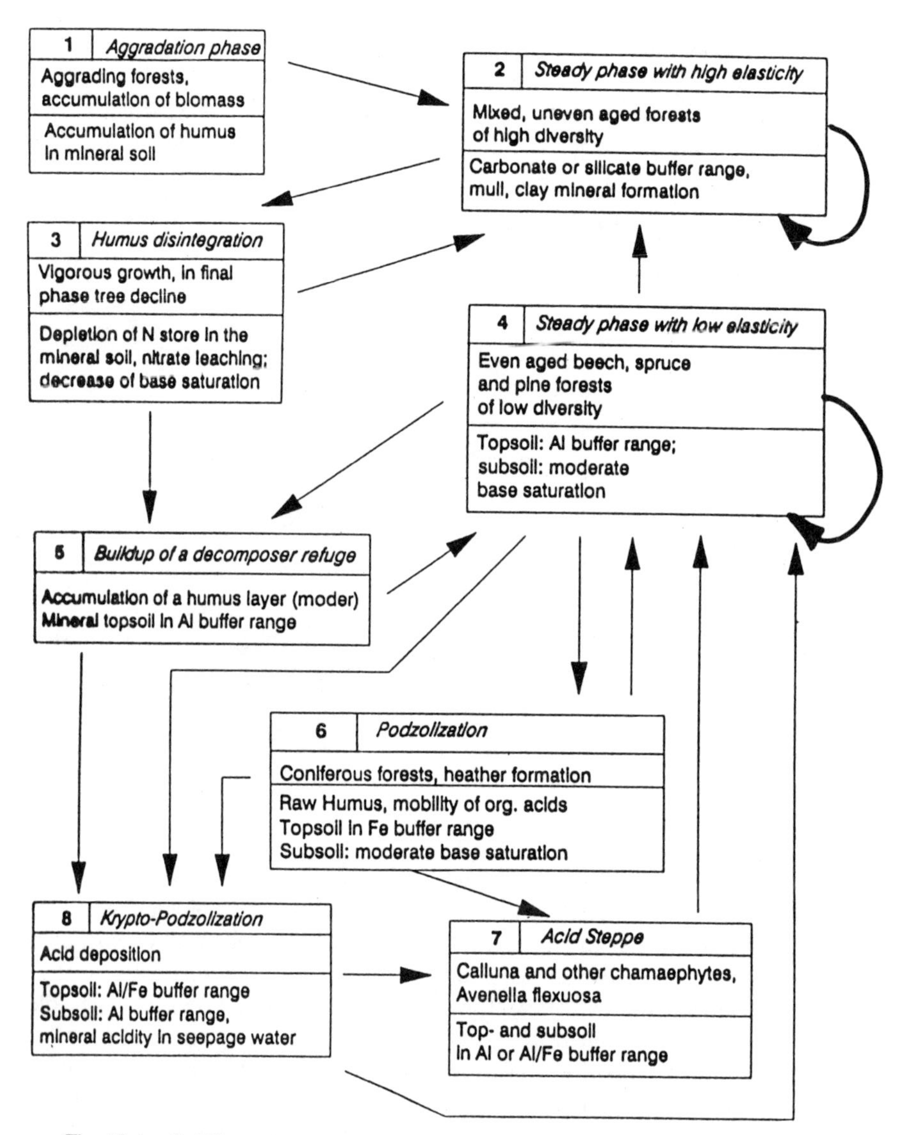

Fig. 10.4. Stability and transition phases of temperate forest ecosystems on aerated soils.

therefore be further subdivided according to the plant association. The senescence of the dominant trees forces the systems into an aggradation phase. In case of high resilience, the species composition may remain unchanged; this is indicated by the ring arrow. If the attractor function of the steady state (the elasticity) is to weak, the systems pass over in a transition phase ($2 \rightarrow 3$, $4 \rightarrow 5$, $4 \rightarrow 6$, see the following section). Systems subjected to cyclic succession (systems of low resilience) return to the original state ($3 \rightarrow 2$, $5 \rightarrow 4$, $6 \rightarrow 4$).

Stability phase with high elasticity (2): The long-term persistence and high resilience of forest ecosystems at high stability enables evolution to develop systems of high species diversity (species richness), habitate diversity (niche structure), and genetic diversity (genetic adaptation). The buffer properties of the soil allow the system to remain close to the "state of rest" that acts as a strong attractor. The spatial heterogeneity of the soil state at the macroscale and the mesoscale, as well as the temporal heterogeneity, are low.

Budget criteria: The output is balanced by atmospheric input and silicate weathering. Deviations from a balanced budget during the stand development play no significant role.

Stability phase with low elasticity (4): The buffer properties of the soil are limited. Extremes of climatic variance, as well as autocatalytic processes during phases of stand development, may therefore lead to phases of nutrient losses and soil acidification. This condition may result in decreased vitality of species. However, the elasticity of the system and the input of nutrients and bases are high enough to keep the system within the attractor range. The vitality of the species and organisms concerned increases again (recovery) before other species can compete successfully.

If the input of basicity by silicate weathering exceeds the loss by leaching, then nutrient storages and base saturation of the soil, and thus the elasticity, increase. Over centuries to millenia the system may then be transferred back to phase 2.

Budget criteria: The output is balanced by atmospheric input and silicate weathering. The lower the elasticity, the weaker is the attractor function of the flux equilibrium.

2.7. Hierarchical Level +3: Succession and Management

Succession is considered to be the replacement of one plant association by another, induced by climate, soil, or the activities of the organisms themselves (Schaefer, 1992). Since succession usually takes place in the course of the regeneration of the dominant tree layer, succession has the same time scale as system renewal. The spatial scale, however, is a landscape that consists of different ecosystems and allows the immigration of species. The ecosystem is constrained at this level by the competitive power of species. If after a change in the conditions (climate, soil state) some species are better adapted to the new environment, succession occurs. Decreases in vitality of dominant tree species and an increasing competition by ground vegetation, especially in the regeneration phase, are the signals from the subordinate level (+2, system renewal). Perturbations (e.g., by fire, wind-throw, or anthropogenic biomass utilization) represent signals from the environment that can also trigger succession.

In managed forests, the change in species composition is man-made and caused, for example, by grazing, burning, thinning, clear felling, planting, and cultural operations. Since no later than the Bronze age, one half of the postglacial quaternary, mankind has had a far reaching influence on forest ecosystems. The unintended triggering of succession by utilization of the forests was followed by the artificial regeneration of tree stands by sowing or planting (for the first time in the fourteenth century in the Nürnberger Reichswald and on a large scale since the nineteenth century). Forest management can initiate the transfer of forest ecosystems into transition phases, but it can also stabilize forest ecosystems, that is, help to maintain them in an attractor range. Afforestation of bare land not only initiates a forest ecosystem, but also the whole process chain that is connected with stand development and which can, depending on the kind of forest, destabilize the system (cf. Section 2.5).

Patterns: According to the ecosystem definition given in Section 1.1, soil changes indicate succession and may often be an early indicator of vegetation changes. The change in species composition of plants is the last stage of a succession process. During this process the ecosystem is in a transition phase (cf. Fig. 10.3). The trajectory on which the system moves may not lead directly to the new attractor range. The forest types of a continent reflect the pattern that succession may lead to, if the climate or the soil is subjected to change (displacement of vegetation zones). The dependence of forest types on the soil conditions shows the significance of changes in the soil conditions needed for succession.

The changes in the storages (increases or decreases) of biomass or necromass (dead wood, litter, or soil organic matter) in the system are a characteristic of transition phases. These characteristics are the consequence of changes in the growth rates of plants or to changes in the population dynamics and activity of secondary producers, mainly decomposers and soil microorganisms. Such changes result in changes of nutrient storages and the acid–base status of the soil that can be calculated applying the material budget equation.

According to the hierarchy theory, transition phases are characterized by the break of the hierarchy: The elasticity is exhausted and the signals from the lower hierarchy levels determine the development of the system.

The transition phases shown in Table 10.2 can be distinguished on the basis of organic matter storage changes in the ecosystem (Ulrich, 1987, 1991).

Aggradation phase (1): If a forest ecosystem is initiated, it starts with an aggradation phase. The extreme case is the development of forest ecosystems after a glaciation on fresh glacial or periglacial sediments. The biomass accumulation should be limited by the availability of nitrogen, this increases slowly by deposition and fixation. The weathering rate of silicates is high as long as there are still easily weatherable silicates present in soil, providing a high nutrient cation input. This means that in the soil the clay content, the storage of mobilizable nutrient cations, and the water capacity increases. Humus and nitrogen are accumulated in soil. The rate of accumulation of basicity in organic matter is balanced by carbonate or silicate weathering, thus the soil does not acidify.

Time scale: In the extreme case mentioned, the phase of increasing elasticity in soil may last for several millenia, that is, it can continue in the stability phase with high elasticity. In central and northern Europe, humans probably destabilized forest ecosystems by grazing and burning when they were still in the phase of increasing soil elasticity.

Patterns: During this phase the soil develops from initial stages to brown earth, which is characterized by a Bv or cambic B horizon (cf. Section 1.1). Species and habitat diversity are increasing. The material balance is characterized by the accumulation of biomass, and of nitrogen and basicity in the biomass, as well as the accumulation of humus, nitrogen, and mobilizable nutrient cations in the mineral soil. The weathering rate of silicates is decreasing according to the consumption of easily weatherable silicates. In calcareous soils, the leaching of bicarbonates is high, as indicated by water hardness ($\sim$ 20 $kmol_c$ ha^{-1} $year^{-1}$), it is supplied by dissolution of carbonate. In noncalcareous soils the leaching of bicarbonates may approach zero if the deeper soil approaches pH 5.0. This result is an important precondition for the accumulation of the nutrient cations released by weathering in the soil. The material budget is thus characterized by a larger input of nutrients (cations and nitrogen compounds) than output.

All other transition phases are characterized by a negative nutrient and base budget: the output exceeds the input, the storages in the soil decrease, and elasticity decreases. The processes involved are different, however, and allow the differentiation of various destabilization phases.

Phase of humus disintegration (3): The material cycle in the ecosystem is decoupled: The rate of decomposition and mineralization exceeds the rate of biomass production. The organic matter and nitrogen storage in the mineral soil that accumulated during phases 1 and 2 are decreasing. This phase can be initiated by the regeneration of dominating trees, by perturbations, and by warming up of the climate. Space and time scale: The process starts patchlike in a tree stand and may then be the driving force for the development of gaps, or it is initiated on a larger space scale by perturbations. It can last for more than a century. It can end either by closure of the material cycle ($3 \rightarrow 2$) or—after acidification of the topsoil to the Al buffer range—by the withdrawal of litter decomposition from the mineral soil ($3 \rightarrow 5$). The loop $2 \rightarrow 3 \rightarrow 2$ may occur at each regeneration and is then an important driving force for soil acidification (i.e., a decrease in elasticity).

Patterns: The ion budget of the soil is characterized by the leaching of nitrates, indicating an equivalent soil acidification. Since the nitrate concentration of soil solution is high, tree growth (forest increment) and biomass production are high. If the change in the acid–base status of the soil exceeds the ability of the tree roots to adapt, a sudden change from high growth to the decline of the dominant trees is possible. Since the humus disintegration involves the organic matter in the whole soil profile including Bv horizons (Bvh), the decline may only concern the deep rooting dominant trees, whereas the ground vegetation is vital and regeneration may take place.

Process (cf. Fig. 10.5): In the case of steady state of soil organic matter, the mineralization of highly polymerized humic substances by fungi is compensated by their new formation via biological humification (Scheffer and Ulrich, 1960). If a soil acidifies, this flux equilibrium is disturbed. The input into the stable compartment is decreasing as a consequence of decreasing bacterial activity and of the inhibition of polymerization by decreasing pH and the formation of organic Al complexes. The

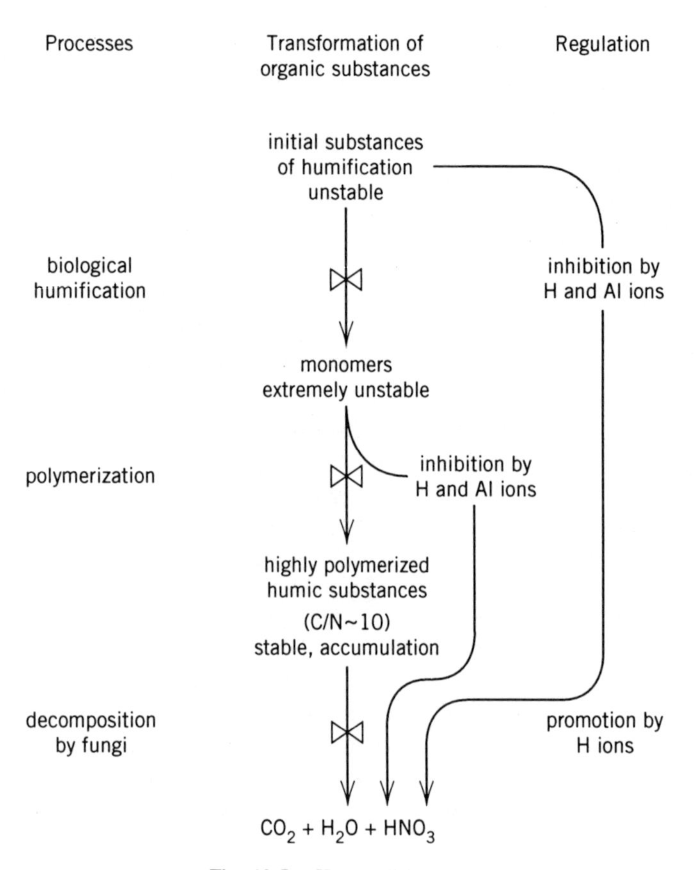

Fig. 10.5. Humus disintegration.

output from the stable soil organic matter compartment is increasing as a consequence of the promotion of decomposition by fungi at low pH (optimum at pH ~ 4.5).

Buildup of decomposer refuge (5): If the acidification of the topsoil (A horizon) has reached low base saturation (Al buffer range), the chemical state of the soil solution changes from alkaline to acidic. This state has serious consequences for the species composition of plants and decomposers. Plants adapted to acid soil conditions also produce a leaf litter that is difficult to decompose, due to a low nitrogen content and also partially to the high content of water soluble phenols. From the soil burrowing species of the decomposers some are lost and others show a decreased activity. Consequently, the leaf litter is no longer mixed into the mineral soil and the accumulation of an organic top layer is initiated. From the point of view of an ecosystem, this phase is called buildup of a decomposer refuge.

Patterns: During this phase, a temporal and spatial decoupling of the material cycle is a characteristic feature. The rate of litter decomposition is lower than the rate of litter production, the accumulation of litter and their decomposition products, together with basicity, takes place at the surface of the mineral soil. This process can follow the phase of humus disintegration. It is also a characteristic feature of the life cycle of even-aged forests on acid soils: The organic top layer is accumulated (transition 4 → 5) in the thicket and pole stage of the stand development. A steady state is reached in the timber tree stage (new attractor: steady phase with low elasticity, 5 → 4). With the opening of the stand (senescence or felling), mineralization may increase and the organic top layer decreases.

Like biomass accumulation litter accumulation in fresh and decomposed state also indicates the accumulation of nutrients and bases outside of the mineral soil, and thus its depletion and acidification. The ion budget of the ecosystem may be balanced, which is not the case for the ion budget of the soil. This number can be calculated applying the material budget equation, if estimates of litter production and its nutrient content are available.

In the case of aggrading stands, the acid load of the mineral soil caused by biomass accumulation and by litter accumulation add up and show a positive feedback: with the maturing of the tree stand the rate of proton production increases (cf. Fig. 10.6). If the adaptability of the tree roots is exceeded, the vitality of the trees may decrease. The risk of such a development is greatest on soils with low elasticity and, with respect to the development of the tree stand, towards the end of the pole stage (bifurcation point of stand development).

The shift of litter decomposition from the mineral soil into a special compartment indicates that mineral soil has unfavorable conditions for the root development of plants. Even acid tolerant species tend to develop a superficial root system that is located mainly in the organic top horizons. Roots and decomposers are an important

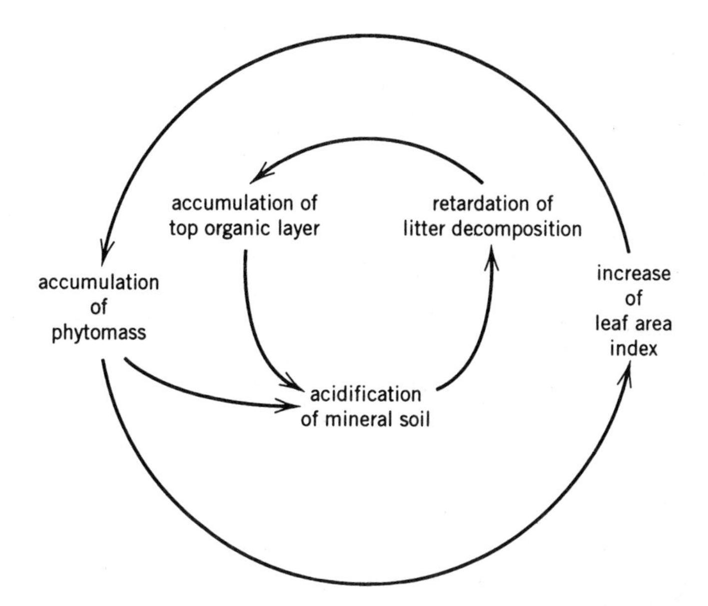

Fig. 10.6. Buildup of a decomposer refuge in aggrading stands.

factor in regulating soil fabric and physical soil conditions such as porosity. The withdrawal of life activities from the subsoil therefore implies adverse changes in soil porosity, aeration capacity, and water conductivity.

Phase of podzolization (6): Podzolization is a soil forming process occurring in aggrading ecosystems on acid soils with plant communities that produce a litter that is difficult to decompose. The heath is a typical podzolizing plant community. Therefore, the process is often naturally connected with the succession of a forest ecosystem to a heath ecosystem (transition 6 → 7). This succession can proceed by a patchlike decline of the tree stand. Due to anthropogenic destabilization of forest ecosystems with low elasticity on sandy soils, this transition occurred on a large spatial scale from northern Germany to Scotland since the Bronze age, especially between 800 and 1800 AD.

Process and Patterns: The proton production in mineral soil by the buildup of biomass and of an organic top layer is enlarged by the blocking of nitrification and the production of soluble organic acids during litter decomposition (Fig. 10.7). Podzolization is thus limited to the aggradation phase of an ecosystem on soils of very low elasticity. If the canopy opens and the plant community (heath or tree stand) declines, the process slows down or stops. On the ecosystem level ion input may balance ion output, but organic matter and basicity is accumulated outside the mineral soil, NH_4-humates and organic acids are leached from the organic top layer into the mineral soil (A horizon). There they increase the proton load and dissolve iron hydroxides. The dissolved organic matter, Al, and Fe are arrested in the mineral soil in a B horizon. In the regeneration phase of the plant community a regradation is possible (6 → 4), if the rate of silicate weathering is high enough to

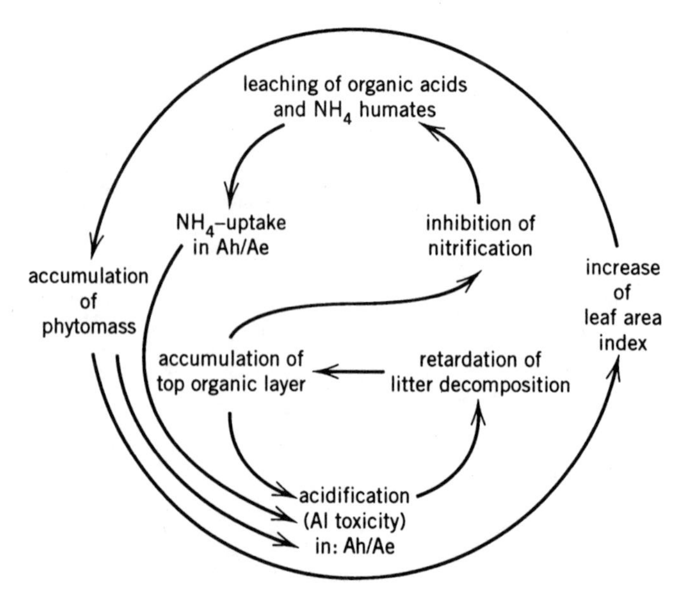

Fig. 10.7. Podzolization. Ah, Ae, surface horizons of mineral soil.

increase nutrient storage and base saturation in mineral soil, and if the nitrogen input allows the development of a more pretentious plant community. Due to nitrogen deposition, this succession is at present characteristic of the heath areas in central Europe.

Cryptopodzolization (8): The term cryptopodzol is used for soils in which Al is mobilized in the entire solum (A and B horizon). Organic matter is stabilized throughout the soil profile by the formation of Al complexes. Such soils were described in the chestnut zone of southern Switzerland (Blaser and Klemmedson, 1987). The same kind of Al stabilization of soil organic matter is a consequence of acid deposition.

Figure 10.8 represents the patterns of humification and humus stabilization. The left hand side indicates the processes in soils of high base saturation. Humus disintegration can lead to a decrease in organic matter and nitrogen storage, the remnants have low nitrogen contents and are stabilized by Al. In this case the organic matter content is usually low and does not show up morphologically. On the

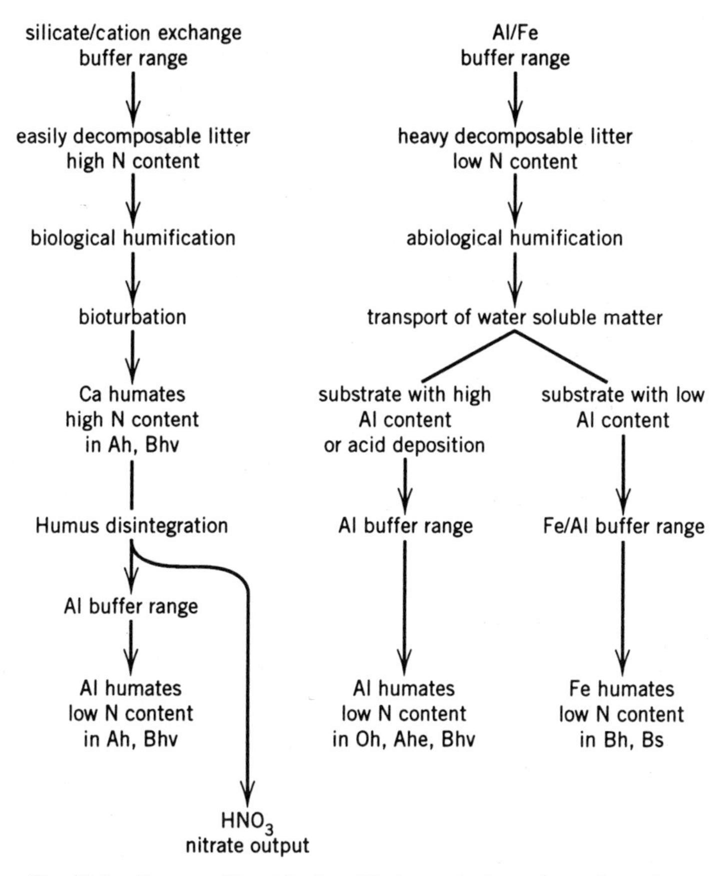

Fig. 10.8. Patterns of humification. Oh, humus horizon of organic top layer.

right-hand side the processes in acid soils are shown. In the rare case of a substrate with high Al content, a site with high precipitation, and the presence of water soluble litter decomposition products, a considerable accumulation of organic matter stabilized by the Al in the subsoil (B horizon) can take place (crytptopodzolization). More widespread is the mobilization of Fe in the acid topsoil (Ae horizon) and the subsequent stabilization of the organic matter as Fe complexes in a Bh horizon of limited extent (podzolization). The mobilization of Al in B horizons by acid deposition begins a development that is comparable to the natural process of cryptopodzolization: Organic matter leached from the topsoil is arrested in the B horizon by Al. The main feature of acid deposition is, however, the leaching of Al ions from the B horizon (cf. the ion budgets of case studies in Ulrich, Chapter 1). This leaching is caused by acid deposition (Ulrich, 1986), which is due to a man-made change in chemical climate starting on a large scale in the middle of the last century.

Patterns: The ion budget is characterized by the input of strong acids (protons and ammonium) accompanied by mobile anions (sulfate and nitrate), and an almost equivalent output of cation acids, especially Al ions, with the seepage water. The organic top layer is densely rooted. There is a high input of mobile Al and Fe into the organic top layer from the mineral soil through root litter (Ulrich, 1989a). Organic matter is accumulating in the organic top layer. Water soluble organic substances are leached into the mineral soil and stabilized in the B horizon. Nitrogen is sequestered together with the organic matter. Thus the nitrogen cycle is interrupted. This decoupling should result in nitrogen deficiency and stunted growth. In fact, this effect can be compensated or even overcompensated for by nitrogen deposition.

All these processes possess a positive feedback mechanism with respect to proton production in the mineral soil. Thus the rate of soil change increases with time up to the bifurcation point of stand development (cf. Section 2.5). The processes may initiate each other and run parallel so that their proton production adds up. Depending on the initial condition, acid deposition could have triggered any of the transition phases shown in Figure 10.4.

From any transition state the ecosystem can return to steady states 2 or 4 (upward leading arrows in Fig. 10.4). If one of these steady states was the starting point, a cyclic succession is indicated. Such transitions are input controlled: chemical heterogeneity in soil decreases, nutrient storage and base saturation increase, and enable the approach of a steady state that functions as attractor. Under natural conditions, silicate weathering is the most important input. Humans learned to use lime and fertilizers as input in order to achieve the same goal.

Probably most forest ecosystems in central Europe are in transition phases. This is due to

- Forest utilization that may go back for centuries and millenia (destabilization).
- Forest management including afforestation of devastated forests and heathlands since the last century (objective: Timber production and stabilization of the forest ecosystem).
- Acid deposition since the middle of the last century (destabilization).

- Nitrogen deposition, slowly increasing since the last century (stabilization) and high rates a few decades ago (increasing risk of destabilization).
- Climate change (may be in effect already and is expected to have great influence in the coming decades).

If mankind had not interfered, central Europe would be almost completely covered by forests. Agricultural sites that are no longer cultivated, thus represent transition phases (succession stages) in the development of forest ecosystems. This succession implies that the subordinate hierarchical levels constrain the superior levels. From the viewpoint of hierarchy theory, this should be a principle that is always valid. Küppers (1991) demonstrated this theory using as an example the succession of hedges along the edges of fields, where the architecture of the plants (branching) becomes the determining factor for succession.

2.8. Hierarchical Level +4: Macroevolution

Evolutionary processes play a role at the level of stand development (+1) as viability selection and, at the level of system renewal (+2), as genetic adaptation. At the level of macroevolution the formation and extinction of species is concerned. The spatial scale is the continent where species are able to migrate. The time scale differs according to the regeneration period of species. The patterns are constituted by the existing species.

3. DESTABILIZATION OF FOREST ECOSYSTEMS BY DEPOSITION OF ACIDITY AND NITROGEN

The results of forest decline research are summarized in this section and the hierarchy theory is used as the order principle. Application of this theory has to be carried out independently for the different components of emission–deposition; the presentation should be considered as an example and does not lay claim to completeness. This theory is limited to the effects of the input of acidity and nitrogen. On the hierarchical levels −1 to −3 the influences on the individual system components are discussed. This presentation is also limited, namely, to the trees and the decomposers. Neither the interactions with other stresses, such as extreme weather conditions or damage causing insects, are included. With respect to the results of forest damage research, reviews by different research groups (FBW, 1989; Horsch et al., 1992; Krupa and Arndt, 1990; Reuther et al., 1991; Schulze et al., 1989; Ulrich 1989a, b) summarize the results achieved.

3.1. Effects of Acid Deposition on the Tree Layer

Table 10.4 shows the effects on the trees of a stepwise break in hierarchy. The acid input comes into effect immediately at levels 0 and −3 (minimal structural units in the sense of Pickett et al., 1989). On the level of the ion cycle the acid deposition

TABLE 10.4. Effects of Acid Deposition on the Plant Subsystem

Level	Process	Filter	New Constraints
+4	Evolution	Competitors	Reduction of gene pools
+3	Succession	Rate of immigration of species Planting, fertilization	Ground vegetation (grasses, heather)
+2	System renewal		
+1	Storage change of biomass: Leaf loss Tree decline	Mobilization of reserves	Canopy closure (open)
0	Material cycle	→ H^+ buffering, crown leaching → Nitrogen deposition Attractor function of flux equilibrium	Flux of water and nutrients to crown
−1	Formation of organs	Fine roots: increased turnover displacement to topsoil	Wood increment Water conducting system
−2	Water/nutrient uptake	Regulation by plant Acid tolerance	Nutrient deficiency Root derangement points
−3	→ Biochemical reaction in leaves	H^+ buffer capacity of apoplast and cells	
−3	→ Soil acidification	Acid buffering Acid accumulation (aluminum sulfates)	Decrease of base saturation Nutrient losses

Long arrows indicate constraints after the break of hierarchy, short arrows indicate effects of deposition of strong acids.

leads via buffering in the leaf to the leaching of cations (K, Mg, Ca, and Mn) from the leaf. Through this the conditions for nutrient uptake (−2) are changed and uptake increased, if the nutrient content of the soil solution is adequate. The effect is thus filtered via the changes in the parameters, which constrain the nutrient uptake.

Another effect is direct damage to the leaf (−3), for example, due to sulfur dioxide (SO_2) and its follow-up products including protons. It was shown that in West Germany, under the present SO_2 concentrations in the air, the neutralization ability of the apoplast and cell is sufficient to be used as a filter to prevent damage.

The soil is another component of interaction, where depending on the buffer range different acid–base reactions take place. Deposited acidity is either neutralized or transformed into weaker acids. In these acid–base reactions an equivalent amount of bases is consumed. The nutrients Ca, Mg, and K, which occur as bases in the soil, are transformed into water soluble neutral salts (e.g., sulfate) and are leached with the seepage water. The base consumption is therefore closely connected with the nutrient depletion. In the soils of central Europe, acid deposition and its buffering in the soil has been under way since the middle of the last century. The filters are the cation exchange equilibria in the soil. Acidification and nutrient depletion have a drastic impact on the soil solution (change from alkaline to acidic) only if the base saturation decreases below around 15%, and only then is the uptake of nutrients inhibited severely. The tree can tolerate a decreasing nutrient supply over decades without showing signs of decline in the above-ground parts (e.g., leaf loss) by selective uptake of nutrients, by the translocation of nutrients within the tree (e.g., from old needles into young), and by adaptation of the growth to the nutrient supply (decrease in radial growth). Acid stress in the soil is filtered in the root system through adaptation, for example, by apoplastic and symplastic tolerance mechanisms. The existing stress nevertheless is visible in the pattern of the content of nutrients and cation acid (especially Al) in leaves and roots, respectively.

Until now, physiological research has failed to clarify the chain of effects causing crown thinning. Should this failure be due to a fundamental methodical problem, it may be possible to apply the method of a correlation of patterns at different stages of the dynamic process, which over a period of years to decades leads to the typical branching pattern of crown thinning. The following interpretation is based on this approach.

On the level of seasonal processes (−1) the process of formation and development of fine roots can be directly influenced by soil changes. Several adaptation reactions operate at this level, which are partly linked to each other by influencing constraining parameters and signals. These reactions can repeatedly put the filtering mechanisms down to level −3 into the circuit. If, for example, due to acid pushes the acid (Al) concentration in the soil solution increases so fast that the tolerance mechanisms of a fine root are not able to react, the meristem (growth tissue) of the root tips can be damaged (signal). Root growth can be continued from secondary meristems (regeneration proficiency as a filter); this process is morphologically manifested (pattern) in an anomaly of root branching (root interference points). It can take weeks to complete the course of such a feedback loop. The lifespan of the fine roots is shortened (signal), which can be balanced by an increased fine root growth (filter). During such pushes the spatial heterogeneity of the chemical soil state increases, with a high seasonal dynamic. The regeneration of fine roots predominantly occurs in those soil layers in which the acid–base and nutrient state reaches the nearest approximation to the optimum (allowing stationary ion uptake). However, the deviation from the optimum can be considerable, and the deciding factor is the relative favorability. In reality, this means that under increasing unfavorable conditions in the subsoil (deep reaching severe acidification of the soil) the fine root system of a tree retreats into the humus rich topsoil (A horizon and organic top layer). The water and nutrient uptake, even with shallow root growth of trees, is secured by an increase in the fine root mass (filter). Under long-term stress (decades) the pattern of the root derangement points continues in the lignified root system interference (Puhe, 1994).

Fine root growth, radial growth including branching and leaf mass, are connected by assimilate production and transport, as well as by the water uptake and transport. In the case of soil born stress (acidification) for assimilate allocation the formation of the fine root system seems to be just as important as leaf formation. An increased assimilate consumption in the fine root system, or a diminished assimilate production (e.g., due to an inhibition of transpiration), leads to decreased radial growth of lignified parts (trunk, coarse roots, branches, and twigs). As a result, the water conducting sapwood area of lignified roots in the trunk and twigs is reduced, impairing the water supply to the leaves, and the level of the material cycle becomes affected with respect to water and nutrients. The inhibition of the water supply to the crown is aggravated by the shift of the fine root system into the upper mineral soil: Only by radial growth can the water conducting system in the lignified roots of the upper mineral soil, cope with the increasing water uptake due to an enhanced fine root formation. On a shorter term the tree is able to increase the flow velocity in the conducting tissue by steeper hydraulic gradients and utilize the water resources in the trunk (a filter at the level of the material cycle). The tree can also concentrate radial growth where a hydraulic bottleneck occurs, such as at the beginning of the crown section, which is manifested in zero growth in the lower trunk (filter on the level of growth processes). If the inhibition of the water supply is not eliminated by these filters, the tree responds with a long-term adaptation (i.e., over years and decades) by adjustment of its transpiring leaf surface to the water supply. In spruce this is realized by the loss of whole shoots causing the formation of typical stress induced branching patterns (cf. Section 2.3). Thereby, level +1 "storage changes in the biomass" is reached. The patterns of the below-ground (roots) and above-ground branching (crown) correspond with each other. However, the modifications in the branching pattern of the roots are ahead of those of the crown by years-to-decades, depending on the efficiency of the filter mechanisms. This type of coupling between the different organs of a tree, which probably include blossom and fruit formation, appears to be characteristic of soil born disturbances, irrespective of the type of disturbance. Both anthropogenic and natural acidification, but also water logging, soil compaction, and anoxia, can lead to the same dynamic in the below- and above-ground branching pattern. A clarification of the causes can attempt to be made by an analysis of the soil chemical and physical patterns, as well as the chemical and anatomical patterns of the root system.

With the phenomenon of crown thinning, particularly after the decline of individual trees, a tree stand loses its function of regulating the microclimate. Now the ecosystem is lead to a bifurcation point where the development either takes the course of renewal and regeneration of the system (persistence), or under the present soil conditions, is dominated by the most competitive species of the ground vegetation to succession. If at this stage the forester provides the tree stand with a second chance through replanting and management, a new forest ecosystem can begin its development. On depleted soils however, the nutrient supply to build up a storage of biomass and an organic top layer may sometimes be insufficient. In such ecosystems the tree stand starts to thin out at the bifurcation point of stand development (cf. Section 2.5) and succession does not set in until some decades later. This finding is exemplified by the yellowing of medium aged spruce stands due to Mg deficiency and by secondary stresses causing other forms of thinning or the collapse of forests in the pole stage.

3.2. Effects of Nitrogen Deposition

The effects of nitrogen deposition are compiled in Table 10.5. In contrast to an input of acidity, an input of nitrogen has immediate effects on the levels of material uptake (−2) and growth processes (−1). Ammonium and nitrate can be taken up through leaves. An increased availability of nitrogen promotes the tissue formation at the places of assimilate allocation, thus it promotes the discharge of assimilates from the chloroplasts and prevents the blocking of photosynthesis. Consequently, higher assimilation rates are possible and growth is increased. Since forest growth is usually limited by lack of nitrogen, nitrogen deposition leads to an increase in

- Radial growth → increase in forest increment.
- Xylem area → improvement of water transport to the canopy. This counteracts the effect of acid deposition (see Section 3.1).
- Height growth → changing the height/diameter ratio with the eventual consequence of mechanical instability of the tree.
- Flowering → consumption of assimilates (→ decrease in xylem area) → Promotion of natural regeneration.
- Fine root mass → improvement of water and nutrient uptake.
- Litter production (via the promotion of ground vegetation) → increase in food resource for decomposers.

These growth increments increase progressively with the improvement of nitrogen nutrition, in so far as they do not compensate each other as seed production and radial growth. Tree stands in the pole stage have the highest growth increments if nitrogen nutrition is improved (level +1, response time decades). These stands can

TABLE 10.5. Effects of Nitrogen Deposition

Level	Process		Effect of NH_x and NO_x Deposition
+3	Succession	(+):	Ground vegetation and fast growing tree species → Change of competition in open timber trees and regenerations
+1	Stand development	(+):	Juvenile phase of stands, biomass → Cation accumulation in biomass, soil acidification
0	Material cycle	(+):	Nutrient cycle
−1	Formation of plant organs	(+):	Diameter increment (→ timber production) Height increment (→ mechanical instability) Xylem area (→ water transport) Seed production (→ natural regeneration) Fine root biomass (→ water/nutrient uptake) Litter production (ground vegetation)
−1	Decomposers	(+):	Bioturbation (depending on humus form)
−1	Phytophages	(+):	Leaf insect damage (due to elevated N contents)
−2	Water/nutrient uptake	(+):	Direct uptake of NH_x and NO_x by leaves In the case of NO_3 nutrition: cation uptake Nitrogen content of leaves
		(−):	In the case of NH_4 nutrition: cation uptake Nutrient relations (N/P, N/K, etc.) in leaves Frost hardiness
−2	Assimilation	(+):	Photosynthesis (assimilate transport) Availability of assimilates
−2	Mineralization	(+):	Denitrification (→ N_2O)
−3	Soil chemical reactions	(+):	In the case of NH_4 deposition: soil acidification In the case of NO_3 leaching: soil acidification

Arrows indicate consequences that follow the primary effects of nitrogen deposition.
(+) = promotion, (−) = inhibition.

accumulate up to 40 kg of N ha^{-1} annually in the aggrading biomass. In timber trees this figure is 10–15 kg of N. Nitrogen deposition has greatly enhanced the growth of young tree stands in central Europe. This deposition is connected with a higher rate of nutrient uptake from soil and indicates that the rate of ecosystem–internal soil acidification is increased in equivalent degree. Nitrogen deposition is considered as the driving force for the development of yellowing of spruce stands, which is a symptom of forest decline in central Europe. Nitrogen promoted growth and soil acidification. If this reaction results in low Mg/Al^{3+} ratios in the soil solution, Mg uptake is inhibited and the yellowing of needles appears as a symptom of Mg deficiency.

In addition, secondary production is promoted by nitrogen input. As far as nitrogen deficiency limits leaf litter decomposition, mineralization and bioturbation can be increased. This finding is relevant especially for organic top layers (humus form moder), which show no signs of podzolization. An increased nitrate concentration in soil solution promotes nitrate leaching and can thus affect ground water quality. The same condition promotes denitrification and the emission of nitrous oxide (N_2O) from soil. Nitrous oxide is a greenhouse gas and contributes to the destruction of the ozone layer. In the case of luxury nitrogen nutrition, the nutrient relationships in plant tissues (N/P, N/K, etc.) change in favor of nitrogen. Consequently, this causes changes in growth relations, a decreased tolerance against phytophages like leaf-eating insects, and diminished frost hardiness.

The negative effects of increased nitrogen inputs finally become evident at the level of succession through the changes in growth between different species. This effect can be manifest through changes in the competition between different tree species or between tree species and species of the shrub and ground vegetation. In mixed stands of European beech (*Fagus sylvatica*) and oak (*Quercus petraea* and *Quercus robur*), beech is promoted. In the renewal phase of beech forests on limestone, the accompanying tree species like *Fraxinus excelsior* and *Acer pseudoplatanus* are favored. The competition by ground vegetation is very marked in spruce ecosystems at higher altitudes (dominant species in ground vegetation is *Calamagrostis villosa*) and in pine forests of the northeastern German flat plain (dominant species is *Calamagrostis epigejos,* Hofmann et al., 1990). These grasses impair the tree stands by water and nutrient uptake, they promote the organic matter accumulation in the organic top layer, and they prevent tree regeneration.

As long as nitrogen is deficient, its input has a stabilizing effect on the forest ecosystem. Nitrogen deposition can mask to a great extent the adverse effects of acid deposition and of ozone (as far as ozone affects photosynthesis and assimilate allocation, Willenbrink and Schatten, 1993). A nitrogen deposition of 10–15 kg of N ha^{-1} annually, covering the requirement for forest increment, can be considered as a stabilizing input into the forest ecosystems of central Europe. The risks of higher nitrogen inputs are the triggering of succession, diminished ground water quality, and the emission of greenhouse gases from forest ecosystems. These risks build up slowly, but their control will become more difficult the later it is attempted. This finding is the consequence of the long-term changes in storages and processes due to nitrogen input.

3.3. Implications of Deposition for Decomposition and Mineralization

The following section is based mostly on the interrelationship of patterns and less on process studies, thus, the causal relationships have to be considered as hypothetical. Depending on the initial state of the soil, acid input induces contrary changes: Either an increase or decrease in the storage of soil organic matter (humus). The effects are compiled in Table 10.6. In this figure, the new constraints after the hierarchy break are differentiated according to the chemical soil state.

In pristine forest ecosystems with mull soils of higher base saturation (silicate or

TABLE 10.6. Effects of Acid Deposition on the Decomposer Subsystem[a]

Level	Process	Filter	New Constraints
+3	Succession	Rate of immigration of species Planting, fertilization	Ground vegetation: Si: nitrophilic flora Al: grasses, heather
+2	System renewal		
+1	Storage change of humus	Change in microclimate due to opening	Si: Nitrogen losses from subsoil (AhBv) Al: Accumulation of organic top layer
0	Material cycle	→ Nitrogen deposition attractor function of flux equilibrium	Decoupling: Si: Nitrate leaching ($+H^+$) Al: N_{org} accumulation ($-H^+$)
−1	Plant	Displacement of fine roots	Si: Root litter in top soil Al: Root litter rich in Al/Fe in Ahe and organic top layer
−1	Decomposers	Displacement according to root litter production	Bioturbation: Si: Decrease Al: Cessation
2	Mineralization	Change in species composition Displacement according to chemical soil state	Si: Humus decomposition in mineral soil by fungi Al: Decrease of microbial activity
−3	→ Soil acidification	Si: H/Calcium exchange in humus Al: Al/Ca exchange in humus	Si: Decrease of base saturation Al: Water soluble humic acids, Al/Fe humates

[a] Si: Soils in silicate/exchange buffer range, Al: soils in Al buffer range.

Long arrows indicate constraints after the break of hierarchy, short arrows indicate processes that are directly affected by nitrogen deposition.

cation exchange buffer range) and relatively high humus contents in the subsoil (Bvh horizons), soil acidification due to deposition (level −3) can initiate humus disintegration (cf. Section 2.7). This finding reflects a positive feedback: Acid deposition has in this case triggered an ecosystem–internal acid production. Base saturation decreases mainly in the subsoil that is intensively rooted by trees, due to the buffering of acid deposition in leaves. Humus disintegration also starts in the subsoil. The retraction of fine roots from the subsoil as a consequence of its acidification diminishes the food resource for decomposers. The decreasing base saturation and the reduced food resource inhibit the activity of deep reaching soil burrowers and result in a population shift of the decomposer chain (level −1). Possible implications for the tree population are (a) a far above average increment as a result of the high nitrogen supply and (b) commonly after warm–dry years severe crown thinning and dying off of individual trees as a result of severe damage to the fine roots in the deeper rooting zone with a secondary effect of enhanced susceptibility to wind-throw. In the case of (b), regeneration can be impaired due to a thriving development of the ground vegetation. Persistence of the same ecosystem may thus be at risk.

In ecosystems with the upper mineral soil layer in the Al buffer range and an organic top layer, the acid input triggers podzolization (cf. Section 2.7). The rate of mineralization by bacteria diminishes, which is reflected in a decrease of microbial biomass and an enhanced respiration (level −2; Anderson, 1992; Wolters and Joergensen, 1991). The decomposers retreat completely into the humic layer (level −1), bioturbation is restricted to the organic top layer, and the humus form (pattern) changes. Due to the processes described in Section 2.7 as cryptopodzolization, accumulation of soil organic matter and nitrogen begins (level +1). The implications of the nitrogen input depend on the base saturation in the Oh horizon and on the type of decomposer chain. If, under favorable base saturation, decomposition and mineralization are limited by nitrogen deficiency, the nitrogen input may activate these processes. Under the condition of cryptopodzolization, however, mineralization is not limited by nitrogen, but by the formation of Al and Fe organic complexes. A kind of flux equilibrium may be reached between (a) nitrogen input as constraint of formation of fine root mass, (b) the input of acidity as constraint of the mobilization of Al and Fe in the mineral topsoil; and (c) the accumulation of mobile and immobile humic substances from root decomposition.

4. DISCUSSION

Any ecosystem analysis is confronted with the problem that for practical reasons only sections of the ecosystem can be considered. The process hierarchy outlined considers only process categories. In order to consider a concrete problem, it is necessary to define the process hierarchy at each level as to which subsystems are involved (e.g., states of the stand development, population, and species organs) and which processes are significant (e.g., removal of biomass, thinning, acid deposition, liming, nitrogen deposition, nitrogen uptake and mineralization, and damage due to bark beetle). As the process categories allow a further differentiation in space and time of single processes within the given reference to space and time, it may be sensible to select additional hierarchical levels for the concrete problems to be considered. This approach corresponds to those suggested by O'Neill et al. (1986) and Pickett et al. (1989). The advantage of the described approach of a continuous process hierarchy is the option to develop concrete hypotheses regarding the effects of a disturbance at level n on all levels. These hypotheses have to specify the constraints exerted by level $n + 1$, the relevant signals of level n, the relevant filtering mechanisms of level $n + 1$, and the change in the constraints exerted by level n. In addition, hypotheses have to be formulated regarding the transmission of the disturbance onto the respective superior ($n + 2, n + 3, \ldots$) and lower levels ($n - 1, n - 2, \ldots$). For ecosystems in the attractor area of a flux equilibrium, the implications of a disturbance in the process hierarchy are limited and may be neglected.

A hierarchical approach offers us the option of following the effects of a special process through the different space and time scales. Reductionist research approaches make it possible to investigate the signals of level $n - 1$ that are effective on level n, and the filter mechanisms on level n. Therefore, it is possible to follow the course of the disturbances on their way through the ecosystem from higher-to-lower levels, as well as in the opposite direction, and determine causal relationships. The signals and the action of the filter mechanism on the respective higher levels are reflected in patterns, as shown in Section 2. These patterns allow the transition between different scales. If, for example, the effect of acid input on trees was determined according to Figure 10.12 with reductionist research approaches, the patterns of the temporal development of emission, of the change in the chemical soil state, of the ion allocation in roots and leaves, of the depth gradients of the root system, of the condition of the crowns of trees, of the element budget of case studies, as well as of the aging and regeneration patterns, can be correlated.

The role of constraining parameters coming into effect on level n (driving variables) can be verified by process models at level $n - 1$ (O'Neill, 1988). Process models of this kind represent holistic approaches in which the variables can be interrelated by strong causal relationships (strong coupling). If several hierarchical levels are to be integrated in system models, simplifying assumptions are required for a coupling; for example, the coupling of patterns by relatively weak relationships (soft coupling). A thorough description and selection of patterns is necessary in order to avoid wrong couplings in verbal models as well as in simulation models.

As pointed out, most effects of a variance in climate and of ecosystem internal processes influence the components of the ecosystem by changing physical and chemical parameters. Such effects start their way through the ecosystem at the level of biochemical and soil chemical reactions (−3) and cause signals at the higher hierarchical levels. At the level of sink and source function of organisms (−2) such signals represent information that is used to regulate physiological processes. Climatic variance, as well as ecosystem internal processes, can also affect higher hierarchical levels directly, for example, the level of seasonal growth processes (−1) by insect damage, or the level of stand development (+1) by wind-throw of the tree layer. Such effects represent disturbances of the ecosystem. A disturbance may thus be defined as an effect that interacts directly at hierarchical levels of -1 or higher. Further examples for externally caused disturbancies are fire and biomass utilization. Ecosystem internally caused disturbancies, such as damage to trees by pathogens, may indicate changes in vitality and competition strength of the populations that are due to a temporal break in hierarchy. All disturbances of the ecosystem that are not filtered, ultimately become effective on the level of stand development. Loss of leaves, yellowing of leaves, premature thinning of timber trees, dying of trees, and the development of gaps in relatively young stands, inevitable shortening of the felling age, extensive perturbations due to wind, snow, insects, failure of natural regeneration, unsatisfactory development of plantations, growth of a grass cover, shift of the age classes toward younger ages—all of these examples are indications that the ecosystem is in a transition phase to a new attractor state, which may no longer represent forest ecosystems. The temporal development of all phenomena described allows us to draw conclusions regarding the change in persistence of the forest ecosystems on a regional scale.

REFERENCES

Anderson T-H (1992): Ber Forschungszentrum Waldökosysteme Univ Göttingen Reihe B 31:154–164.

Augustin S (1992): Mikrobielle Stofftransformationen in Bodenaggregaten. Ber Forschungszentrum Waldökosysteme Univ Göttingen A 85:152 S.

Bergmann W (1988): Ernährungsstörungen bei Kulturpflanzen—Entstehung, visuelle und analytische Diagnose. Fischer Verlag Jena.

Blaser P, Klemmedson JO (1987): Die Bedeutung von hohen Aluminiumgehalten für die Humusanreicherung in sauren Waldböden. Z Pflanzenernaehr Bodenkd. 150:334–341.

Ellenberg H (1978): Vegetation Mitteleuropas mit den Alpen in ökologischer Sicht. Ulmer Stuttgart.

FBW (Forschungsbeirat Waldschäden/Luftverunreinigungen) (1989): Dritter Bericht. Kernforschungszentrum Karlsruhe.

Godbold DL (1991): Die Wirkung von Aluminium und Schwermetallen auf Picea abies Sämlinge. Schriften Forstl Fak Univ Göttingen 104:156 S.

Gruber F (1992): Dynamik und Regeneration der Gehölze. Ber Forschungszentrum Waldökosysteme Univ Göttingen A 86:420 S.

Haken H (1981): Erfolgsgeheimnisse der Natur. Deutsche Verlags-Anstalt Stuttgart 255 S.

Hartmann FK, Jahn G (1967): Waldgesellschaften des mitteleuropäischen Gebirgsraumes nördlich der Alpen. Stuttgart.

Hauhs M (1985): Wasser- und Stoffhaushalt im Einzugsgebiet der Langen Bramke (Harz). Ber Forschungszentrum Waldökosysteme Univ Göttingen 17:206 S.

Hildebrand EE (1990): Die Bedeutung der Bodenstruktur für die Waldernährung, dargestellt am Beispiel des Kaliums. Forstwiss Centralbl 109:2–12.

Hildebrand EE (1992): Die chemische Untersuchung ungestört gelagerter Waldbodenproben—Methoden und Informationsgewinn. Kernforschungszentrum Karlsruhe PEF-Ber.

Hofmann G, Heinsdorf D, Krauss H-H (1990): Wirkung atmogener Stickstoffeinträge auf Produktivität und Stabilität von Kiefern-Forstökosystemen. Beitr Forstwirtsch 24:59–73.

Holling CS (1973): Resilience and stability of ecological systems. Annu Rev Ecol Syst 4:1–23.

Horn R (1987): Die Bedeutung der Aggregierung für die Nährstoffsorption im Boden. Z Pflanzenernaehr Bodenkd 150:13–16.

Horsch F, Filby WG, Fund N, Gross S, Kändler G, Reinhardt W (Eds.) (1992): Projekt Europäisches Forschungszentrum für Massnahmen der Luftreinhaltung, 8. Statuskolloquium. Kernforschungszentrum Karlsruhe, 570 pp.

Krupa SV, Arndt U (Eds.) (1990): The Hohenheim long-term experiment: Effects of ozone, sulphur dioxide and simulated acidic precipitation on tree species in a microcosm. Environ Pollut 68:193–478.

Küppers M (1991): Die Bedeutung des Wechselspiels von Photosynthese, Blattpopulation und pflanzlicher Architektur für Wachstum und Konkurrenzkraft. In Schmid B, Stöcklin J (Hrsgb) Populationsbiologie der Pflanzen. Basel: Birkhäuser Verlag. S. 165–182.

Lenz R, Schall P (1991): Theorie und Modellierung von Waldschadensprozessen im Fichtelgebirge—ihre hierarchische Strukturierung und technische Anwendung. Verh Ges Ökol 19:647–661.

Ludyck G (1977): Theorie dynamischer systeme. Berlin: Elitera Verlag.

Manderscheid B (1991): Die Simulation des Wasserhaushaltes als Teil der Ökosystemforschung. Ber Forschungszentrum Waldökosysteme Univ Göttingen B 24:211–242 (1991).

Mayer H (1984): Wälder Europas. New York: Stuttgart Fischer.

Meiwes K-J, Hauhs M, Gerke H, Asche N, Matzner E, Lamersdorf N (1984): Die Erfassung des Stoffkreislaufs in Waldökosystemen. Ber Forschungszentrum Waldökosysteme Univ Göttingen 7: 68–142.

Müller F (1992): Hierarchical approaches to ecosystem theory. Ecol Modelling 63:215–242.

Murach D, Wiedemann H (1988): Dynamik und chemische Zusammensetzung der Feinwurzeln von Waldbäumen als Mass für die Gefährdung von Waldökosystemen durch toxische Luftverunreinigungen. Ber Forschungszentrum Waldökosysteme Univ Göttingen B 10:287 S.

Nye PH, Tinker PB (1977): Solute movement in the soil-root system. Oxford, London: Blackwell 342 pp.

O'Neill RV (1988): Hierarchy theory and global change. In Rosswall T, Woodmansee G, Risser PG (eds): Scales and Global Change. London: Wiley. pp. 29–45.

O'Neill RV, DeAngelis DL, Waide JB, Allen TFH (1986): A hierarchical concept of ecosystems. Princeton, New Jersey: 253 pp.

Pickett STA, Kolassa J, Arnesto JJ, Collins SL (1989): The ecological concept of disturbance and its expression at various hierarchical levels. Oikos 54:129–136.

Prigogine I (1979): Vom Sein zum Werden—Zeit und Komplexität in den Naturwissenschaften. München: Piper Verlag. 261 pp.

Puhe J (1994): Die Wurzelentwicklung der Fichte (*Picea abies* [L.] Karst.) bei unterschiedlichen chemischen Bodenbedingungen. Ber Forschungszentrum Waldökosysteme Univ Göttingen A 108:128 pp.

Raben G (1988): Untersuchungen zur raumzeitlichen Entwicklung boden- und wurzelchemischer Stressparameter und deren Einfluss auf die Feinwurzelentwicklung in bodensauren Waldgesellschaften des Hils. Ber Forschungszentrum Waldökosysteme Univ Göttingen A 38:253 S.

Remmert H (1991): The mosaic-cycle concept of ecosystems—an overview. Ecological Studies 85. Berlin, Heidelberg: Springer pp. 1–21.

Reuther M, Kirchner M, Kirchinger E, Reiter H, Rösel K, Pfeifer U (1991): Expertentagung Waldschadensforschung im östlichen Mitteleuropa und in Bayern. Neuherberg: GSF Forschungszentrum 616 pp.

Roloff A (1989): Kronenentwicklung und Vitalitätsbeurteilung ausgewählter Baumarten der gemässigten Breiten. Schriften Forstl Fak Univ Göttingen 93:258 S.

Schaefer M (1991): The animal community: diversity and resources. In Röhrig E, Ulrich B (eds) Temperate Deciduous Forests. Amsterdam The Netherlands: Elsevier pp 51–120.

Schaefer M (1992): Wörterbücher der Biologie: Ökologie. Stuttgart, Germany: Fischer Verlag 3. Aufl.

Scheffer F, Ulrich B (1960): Humus und Humusdüngung I. Morphologie, Biologie, Chemie und Dynamik des Humus. Stuttgart, Germany: Enke Verlag 266 pp.

Schulze ED, Lange OL, Oren R (eds) (1989): Forest decline and air pollution. Ecological Studies 77: Berlin, Heidelberg, New York: Springer. 475 pp.

Steiger BV, Baccini P (1990): Regionale Stoffbilanzierung von landwirtschaftlichen Böden mit messbarem Ein- und Austrag. Ber 38 Nat Forschungsprogramm Boden, Liebefeld-Bern.

Sverdrup H, Warfvinge PG (1988): Chemical weathering of minerals in the Gardsjön catchment in relation to a model based on laboratory rate coefficients. In Nilsson J, Grennfelt P (eds): Critical loads for sulphur and nitrogen. Miljorapport 15:131–149, Nordic Counsil of Ministers Copenhagen.

Ulrich B (1981a): Ökologische Gruppierung von Böden nach ihrem chemischen Bodenzustand. Z Pflanzenernaehr Bodenkd 144:289–305.

Ulrich B (1981b): Theoretische Betrachtung des Ionenkreislaufs in Waldökosystemen. Z Pflanzenernaehr Bodenkd 144:647–659.

Ulrich B (1986): Natural and anthropogenic components of soil acidification. Z Pflanzenernaehr Bodenkd 149:702–717.

Ulrich B (1987): Stability, elasticity, and resilience of terrestrial ecosystems with respect to matter balance. Ecological Studies 61. Berlin, Heidelberg, Germany: Springer. pp. 11–49.

Ulrich B (1989a): Effects of acidic precipitation on forest ecosystems in Europe. In Adriano DC, Johnson AH (eds): Acidic Precipitation Vol. 2 Biological and Ecological Effects. New York, Berlin, Heidelberg: Springer. pp 189–272.

Ulrich B (ed) (1989b): Internationaler Kongress Waldschadensforschung: Wissensstand und Perspektiven. Vorträge (2 Bände), Poster Kurzfassungen (2 Bände). Kernforschungszentrum Karlsruhe.

Ulrich B (1991): An ecosystem approach to soil acidification. In Ulrich B, Sumner ME (eds): Soil Acidity, Berlin, Heidelberg, New York: Springer pp. 28–79.

Ulrich B (1992): Forest ecosystem theory based on material balance. Ecol Modelling 63:163–183.

Webster JR, Waide JB, Patten BC (1975): Nutrient recycling and the stability of ecosystems. In Howell FG, Gentry JB, Smith MH (eds): Mineral Cycling in southeastern Evcosystems. ERDA Sympos Series (CONF74D513), pp. 1–27.

Willenbrink J, Schatten Th (1993): CO_2-Fixierung und Assimilatverteilung in Fichten unter Langzeitbegasung mit Ozon. Forstwiss Centralbl 112:50–56.

Wolters V (1989): Die Zersetzernahrungskette von Buchenwäldern. Untersuchungen zur ökosystemaren Bedeutung der Interaktionen zwischen Bodentieren und Mikroflora. Verh Ges Ökol 18:213–219.

Wolters V (1991): Soil invertebrates—effects on nutrient turnover and soil structure. Z Pflanzenernaehr Bodenkd 154:389–402.

Wolters V, Joergensen RG (1991): Microbial carbon turnover in beech forest soils at different stages of acidification. Soil Biol Biochem 23:897–902.

INDEX